T0306081

COSMIC MASERS - FROM OH TO H_0

IAU SYMPOSIUM 287

COVER ILLUSTRATION:

A picture of Table Montain, Cape Town, taken from Robben Island.

This iconic symbol of South Africa was recently voted one of the 'New 7 Wonders of the World'.

IAU SYMPOSIUM PROCEEDINGS SERIES

Chief Editor

IAN F. CORBETT, IAU General Secretary
icorbett@eso.org

Editor

THIERRY MONTMERLE, IAU Assistant General Secretary
Institut d'Astrophysique de Paris,
98bis, Bd Arago, 75014 Paris, France
montmerle@iap.fr

INTERNATIONAL ASTRONOMICAL UNION

UNION ASTRONOMIQUE INTERNATIONALE

COSMIC MASERS - FROM OH TO H$_0$

PROCEEDINGS OF THE 287th SYMPOSIUM OF THE INTERNATIONAL ASTRONOMICAL UNION HELD IN STELLENBOSCH, SOUTH AFRICA JANUARY 29 - FEBRUARY 3, 2012

Edited by

ROY S. BOOTH
SA SKA Project, SOUTH AFRICA

ELIZABETH M. L. HUMPHREYS
ESO Garching, GERMANY

and

WOUTER H. T. VLEMMINGS
Onsala Space Observatory, Chalmers University of Technology, SWEDEN

CAMBRIDGE
UNIVERSITY PRESS

University Printing House, Cambridge CB2 8BS, United Kingdom

One Liberty Plaza, 20th Floor, New York, NY 10006, USA

477 Williamstown Road, Port Melbourne, VIC 3207, Australia

314-321, 3rd Floor, Plot 3, Splendor Forum, Jasola District Centre, New Delhi - 110025, India

103 Penang Road, #05-06/07, Visioncrest Commercial, Singapore 238467

Cambridge University Press is part of the University of Cambridge.

It furthers the University's mission by disseminating knowledge in the pursuit of education, learning and research at the highest international levels of excellence.

www.cambridge.org
Information on this title: www.cambridge.org/9781107032842

First published 2012

A catalogue record for this publication is available from the British Library

ISBN 978-1-107-03284-2 Hardback

Table of Contents

Preface xii

Organizing committee xiii

Conference photograph xiv

Conference participants xv

Tribute to Yolanda Gómez Castellano xvii

T1: Advances in Maser Theory *Chair: Roy Booth*

Advances in Maser Theory 3
    V. Strelnitski

Modelling of Cosmic Molecular Masers: Introduction to a Computation Cookbook   13
    A. M. Sobolev & M. D. Gray

A detailed trace of the pump for 1720-MHz OH masers in SNRs 23
    M. D. Gray

T2: Polarization and magnetic fields *Chair: Athol Kemball*

Maser polarization and magnetic fields 31
    W. H. T. Vlemmings

Polarization of Class I methanol (CH_3OH) masers 41
    A. P. Sarma

Polarization of the Recombination Line Maser in MWC349 49
    C. Thum, D. Morris, & H. Wiesemeyer

VLBA SiO maser observations of the OH/IR star OH 44.8-2.3: magnetic field
    and morphology 54
    N. Amiri, W. H. T. Vlemmings, A. J. Kemball & H. J. van Langevelde

Linear polarization of hydroxyl masers in circumstellar envelope outer regions . . 59
    P. Wolak, M. Szymczak & E. Gérard

Maser polarization with ALMA 64
    A. F. Pérez-Sánchez & W. Vlemmings

High resolution magnetic field measurements in high-mass star-forming regions
    using masers 69
    G. Surcis, W. H. T. Vlemmings, H. J. van Langevelde & B. H. Kramer

The magnetic field of IRAS 16293-2422 as traced by shock-induced H_2O masers   74
    F. O. Alves, W. H. T. Vlemmings, J. M. Girart & J. M. Torrelles

Water Maser Emission Around Low/Intermediate Mass Evolved Stars 79
    M. L. Leal-Ferreira, W. H. T. Vlemmings, P. J. Diamond, A. Kemball,
    N. Amiri & J.-F. Desmurs

Observational tests of SiO maser polarisation models 81
    L. L. Richter, A. J. Kemball & J. L. Jonas

T3a: Star formation: maser variability *Chair: Philip Diamond*

Variability of Class II methanol masers in massive star forming regions.   85
   S. Goedhart, M. Gaylard, & J. van der Walt

Binary systems: implications for outflows & periodicities relevant to masers   93
   N. K. Singh & A. A. Deshpande

Intermittent maser flare around the high-mass young stellar object G353.273+0.641   98
   K. Motogi, K. Sorai, K. Fujisawa, K. Sugiyama & M. Honma

VERA Observations of the H_2O Maser Burst in Orion KL   103
   T. Hirota, M. Tsuboi, K. Fujisawa, M. Honma, N. Kawaguchi, M. K. Kim,
   H. Kobayashi, H. Imai, T. Omodaka, K. M. Shibata, T. Shimoikura, &
   Y. Yonekura

The variability of cosmic methanol masers in massive star-forming regions   108
   J. P. Maswanganye & M. J. Gaylard

Chasing the flare in Orion KL: Observations of the 22 GHz H_2O Masers at
   HartRAO.   110
   S. Otto & M. J. Gaylard

On the Methanol masers in G9.62+0.20E: Preliminary colliding-wind binary (CWB)
   calculations   112
   S. P. van den Heever, D. J. van der Walt, J. M. Pittard & M. G. Hoare

T3b: Star formation masers *Chair: Crystal Brogan*

Masers in star forming regions   117
   A. Bartkiewicz & H. J. van Langevelde

Masers in GLIMPSE Extended Green Objects (EGOs)   127
   C. J. Cyganowski, C. L. Brogan, T. R. Hunter, E. Churchwell, J. Koda,
   E. Rosolowsky, S. Towers, B. Whitney, & Q. Zhang

44 GHz Methanol Maser Surveys   133
   S. E. Kurtz

A Highly-collimated Water Maser Bipolar Outflow in the Cepheus A HW3d
   Massive Protostellar Object.   141
   J. O. Chibueze, H. Imai, D. Tafoya, T. Omodaka, O. Kameya, T. Hirota,
   S.-N. Chong, & J. M. Torrelles

Methanol masers and millimetre lines: a common origin in protostellar envelopes   146
   K. J. E. Torstensson, H. J. van Langevelde, F. F. S. van der Tak,
   W. H. T. Vlemmings, L. E. Kristensen, S. Bourke & A. Barkiewicz

The infrared environment of methanol maser rings at high spatial resolution . . .   151
   J. M. De Buizer, A. Bartkiewicz & M. Szymczak

Masers as evolutionary tracers of high-mass star formation.   156
   S. L. Breen & S. P. Ellingsen

Class I methanol masers in low-mass star formation regions   161
   S. V. Kalenskii, V. I. Slysh, L. E. B. Johansson, P. Bergman, S. Kurtz,
   P. Hofner, & C. M. Walmsley

Dynamical detection of a magnetocentrifugal wind driven by a 20 $M_\odot$ YSO 166
    L. J. Greenhill, C. Goddi, C. J. Chandler, E. M. L. Humphreys, &
    L. D. Matthews

The W51 Main/South SFR complex seen through 6-GHz OH and methanol masers   171
    S. Etoka, M. D. Gray & G. A. Fuller

What is happening in G357.96-0.16? 176
    T. R. Britton, M. A. Voronkov, & V. A. Moss

Infrared characteristics of sources associated with OH, H_2O, SiO and CH_3OH
   masers 178
    J. Esimbek, J. J. Zhou, G. Wu & X. D. Tang

Massive star-formation toward G28.87+0.07 180
    J. J. Li, L. Moscadelli, R. Cesaroni, R. S. Furuya, Y. Xu, T. Usuda,
    K. M. Menten, M. Pestalozzi, D. Eliav, & E. Schisano

High-mass Star Formation in the Regions IRAS 19217+1651 and 23151+5912 . . 182
    V. Migenes, I. T. Rodríguez & M. A. Trinidad

325 GHz Water Masers in Orion Source I 184
    F. Niederhofer, E. Humphreys, C. Goddi & L. J. Greenhill

10 years of 12.2 GHz methanol maser VLBI observations towards NGC 7538 IRS1 N:
   proper motions and maser saturation 186
    M. Pestalozzi, A. Jerkstrand & J. Conway

Internal Proper Motion of 6.7 GHz Methanol Masers in Ultra Compact HII Region
   S269 188
    S. Sawada-Satoh, K. Fujisawa, K. Sugiyama, K. Wajima & M. Honma

The radial velocity acceleration of the 6.7 GHz methanol maser in Mon R2 IRS3 190
    K. Sugiyama, K. Fujisawa, N. Shino, & A. Doi

A Circumstellar Disk toward the High-mass Star-forming Region IRAS 23033+5951 192
    M. A. Trinidad, T. Rodríguez & V. Migenes

Molecular outflows toward methanol masers: detection techniques and their prop-
   erties 194
    H. M. de Villiers, M. A. Thompson, A. Chrysostomou & D. J. van der Walt

T4: Stellar Masers *Chair: Elizabeth Humphreys*

Masers in evolved star winds 199
    A. M. S. Richards

Radio and IR interferometry of SiO maser stars 209
    M. Wittkowski, D. A. Boboltz, M. D. Gray, E. M. L. Humphreys,
    I. Karovicova, & M. Scholz

Maser emission during post-AGB evolution 217
    J.-F. Desmurs

Water Fountains in Pre-Planetary Nebulae: The Case of IRAS16342−3814 225
    M. Claussen, R. Sahai, M. Morris, & H. Rogers

The first water fountain in a planetary nebula 230
    O. Suárez, J. F. Gómez, P. Bendjoya, L. F. Miranda, M. A. Guerrero,
    G. Ramos-Larios, J. R. Rizzo, & L. Uscanga

Polarization properties of R Cas SiO masers.............................. 235
   K. A. Assaf, P. J. Diamond, A. M. S. Richards & M. D. Gray

The final 112-frame movie of the 43 GHz SiO masers around the Mira Variable TX
Cam.. 240
   I. Gonidakis, P. J. Diamond & A. J. Kemball

High Resolution Radio and IR Observations of AGB Stars................... 245
   W. Cotton, G. Perrin, R. Millan-Gabet, O. Delaa, & B. Mennesson

OH mainline maser polarisation properties of post-AGB stars 250
   J. M. Chapman, I. Gonidakis, R. M. Deacon & A. Green

Preliminary results on SiO v=3 J=1–0 maser emission from AGB stars 252
   J.-F. Desmurs, V. Bujarrabal, M. Lindqvist, J. Alcolea, R. Soria-Ruiz, &
   P. Bergman

1612 MHz OH maser monitoring with the Nançay Radio Telescope........... 254
   D. Engels, E. Gérard, & N. Hallet

The Hamburg Database of Circumstellar OH Masers....................... 256
   D. Engels

Imaging the water masers toward the H_2O-PN IRAS 18061−2505 258
   Y. Gómez, D. Tafoya, O. Suárez, J. F. Gómez, L. F. Miranda, G. Anglada,
   J. M. Torrelles, & R. Vázquez

TWINKLING STARS The disappearing SiO masers of W Aql............... 260
   S. Ramstedt, W. Vlemmings, S. Mohamed, Y. K. Choi & H. Olofsson

T5: Maser Surveys *Chair: Jessica Chapman*

SiO Maser Surveys of Nearby Miras and their Kinematics in the Galaxy....... 265
   S. Deguchi

Water maser follow-up of the Methanol Multi-Beam Survey................. 275
   A. Titmarsh, S. Ellingsen, S. Breen, J. Caswell & M. Voronkov

Identification of Class I Methanol Masers with Objects of Near and Mid-Infrared
Bands and the Third Version of the Class I Methanol Maser (MMI) Catalog 280
   O. Bayandina, I. Val'tts & G. Larionov

25 GHz methanol masers in regions of massive star formation 282
   T. R. Britton & M. A. Voronkov

44-GHz class I methanol maser survey towards 6.7-GHz class II methanol masers 284
   D.-Y. Byun, K.-T. Kim & J.-H. Bae

Water Masers Toward Star-Forming Regions in the Bolocam Galactic Plane Survey 286
   M. K. Dunham & The BGPS Team

The VLBI mapping survey of the 6.7 GHz methanol masers with the JVN/EAVN 288
   K. Fujisawa, K. Hachisuka, K. Sugiyama, A. Doi, M. Honma, Y. Yonekura,
   T. Hirota, S. Sawada-Satoh, Y. Murata, K. Motogi, H. Ogawa, X. Chen,
   K.-T. Kim & Z.-Q. Shen

Simultaneous observations of SiO and H_2O masers toward known stellar SiO and/or H_2O maser sources 290
  J. Kim, S.-H. Cho & S. J. Kim

SHOOTING STARS Masers from red giants 292
  S. Ramstedt, W. Vlemmings, E. Humphreys & F. Alves

New OH Observations toward Northern Class I Methanol Masers 294
  I. E. Val'tts, I. D. Litovchenko , O. S. Bayandina, A. V. Alakoz,
  G. M. Larionov, D. V. Mukha, A. S. Nabatov, A. A. Konovalenko,
  V. V. Zakharenko, E. V. Alekseev, V. S. Nikolaenko, V. F. Kulishenko &
  S. A. Odincov

22 GHz Water Maser Survey of the Xinjiang Astronomical Observatory 296
  J.-J. Zhou, J. Esimbek & G. Wu

T6: Cosmology and the Hubble constant *Chair: Willem Baan*

Cosmology and the Hubble Constant: On the Megamaser Cosmology Project (MCP) 301
  C. Henkel, J. A. Braatz, M. J. Reid, J. J. Condon, K. Y. Lo,
  C. M. V. Impellizzeri & C. Y. Kuo

Mrk 1419 - a new distance determination 311
  C. M. V. Impellizzeri, J. A. Braatz, C.-Y. Kuo, M. J. Reid, K. Y. Lo,
  C. Henkel & J. J. Condon

Optical Properties of the Host Galaxies of Extragalactic Nuclear H_2O Masers . . 316
  I. Zaw, G. Zhu, M. Blanton, & L. J. Greenhill

T7: AGN and megamasers *Chair: Lincoln Greenhill*

AGN and Megamasers 323
  A. Tarchi

Masers in Starburst Galaxies 333
  J. Darling

Long term Arecibo monitoring of the water megamaser in MG J0414+0534 340
  P. Castangia, C. M. V. Impellizzeri, J. P. McKean, C. Henkel,
  A. Brunthaler, A. L. Roy, & O. Wucknitz

Searching for new OH megamasers out to redshifts $z > 1$ 345
  K. W. Willett

Expectations of maser studies with FAST 350
  J. S. Zhang, D. Li & J. Z. Wang

The origin of Keplerian megamaser disks. 354
  M. Wardle & F. Yusef-Zadeh

T8: Maser Astrometry *Chair: Huib Jan van Langevelde*

Maser Astrometry: from Galactic Structure to Local Group Cosmology 359
  M. J. Reid

Methanol Maser Parallaxes and Proper Motions.......................... 368
  Y. Xu, M. J. Reid, L. Moscadelli, K. M. Menten, X. W. Zheng,
  A. Brunthaler, B. Zhang, K. L. J. Rygl, J. J. Li, & A. Sanna

VLBI multi-epoch water maser observations toward massive protostars........ 377
  J. M. Torrelles, J. F. Gómez, N. A. Patel, S. Curiel, G. Anglada, &
  R. Estalella

Maser astrometry with VERA and Galactic structure 386
  M. Honma, T. Nagayama, T. Hirota, N. Matsumoto, N. Sakai,
  N. Kawaguchi & VERA project members

Astrometry of Galactic Star-Forming Regions ON1 and ON2N with VERA 391
  T. Nagayama & VERA project members

VLBI maser kinematics in high-mass SFRs: G23.01–0.41................... 396
  A. Sanna, L. Moscadelli, R. Cesaroni & C. Goddi

3D velocity fields from methanol and water masers in an intermediate-mass protostar 401
  C. Goddi, L. Moscadelli & A. Sanna

Trigonometric Parallax of the Protoplanetary Nebula OH 231.8+4.2 407
  Y. K. Choi, A. Brunthaler, K. M. Menten & M. J. Reid

Astrometry of water masers in post-AGB stars.......................... 411
  H. Imai & VERA collaboration

Relative parallaxes in the massive star forming region W33 413
  K. Immer, M. J. Reid & K. M. Menten

3-Dimensional kinematics of water/SiO masers in Orion-KL................ 415
  M. K. Kim, T. Hirota, K. Hideyuki & VERA project team members

Annual Parallax Measurements of an Infrared Dark Cloud MSXDC G034.43+00.24   417
  T. Kurayama

The bar effect in the galactic gas motions traced by 6.7 GHz methanol maser
  sources with VERA 419
  N. Matsumoto, M. Honma & VERA project members

Mass distribution of the Galaxy with VERA............................. 421
  N. Sakai, M. Honma, H. Nakanishi, H. Sakanoue, T. Kurayama & VERA
  project members

Distance and Maser Outflows of the Galactic Star-forming Region W51 Main/South   423
  M. Sato, M. J. Reid, A. Brunthaler & K. M. Menten

Trigonometric Parallax of RCW 122 425
  Y. W. Wu, Y. Xu, K. M. Menten, X. W. Zheng & M. J. Reid

Distance and Size of the Red Hypergiant NML Cyg....................... 427
  B. Zhang, M. J. Reid, K. M. Menten, X. W. Zheng & A. Brunthaler

T9: New masers and further developments in maser physics
Chair: Mark Reid

New class I methanol masers. 433
    M. A. Voronkov, J. L. Caswell, S. P. Ellingsen, S. L. Breen, T. R. Britton,
    J. A. Green, A. M. Sobolev & A. J. Walsh

OH Masers and Supernova Remnants 441
    M. Wardle & K. McDonnell

Class I Methanol Masers in the Galactic Center. 449
    L. O. Sjouwerman & Y. M. Pihlström

Radio Recombination Line Maser Objects: New Detections with the SMA 455
    I. Jiménez-Serra

Unveiling the kinematics of the disk and the ionized stellar wind of the massive
    star MWC349A through RRL masers. 460
    A. Báez-Rubio & J. Martín-Pintado

Intrinsic Sizes of the W3 (OH) Masers via Short Time Scale Variability 465
    T. Laskar, W. M. Goss & B. A. Zauderer

OH Maser sources in W49N: probing differential anisotropic scattering with
    Zeeman pairs. 470
    A. A. Deshpande, W. M. Goss & J. E. Mendoza-Torres

T10: Masers and the impact of new facilities *Chair: Wouter*
Vlemmings

Maser observations with new instruments 477
    A. Wootten

MeerKAT and its potential for Cosmic MASER Research. 483
    R. Booth, S. Goedhart & J. Jonas

KVN Single-dish Water and Methanol Maser Line Surveys of Galactic YSOs. . . 488
    K.-T. Kim, D.-Y. Byun, J.-H. Bae, W.-J. Kim, H.-W. Kang, C. S. Oh, &
    S.-Y. Youn

Methanol masers in the Herschel era Putting them in the star formation context 492
    M. Pestalozzi

Early results from a diagnostic 1.3 cm survey of massive young protostars 497
    C. L. Brogan, T. R. Hunter, C. J. Cyganowski, R. Indebetouw, R. Friesen &
    C. Chandler

EVLA imaging of the water masers in the massive protostellar cluster NGC6334I 502
    T. R. Hunter & C. L. Brogan

OH maser observations using the Russian interferometric network "Quasar" in
    preparation for scientific observations with the space mission RadioAstron. 504
    I. D. Litovchenko, A. V. Alakoz, V. I. Kostenko, S. F. Lihachev,
    A. M. Finkelstein & A. V. Ipatov

IAU (Maser) Symposium 287 Summary. 506
    Karl M. Menten

Author index 516

Preface

The fourth IAU Symposium on Astronomical *masers*, IAUS 287, entitled *Cosmic Masers-from OH to H_0* was held in South Africa from January 29 February 3. The venue was the excellent Wallenberg Conference Centre in the beautiful old town of Stellenbosch nestling in the foothills of one of the countrys foremost wine districts.

The meeting was opened by Dr. Bernie Fanaroff, Director of the South African SKA Project, who pointed out that human-kind had its origins in Africa and learned basic skills before migration and continental drift led to the population of the rest of the World.

Despite a strenuous programme with lively discussions, the participants found time to visit the Cape Town Water Front and take a boat trip to historical Robben Island, where the present political structure was formulated in the mid 90s. They also enjoyed an African evening at a local hospice, where they sampled truly African food, song and dance.

Interstellar hydroxyl *masers* were first reported in 1965, at the beginning of an avalanche of detections of interstellar molecules and the emergence of the science of Cosmo-chemistry. Many molecules are now found to exhibit polarized (circular and linear) stimulated emission in a variety of astrophysical environments and the hierarchy extends from the interstellar medium and regions of stellar formation through winds around evolved stars and even the compressed gas surrounding Supernova remnants. *masers* are also detected in other galaxies as analogues of Galactic *masers*, and in the dense molecular regions near AGN. Most recently, *masers* have been detected in Comets and other regions of the Solar System. In all cases they have proved to be important diagnostics of their associated regions, pinpointing dense gas, its motions and the local magnetic field structure in great detail.

This phenomenon has proved to be invaluable in astrometry and the details of Galactic structure and rotation have been refined through precise observation of *masers*. The same properties of *masers* have been used to great effect to measure the study of Galactic nuclei and because of the precise measurement of position and velocity, water *masers* have enabled the unambiguous determination of the Hubble Constant at small red-shifts.

Recent *maser* surveys, and measurements of the above phenomena were discussed at the Stellenbosch meeting along with new detections of known *masers* and detections of new *maser* species. Discussion extended from reviews of the theory of the phenomenon, *maser* properties such as variability and polarisation, to its use in astrometry and in cosmology.

More than 110 participants, including some 25 students (many securing IAU grants) attended the meeting. There were 68 oral contributions, including 8 by students and many posters, displayed all week and discussed in 3 separate poster sessions. In general the meeting followed its predecessor with 10 different scientific themes, each with a review talk, one or two invited talks and contributed talks.

Finally we thank our sponsors the South African SKA project and the Natural Science Research Council, NRF.

Roy Booth, Liz Humphreys & Wouter Vlemmings

THE ORGANIZING COMMITTEE

Scientific

A. Bartkiewicz (Poland)
R.S. Booth (Chair, South Africa)
V. Bujarrabal (Spain)
J.M. Chapman (Australia)
M. Elitzur (USA)
S.P. Ellingsen (Australia)
Y. Gómez (Mexico)
M.D. Gray (UK)

M. Honma (Japan)
E.M.L. Humphreys (co-Chair, Germany)
A.J. Kemball (USA)
K.-T. Kim (Taiwan)
H.J. van Langevelde (Netherlands)
J.M. Moran (USA)
W.H.T. Vlemmings (co-Chair, Sweden)

Local

K. de Boer
R.S. Booth (Chair)
S. Fishley
M.J. Gaylard
S. Goedhart (co-Chair)
R. Hames

M. Oozeer
S. Passmoor
A. Schroeder
D.J. van der Walt
P.A. Whitelock

Acknowledgements

The symposium was sponsored and supported by the IAU Divisions I (Fundamental Astronomy), VI (Interstellar Matter), VII (Galactic System), and X (Radio Astronomy); and by the IAU Commission No. 40 (Radio Astronomy).

Funding by the
National Research Foundation (NRF),
and
the South African SKA project
are gratefully acknowledged.

Participants

Khudhair Abbas Assaf **Al-Muntafki**, School of Physics & Astronomy, Manchester University, UK   kam@jb.man.ac.uk
Alexey **Alakoz**, Astro Space Center of the P.N. Lebedev Physical Institute, Moscow, Russia   l-sha@yandex.ru
Felipe **Alves**, Argelander Institut for Astronomie, Bonn, Germany   falves@astro.uni-bonn.de
Nikta **Amiri**, Leiden Observatory/JIVE, Leiden, Netherlands   amiri@strw.leidenuniv.nl
Tao **An**, Shanghai Astronomical Observatory, Shanghai, China   antao@shao.ac.cn
Bernard Duah **Asabere**, Physics Department, University of Johannesburg, South Africa   b.asabere@rocketmail.com
Willem **Baan**, ASTRON, Dwingeloo, Netherlands   baan@astron.nl
Alejandro **Baez Rubio**, Centro de Astrobiologia (CSIC-INTA), Madrid, Spain   baezra@cab.inta-csic.es
Olga **Bayandina**, Astro Space Center of the Lebedev Physical Institute of RAS, Moscow, Russia   Bayandix@yandex.ru
Per **Bergman**, Onsala Space Observatory, Chalmers University, Sweden   per.bergman@chalmers.se
Roy S. **Booth**, SA SKA, Cape Town, South Africa   rbooth@ska.ac.za
Shari **Breen**, CSIRO Astronomy and Space science, NSW, Australia   Shari.Breen@csiro.au
Tui Rose **Britton**, Dept. of Physics and Astronomy, Macquarie University, NSW, Australia   tui.britton@csiro.au
Crystal **Brogan**, NRAO, Charlottesville, VA, USA   cbrogan@nrao.edu
Do-Young **Byun**, Korea Astronomy and Space Science Institute, Seoul, South Korea   bdy@kasi.re.kr
Jessica **Chapman**, CSIRO Astronomy & Space Science, NSW, Australia   Jessica.Chapman@csiro.au
James Okwe **Chibueze**, Kagoshima University, Kagoshima, Japan   james.chibueze@gmail.com
Se-Hyung **Cho**, Korea Astronomy and Space Science Institute, Seoul, South Korea   cho@kasi.re.kr
Yoon Kyung **Choi**, Max-Planck-Institut fuer Radioastronomie, Bonn, Germany   ykchoi@mpifr.de
Mark **Claussen**, NRAO, Socorro, NM, USA   mclausse@nrao.edu
Willliam **Cotton**, NRAO, Charlottesville, VA, USA   bcotton@nrao.edu
Claudia **Cyganowski**, Harvard/SAO CfA, Cambridge, MA, USA   ccyganow@cfa.harvard.edu
Jeremy **Darling**, University of Colorado, Boulder, CO, USA   jdarling@colorado.edu
James **De Buizer**, SOFIA/USRA, NASA Ames Research Center, CA, USA   jdebuizer@sofia.usra.edu
Eduardo **de la Fuente**, CUCEI, Universidad de Guadalajara, Mexico   edfuente@gmail.com
Lientjie **de Villiers**, Centre for Astrophysics Research, University of Hertfordshire, UK   lientjiedv@gmail.com
Shuji **Deguchi**, NRO/NAOJ, Nagano, Japan   deguchi@nro.nao.ac.jp
Avinash **Deshpande**, Raman Research Institute, Bangalore, India   desh@rri.res.in
Jean-Francois **Desmurs**, OAN, Madrid, Spain   desmurs@oan.es
Phil **Diamond**, CSIRO Astronomy and Space Science, NSW, Australia   phil.diamond@csiro.au
Miranda **Dunham**, Yale University, New Haven, CT, USA   miranda.dunham@yale.edu
Dieter **Engels**, Hamburger Sternwarte, Hamburg, Germany   dengels@hs.uni-hamburg.de
Jarken **Esimbek**, Xinjiang Astronomical Observatory, Xinjiang, China   jarken@xao.ac.cn
Sandra **Etoka**, JBCA, University of Manchester, UK   sandra.etoka@googlemail.com
Michael **Feast**, SAAO/UCT, University of Cape Town, South Africa   mwf@ast.uct.ac.za
Kenta **Fujisawa**, Yamaguchi University, Yamaguchi, Japan   kenta@yamaguchi-u.ac.jp
Michael J. **Gaylard**, HartRAO, Krugersdorp, South Africa   mike@hartrao.ac.za
Ciriaco **Goddi**, ESO, Garching, Germany   cgoddi@eso.org
Sharmila **Goedhart**, SKA SA, Cape Town, South Africa   sharmila@ska.ac.za
Ioannis **Gonidakis**, CSIRO, NSW, Australia   ioannis.gonidakis@csiro.au
Malcolm **Gray**, JBCA, University of Manchester, UK   Malcolm.Gray@manchester.ac.uk
Lincoln **Greenhill**, Harvard/Smithsonian CfA, Cambridge, MA, USA   greenhill@cfa.harvard.edu
Daniel **Harris**, SAO, Cambridge, MA, USA   harris.d.e.42@gmail.com
Christian **Henkel**, MPIfR, Bonn, Germany   chenkel@mpifr-bonn.mpg.de
Tomoya **Hirota**, National Astronomical Observatory of Japan, Tokyo, Japan   tomoya.hirota@nao.ac.jp
Mareki **Honma**, NAOJ, Japan   mareki.honma@nao.ac.jp
Liz **Humphreys**, ESO, Garching, Germany   ehumphre@eso.org
Todd **Hunter**, NRAO, Charlottesville, VA, USA   thunter@nrao.edu
Hiroshi **Imai**, Kagoshima University, Kagoshima, Japan   hiroimai@sci.kagoshima-u.ac.jp
Katharina **Immer**, MPIfR, Bonn, Germany   kimmer@mpifr-bonn.mpg.de
Violette **Impellizzeri**, ALMA, Santiago, Chile   vimpelli@alma.cl
Izaskun **Jimenez-Serra**, Harvard-Smithsonian Center for Astrophysics, MA, USA   ijimenez-serra@cfa.harvard.edu
Norio **Kaifu**, NAOJ, Japan   noriokaifu@aol.com
Sergei **Kalenskii**, Lebedev Physical Institute, Astro Space Center, Russia   kalensky@asc.rssi.ru
Athol **Kemball**, University of Illinois at Urbana-Champaign, IL, USA   akemball@illinois.edu
Jaeheon **Kim**, Yonsei University Observatory/KASI/Kyung Hee University, South Korea   jhkim@kasi.re.kr
Mikyoung **Kim**, NAOJ, Japan   mikyoung.kim@nao.ac.jp
Kee-Tae **Kim**, Korea Astronomy & Space Science Institute (KASI), South Korea   ktkim@kasi.re.kr
Tomoharu **Kurayama**, Kagoshima University, Japan   kurayama@sci.kagoshima-u.ac.jp
Stan **Kurtz**, UNAM, Mexico   s.kurtz@crya.unam.mx
Tanmoy **Laskar**, Harvard University, Cambridge, MA, USA   tanmoylaskar@gmail.com
Marcelo **Leal-Ferreira**, Argelander-Institut fur Astronomie, Germany   ferreira@astro.uni-bonn.de
Jingjing **Li**, Purple Mountain Observatory, China   jjli@pmo.ac.cn
Michael **Lindqvist**, Onsala Space Observatory, Chalmers University, Sweden   Michael.Lindqvist@chalmers.se
Ivan **Litovchenko**, AstroSpace Center of Lebedev Physical Institute, Moscow, Russia   grosh@asc.rssi.ru
Jabulani Paul **Maswanganye**, HartRAO/ Wits University, Krugersdorp, South Africa   pop7paul@gmail.com
Naoko **Matsumoto**, National Astronomical Observatory of Japan, Japan   naoko.matsumoto@nao.ac.jp
Karl **Menten**, MPIfR, Bonn, Germany   kmenten@mpifr.de
Victor **Migenes**, Brigham Young University, UT, USA   vmigenes@byu.edu
David **Morris**, IRAM, France   morris@iram.fr
Kazuhito **Motogi**, Hokkaido University, Japan   motogi@astro1.sci.hokudai.ac.jp
Takumi **Nagayama** NAOJ, Japan   takumi.nagayama@nao.ac.jp
Liudmila **Nazarova**, EAAS, Moscow, Russia   lsnazarova@rambler.ru
Florian **Niederhofer**, ESO, Garching, Germany   floniederhofer@gmx.de
Hans **Olofsson**, Onsala Space Observatory, Chalmers University, Sweden   hans.olofsson@chalmers.se
Nadeem **Oozeer**, SKA SA, South Africa   nadeem@ska.ac.za
Sunelle **Otto**, HartRAO, Krugersdorp, South Africa   sunelle@hartrao.ac.za
Tomoaki **Oyama**, National Astronomical Observatory of Japan, Japan   t.oyama@nao.ac.jp
Sean **Passmoor**, SKA SA, South Africa   sean@ska.ac.za
Andres **Perez Sanchez**, AIfA, Bonn, Germany   aperez@astro.uni-bonn.de
Michele **Pestalozzi**, IFSI-INAF, Rome, Italy   michele.pestalozzi@gmail.com
Nurur **Raman**, University of Johannesburg, Johannesburg, South Africa   nrahman@uj.ac.za
Sofia **Ramstedt**, Argelander Institute for Astronomy, Bonn, Germany   sofia@astro.uni-bonn.de
Mark **Reid**, Harvard-Smithsonian CfA, Cambridge, MA, USA   reid@cfa.harvard.edu
Anita M. S. **Richards**, UK ARC Node, JBCA, University of Manchester, UK   amsr@jb.man.ac.uk
Laura **Richter**, Rhodes University, South Africa   llrichter@gmail.com

Nobuyuki **Sakai**, Graduate University for Advanced Studies, Japan
Alberto **Sanna**, MPIfR, Bonn, Germany
Anuj **Sarma**, DePaul University, Chicago, IL, USA
Mayumi **Sato**, MPIfR, Bonn, Germany
Satoko **Sawada-Satoh**, Mizusawa VLBI Observatory, NAOJ Japan
Nishant **Singh**, Raman Research Institute, Bangalore, India
Lorant **Sjouwerman**, NRAO, Socorro, NM, USA
Rupert **Spann**, SKA SA, South Africa
Vladimir **Strelnitski**, Maria Mitchell Observatory, Nantucket, MA, USA
Olga **Suarez**, Observatoire de la Cote d'Azur, Nice, France
Koichiro **Sugiyama**, Yamaguchi University, Yamaguchi, Japan
Gabriele **Surcis**, JIVE, Dwingeloo, Netherlands
Andrea **Tarchi**, INAF-Osservatorio Astronomico di Cagliari, Italy
Clemens **Thum**, IRAM, Granada, Spain
Anita **Titmarsh**, UTas/CASS, Tasmania, Australia
Jose-Maria **Torrelles**, ICE(CSIC)-UB/IEEC, Barcelona, Spain
Karl Johan Erik **Torstensson**, JIVE/Leiden Observatory, Netherlands
Miguel Angel **Trinidad**, Guanjuato University, Guanjuato, Mexico
Irina **Val'tts**, Astro Space Center of the Lebedev Physical Institute, Russia
Stefanus Petrus **van den Heever**, North West University, South Africa
Johan **van der Walt**, North-West University, South Africa
Huib **van Langevelde**, JIVE/Leiden, Dwingeloo, Netherlands
Wouter **Vlemmings**, Chalmers University, Sweden
Maxim **Voronkov**, CSIRO Astronomy & Space Science, Australia
Andrew **Walsh**, James Cook University, Australia
Junzhi **Wang**, Department of Astronomy, Nanjing University, China
Mark **Wardle**, Macquarie University, Australia
Patricia **Whitelock**, SAAO/UCT, South Africa
Kyle **Willett**, University of Minnesota, MN, USA
Markus **Wittkowski**, ESO, Garching, Germany
Pawel **Wolak**, Nicolaus Copernicus University, Poland
Alwyn **Wootten**, NRAO, Charlottesville, VA, USA
Yuanwei **Wu**, MPIfR, Bonn, Germany
Ye **Xu**, Purple Mountain Observatory, Nanjing, China
Ingyin **Zaw**, New York University Abu Dhabi, New York, USA
Jiangshui **Zhang**, Center for Astrophysics, Guangzhou University, China
Bo **Zhang**, MPIfR, Bonn, Germany
Jianjun **Zhou**, Xinjiang Astronomical Observatory, Xinjiang, China

nobuyuki.sakai@nao.ac.jp
asanna@mpifr-bonn.mpg.de
asarma@depaul.edu
msato@mpifr-bonn.mpg.de
satoko.ss@nao.ac.jp
nishant@rri.res.in
lsjouwer@nrao.edu
rupert.spann@ska.ac.za
vladimir@mmo.org
olga.suarez@oca.eu
koichiro@yamaguchi-u.ac.jp
gabsurcis2@yahoo.it
atarchi@oa-cagliari.inaf.it
thum@iram.fr
anitat@postoffice.utas.edu.au
torrelles@ieec.cat
kalle@strw.leidenuniv.nl
trinidad@astro.ugto.mx
ivaltts@asc.rssi.ru
13077724@nwu.ac.za
johan.vanderwalt@nwu.ac.za
langevelde@jive.nl
wouter.vlemmings@chalmers.se
Maxim.Voronkov@csiro.au
andrew.walsh@jcu.edu.au
junzhiwang@nju.edu.cn
mark.wardle@mq.edu.au
paw@saao.ac.za
willett@physics.umn.edu
mwittkow@eso.org
wolak@astro.uni.torun.pl
awootten@nrao.edu
ywei@mpifr-bonn.mpg.de
xuye@pmo.ac.cn
iz6@nyu.edu
jszhang@gzhu.edu.cn
bzhang@mpifr.de
zhoujj@xao.ac.cn

Tribute to Yolanda Gómez Castellano

The proceedings of IAU Symposium 287 are dedicated to the memory of Yolanda Gómez Castellano. Yolanda died on 16 February 2012, shortly after the symposium, at the age of forty-nine. She had struggled with cancer for a number of years, and her illness took a sudden, and unexpected, turn for the worse just weeks before the meeting. She was a member of the Scientific Organizing Committee of the symposium, and had planned to attend the meeting, canceling her travel plans only days before her scheduled departure for South Africa. Despite the state of her health, Yolanda still had sufficient energy to fulfill her wish to participate in the conference by sending her contribution as a poster (*Imaging the water masers toward the H_2O-PN IRAS 18061-2505*) through Mexican colleagues.

Yolanda's professional career was based at the National Autonomous University of Mexico (UNAM). Her professional degrees (in physics) were all obtained from the UNAM, although her doctoral research (*A study in Radio Frequencies of Young Planetary Nebulae and OH/IR Stars*) was carried out at the Harvard-Smithsonian Center for Astrophysics. Upon her return to Mexico from Boston, she worked side-by-side with her husband, Luis Felipe Rodríguez, to establish a Mexican presence in the international radio astronomy community.

Yolanda worked for more than twenty years at the UNAM; she began her research career at the Institute for Astronomy at the Mexico City campus. Later, she was a founding member of the Center for Radio Astronomy and Astrophysics (CRyA) at the Morelia Campus of the UNAM, where she was a full professor.

Yolanda's primary research area was the study of gaseous nebulae, both young planetary nebulae associated with evolved stars and compact HII regions associated with young massive stars. She also did significant work on protoplanetary disks and jets associated with young stellar objects. Her research resulted in more than 65 published papers with numerous important contributions. Among her most notable discoveries was the first detection of water vapor around a planetary nebula. This was unexpected because the radiation field around these stars should destroy any nearby molecules. She was actively pursuing this fascinating line of research at the time of her death and in fact this was the topic of her contribution at the Symposium. In addition, Yolanda was a generous and cheerful teacher, always willing to help and guide her students. Her enthusiasm led many of her students to become astronomers: without exception her former masters and doctoral students have joined the research community, and are now independent researchers.

She was a firm believer that science is for everyone — not just for specialists but also for the general public. Consequently, Yolanda not only assumed an active role in the postgraduate astronomy program of the UNAM but she also taught undergraduate physics at the state university of Michoacán, in addition to giving courses in astronomy in several Latin American countries. Particularly notable was Yolanda's extensive involvement in public outreach. She gave innumerable talks, workshops, radio and television interviews and published more than 40 popular science articles. She had a key role in the International Year of Astronomy as celebrated in Mexico, organizing all of the events of the CRyA and serving as president of the state organizing committee. The 350 public outreach events that took place that year attracted more than 100,000 participants. The highlight was the *Noche de las Estrellas* at the prehispanic ruins in Tzintzuntzan, Michoacán, which an estimated 10,000 people attended. In addition, Yolanda contributed to the development of teaching materials for popularizing astronomy and the design of astronomical displays and activities for science museums. Thanks to the warmth, enthusiasm and dedication with which Yolanda took astronomy to the public, the state government of Michoacán awarded her the 2008 State Prize for Public Outreach of Science and Technology.

Although perhaps little recognized by her professional colleagues, Yolanda was a avid lover of the arts. Not only did she come from a very musically-talented family, but she was also a devoted follower of the visual arts, decorating her home with outstanding works by national and regional Mexican artists. Moreover, she herself had a creative, artistic side: the logo of the CRyA — a radio telescope depicted in the style of Oriental calligraphy — was of her own design.

Those of us who knew Yolanda personally will miss her ready smile and the cheerfulness that accompanied her wherever she went. She will be greatly missed, by her former students, by her colleagues and friends in Morelia, in Mexico, and around the world, and most of all by her family, in particular by her children, Vicente and Cecilia, and her husband Luis Felipe, to whom we extend the warmest embrace.

Rest in Peace, Yolanda.

S. Kurtz & J.M. Torrelles

T1:

Advances in Maser Theory *Chair: Roy Booth*

Cosmic Masers - from OH to H_0
Proceedings IAU Symposium No. 287, 2012
R.S. Booth, E.M.L. Humphreys & W.H.T. Vlemmings, eds.

© International Astronomical Union 2012
doi:10.1017/S1743921312006576

Advances in Maser Theory

Vladimir Strelnitski

Maria Mitchell Observatory, 4 Vestal Street, Nantucket, MA 02554
email: `vladimir@mmo.org`

Abstract. The groundwork of the cosmic maser theory was laid four decades ago. The elapsed time, including the few years after the last IAU symposium dedicated to masers, did not add much to the fundamentals. In this review, I will summarize some cornerstones of the theory, with an emphasis on issues that don't seem to have received due attention in the past. I will also comment on some new developments.

1. Introduction

Before discussing the theory of astrophysical masers, I would like for us to remember a person who was especially knowledgeable and productive in this field during the past three decades, Professor Bill Watson. Unfortunately, Bill passed away between this and the previous maser symposium. He worked actively in the field up until his premature death in 2009.

The foundations of the theory of maser amplification in astrophysical environments were developed soon after the discovery of the first cosmic masers – in the late 1960s and early 1970s. Seminal papers were published by Marvin Litvak and by Peter Goldreich with colleagues. Later, the theory was elaborated upon and concretized by other authors, but qualitatively new ideas appeared rarer and rarer. When I started preparing this review, I realized that *advances* in cosmic maser theory during the four years passed after the Australia symposium have been modest at best. This explains my decision to dedicate most of the presentation to revisiting the basics of the theory. By doing so, I will emphasize the points that, in my opinion, have not been given enough attention in the past and yet have a good potential to provide deeper insight into the physics of masers. I will also comment on some new developments, especially those that can illustrate the application of the classical aspects of the theory.

I hope that this approach will help us create a common theoretical context for further discussions. I also hope that this review will help practitioners in this field, especially the young researchers, to make fast order-of-magnitude estimates of physical parameters of the maser environment while planning or interpreting observations and that it will help them to avoid some typical mistakes.

2. Population Inversion and Masing as non-LTE Phenomena

Masing in a spectral line requires the fulfillment of two conditions: (1) population inversion, and (2) the absolute value of the optical depth in the line (the 'gain' of the maser) to be > 1. Maser radiation is an extreme case of non-LTE radiation. What is its place among other observable non-LTE phenomena? There are two useful approaches to answer this question.

The first approach (e.g. Strelnitski, Ponomarev & Smith (1996)) is based on the notion of excitation temperature:

3

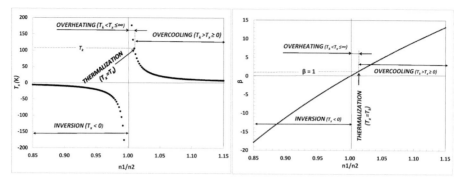

Figure 1. (left) Excitation states of a transition described by the T_x parameter. (right) Excitation states of a transition described by the β parameter.

$$T_x = \frac{h\nu}{k} [\ln(\frac{n_1}{n_2})]^{-1} \, , \qquad (2.1)$$

where $n_1 = N_1/g_1$ and $n_2 = N_2/g_2$ are the populations, per a degenerate sublevel, of the lower and upper levels of the transition, g_1 and g_2 are the statistical weights, and the other symbols have their usual meaning. Fig. 1(left) shows T_x as a function of n_1/n_2 for a transition with an arbitrary frequency ($\nu = 22.2$ GHz). The function has a singular point at $n_1 = n_2$, jumping from $+\infty$ to $-\infty$ when n_1/n_2 changes from > 1 to < 1.

While T_x is only a useful parameter, the kinetic temperature of the gas, T_k, describing the energetic state of the translational degrees of freedom, has a meaningful thermodynamical sense, because translational motions are 'maxwellized' in most astrophysical environments. If collisions control the populations of levels 1 and 2, T_x approaches T_k, and the transition 1-2 is said to be 'thermalized' (see more in section 4). T_k is just a point on the graph of T_x (its value in Fig. 1(left) was arbitrarily set to 106 K). It is natural to call the state of the transition with $+\infty \geqslant T_x \geqslant T_k$ 'overheating' and the state with $T_k \geqslant T_x \geqslant 0$ – 'overcooling.' The former causes an enhanced emission and the latter – an enhanced absorption in the line.

The other way to describe the excitation states of a transition is via the parameter

$$\beta_{12} \equiv \frac{1 - \frac{b_2}{b_1}\exp(-\frac{h\nu}{kT_k})}{1 - \exp(-\frac{h\nu}{kT_k})} = \frac{1 - \exp(-\frac{h\nu}{kT_x})}{1 - \exp(-\frac{h\nu}{kT_k})} \, , \qquad (2.2)$$

first introduced by Brocklehurst & Siton (1972). The coefficients b_1, b_2 are Menzel's coefficients of departure from LTE populations (Menzel (1937)). The graph of β as a function of n_1/n_2 is shown in Fig. 1(right). This function has no singularities, it goes smoothly through 0 when n_1/n_2 passes from > 1 to < 1 and it equals $+1$ at the point of thermalization ($T_x = T_k$). Like T_x, the β function is positive for ($n_1/n_2 \geqslant 1$) and negative for the population inversion. However, unlike T_x, which is a single function for each transition, there is an infinite number of β functions for each transition, corresponding to different values of T_k. The plot in Fig. 1(right) is for the same $\nu = 22.2$ GHz and $T_k = 106$ K as in Fig. 1(left).

An important methodological conclusion from this analysis is that overcooling of a transition ($T_x < T_k$; $\beta > 1$) is not 'anti-inversion' and the enhanced absorption caused by overcooling is not an 'anti-maser' effect, as they are frequently called. Inversion and maser are *absolute* phenomena, whereas the occurrence of overcooling and the ensuing enhanced absorption depend on the value of kinetic temperature. The opposite to overcooling of a transition is its overheating, not inversion. Unlike the 'relative' non-LTE phenomena, masing can, in principle, be accompanied by specific modifications of the radiation field and its interaction with matter. These effects include the line and beam narrowing; competition between spatial modes for pumping (which may make the apparent geometry of maser spots different from the actual geometry of the active region); the exertion of a 'negative' radiation pressure upon the medium; and the transition of the maser to the regime of 'oscillations'(as distinguished from the one-pass, travelling-wave amplification).

3. Maser as a Quantum Heat Engine

It is easily shown that a steady-state inversion of populations is impossible in a *two-level* quantum system. Whatever the number and temperatures of the energy reservoirs interacting with such a system, its excitation temperature will always be between the temperatures of the reservoirs – kinetic temperature, T_k, for collisional reservoirs and brightness temperature, T_b [see equation (4.4) below], for radiative reservoirs. Since both T_k and T_b are essentially positive, T_x of a two-level system will always be positive.

At least one extra level is needed in order to create the 'pumping' – a net population transfer from the lower to the upper maser level, which can cause and sustain population inversion. From the point of view of thermodynamics, any steady-state pumping mechanism is a cyclic quantum 'heat engine,' with some transition(s) of the cycle taking energy from high-temperature reservoirs ('sources') and some other transitions giving energy to low-temperature reservoirs ('sinks'). The source and sink are equally important for pumping. A correct description of a pumping mechanism should therefore refer to the types of *both* of them, not just to the type of the source, as it is frequently presented. Considering the elementary processes responsible for pumping – collisional (C), radiative (R), or chemical (X), there are nine possible source-sink combinations:

$$\begin{array}{ccc} RR & RC & RX \\ CR & CC & CX \\ XR & XC & XX \end{array}$$

The pumping is, respectively, 'radiative-radiative,' 'radiative-collisional,' etc.

The thermodynamical approach to maser physics is a helpful tool. A few examples of its application include: the idea of the (counter-intuitive) collisional-collisional (CC) pumping (Strelnitski (1984), Kylafis & Norman (1987), Elitzur & Fuqua (1989)); the account for the rates of *both* the source and the sink processes in a correct evaluation of the upper limit of maser luminosity (e.g. Strelnitski (1984)); the idea of the photon sink on the cold dust as a boost of the CR pumping rate and as a possible cause of variability of masers due to variations of the dust temperature (Strelnitski (1981)).

4. Thermalization

When the collisional transition probability, in a *two-level* system surpasses the Einstein coefficient of spontaneous emission,

$$C_{21} \equiv q_{21} N_c \gtrsim A_{21}, \tag{4.1}$$

T_x approaches T_k – the transition is 'thermalized' (see section 1). Here $q_{21} = \langle v\sigma \rangle$, v is the relative velocity of the target and collider, $\sigma(v)$ is the collision cross-section, and the angle brackets signify averaging over the velocity distribution. The condition (4.1) translates into a condition for the collider's number density: $N_c \gtrsim A_{21}/q_{21}$.

One may expect that sufficiently frequent collisions between the maser levels would do the same – bring the (negative) excitation temperature of the transition to the (positive) kinetic temperature, i.e. destroy the population inversion. However, as argued in section 2, a maser is, by necessity, a system of *more than two* levels. The use of the naïve two-level thermalization condition (4.1) to evaluate the upper limit of an allowable gas density in a maser may lead to large mistakes.

Actually, the frequently used condition (4.1) requires some elaboration even in its application to a two-level system. This condition follows from the obvious equation for the ratio of the level populations in a steady-state two-level system:

$$\frac{N_2}{N_1} = \frac{C_{12} + B_{12}J_{12}}{A_{21} + C_{21} + B_{21}J_{12}} , \tag{4.2}$$

where B_{12} and B_{21} are the Einstein coefficients for stimulated absorption and emission, respectively, and J_{12} is the radiation intensity averaged over directions and frequencies within the line. When $C_{21} >> A_{21}$ *and* $B_{21}J_{21}$, it follows from equation (4.2) and the well-known relation between C_{12} and C_{21}:

$$\frac{N_2}{N_1} = \frac{g_2}{g_1}e^{-\frac{h\nu}{kT_k}} \tag{4.3}$$

– the Boltzmann population ratio with $T_x = T_k$, which means thermalization by collisions.

However, the additional condition $C_{21} >> B_{21}J_{21}$ can be unquestionably fulfilled only when $J_{12} \to 0$ – in an optically thin medium without external radiation. If $C_{21} << A_{21}$ in an optically thin system, and the external radiation is strong enough to also make $C_{21} << B_{21}J_{21}$, then, using the well-known relation $g_1B_{12} = g_2B_{21}$ and the definition of the brightness temperature (based on the Planck equation),

$$T_r = \frac{h\nu}{k}\left[\ln\left(\frac{2h\nu^3}{c^2 J_{12}} + 1\right)\right]^{-1}, \tag{4.4}$$

we get from equation (4.2):

$$\frac{N_2}{N_1} = \frac{g_2}{g_1}e^{-\frac{h\nu}{kT_r}}, \tag{4.5}$$

i.e. the Boltzmann distribution with $T_x = T_r$. One can consider this case as 'thermalization' of the transition by external radiation.

In an optically thick medium (the smallest optical depth $\tau_{min} >> 1$), the trapped line photons and the collisions work in concert toward thermalization, and T_x (and T_r) approach T_k at lower densities than predicted by condition (4.1), namely, at $N_c \sim A_{21}/(q_{21} \cdot \tau_{min})$ [see e.g. Avrett & Hummer (1965)].

In a maser, the rate of thermalizing collisions should be compared not with the spontaneous emission rate A_{21} but with the rate of pumping. It is customary to describe the rate of pumping by four phenomenological coefficients: Λ_1, Λ_2 $(\mathrm{cm^{-3}s^{-1}})$ – the coefficients of population supply to the maser levels from other levels, and Γ_1, Γ_2 $(\mathrm{s^{-1}})$ – the population decay rates. Assuming, for simplicity, that $\Gamma_1 = \Gamma_2 \equiv \Gamma$ and $C_{12} = C_{21} \equiv C$, one can present the (unsaturated) population *difference* (the quantity that determines the maser gain) as

$$\Delta N_0(\mathrm{cm^{-3}}) \equiv (N_2 - N_1)_0 \approx \frac{\Lambda_2 - \Lambda_1}{\Gamma + C}. \tag{4.6}$$

By formal analogy with the two-level system, it may appear that 'thermalizaiton' of the maser occurs when C exceeds Γ. However, equation (4.6) is not analogous to equation (4.2). If pumping involves collisions, the quenching of the maser can be postponed to a much higher density than determined by $C \sim \Gamma$. For example, if the population supply involves collisions, then Λ_2 and Λ_1, and thus the whole numerator in equation (4.6) will grow with N_c, which would prevent the decrease of ΔN_0 after C becomes $\sim \Gamma$. A good example of this case is presented by the H masers (section 6).

The extreme case is the CC pumping in which the source and the sink are due to collisions with two kinds of particles with different kinetic temperatures (section 3). In this case, all the terms in the right-hand side of equation (4.6) are proportional to the gas density, and, as long as the relative abundances of the two kinds of colliders and their temperature difference are sustained, there is *no* limiting ('thermalizing') density at all.

5. Saturation

When the rate of radiative transitions stimulated by the growing maser radiation becomes comparable with the pumping rate, the equation for the population difference can be presented in the form:

$$\Delta N \approx \frac{\Delta N_0}{1 + \frac{B_{21}}{\Gamma + C}} \approx \frac{\Delta N_0}{1 + \frac{J_{12}}{J_s}}, \tag{5.1}$$

where ΔN_0 is the unsaturated population difference given by equation (4.6), and

$$J_s \equiv \frac{\Gamma + C}{B_{21}} \tag{5.2}$$

is the 'saturation intensity.'

In a maser, $dI \propto \Delta N \cdot I \cdot ds$, where ds is the differential of the path length. In an unsaturated maser, $\Delta N \equiv \Delta N_0$ does not depend on I, which results in $dI/I = \text{const} \cdot ds$ and thus an exponential growth of I. Under strong saturation, when $J_{12} >> J_s$, we can ignore unity in the denominator of equation (5.1); ΔN becomes inversely proportional to J_{12}, and thus to the central intensity of the line, I. The dependence of dI on I cancels. Combining equations (4.7), (5.1) and (5.2), we get: $dI = \text{const} \cdot (\Lambda_2 - \Lambda_1) \cdot ds$ – a linear amplification proportional to the pump rate.

The fully saturation regime is the one of maximum usage of available pumping, with the release of a maser photon for each pumping cycle (in an unsaturated regime, most of the pumping cycles are closed by a collisional relaxation of transition 2-1). This fact is used to estimate the capability of a theoretical pumping mechanism to secure the observed flux density, S, in the masing line. A useful relation is (Strelnitski (1984)):

$$\left(\frac{N_1 \Delta \Gamma}{\text{cm}^{-3} \, \text{s}^{-1}} \right) \gtrsim 10^1 \left(\frac{l}{\text{A.U.}} \right)^{-3} \left(\frac{D}{\text{kpc}} \right)^2 \left(\frac{S}{\text{Jy}} \right) \left(\frac{\Delta \nu / \nu}{10^{-6}} \right), \tag{5.3}$$

where $N_1 \Delta \Gamma \approx \Delta \Lambda$ is the pump rate; l is the amplification length along the line of sight; D is the distance to the source; and $\Delta \nu$, ν are the line's width and central frequency. Note that equation (5.3) contains only one geometrical parameter of the maser – its length along the line of sight, whose probable upper limit can often be estimated from the interferometrical map of the source (e.g. Strelnitski (1984)). This equation is valid, within an order of magnitude, for any geometry elongated along the line of sight. In the extreme (and improbable) case of purely spherical geometry, the coefficient 10^1 should be replaced by 10^2.

6. Hydrogen Masers and Lasers

The first mm and submm masers on hydrogen recombination lines were discovered at the end of the 1980s in the emission-line star MWC 349 (Martín-Pintado *et al.* (1989)). Later, the proof of masing (formally, already 'lasing') in this star was extended to the

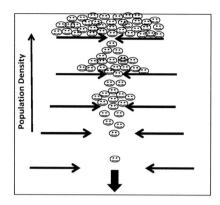 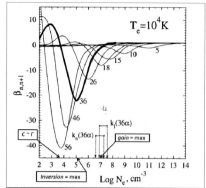

Figure 2. (left) To the formation of population inversion in recombining H atoms. (right) $\beta(N_e)$ for hydrogen recombination lines. The three arrows indicate the values of $\log N_e$ for which the H36α line has (1) $C \sim \Gamma$; (2) maximum inversion; and (3) maximum maser gain. Based on Storey & Hummer (1995) and Strelnitski, Ponomarev & Smith (1996).

IR domain (Strelnitski *et al.* (1996); Thum *et al.* (1998)). The observational aspects of hydrogen masers and lasers were reviewed by Martín-Pintado (2002). Here I will comment on some theoretical moments.

D. Menzel, the person who, apparently, was the first to understand the very possibility of radiation amplification in an inverted quantum system (Menzel (1937)), was also one of the first to create and... overlook a numerical model of hydrogen recombination line masers, discovered a half-century later. Use the above equation (2.2) to check that in Table 6 of his paper with Baker (Baker & Menzel (1938)), dedicated to the calculations of the theoretical Balmer decrements in gaseous nebulae, all the populations on levels $n > 5$ are *inverted* at $T_e = 10000$ K.

At low densities, typical for large HII regions, the inversion of populations across a wide group of Rydberg levels naturally arises as a result of spontaneous cascade after recombination to high-n levels. The rate of spontaneous decay steeply increases with decreasing n (approximately as n^{-5}). The analogy with a crowd rushing out from a building through a sequence of ever broadening doors (Fig. 2, left) may help understand why this leads to population inversion.

At higher densities, collisions with electrons and ions take part in the pumping. They redistribute the populations so that, depending on the density, the inversion for some group of levels increases, as compared with the purely radiative case, and for another group it decreases and can even be replaced by overcooling. Fig. 2(right) reproduces the results of calculations of the β parameter, as a function of electron density, for hydrogen α-lines with n from 5 to 56, for $T_e = 10^4$ K. At $N_e \approx 10^4$ cm^{-3}, for example, the lines around H55α attain maximum inversion (maximum absolute value of $\beta < 0$), whereas the lines around H25α are subject to maximum overcooling ($\beta > 1$). The shift of maser 'thermalization' to higher densities (see section 4) is illustrated with the H36α line. The 'two-level' condition of thermalization ($C \sim \Gamma$) is attained at $N_e \sim 5 \times 10^3$ cm^{-3}, but the inversion and the maser gain reach their maxima at much higher densities, $\sim 10^5$ and 10^7 cm^{-3}, respectively.

The possibility of maser amplification in high-n hydrogen recombination lines during the cosmological epochs of recombination and reionization was first considered by Spaans & Norman (1997). More recent investigation showed that observable effects from the epoch of recombination are improbable, but masing of the giant ionized clouds of the first galaxies, at $z \sim 10$, in the hydrogen lines whose frequencies, after red shift, fall into mm

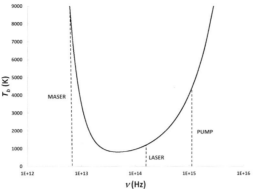

Figure 3. Lasers *vs.* masers

and short-cm domains may make them detectable with the modern radio-astronomical facilities (Loeb & Strelnitski, in preparation).

7. **Astrophysical Lasers** *versus* **Masers**

The broad inventory of masing and lasing hydrogen lines in MWC 349 helped understand why astrophysical masers are so common as compared with lasers. One of the possible reasons is connected with saturation (Strelnitski, Smith & Ponomarev (1996)).

Another way to understand the rareness of lasers (Messenger & Strelnitski (2010)) is based on the plausible assumption that the pumping cycles in most astrophysical environments contain at least one radiative step. The condition of maximum one maser photon per one pumping cycle (section 5) leads then to the conclusion that the *intensity* of the masing/lasing line must always be less or, at best, equal to the intensity of the pumping (source or sink) radiation. The dependence of brightness temperature on frequency for radiation of *fixed intensity* is shown in Fig. 3. The two branches of the curve correspond, asymptotically, to the Wien (laser) and Rayleigh-Jeans (maser) domains. Considering that the frequency of pumping is always higher than that of the maser or laser, Fig. 3 shows that, with the same pumping frequency, masers can achieve very high brightness temperatures, while the brightness temperature of lasers should, typically, be *lower* than that of pumping, which, observationally, creates a vicious circle. The probability of detecting high-gain astrophysical lasers is therefore low. This was recently pointed out also by Salzmann & Takabe (2011) in connection with the putative astrophysical X-ray lasers.

8. **Polarization**

Polarization of maser radiation, with its potential of extracting important information about the magnetic field in the maser's environment, remains one of the most controversial and 'polarizing' topics of cosmic maser theory.

The 'magnetic' interpretation of observed polarization was first addressed in the seminal paper of Goldreich, Keeley & Kwan (1973; hereafter 'GKK'). They considered the simplest case of 1-dimensional maser permeated by a homogeneous magnetic field B; the lowest angular momentum, $J = 1 - 0$, transition; and an isotropic pumping. Using an analytic, semi-classical approach, they obtained results for various combinations of four key parameters: the bandwidth of radiation, $\Delta\omega$; the Zeeman splitting $g\Omega$ (where g is the Landé g value for the upper state and $\Omega = eB/mc$ is the girofrequency); the population decay rate, Γ; and the stimulated emission rate, $R \equiv BJ$.

For paramagnetic molecules, such as OH, the Zeeman splitting $g\Omega$ in the magnetic fields of the expected strength is greater than the maser line width, which leads to the well anticipated result of a Zeeman pattern. The strength of the magnetic field can be directly determined from the observed separation of the circularly polarized σ components.

More complicated are the results for non-paramagnetic molecules, such as SiO, H_2O or CH_3OH , for which $g\Omega < \Delta\omega$ for any plausible values of the magnetic field strength. Analytical solutions were obtained by GKK only for linear polarization. One of the most cited results is that in a completely saturated maser ($R \gg \Gamma$) and strong enough field $[g\Omega > (R\Gamma)^{1/2}]$ the fractional linear polarization – the ratio of the Stokes parameters $Q = I_x - I_y$ and $I = I_x + I_y$, is

$$\frac{Q}{I} = \frac{3\sin^2\theta - 2}{3\sin^2\theta} \quad \text{for } \sin^2\theta \geqslant 1/3 \ (\theta \geqslant 35°);$$

$$\frac{Q}{I} = -1 \quad \text{for } \sin^2\theta \leqslant 1/3 \ (\theta \leqslant 35°),$$

where θ is the angle between the direction of the magnetic field and the propagation direction. Two important conclusions follow from this result: (1) linear polarization indicates the direction of the projection of the magnetic field onto the plane of the sky: it is either parallel or perpendicular to this projection; the change of the sign occurs at $\sin^2\theta = 2/3$ ($\theta = 54.7°$, the so called 'van Vleck angle'); and (2) when the direction of the magnetic field is close to the direction of the line of sight ($\theta \leqslant 35°$), the linear polarization reaches 100%.

Watson and co-authors revisited the problem in several papers (e.g. Watson & Western 1984; Watson & Wyld 2001) using the same formalism as GKK but obtaining solutions numerically. They showed, in particular, that the expected degree of linear polarization decreases with the decreasing degree of maser saturation and with the increasing angular momentum of the maser levels.

Detection of circular polarization in the maser radiation of non-paramagnetic molecules (SiO, H_2O) by the end of 1980s (Barvainis, McIntosh & Predmore (1987); Fiebig & Güsten (1989)) required a theory which would allow for an assessment of the magnetic field strength from the observed Stokes V parameter. The first interpretation was based on the theory for non-masing lines, which seems to be adequate for an unsaturated maser. According to this theory, V is proportional to $B\cos\theta$, and to the derivative of the intensity over frequency within the line:

$$V = \text{const} \cdot B\cos\theta \cdot \frac{dI}{d\nu}, \tag{8.1}$$

where the constant depends on the quantum-mechanical properties of the transition.

Nedoluha & Watson (1992) were the first to extend this classical theory to the case of a maser – a partially saturated H_2O maser. They came to the conclusion that, as long as $g\Omega \gg R$, the error in the strength of the magnetic field obtained with the non-maser theory is less than a factor of 2, but it may be much greater if this condition is not fulfilled. Another important difference from the non-maser theory is the dependence of V on θ. The deviation from the classical $\cos\theta$ dependence increases with the intensity of the maser (Watson & Wyld 2001). For high intensities and saturation, the maximum V is achieved at $\cos\theta \ll 1$, closer to the perpendicular orientation of the field to the line of sight.

A persistent problem with the purely magnetic explanations of maser polarization is that, in some cases, the assessed magnetic field is implausibly strong. For example,

Deguchi & Watson (1986) showed that the high degree of linear polarization in some strong H_2O masers (such as the famous 'super-maser' in Orion) requires magnetic fields of up to $\sim 1G$, whereas the subsequently measured circular polarization of H_2O emission led to the fields ~ 10 mG (Fiebig & Güsten (1989); Nedoluha & Watson (1992)). Unlikely strong magnetic fields, $B \sim 10 - 100$ G, were indicated by the circular polarization of SiO maser in several late-type stars (Barvainis, McIntosh & Predmore (1987)). More recently, the magnetic interpretation of circular polarization observed in several 6.7 GHz and other methanol masers with the use of the Landé factors extrapolated from the only known laboratory measurements by Jen (1951)) for the 25 GHz transitions, with the correction for the old, recently revealed order-of-magnitude arithmetical error in the extrapolation, led to implausibly high values of the magnetic fields (e.g. Vlemmings, Torres & Dodson (2011); Fish *et al.* (2011)).

These cases urge the researchers to address the studies of non-magnetic or non-Zeeman mechanisms for polarization. I will briefly comment on a few of them.

The capability of *anisotropic pumping* to produce linear polarization has been discussed in many papers, starting in 1960s. For a recent application to unsaturated masers, see, for example, Asensio Ramos, Landi Degl'Innocenti & Trujillo Bueno (2005). Western & Watson (1983) showed that linear polarization can be produced by the competition between intersecting rays of a saturated maser in a non-spherical medium or a medium with velocity gradients, if, in addition, there is a lack of axial symmetry along the line of sight.

Variations of the orientation of a (relatively weak) magnetic field along the line of sight can cause circular polarization of the output radiation, if the input radiation is linearly polarized (Wiebe & Watson (1998)). In a maser with $R \sim g\Omega$, circular polarization can appear because of the change of the quantization axis from the direction of the magnetic field to the direction of propagation (Nedoluha & Watson (1994)).

Problems with the theories of maser polarization were strongly emphasized by Elitzur, including the acknowledgment of 'incompleteness' of his own, alternative analytical approach that he developed in a series of papers in the 1990s. The references and the summary of his views can be found in Elitzur (2002) and Elitzur (2007). In particular, he put into question the validity of the results obtained by numerical simulations based on GKK theory. References, including the papers with counter-arguments by Watson and co-authors can be found in Dinh-V-Trung (2009). The latter author performed a computer modeling of maser amplification in the presence of a strong magnetic field, proceeding from random realizations of broadband seed radiation for the maser. Averaging over the ensemble of emerging maser radiation led to the polarization parameters that are in general agreement with the old numerical simulations based on GKK theory, which the author considers to be a confirmation of the validity of the old numerical modeling. The summary of Watson's views on the problem of polarization is presented in his last review (Watson (2009)).

The best remedy from the uncertainties of the current theories of cosmic maser polarization is, probably, a complex approach, when each significant conclusion (such as the strength and structure of the magnetic field) is verified by more than one independent method.

9. Concluding Remarks:

- 'Anti-inversion' and 'anti-maser' don't have physical meaning.
- Pumping mechanisms must be identified by *both* their source *and* sink.
- Be careful while assessing when a maser 'thermalizes'!

- Saturation is the key to most maser problems.
- First numerical models of H masers were created a half-century before their discovery (without computers and without understanding the result).
- Beware of land mines while entering the field of maser polarization!

Acknowledgments: The preparation of this review and the participation of the author in the Symposium were supported by the IAU and the Maria Mitchell Association, which is gratefully acknowledged.

References

Asensio Ramos, A., Landi Degl'Innocenti, E., & Trujillo Bueno, J. 2005, *ApJ* 625, 985
Avrett, E. H. & Hummer, D. G. 1965, *MNRAS* 130, 295
Baker, J. G. & Menzel, D. H. 1938, *ApJ* 88, 52
Barvainis, R., McIntosh, G., & Predmore, C. R. 1987, *Nature* 329, 613
Brocklehurst, M. & Siton, M. J. 1972, *MNRAS* 157, 179
Deguchi, S. & Watson, W. D. 1986, *ApJ* 302, 750
Elitzur, M. 2002, in: V. Migenes & M. Reid, (eds.), *Cosmic Masers: From Protostars to Black-holes*, Proc. IAU Symposium No.206 (San Francisco: ASP), p. 452
Elitzur, M. 2007, in: J. Chapman & W. Baan (eds.), *Astrophysical Masers and their Environments*, Proc. IAU Symposium No.242 (San Francisco: ASP), p. 7
Elitzur, M. & Fuqua 1989, *ApJ* (Letters) 347, L35
Fiebig & Güsten 1989, *AA* 214, 333
Fish, V., Muehlbrad, T. C., Pratap, P., Sjouwerman, L. O., Strelnitski, V., Pihlström, Y. M., & Bourke, T. M. 2011, *ApJ* 729, 14
Goldreich, P., Keeley, D. A., & Kwan, J. Y. 1973, *ApJ* 179, 111
Jen, C. K. 1951, *Phys. Rev.* 81, 197
Kylafis, N. D. & Norman, C. 1987, *ApJ* 323, 346
Litvak, M. M. 1970, *Phys. Rev. A* 2, 2107
Martín-Pintado, J. 2002, in: V. Migenes & M. Reid (eds.), *Cosmic Masers: From Protostars to Blackholes*, Proc. IAU Symposium No.206 (San Francisco: ASP), p. 226
Martín-Pintado, J., Bachiller, R., Thum, C., & Walmsley, M. 1989, *AA* 215, L13
Menzel, D. H. 1937, *ApJ* 85, 33
Messenger, S. J. & Strelnitski, V. 2010, *MNRAS* 404, 1545
Nedoluha, G. E. & Watson, W. D. 1992, *ApJ* 384, 185
Nedoluha, G. E. & Watson, W. D. 1994, *ApJ* 423, 394
Salzmann, D. & Takabe, H. 2011, *PASJ* 63, 727
Spaans, M. & Norman, C. A. 1997, *ApJ* 488, 27
Storey, P. J. & Hummer, D. G. 1995, *MNRAS* 272, 41
Strelnitski, V. 1981, *Soviet Astron.* (Letters) 7, 223
Strelnitski, V. 1984, *MNRAS* 207, 339
Strelnitski, V., Ponomarev, V., & Smith, H. 1996, *ApJ* 470, 1118
Strelnitski, V., Haas, M. R., Smith, H. A., Erickson, E. F., Colgan, S. W. J., & Hollenbach, D. J. 1996, *Science* 272, 1459
Strelnitski, V., Smith, H., & Ponomarev, V. 1996, *ApJ* 470, 1134
Thum, C., Martin-Pintado, J., Quirrenbach, A., & Matthews, H. E. 1998, *AA* 333, L63
Vlemmings, W. H. T., Torres, R. M., & Dodson, R., 2011 *AA* 529, 95
Watson, W. D. 2002, in: V. Migenes & M. Reid (eds.), *Cosmic Masers: From Protostars to Blackholes*, Proc. IAU Symposium No. 206 (San Francisco: ASP), p. 464
Watson, W. D. 2009, *Rev.Mexicana AyA* (Serie de Conferencias) 36, 113
Watson & Wyld 2001, *ApJ* (Letters) 558, L55
Western & Watson, W. D. 1983, *ApJ* 275, 195
Wiebe & Watson, W. D. 1998, *ApJ* (Letters) 503, L71

Cosmic Masers - from OH to H_0
Proceedings IAU Symposium No. 287, 2012
R.S. Booth, E.M.L. Humphreys & W.H.T. Vlemmings, eds.
© International Astronomical Union 2012
doi:10.1017/S1743921312006588

Modelling of Cosmic Molecular Masers: Introduction to a Computation Cookbook

Andrej M. Sobolev[1] and Malcolm D. Gray[2]

[1] Ural Federal University,
Lenin Ave. 51, Ekaterinburg 620000, Russia
email: Andrej.Sobolev@usu.ru

[2] University of Manchester, Manchester M13 9PL, UK
email: Malcolm.Gray@manchester.ac.uk

Abstract. Numerical modeling of molecular masers is necessary in order to understand their nature and diagnostic capabilities. Model construction requires elaboration of a basic description which allows computation, that is a definition of the parameter space and basic physical relations. Usually, this requires additional thorough studies that can consist of the following stages/parts: relevant molecular spectroscopy and collisional rate coefficients; conditions in and around the masing region (that part of space where population inversion is realized); geometry and size of the masing region (including the question of whether maser spots are discrete clumps or line-of-sight correlations in a much bigger region) and propagation of maser radiation. Output of the maser computer modeling can have the following forms: exploration of parameter space (where do inversions appear in particular maser transitions and their combinations, which parameter values describe a 'typical' source, and so on); modeling of individual sources (line flux ratios, spectra, images and their variability); analysis of the pumping mechanism; predictions (new maser transitions, correlations in variability of different maser transitions, and the like). Described schemes (constituents and hierarchy) of the model input and output are based mainly on the experience of the authors and make no claim to be dogmatic.

Keywords. interstellar matter, molecules, masers, numerical modelling

1. Introduction

Creating maser models makes it possible to justify the use of masers as sensitive tracers of specific objects (young stellar objects, late-type stars, active galactic nuclei, for example), to use them as tools (to study kinematics, measure physical parameters including magnetic fields, to make direct measurements of distance, and to study the structure of the Galaxy, to name just some), and to study masers as a unique phenomenon which is interesting in itself.

Modelling allows us to find likely/optimum conditions in the maser sources, interpret spectra and maps of generic sources and individual objects, reveal relations between maser spots and physical condensations, understand variability, predict likely new masers, reveal detailed pumping schemes and resolve propagation problems and so on.

In many cases modelling of cosmic molecular masers should be numerical. Analytical estimates often can carry only qualitative character for at least two reasons. The first reason is connected with the nature of the maser phenomenon, and in particular the complexity of the pumping and saturation schemes in real molecules: maser amplification happens due to the existence of the population inversion which is created by the pumping mechanism. For the brightest masers it was shown that these mechanisms involve very large numbers of levels and transitions (see, for example, Sobolev (1989) for H_2O masers, Sobolev & Deguchi (1994a) for CH_3OH masers and Gray (2007) for OH masers). Typically the figures are hundreds of levels and thousands of transitions. We note that here, and further on, we are trying to cite the historically earliest papers in accessible journals; later works refer to them. Given the likely numbers of levels and transitions, analytical techniques must almost always give way to the processing power of the computer. The second reason is connected with the complexity of the astrophysical maser environments. Simple geometrical models can describe these systems only approximately. The presence of inversions itself requires non-equilibrium conditions, where there is an interplay between different constituents of the matter and radiation, magnetic and velocity fields. The masing gas is often subject to large-scale velocity gradients, and the medium is normally highly turbulent at smaller scales. For some specific examples, see Deguchi (1982) and Humphreys *et al.* (1996) for circumstellar masers, Sobolev, Wallin & Watson (1998) for interstellar masers and Wallin, Watson & Wyld (1998) for masers in the accretion disks of active galactic nuclei. Additionally, in some cases, maser amplification can be restricted by the capacity of the medium to carry emission of extreme brightness (the phenomenon known as saturation) and extremely bright emission affects the medium through which it propagates.

In this paper we will try to outline the basics of computational modelling of astrophysical molecular masers in the most widespread case of stationary, that is steady-state, masers which arise due to a pumping mechanism creating inversion of population numbers for the levels of one or more maser transitions.

2. Model input

The first essential stage of model construction is assembly of all the relevant input data and relations that are necessary for construction of a computational model.

Scheme of levels. Astrophysical molecular masers appear because of the operation of the pumping mechanism. In the majority of cases, this consists of the combined action of numerous pumping cycles. Links of these cycles connect energy levels of the molecule and correspond to quantum-mechanical transitions, usually driven by radiative and/or collisional processes. It is very important to know which levels are essential for operation of the pumping mechanism. This often requires special consideration, based on the knowledge of relevant molecular spectroscopy and physical conditions in the maser source and its environment which in turn controls the transition rates. It is generally thought that the scheme of energy levels used for computations can be limited to those levels which are expected to have high population numbers (numbers of molecules in the state corresponding to the particular energy level). In some cases this can be erroneous, and it is recommended to include also levels which are connected with the highly populated levels by the set of transitions of high probability (that is possessing large radiative or collisional rates). This statement is illustrated by the example of class II methanol maser pumping where the sparsely populated torsionally excited levels play the role of extremely important transients: in computations of Sobolev & Deguchi (1994b), neglecting torsionally excited levels in a representative model turned strong maser emission into absorption. It is difficult to make recommendations on how to construct adequate schemes of levels for the cases when available molecular spectroscopy data is insufficient. But for the case of sufficient spectroscopic data one can constrain an adequate scheme of levels using an approach similar to described in Sobolev (1989) and Sobolev & Deguchi (1994b).

Transition rate coefficients. The first numerical models appeared just after reliable estimates for the rates of transitions became available (see de Jong (1973), Shmeld *et al.* (1976) and Bolgova *et al.* (1988) for different pumping mechanisms of interstellar H_2O

masers, Deguchi (1977) for circumstellar H_2O masers, Cesaroni & Walmsley (1991) and Gray *et al.* (1991) for OH masers in various sources of different nature, Sobolev & Strelnitskii (1983) and Sobolev & Deguchi (1994b) for CH_3OH masers of different classes, Deguchi & Iguchi (1976) for circumstellar SiO masers). It is worth pointing out that the absence of essential data on molecular excitation forced early enthusiasts of maser research to use (and sometimes invent) approximations for the necessary values. Most of the papers mentioned above used different guesses on the rules and values of collisional rates. These guesses were not purely serendipitous and were based on deep knowledge of molecular spectroscopy and physics of collisional excitation. Moreover, stability of conclusions was usually verified by comparison of results obtained with different models for the transition rates. Later examination (see for example Cragg *et al.* (2005) for CH_3OH maser models) with the use of refined data have shown that the qualitative conclusions of the previous modelling remained unchanged, though some numbers, for example the density threshold for collisional quenching of class II CH_3OH masers, altered by an order of magnitude.

Conditions in and around masing regions. The next essential part of the input consists of a description of the physical parameters in the masing region (internal conditions), and in its nearby environment (external conditions). Important parameters determining internal conditions include the kinetic temperature, abundance of the maser molecule and number density of the medium – all of which affect collisional and radiative rates – and the velocity field, and the temperature of any dust present, which affect radiation transfer specifically. Examples of the second (external) type are usually restricted to radiation fields (typically from a nearby protostellar object in the case of a star-forming region maser). Population transfer rates are usually represented as products of a (first-order) rate-coefficient and a fractional level population. For the case of collisional rates, the first order rate coefficient is in turn a product of the number density of a collision partner and a second-order rate coefficient. The functional forms of such coefficients are relatively well known (see, for example, the relevant theory in Flower (2003)), and are normally determined by the kinetic temperatures of collisional partners, usually molecular hydrogen and helium (see, e.g. Cragg *et al.* (2005) for the models of class II CH_3OH masers). In some cases other essential collisional partners (e.g. electrons in Bolgova *et al.* (1988)) were introduced. In contrast, functions of physical parameters associated with the radiative rates are very complicated, and usually require long computations. This comes about because evaluation of radiative rates requires solution of the radiative transfer problem, the complexity of which will be discussed in the special subsection below. Radiative rates are determined by emission from external and internal sources and by extinction (absorption and scattering) within the region where the maser is formed. Knowledge of the radiation field parameters is required for all important pumping lines in addition to the maser line itself. Here, we provide some explanatory notes on the physical parameters that determine the effective rate coefficients of radiative fields. Emission and absorption in transitions of the masing molecule itself were shown to be important by calculations in some of the earliest models (see, for example, de Jong (1973)). In the simplest and most common case, the pumping is realized via collisional excitation and radiative decay, followed by escape of the photons from the maser formation region. Later it was established that absorption by internal dust can create a maser even in the case where high optical depth in the pumping lines, within the object, prevents escape of photons (see, for example, Shmeld *et al.* (1976)). In some cases, this internal dust plays the role of a source of energy for the pumping, as in Voronkov *et al.* (2005), so the role of internal dust is twofold. Dust emission from the maser environment can be the main source of energy in the pumping mechanism. Good examples of this are presented by the strong class II methanol masers and OH masers (Sobolev & Deguchi (1994b), Cragg *et al.* (2005) and Gray (2001)). It is important to mention that the properties of the dust grains (sizes, composition, an so on) influence its spectrum. These differences can affect maser line intensities considerably (see Ostrovskii & Sobolev (2002), Gray (2001)). Apart from the dust, the role of external sources of pumping radiation can be taken by optically thick HII regions (Slysh *et al.* (2002)). This is most efficiently done by the youngest sources, hypercompact HII regions, which are bright at sufficiently high frequencies (Sobolev *et al.* (2007)). External sources of radiation also play an important role as the background that is amplified by the masers (Cooke & Elitzur (1985), Sobolev & Deguchi (1994b) and Gray (2001)). It should be noted that the maser brightness is a nonlinear function of the background intensity. It was shown that in some cases background emission that gives considerable radiation density at the frequency of a maser transition can produce maser saturation effects (see Sobolev & Deguchi (1994b) and Sobolev *et al.* (1997a)). The last parameter of the maser formation region which we would like to mention here has a chemical nature. It is the molecular abundance of the maser molecule, which is very important because it determines number of masing molecules in the line of sight, and, hence, the brightness of the maser emission. Chemical questions are very important because explaining the strongest masers often requires an assumption of a very high abundance of the masing molecule. Such high abundances put restrictions on the physical state and evolution of the region. Chemical considerations are also very important when one considers relative intensities of masers in transitions of different maser species (see the analysis of coexistence of OH and CH_3OH masers in Hartquist *et al.* (1995) and Cragg *et al.* (2004)).

Geometry, kinematics and size of the masing region. The geometry of the maser region strongly affects the solution of the radiative transfer problem in both its aspects: calculation of internal radiative fields (the necessity of which were considered in the previous subsection) and calculation of the properties of the emergent radiation. The geometry of masing objects at the scale of, say, VLBI resolution is usually difficult to establish. However, on the larger scale of a formation region, consisting of many individual masing objects or features, it is often possible to suggest a suitable geometry for modelling, often from the evidence of observations of the source in non-maser lines, particularly in lines of CO, that trace low-velocity molecular outflows, and in lines of shock-tracing molecules, for example SiO and molecular hydrogen. There are various forms commonly used for modelling: megamasers typically have an overall disk or toroidal structure (Lockett & Elitzur (2008) and the clumpy model by Wallin *et al.* (1998)), and a similar geometry is applicable to some types of star-forming region masers (see, for example, the model in Gallimore *et al.* (2003)). Other star-forming region masers have been placed at the edges of an outflow, as in models by Mac Low *et al.* (1994) and Sobolev *et al.* (1998)). Masers associated with shocks, for example OH 1720-MHz masers in SNRs and, in particular, water masers in outflows are often treated in a slab geometry with the slabs arranged parallel to the shock front (for example, Yates *et al.* (1997)). For masers associated with the shell at the edge of an HII region, slab models have also been used, but introducing clumps is sometimes necessary (Sutton *et al.* (2001)). For stellar masers, the obvious geometry is spherical, since these masers form in shells in the circumstellar envelope. However, for certain supergiants and post-AGB objects, the spherical approximation is poor, and the model should probably rely on a different geometry including clumps (for example Humphreys *et al.* (1996)). Processes of radiative transfer are tightly related to velocity fields within the region. It is well known that monotonic large velocity gradients lead to localization of the radiative transfer problem, which simplifies calculations and reduces the number of molecules contributing to maser emission. Non-monotonic and irregular fields result in interactions between spatially differentiated parts of the object which greatly affects maser spectra and images, as in Deguchi (1982), Humphreys *et al.* (1996) and Sobolev, Wallin & Watson (1998). Velocity fields cause non-local line overlaps which can play an extremely important role in the pumping of some maser species (see OH maser models in Cesaroni & Walmsley (1991) and Pavlakis & Kylafis (1996)). It should be noted that the observational data on maser spot kinematics (radial velocities and proper motions) provides tight constraints for the computational models, but only in rare cases does it provide definitive association with a certain type of astrophysical object, such as the accretion disk in NGC 4258 described in Greenhill *et al.* (1995), or the outflow in the BN/KL region in Orion, described in Greenhill *et al.* (1998). Often interpretation is not unique (see the discussion of the disk versus outflow model in Shchekinov & Sobolev (2004) and various models in van der Walt *et al.* (2007)). Combined computations of kinematics and excitation are necessary to distinguish between the various possibilities.

Discrete clumps or correlations? We would like to draw attention to one basic problem relevant to the geometry and size of the maser formation region. The question is often polarized and reduced to two extreme cases when 1) maser spots are assumed to be discrete clumps (density inhomogeneities) with sizes comparable to the sizes of the maser spots, and 2) maser spots are assumed to be correlations in the distributions of some physical parameters within an orders of magnitude more extended maser formation region (possibilities include correlations in velocity fields suggested in Deguchi (1982) and Sobolev, Wallin & Watson (1998) or alignment of the structures in the line of sight suggested in Deguchi & Watson (1989)). The importance of this question is directly related to interpretation of the motion of maser spots and trigonometric parallax measurements (see, for example, Bloemhof *et al.* (1996) and Sobolev *et al.* (2008)). Small angular sizes of maser spots, and normally clear distinction of the spots from each other, suggest that the maser spots are formed in discrete clumps. This view is supported by a number of observational facts. For example, the ordered angular and spatial spot distribution, and the orientation in general agreement with the proper motions, support a kinematical interpretation for velocities of the 6-GHz CH_3OH maser spots in several sources (see, for example, Moscadelli *et al.* (2011)). The correlation interpretation is supported by the detection of extended maser structures, with sizes orders of magnitude greater than those of the maser spots themselves (such observations are described in Gwinn (1994), Minier *et al.* (2002), Harvey-Smith & Cohen (2006) and interpretations in Sobolev *et al.* (2003)). Observations indicate that there can be numerous maser spots within one extended maser structure. It is instructive to consider the situation in the context of such a well-studied region as W3(OH). Bloemhof *et al.* (1996) measured intrinsic morphologies of interstellar OH maser spots in this source at two distinct epochs. They found that maser spot shapes persist with time, and came to the following conclusion: 'These observations provide the first direct evidence that the motions measured are due to actual physical movement of discrete clumps of maser-emitting matter, rather than to some sort of nonkinematic effect, such as traveling excitation phenomena or chance realignments of coherency paths through the masing gas. The kinematic assumption is crucial to astrophysical applications of maser proper-motion measurements, including distance determinations and studies of source dynamics.' Xu *et al.* (2006) measured trigonometric parallaxes for numerous 6-GHz CH_3OH maser spots, and have found that the measured values are in close internal agreement. This provides support to the kinematic (discrete clump) hypothesis because the 6-GHz CH_3OH masers in W3(OH) are distributed within the same region as those of OH (Menten *et al.* (1988)). At the same time, considering BIMA data on the numerous maser and non-maser CH_3OH lines in W3(OH) Sobolev *et al.* (2005) concluded that the 6-GHz CH_3OH maser formation region in this source has a size of order 1 arcsecond, which is about 3 orders of magnitude greater than that of the 6.7-GHz maser spots, and corresponds to the total extent of the region where the strong class II methanol masers are distributed. This conclusion was later supported by the observational discovery of large-scale methanol and hydroxyl maser filaments in W3(OH) (Harvey-Smith & Cohen (2006)). Computational modelling has shown that the model comprising a uniform masing layer and a turbulent velocity field (that is spots appear due to velocity correlations) has good potential to reproduce observational data following the time evolution of the maser images (Sobolev *et al.* (2003)), and does not prevent measurements of trigonometric parallaxes (Sobolev *et al.* (2008)). Of course these models are greatly simplified but they have shown the potential of the correlation interpretation. The actual situation is likely to be in between these extremes, and the maser images and spectra, as well as their evolution, are probably determined both by relatively well-delineated clumps (density inhomogeneities) and correlations (coincidences of velocities and other physical parameters in the line of sight). In conclusion, though at present this complicated situation cannot be resolved both by observational and computational means, it should be noted that the kinematic interpretation of the maser spot velocities should not be considered as contradictory to existence of correlations within extended masing regions.

Radiative transfer. The solution to any one model in a parameter-space search consists of a set of population numbers for the masing molecule at each discretized position within the model geometry. For example, in a uniform spherical model, the solution is completely specified by one set of population numbers at each modeled radius. To obtain such a solution requires solving the combined radiative transfer and kinetic master (or statistical equilibrium) equations. The majority of models have used different methods that allow us to localize the radiative transfer problem (see, for example, de Jong (1973), Shmeld *et al.* (1976) and Deguchi & Iguchi (1976)). The large velocity gradient (LVG) approximation was most widely used. The geometry in it is apparent only through the form of the LVG escape probabilities that have to be solved for, with the population numbers, at a single representative point. Self-consistent population numbers and escape probabilities are achieved iteratively. More sophisticated models typically rely on a radiative transfer algorithm such as accelerated lambda iteration (ALI) or coupled escape probability (CEP) to provide sets of populations with spatial resolution. The ALI method as applied to molecular line transfer (Scharmer & Carlsson (1985)) relies on a solution of modest accuracy from ordinary lambda-iteration, that is an iterative solution of the radiative transfer and kinetic master problems from a starting solution, typically with Boltzmann population numbers. The ALI theory introduces perturbations on all population and radiative variables, and constructs a set of equations for these using an approximate, usually local, lambda operator. These perturbations are then iterated to zero against an error vector based on the exact lambda operator. ALI has the advantage of only ever having to solve linear equations, but it remains necessary to compute various radiation integrals over frequency and direction. This method was successfully applied to the interpretation of the OH masers (Gray (2001)). It is available in slab, spherical and perhaps other geometries. The Monte Carlo approach was suggested by Bernes (1979). A great advantage of this method is the possibility of solving radiative transfer problems in the maser and pumping lines simultaneously. It was applied to interpret the 25-GHz CH_3OH masers in Orion by Sobolev & Strelnitskii (1983). This method is applicable for any geometry (see, for example, Juvela (1998)), but requires great computational time. This is the main reason why this method has not been widely used for maser modeling. There has been recent renewed interest in integral equation (direct non-linear) methods, and particularly the CEP implementation (Elitzur & Asensio Ramos (2006)). In such methods, the radiation mean intensity is eliminated analytically, yielding a set of integral equations in the unknown population numbers. In a discrete model, the integral equations become non-linear algebraic equations in the population numbers. A spherical version of the CEP algorithm has recently been implemented (Yun *et al.* (2009)). Acceleration of convergence of the complicated models introduced above can often be significantly improved by the use of convergence accelerators that use knowledge of past iterates to make an improved estimate of the most recent. Two accelerators worthy of attention are Ng's method (Ng (1974)) and the ORTHOMIN accelerator (Vinsome (1976)).

Propagation of maser radiation. The propagation of the maser radiation is directly related to the formation of images and affects their intensities. Even in the simplest models, where the maser radiation is unsaturated and the radiative transfer problem is not fully resolved, this can require considerable computational effort (see the models of the turbulent medium in Sobolev, Wallin & Watson (1998) and the circumstellar envelope in Humphreys *et al.* (1996)). Further, the propagation of the maser radiation introduces additional problems that do not usually apply to the transfer of, for example, the pumping radiation in the same model. One of these is that many radiative transfer codes are not constructed to function properly when the optical depth and source function become negative, as they will for masers. There is also the geometrical problem that the maser radiation is tightly beamed, so that a better angular resolution is required for the maser rays than for the rays of the non-maser radiation. A more fundamental consideration is that maser radiation is expected to retain some coherence properties that cause a departure from the Gaussian statistics of ordinary radiation. To model this behaviour correctly, maser saturation must be considered in the semi-classical approximation, where the molecular response is treated quantum-mechanically through density-matrix theory. Semi-classical models of an ideal 1-D maser with full semi-classical coupling have been considered by Menegozzi & Lamb (1978), and in the more recent work by Dinh-v-Trung (2009a); Dinh-v-Trung (2009b), where the latter work includes polarization. An approximate semi-classical model that allows propagation of the power spectrum, rather than many realizations of the electric field has also been devised (Field & Gray (1988)). Numerical solution of the full equations for the frequency-dependent radiative transport that includes the thermal motion of the molecules is very complicated. Until now it was carried out only in Watson & Wyld (2003) for the simple geometries of isolated spheres and thin disks viewed edge-on. Another complication in computation of propagating maser radiation arises in the presence of magnetic fields causing splitting and polarization of the maser lines. This is especially important for OH masers, modelled in Watson *et al.* (2004).

3. Model output

The outcome of the modelling has different forms depending on the task of research study. It can be exploratory or address some well-defined question. This section describes some types of model output used in our research.

Exploration of parameter space. One of the primary tasks of the modelling is search for the ranges of parameters where the model can reproduce observed values of maser fluxes, flux ratios (when applicable) and other characteristics. The goal is achieved by exploration of a parameter space which can be done in several different ways. Model calculations can require considerable computer time. To gain a broad view, it is useful to study the effects on the output parameters of varying certain of the model input parameters one-at-a-time about chosen standard conditions. The other model parameters remain fixed at the standard values in all runs. This approach was used in, among other works, Pavlakis & Kylafis (1996) and Sobolev et al. (1997a) with considerable success. However, in some cases choice of a single or even a few representative models is not possible, and it is necessary to explore wide ranges of parameter space. In such a situation it is common to calculate a grid of models with relevant detailing. The number of maser model parameters usually exceeds 3, and presentation of results in the form of cross-sections looks adequate. Rather sophisticated ways of presentation of 3D cross-sections can be found, for example, in Juvela (1998). In many cases conciseness, combined with sufficient clarity can be achieved following the style of presentation used in Cragg et al. (2002), where the brightness temperature was shown as a set of contours in a set of cross-sections with coordinates representing 2 parameters (hydrogen number density and specific column density of the masing molecule). Exploration of the parameter space is used when it is necessary to constrain parameter domains of a generic source that represents common characteristics of some masing object (Cragg et al. (2004)). In other circumstances, it may be preferable to create models of individual objects (next subsection), or to study pumping regimes (see example in Sobolev et al. (2007)).

Modelling of individual sources. Creating models of the most-studied and prototypic sources is one of the basic tasks of computational modelling. The models are designed to explain one or more sets of observational data. One type of modelling studies is devoted to searching for the set of parameters providing the best (or at least an acceptable) fit to observational data. Many computational models based on pumping analysis succeeded in reproducing observed line flux ratios of individual objects (CH$_3$OH masers in W3(OH) by Sutton et al. (2001), G345.01+1.79 and NGC 6334F by Cragg et al. (2001), OH masers in the magnetic-field specific model of W75N by Gray et al. (2003)). Other models managed to obtain fits to maser kinematics (H$_2$O masers in W49N by Mac Low et al. (1994) and in L1287 by Fiebig (1997)). Another type of modelling studies aims to reproduce general characteristics of the observational data because searching for exact coincidence is impossible, for example because of the stochastic character of the emission. Reproducing spectra and images for CH$_3$OH masers in Ori-KL was done by Sobolev et al. (1998), for CH$_3$OH masers in W3(OH) by Sobolev et al. (2005), and for SiO masers in o Cet by Gray et al. (2009). Maser variability was addressed in Sobolev et al. (2003) and Gray (2005).

Analysis of the pumping mechanism. Analysis of the nature of the maser phenomenon requires simulation of the pumping mechanism. As discussed above, such mechanisms consist of the combined action of numerous pumping cycles. Heating cycles deliver molecules from the excitation state corresponding to the lower energy level (briefly, lower maser level) of the maser transition, up to the upper maser level. Cooling cycles have opposite starting and final points. It is instructive to describe pumping as a heat engine as suggested in Strelnitskii (1988, and this volume). In this approach, transitions of the upward links of the pumping cycles represent the source of the pumping energy, and downward links provide the sink of energy. Numerical definition of the pumping cycle efficiency, and means to draw conclusions about the thermodynamic nature of these links were introduced in Sobolev (1986). This provided grounds for thermodynamic methods of analysis of the pumping mechanism based on degrees of participation of the cycles of certain thermodynamic types described in Sobolev et al. (1989), and for the methods of tracing pumping routes described in Sobolev & Deguchi (1994a) and Gray (2007). These methods were successfully applied to analysis of the pumping of CH$_3$OH masers (Sobolev & Deguchi (1994b), Sutton et al. (2001)) and OH masers (Gray (2007, and this volume)). Examples of detailed analysis of behaviour of the characteristics of the pumping cycles for masers under different degrees of saturation is presented in Sobolev & Deguchi (1994a).

Predictions. An attractive point in maser computations is their predictive potential. The papers Sobolev *et al.* (1997b) and Johns *et al.* (1998: substituted CH_3OH species) were devoted to extraction of data on the maser candidates from a number of models of class II CH_3OH masers. Some predictions were observationally confirmed later (for example Cragg *et al.* (2001), Sutton *et al.* (2001), Voronkov *et al.* (2002), and Sobolev *et al.* (2002)). It is noteworthy that some of the successful predictions did not follow the phenomenological series rules described in Menten (1991) and Sobolev (1993). The situation regarding predictions of class I CH_3OH maser candidates is quite similar, and is described in more detail in the contribution by Voronkov *et al.* (this volume). Modelling also allows us to predict correlations in variability of different maser transitions and their response to evolution of the pumping conditions (see, for example, Gray (2005)).

4. Acknowledgements

AMS thanks the Russian Foundation for Basic Research (grant 10-02-00589-a) and the Russian federal task program 'Research and operations on priority directions of development of the science and technology complex of Russia for 2007–2012' (state contract 16.518.11.7074). The authors are grateful to V.S. Strelnitski for fruitful discussion.

References

Bernes, C. 1979, *Astron. Astrophys.*, 73, 67
Bloemhof, E. E., Moran, J. M., & Reid, M. J. 1996, *ApJ*, 467, 117
Bolgova, G. T., Makarov, S. V., & Sobolev, A. M. 1988, *Astrophysics*, 28, 239
Cesaroni, R. & Walmsley, C. M. 1991, *Astron. Astrophys.*, 241, 537
Cooke, B. & Elitzur, M. 1985, *ApJ*, 295, 175
Cragg, D. M., Sobolev, A. M., Ellingsen, S. P., *et al.* 2001, *MNRAS*, 323, 939
Cragg, D. M., Sobolev, A. M., & Godfrey, P. D. 2002, *MNRAS*, 331, 521
Cragg, D. M., Sobolev, A. M., Caswell, *et al.* 2004, *MNRAS*, 351, 1327
Cragg, D. M., Sobolev, A. M., & Godfrey, P. D. 2005, *MNRAS*, 360, 533
Deguchi, S. & Iguchi, T. 1976, *PASJ*, 28, 307
Deguchi, S. 1977, *PASJ*, 29, 669
Deguchi, S. 1982, *PASJ*, 29, 669
Deguchi, S. & Watson, W. D. 1989, *ApJ*, 340, 17
de Jong, T. 1973, *Astron. Astrophys.*, 26, 297
Dinh-v-Trung 2009, *MNRAS*, 396, 2319
Dinh-v-Trung 2009, *MNRAS*, 399, 1495
Elitzur, M. & Asensio Ramos, A. 2006, *MNRAS*, 365, 779
Fiebig, D. 1997, *Astron. Astrophys.*, 758, 770
Field, D. & Gray, M. D. 1988, *MNRAS*, 234, 353
Flower, D. R. 2003, Molecular Collisions in the Interstellar Medium, pp. 145. ISBN 0521545749. Cambridge, UK: Cambridge University Press, December 2003.
Gallimore, J. F., Cool, R. J., Thornley, M. D., & McMullin, J. 2003, *ApJ*, 586, 306
Gray, M. D., Doel, R. C., & Field, D. 1991, *MNRAS*, 252, 307
Gray, M. D. 2001, *MNRAS*, 324, 57
Gray, M. D., Hutawarakorn, B., & Cohen, R. J. 2003, *MNRAS*, 343, 1067
Gray, M. D. 2005, *Astrophys.Space Sci.*, 295, 309
Gray, M. D. 2007, *MNRAS*, 375, 477
Gray, M. D., Wittkowski, M., Scholz, M., *et al.* 2009, *MNRAS*, 394, 51
Greenhill, L. J., Jiang, D. R., Moran, J. M., *et al.* 1995, *ApJ*, 440, 619

Greenhill, L. J., Gwinn, C. R., Schwartz, C., *et al.* 1998, *Nature*, 396, 650

Gwinn, C. 1994, *ApJ*, 431, L123

Hartquist, T. W., Menten, K. M., Lepp, S., & Dalgarno, A. 1995, *MNRAS*, 272, 184

Harvey-Smith, L. & Cohen, R. J. 2006, *MNRAS*, 371, 1550

Humphreys, E. L. M., Gray, M. D., Yates, J. A., *et al.* 1996, *MNRAS*, 282, 1359

Johns, K. P., Cragg, D. M., Godfrey, P. D., & Sobolev, A. M. 1997, *MNRAS*, 300, 999

Juvela, M. 1998, *Astron. Astrophys.*, 329, 659

Lockett, P. & Elitzur, M. 2008, *ApJ*, 677, 985

Mac Low, M.-M., Elitzur, M., Stone, J. M., & Koenigl, A. 1994, *ApJ*, 427, 914

Menegozzi, L. N. & Lamb, Jr., W. E. 1978, *Phys. Rev. A*, 17, 701

Menten, K. M. 1991, *ASPC*, 16, 119

Menten, K. M., Johnston, K. J., Wadiak, E. J., *et al.* 1988, *ApJ*, 381, L41

Minier, V., Booth, R. S., & Conway, J. E. 2002, *Astron. Astrophys.*, 383, 614

Moscadelli, L., Sanna, A., & Goddi, C. 2011, *Astron. Astrophys.*, 536, 38

Ng, K.-C. 1974, *JCP*, 61, 2680

Ostrovskii, A. B. & Sobolev, A. M., 2002, *IAUS*, 206, 183

Pavlakis, K. G. & Kylafis, N. D. 1996, *ApJ*, 467, 300

Scharmer, G. B. & Carlsson, M. 1985, *J. Comp. Phys.*, 59, 56

Shchekinov, Yu.A. & Sobolev, A. M. 2004, *Astron. Astrophys.*, 418, 1045

Shmeld, I. K., Strel'nitskii, V. S., & Muzylev, V. V. 1976, *Soviet Astronomy*, 20, 411

Slysh, V. I., Kalenskii, S. V., & Val'tts, I. E. 2002, *Astron. Reports*, 46, 49

Sobolev, A. M. & Strelnitskii, V. S. 1983, *Soviet Astronomy Letters*, 9, 12

Sobolev, A. M. 1986, *Soviet Astronomy*, 30, 399

Sobolev, A. M. 1989, *Astron.Nachr.*, 310, 343

Sobolev, A. M. 1993, *LNP*, 412, 215

Sobolev, A. M. & Deguchi, S. 1994a, *ApJ*, 433, 719

Sobolev, A. M. & Deguchi, S. 1994b, *Astron. Astrophys.*, 291, 569

Sobolev, A. M., Cragg, D. M., & Godfrey, P. D. 1997, *Astron. Astrophys.*, 324, 211

Sobolev, A. M., Cragg, D. M., & Godfrey, P. D. 1997, *MNRAS* (Letters), 288, 39

Sobolev, A. M., Wallin, B. K., & Watson, W. D. 1998, *ApJ*, 498, 763

Sobolev, A. M., Ostrovskii, A. B., Malyshev, A. V., *et al.* 2002, *IAU Symposium*, 206, 179

Sobolev, A. M., Watson, W. D., & Okorokov, V. A. 2003, *ApJ*, 590, 333

Sobolev, A. M., Sutton, E. C., Cragg, D. M., & Godfrey, P. D. 2005, *Astrophys.Space Sci.*, 295, 189

Sobolev, A. M., Cragg, D. M., Ellingsen, S. P., *et al.* 2007, *IAUS*, 242, 81

Sobolev, A. M., Sutton, E. C., Watson, W. D., *et al.* 2008, *Radiophys. and Radioastron.*, 13, S76

Strelnitskii, V. S. 1988, *Proc.IAU Symp.*, 129, 239

Sutton, E. C., Sobolev, A. M., Ellingsen, S. P., *et al.* 2001, *ApJ*, 554, 173

van der Walt, D. J., Sobolev, A. M., & Butner, H. 2007, *Astron. Astrophys.*, 464, 1015

Vinsome, P. K. W. 1976, *4th SPE Symp. on Reservoir Simul. Soc. of Petroleum Engineers*, 4, 149

Voronkov, M. A., Austin, M. C., & Sobolev, A. M. 2002, *Astron. Astrophys.*, 387, 310

Voronkov, M. A., Sobolev, A. M., Ellingsen S. P., *et al.* 2005, *Astrophys.Space Sci.*, 295, 217

Wallin, B. K., Watson, W. D., & Wyld, H. W. 1998, *ApJ*, 495, 774

Watson, W. D. & Wyld, H. W. 2003, *ApJ*, 598, 357

Watson, W. D., Wiebe, D. S., McKinney, J. C., & Gammie, C. F. 2004, *ApJ*, 604, 707

Xu, Y., Reid, M. J., Zheng, X. W., & Menten, K. M. 2006, *Science*, 311, 54

Yates, J. A., Field, D., & Gray, M. D. 1997, *MNRAS*, 285, 303

Yun, Y. J., Park, Y.-S., & Lee, S. H. 2009, *Astron. Astrophys.*, 507, 1785

Cosmic Masers - from OH to H_0
Proceedings IAU Symposium No. 287, 2012
R.S. Booth, E.M.L. Humphreys & W.H.T. Vlemmings, eds.

© International Astronomical Union 2012
doi:10.1017/S174392131200659X

A detailed trace of the pump for 1720-MHz OH masers in SNRs

M. D. Gray

Jodrell Bank Centre for Astrophysics,
School of Physics and Astronomy,
Alan Turing Building, University of Manchester,
Oxford Road, Manchester, M13 9PL, United Kingdom
email: Malcolm.Gray@manchester.ac.uk

Abstract. I have analysed the pumping scheme of 1720-MHz OH masers in great detail.

Keywords. masers, molecular processes, radiative transfer, methods: numerical, stars: winds, outflows, ISM: supernova remnants, radio lines: ISM

1. Introduction

The only OH maser line observed towards supernova remnants (SNRs) is the 1720-MHz outer satellite line from the rovibrational ground state. Only about 10 per cent of Galactic SNRs support 1720-MHz OH masers (see, for example, Hoffman *et al.* (2005)). The SNRs that do support 1720-MHz masers have the following properties: they are old ($>10^4$ yr), placing them in the radiative phase of expansion with non-dissociative C-type shocks expanding at $\lesssim 60 \, \mathrm{km \, s^{-1}}$, and they expand into molecular pre-shock gas.

The SNR masers themselves form in the dense post-shock gas, usually in a number of groups (see, for example, Frail & Mitchell (1998)). At VLBI resolution, the brightest maser cores are observed to be elongated parallel to the shock front, as expected, and have scale sizes of order 100 AU. Brightness temperatures of the maser emission from these cores of order 10^8-10^9 K suggest moderate saturation. An example of a maser-supporting SNR is W44, and this object is shown in Fig. 1 with zones of maser emission ringed.

A pumping model was developed by Elitzur (1976), involving radiative transitions and hard-sphere collisions. Large inversions at 1720 MHz could only be produced if excitation to the $^2\Pi_{3/2}, J = 5/2$ rotational state predominated over excitation to rotational states in the $^2\Pi_{1/2}$ ladder, and if the radiative decays to the ground rotational state were optically thick. This situation leads to inversion because the upper state of the 1720-MHz transition, which has $F = 2$, can receive population radiatively from two hyperfine levels in $J = 5/2$ (with $F = 3$ and 2), whilst the lower state, with $F = 1$, can receive population from only one $J = 5/2$ level (with $F = 2$). This pumping model was revisited by Lockett *et al.* (1999), using more modern rate coefficients for collisions between OH and both ortho- and para-H_2, with results that generally supported the earlier hard-sphere model. Large 1720-MHz inversions required the following conditions: $n_{H_2} = 10^4 - 5 \times 10^5 \, \mathrm{cm^{-3}}$, $T_K = 50 - 125 \, \mathrm{K}$, $T_d < 50 \, \mathrm{K}$ and an OH column density of $10^{16} - 10^{17} \, \mathrm{cm^{-2}}$.

2. Computational Model

To make the detailed trace of the population flow, and for comparison with the earlier models introduced above, I obtained an accelerated lambda iteration (ALI) solution to the OH pumping problem in a slab geometry. Apart from a velocity shift of $10 \, \mathrm{km \, s^{-1}}$

to simulate the C-shock, the slab was uniform over an extent of 2×10^{16} cm, with the following conditions: $n_{H_2} = 3 \times 10^4$ cm^{-3}, $T_K = 105$ K, $T_d = 10$ K and an OH abundance of 10^{-4}. No microturbulence was included: it has a very detrimental effect on the 1720-MHz inversion. Line overlaps were found between the pairs of transitions 6-2, 5-2 and between 8-4, 7-4. See Fig. 2 for the energy level numbering system. The model produced a maximum 1720-MHz inversion of 0.076 cm^{-3} at a depth of 4.71×10^{14} cm (measured from the side of the slab more remote from the site of the SN).

3. Population Tracing

The population tracing method, implemented through the computer code TRACER, uses a log of matrix operations to restore information usually lost in numerical solutions. For details see Gray (2007) and Gray *et al.* (2005). If the important all-process rate coefficients form a small subset of the total number used in the model (and there is no guarantee of this) then it is often possible to express the inversion derived from a small residual matrix in terms of original coefficients. For example, the 1720-MHz inversion,

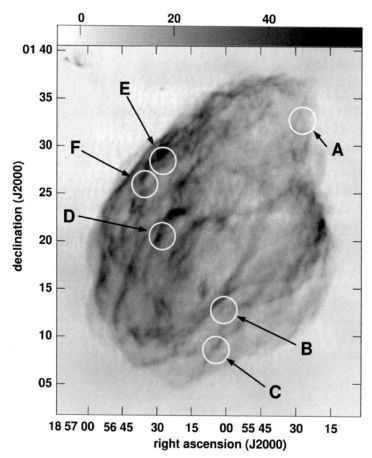

Figure 1. Maser zones (circled) in the SNR W44 superimposed on a continuum image at 1442 MHz (Hoffman *et al.* 2005).

expressed in terms of the coefficients of a residual 4×4 matrix is,

$$\Delta\rho_{41} = \frac{N\delta}{Dk_{44}} \left\{ \frac{k_{14}}{g_4} - \frac{k_{41}}{g_1} + \frac{k_{33}}{\delta} \left(\frac{k_{12}k_{24}}{g_4} - \frac{k_{42}k_{21}}{g_1} \right) + \frac{k_{22}}{\delta} \left(\frac{k_{13}k_{34}}{g_4} - \frac{k_{43}k_{31}}{g_1} \right) \right.$$
$$\left. + \frac{k_{12}k_{23}k_{34}}{g_4\delta} - \frac{k_{43}k_{32}k_{21}}{g_1\delta} + \frac{k_{13}k_{32}k_{24}}{g_4\delta} - \frac{k_{42}k_{23}k_{31}}{g_1\delta} \right\} \tag{3.1}$$

where g_x is the statistical weight of level x, $\delta = k_{22}k_{33} - k_{23}k_{32}$ and k_{xy} is the all-process rate coefficient for transfer of population from level x to level y, unless $x = y$, when k_{xx} represents the rate of transfer from level x to all other levels. Note that neither δ nor the expression N/D can change the sign of the inversion. The object of the trace is to expand these coefficients in terms of less-processed forms until the most important coefficients of the original 36×36 matrix are reached.

Note that the rate coefficients are grouped in antagonistic pairs in eq.(3.1), comprising a forward (pumping) term, and a reverse (negative, or anti-pumping) term. I refer to the first of these as the group A set of routes, based on the TRACER expansion of $(k_{14}/g_4) - (k_{41}/g_1)$. Group A routes can only contain levels 1, 4, and levels higher than 4. The next group, B, is the first term in round brackets in eq.(3.1), and is formed of routes that involve levels 1,2,4 and higher levels. The remaining groups are labelled C, D and E in the order that they appear in eq. (3.1).

4. Results

The dominant pumping route found from the trace is shown in Fig. 2; the reverse route is omitted for clarity. Perhaps surprisingly, this route comes from the B-group, involving level 2 as well as levels 1, 4 and the higher levels shown. Apart from the small contribution that pumps via the $^2\Pi_{3/2}, J = 7/2$ levels, the route involves only the ground state and $^2\Pi_{3/2}, J = 5/2$, as expected. The routes shown contribute 63 per cent of the total 1720-MHz inversion under the conditions of the model. The main route comprises three parts: a mainly radiative excitation from level 1 to level 5, followed by a radiative decay to level 2; both these transitions are significantly optically thick, but do not overlap significantly. Finally, excess population in level 2 is moved to level 4 by a rapid collisional transfer across the ground-state lambda-doublet. Additional variations from the B-group involving level 6 and the other half of the $^2\Pi_{3/2}, J = 5/2$ lambda doublet (levels 7 and 8) raise the inverting effect of group B to 84 per cent of the total. The group-D routes, involving level 3 in addition to those involved in B, are also significantly inverting, and, combined with the B-routes, are responsible for over 99 per cent of the inversion. Of the remainder, the A-routes are anti-inverting overall, whilst groups C and E make very small positive contributions to the inversion.

5. Discussion

The traced route in Fig. 2 is similar to that derived from earlier models in that it involves the $^2\Pi_{3/2}$ ladder almost exclusively, and makes particularly strong use of the $J = 5/2$ lambda-doublet. However, it differs in using a radiative excitation and final collisional transfer within the ground-state lambda doublet instead of a collisional excitation to either level 7 or level 8. The basic reason for this is that, under the conditions of the present model, the upward collisional rate coefficients are comparatively slow compared to the radiative excitation rate from level 1 to level 5. For example, the respective fully-traced back first-order collisional rate coefficients for the 1-8 and 1-7 transitions are

1.9×10^{-7} Hz and 4.2×10^{-7} Hz. By contrast, the all-process rate-coefficient k_{15}, also fully traced back, is 0.0184 Hz. This leads to an overall pumping rate for the route in Fig. 2 that is of order 50 times faster than those via level 7 and 8, both of which form part of the A-group routes. It should be remembered, however, that the present model is close to the minimum value for overall H_2 number density, and higher densities would make the collisional excitations more efficient. It should also be remembered that the final collisional cross-doublet transfer is faster than the excitation to either level 7 or level 8: k_{24} $=5.72\times10^{-6}$ Hz, but the reverse rate coefficient is the same, so that the necessary predominance of the plotted inverting route and the (omitted) reverse route cannot depend on this final link. In fact, this is borne out by inspecting the groups $k_{15}k_{52}$ and $k_{25}k_{51}$. The former product, part of the pumping route, is the faster because it has the excitation via the stronger transition: 1-5 (which changes F), and the decay via the weaker, taking advantage of the greater optical depth in 1-5, and of the overlap of 5-2 with the stronger 6-2 transition.

6. A more realistic model

The model discussed above, to which TRACER was applied, is a fairly crude representation of a SNR maser zone. In a realistic model, the kinetic temperature, density and OH abundance are likely to vary significantly with depth (that is distance measured from the

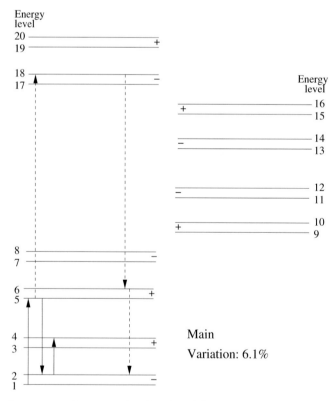

Figure 2. The fastest pumping route for the 1720-MHz inversion. Reverse routes are omitted for clarity. Level numbers are shown, ordered by increasing energy, as are parity designations $(+/-)$. Solid arrows show the main route, whilst the dashed arrows indicate a variation that is effective at the level of 6.1 per cent with respect to the main route.

shock front towards the site of the SN). A model that I considered was the set of C-shock solutions by Le Bourlot *et al.* (2002). However, adapting these for the 1720-MHz OH maser system proved problematic because the range of depth where the OH abundance is adequate to form a maser is rather narrow, and has temperatures significantly higher than those associated with 1720-MHz SNR masers.

The remedy is to include the SNR-specific processes of cosmic-ray and X-ray irradiation of the model cloud Wardle (1999). Energetic electrons produced by this irradiation excite H_2 molecules collisionally, with resultant UV-emission. This ultra-violet radiation is sufficient to generate OH from water by photodissociation at large depths, where the temperature is suitable for OH masers to form.

7. Conclusion

A simple model of a SNR maser zone has been constructed, and the pump of the 1720-MHz OH maser in this type of source has been traced in detail. The results are broadly in agreement with earlier work: the pumping routes are almost exclusively confined to the $^2\Pi_{3/2}$ ladder of levels, and to the ground and first excited rotational states in particular. The pump routes rely on radiative decay via optically thick transitions from $J = 5/2$ levels to $J = 3/2$ levels. However, under the single set of conditions so far studied, the initial excitation from level 1 is radiatively dominated, whilst the pump route is completed by a collisional transfer across the arms of the ground-state lambda doublet. This conclusion could change significantly at higher densities where the collisional excitations to $J = 5/2$ are faster.

References

Elitzur, M. 1976, *ApJ*, 203, 124

Frail, D. A. & Mitchell, G. F. 1998, *ApJ*, 508, 690

Gray, M. D. 2007, *MNRAS*, 375, 477

Gray, M. D., Howe, D. A., & Lewis, B. M. 2005, *MNRAS*, 364, 783

Hoffman, I. M., Goss, W. M., Brogan, C. L., & Claussen, M. J. 2005, *ApJ*, 627, 803

Le Bourlot, J., Pineau des Forêts, G., Flower D. R., & Cabrit, S. 2002, *MNRAS*, 332, 985

Lockett, P., Gauthier, E., & Elitzur, M. 1999, *ApJ*, 511, 235

Wardle, M. 1999, *ApJ*, 525, L101

Lively discussions during poster sessions.
Proceeding pictures by Kalle Torstensson.

T2:

Polarization and magnetic fields *Chair: Athol Kemball*

Cosmic Masers - from OH to H_0
Proceedings IAU Symposium No. 287, 2012   © International Astronomical Union 2012
R.S. Booth, E.M.L. Humphreys & W.H.T. Vlemmings, eds.   doi:10.1017/S1743921312006606

Maser polarization and magnetic fields

W. H. T. Vlemmings

Department of Earth and Space Sciences, Chalmers University of Technology, Onsala Space
Observatory, SE-439 92 Onsala, Sweden

Abstract. Maser polarization observations can reveal unique information on the magnetic field
strength and structure for a large number of very different astronomical objects. As the different masers for which polarization is measured, such as silicon-monoxide, water, hydroxil and
methanol, probe different physical conditions, the masers can even be used to determine for
example the relation between magnetic field and density. In particular, maser polarization observations have improved our understanding of the magnetic field strength in, among others, the
envelopes around evolved stars, Planetary Nebulae (PNe), massive star forming regions, supernova remnants and megamaser galaxies. This review presents an overview of maser polarization
observations and magnetic field determinations of the last several years and discusses some of
the theoretical considerations needed for a proper maser polarization analysis.

Keywords. masers, polarization, magnetic fields

1. Introduction

Because of their compactness, their high brightness and the fact that they occur in a
wide variety of astrophysical environments, masers are excellent astrophysical probes. As
the masers are often highly linearly and circularly polarized, polarization observations
add fundamental information on both the masing process (such as pumping and level of
maser saturation) and the physical conditions in the masing gas. With a detailed theory
of maser polarization propagation, the observations can yield the strength of the magnetic
field along the maser line of sight and the two-dimensional (or even three-dimensional)
field structure. Consequently, they allow for a determination of the dynamical importance
of the magnetic field. And, since masers are specifically suited for high angular resolution
observations with interferometry instruments, the polarization observations can probe the
magnetic field properties at unprecedented small scales.

In the last several years, maser polarization observations have been used to determine
the strength and structure of magnetic field in the circumstellar envelopes (CSE) of
Asymptotic Giant Branch (AGB) stars, high-mass star forming regions and supernova
remnants (SNRs), while observations of the magnetic field strength in megamasers have
also become possible. Here, I will mainly discuss results obtained since the last maser IAU
symposium (IAU 242, 2007, eds. Chapman & Baan), focusing on the maser transitions
in the radio-wavelength regime. In particular, in 2008, the first measurement of 6.7 GHz
methanol maser circular polarization was made (Vlemmings 2008) and the considerations
of its analysis are discussed. In §2, the theoretical background of maser polarization are
briefly described. Recent magnetic field measurements and their consequences for related
astrophysical problems involving evolved stars and high-mass star forming regions are
presented in §3 and §4 respectively. §5 highlights a few further maser polarization results
and finally §6 presents perspectives for future instruments and observations.

2. Background and considerations

There exists extensive literature on the theory of maser polarization, as polarization during maser amplification differs from the regular thermal emission case due to the stimulated emission process and a range of other properties of the masing process that can influence the radiation polarization characteristics. The main theoretical problem is framed by constructing the density matrix evolution and radiative transfer equations for the maser emission including the Zeeman terms (Goldreich *et al.* 1973). The Zeeman effect occurs when the degeneracy of magnetic substates is broken under the influence of a magnetic field. The magnitude of the Zeeman effect is significantly different for paramagnetic (e.g. OH) and non-paramagnetic molecules (e.g. SiO, H_2O and methanol), due to the ratio between the Bohr magneton ($\mu_B = e\hbar/2m_e c$) and the nuclear magneton ($\mu_N = e\hbar/2m_n c$). In these expressions e is the electron charge, $\hbar$ is the Planck constant and c the speed of light. The ratio of the two ($\mu_B/\mu_N \approx 10^3$) is determined by the ratio of the electron mass (m_e) and the nucleon mass (m_N) and implies three orders of magnitudes larger Zeeman splitting for a similar magnetic field strength in the case of paramagnetic molecules compared to the non-paramagnetic ones.

A fundamental difference in the treatment of polarized maser emission exists between cases when the magnetic transitions overlap in frequency or when they are well separated. This can be defined by the splitting ratio $r_Z = \Delta\nu_Z/\Delta\nu_D$, where $\Delta\nu_Z$ is the Zeeman splitting and $\Delta\nu_D$ the Doppler line width. Typically, $r_Z \gtrsim 1$ for the paramagnetic molecule OH, while $r_Z < 1$ for the other, non-paramagnetic maser species. In the case of $r_Z > 1$ there are no theoretical ambiguities and the Zeeman components are well separated and resolved. The magnetic transitions $\Delta m_F = \pm 1$ give rise to the $\sigma^\pm$ components, circularly polarized perpendicular to the magnetic field B. The transition $\Delta m_F = 0$ gives rise to the π component, linearly polarized along B. For an arbitrary angle between B and the maser propagation direction θ, the resultant components are elliptically polarized for $\theta < \pi/2$, and linearly polarized for $\theta = \pi/2$. The observed splitting of the Zeeman components directly gives the magnetic field strength $B\cos\theta$.

In the case of $r_Z < 1$, the Zeeman components overlap. The derived B strengths when $r_Z < 1$ do not only depend on the circular polarization fraction but also depend on the maser saturation level. Especially for masers that are saturated, simple assumptions with a fixed proportionality between circular polarization and magnetic field strength can lead to B being overestimated by up to a factor of 4. Theoretical work has been done both analytically and numerically with different implications for the derived magnetic field strengths (e.g.Watson 1994, Elitzur 1996, and references therein). A comparison of the different polarization theories is given by Gray (2003), who finds the numerical models can be used to accurately describe the maser polarization. In even more recent work (Dinh-V-Trung 2009) has revisited the theoretical maser calculations from first principles, avoiding the typically made assumptions that the maser radiation field is stationary and that the different spectral components are not correlated. Dinh-V-Trung (2009) finds that the numerical calculation of maser polarization (e.g.Watson 1994) are sufficient in the unsaturated and partially saturated regime.

In both Zeeman splitting cases ($r_Z < 1$ and $r_Z > 1$) there are several other properties of the maser and its surrounding medium that need to be taken into account when interpreting polarization observations. Especially at the low frequencies, Faraday rotation can make a direct connection between the polarization angle and the B-field uncertain. External Faraday rotation can cause significant vector rotation originating along the line of sight to the maser source. For instance at 1.6 GHz, a typical interstellar electron density and magnetic field strength can cause up to a full $\sim 180°$ rotation towards the W3(OH) star forming region. Additionally, internal Faraday rotation can alter the polarization characteristics of individual maser features in a source in different ways, possibly destroying any large scale structure in the linear polarization measurements (e.g. Fish & Reid 2006).

2.1. *On the Zeeman splitting of methanol masers*

As mentioned above, the interpretation of maser polarization depends critically on the Zeeman frequency shift in relation to the maser saturation level. The recent detection of the circular polarization of 6.7 GHz methanol masers (e.g. Vlemmings 2008; Fig. 1), has led to the need of a careful evaluation of the relevant methanol Zeeman frequency shift (Vlemmings *et al.* 2011b).

The methanol molecule is a non-paramagnetic molecule and as a result the Zeeman splitting under the influence of a magnetic field is extremely small. The split energy, ΔE_Z, of an energy level under the influence of a magnetic field, B, can be described as $\Delta E_z = g_L \mu_N M_J B$, where M_J denotes the magnetic quantum number for the rotational transition described with the total angular momentum quantum number J, B is the magnetic field strength in units of Tesla ($= 10^4$ G), μ_N is the nuclear magneton and g_L is the Landé g-factor. The Zeeman effect is determined by the Landé g-factor, which needs to be determined from laboratory spectroscopy.

In most previous publications of methanol polarization, the g-factor used to determine the magnetic field strength was based on laboratory measurements performed many years ago on a number of methanol transitions near 25 GHz methanol (Jen 1951). However, there are several caveats regarding these measurements. Firstly, g_L is an average of the true g-factor of several interacting states. Additionally, the measurements are classified as preliminary, and the exact transitions that were used are not specified. The observations were done on poorly identified transitions around 25 GHz with $\Delta J = 0$ and $K = 2 - 1$, which likely indicates it concerns the E1-type methanol maser. It is thus not impossible that an extrapolation to the 6.7 GHz $5_1 - 6_0$ A$^+$ methanol transition and others transitions with different ΔJ and quantum number K is invalid.

While it is thus unclear if the g-factor determined in 1951 can be used for the 6.7 GHz methanol maser, it is the only estimate available to us at the moment. For the Zeeman splitting coefficient for the $5_1 - 6_0$ A$^+$ 6.7 GHz methanol maser transition extrapolated from the laboratory measurements this implies 0.005 km s^{-1} G^{-1}, an order of magnitude smaller than previously used and implying unlikely high magnetic fields of ~ 100 mG in the methanol maser region of massive star forming regions. Based on comparison with 6 GHz OH masers at similar densities, for which typical fields measured ar of order ~ 10 mG, the true g-factor of methanol could be an order of magnitude larger.

Alternatively, the methanol circular polarization can be caused by non-Zeeman effects. In Vlemmings *et al.* (2011b) various different effects were investigated and ruled unlikely. The case of the rotation of the axis of symmetry for the molecular quantum states deserves some more detail here. This can occur when, as the maser brightness increases while it becomes more saturated, the rate for maser stimulated emission R becomes larger than the Zeeman frequency shift $g\Omega$. While $g\Omega \gg R$, the magnetic field direction is the quantization axis. Then, when R becomes larger than $g\Omega$, the molecules interact more strongly with the radiation field than with the magnetic field and the quantization axis changes towards the maser propagation direction. This change will cause an intensity-dependent circular polarization that mimics the regular Zeeman splitting.

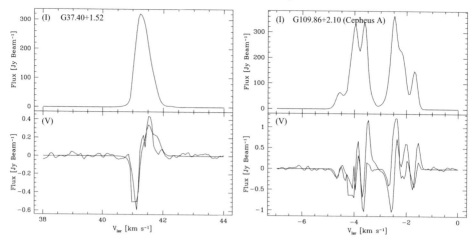

Figure 1. From Vlemmings (2008), Total intensity and circular polarization spectrum for G37.40+1.52 (left) and G109.86+2.10 (Cepheus A; right). The thick solid line in the bottom panel is best fit fractional total power derivative to the circular polarization spectrum.

From the methanol maser Zeeman splitting coefficient given above, $g\Omega \approx 0.1B[\mathrm{mG}]\,\mathrm{s}^{-1}$ for the 6.7 GHz methanol maser. The rate for stimulated emission can be estimated using:

$$R \simeq AkT_{\mathrm{b}}\Delta\Omega/4\pi h\nu. \qquad (2.1)$$

Here A is the Einstein coefficient for the maser transition, which is equal to $0.1532 \times 10^{-8}\,\mathrm{s}^{-1}$, and k and h are the Boltzmann and Planck constants respectively. The maser frequency is denoted by ν, and T_{b} and $\Delta\Omega$ are the maser brightness temperature and beaming solid angle. observations indicate typically $T_{\mathrm{b}} \lesssim 10^{10}$ K. The beaming angle $\Delta\Omega$ is harder to estimate and decreases rapidly with increasing maser saturation level. If we very conservatively assume a maser beaming angle of $\Delta\Omega \approx 10^{-2}$, the typical maser stimulated emission $R \sim 0.04\,\mathrm{s}^{-1}$. Thus, for a typical field strength of ~ 10 mG, $g\Omega/R > 25$ and only for the most saturated masers would we expect the non-Zeeman effect to be applicable.

3. Evolved stars and planetary nebulae

Maser polarization observations are the predominant source of information about the role of magnetic fields during the late stages of stellar evolution. Most observations have focused on the masers in the CSEs of AGB stars, as OH, H_2O and SiO masers are fairly common in these sources. However, polarization observations of masers around post-AGB stars and (Proto-)PNe are becoming more common as more such sources with maser emission are found.

3.1. AGB stars & Supergiants

Recent polarization observations of CSE 1.6 GHz OH masers confirm a regular structure and few milliGauss magnetic field strength in AGB OH maser envelopes (Wolak et al. 2012) as found in previous observations. Recent years have also seen an increase in 22 GHz H_2O and 43 GHz SiO maser observations of Mira variables, OH/IR stars and supergiants. As the different maser species typically occur in different regions of the CSE, combining observations of all three species allows us to form a more complete picture of the magnetic field throughout the entire envelope (Fig. 2). Close to the central star, SiO maser linear

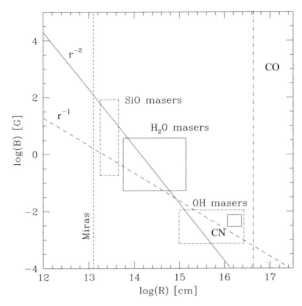

Figure 2. Magnetic field strength vs. radius relation as indicated by current maser polarization observation of a number of Mira stars. The boxes show the range of observed magnetic field strengths derived from the observations of SiO, H_2O and OH masers and thermal CN. The thick solid and dashed lines indicate an r^{-2} solar-type and r^{-1} toroidal magnetic field configuration. The vertical dashed line indicates the stellar surface. CO polarization observations will uniquely probe the outer edge of the envelope (vertical dashed dotted line).

polarization reveals an ordered B-field with a linear polarization fraction ranging up to $m_l \sim 100\%$ (e.g. Cotton *et al.* 2011, Amiri *et al.* 2012). A large single dish survey of SiO maser polarization revealed an average field strength of 3.5 G when assuming a regular Zeeman origin of the polarization, indicating a dynamically important B-field (Herpin *et al.* 2006). The observations find no specific support for other (non-Zeeman) interpretations of the polarization (e.g. Richter *et al.*, this proceedings).

Further out in the envelope, also 22 GHz H_2O maser measurements reveal significant B-fields, both around Miras and supergiants (Vlemmings *et al.* 2005, and references therein). The measured field strength is typically of the order of $\sim 100 - 300$ mG but can be up to several Gauss. The strongest field strengths are found around Mira variables, consistent with the H_2O masers occurring closer to the star. As no linear polarization has been detected thus far, describing the magnetic field shape is difficult. However, for the Supergiant VX Sgr, the complex maser structure reveals an ordered field reversal across the maser region consistent with a dipole B-field. Interestingly, the orientation of the field determined from the H_2O maser polarization is similar to the orientation of the outflow determined from other H_2O maser observations as well as the orientation of a dipole field determine from OH maser polarization. It is even similar to the direction of the SiO maser polarization measured with the Submillimeter Array (SMA) (Fig.2; Vlemmings *et al.* 2011a).

3.2. *Proto-Planetary nebulae*

By now, a number of P-PNe have had their magnetic fields measured using OH and H_2O masers. Similar to the magnetic field strengths around their progenitor stars, the P-PNe fields are ~ 1 mG in the OH maser region (e.g. Etoka & Diamond 2010). Single dish surveys reveal linear and circular polarization in respectively $\sim 50\%$ and $\sim 75\%$ of the

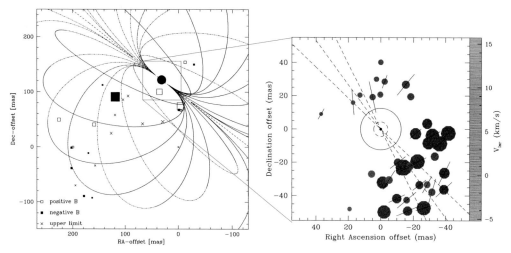

Figure 3. (left) The dipole magnetic field of the supergiant VX Sgr as determined from a fit to the H_2O maser magnetic field observations (Vlemmings *et al.* 2005). (right) Positions and polarization of the VX Sgr $v = 0, J = 5 - 4$ ^{29}SiO masers observed with the SMA (Vlemmings *et al.* 2011a). The masers spots are plotted with respect to the peak of the continuum emission. The black vectors are the observed polarization vectors scaled linearly according to polarization fraction. The long dashed inner circle indicates the star and the solid circle indicates the location of the 43 GHz SiO masers. The short dashed circle indicates the minimum radius of the ^{28}SiO masers. The dashed lines indicate the position angle and its uncertainty of the inferred orientation of the dipole magnetic field of VX Sgr observed using H_2O and OH masers(Vlemmings *et al.* 2005, Szymczak *et al.* 2001).

sources, dependent on the OH maser line (polarization is more common in the 1612 MHz OH satellite line than in the 1665 and 1667 main line masers). The polarization fraction is typically less than 15% (Szymczak & Gérard 2004).

A very small fraction of the Post-AGB/P-PNe maser stars show highly collimated H_2O maser jets (see e.g. Desmurs 2012; these proceedings). These so-called water-fountain sources are likely the progenitors of bipolar PNe and there are indications that they evolve from fairly high-mass AGB stars. The archetype of this class is W43A and polarization observations have revealed that the maser jet is magnetically collimated (Vlemmings *et al.* 2006). In addition to the observations of W43A, recent Australia Telescope Compact Array (ATCA) observations of the likely water-fountain source IRAS 15445-5449 (Pérez-Sánchez *et al.* 2011) also indicate a magnetic H_2O maser jet. The first measurement of the magnetic field strength within a few tens of AU of the binary post-AGB Rotten Egg Nebula, in the H_2O maser region entrained by the fast bipolar outflow, was also recently presented (Leal-Ferreira *et al.* 2012).

3.3. *Planetary Nebulae*

There are only a handful of PNe known which show maser emission, and even less of these have masers that are strong enough to provide B-field measurements from polarization observations. One of the sources that shows both OH and H_2O maser emission is the very young PNe K3-35. In this source, the OH masers indicate a B-field of a ~ 0.9 mG at 150 AU from the central object (Gómez *et al.* 2009).

3.4. *Summary*

Fig. 2 gives a summary of the current magnetic field measurements in CSEs of Mira stars. Although the exact relation between the magnetic field strength and distance

Table 1. Energy densities in AGB envelopes

		‖ Photosphere	SiO	H_2O	OH
B	[G]	$\sim 50?$	~ 3.5	~ 0.3	~ 0.003
R	[AU]	-	~ 3	~ 25	~ 500
		-	$[2-4]$	$[5-50]$	$[100-10.000]$
V_{exp}	[km s^{-1}]	~ 5	~ 5	~ 8	~ 10
n_{H_2}	[cm^{-3}]	$\sim 10^{14}$	$\sim 10^{10}$	$\sim 10^8$	$\sim 10^6$
T	[K]	~ 2500	~ 1300	~ 500	~ 300
$B^2/8\pi$	[dyne cm^{-2}]	$10^{+2.0}?$	$10^{+0.1}$	$10^{-2.4}$	$10^{-6.4}$
nKT	[dyne cm^{-2}]	$10^{+1.5}$	$10^{-2.8}$	$10^{-5.2}$	$10^{-7.4}$
ρV_{exp}^2	[dyne cm^{-2}]	$10^{+1.5}$	$10^{-2.5}$	$10^{-4.1}$	$10^{-5.9}$
V_A	[km s^{-1}]	~ 15	~ 100	~ 300	~ 8

to the central star remains uncertain, the field strengths are obvious strong enough to dynamically influence the shaping of the outflow and help shape asymmetric PNe. To properly determine the possible effect of the magnetic fields, it is illustrative to study the approximate ratios of the magnetic, thermal and kinematic energies contained in the stellar wind. In Table 1 I list these energies along with the Alfvén velocities and typical temperature, velocity and temperature parameters in the envelope of AGB stars. While many values are quite uncertain, as the masers that are used to probe them can exist in a fairly large range of conditions, it seems that the magnetic energy dominates out to $\sim 50 - 100$ AU in the circumstellar envelope.

4. High-mass star formation

Star forming regions, and especially those forming high-mass stars, often contain a wide variety of maser species tracing many different density and temperature regimes. As was the case for evolved stars, most information on the small scale magnetic fields comes from maser polarization observations, dominated by OH maser measurements but also with an increasing number of H_2O and methanol maser observations.

4.1. OH masers

There exists a large number of OH maser polarization measurements, probing densities from $\sim 10^5 - 10^8$ cm^{-3}. OH masers are often strongly polarized and as the Zeeman splitting is large, observations of the separate σ^+ and σ^- components directly yields a magnetic field strength. Observations of the 100% linearly polarized π-component are extremely rare however, likely due to magnetic beaming and the overlap of several differently polarized masers along the line of sight (Fish & Reid 2006). The measured field strengths are typically around ~ 1 mG. Most of the recent OH maser work has been performed by Caswell *et al.* (2009, 2011), including observations of the polarization of 6 GHz OH masers. These results seem to imply a relation between the Galactic magnetic field and that measured in the OH maser region.

4.2. H_2O masers

After the first discovery of interstellar H_2O maser Zeeman splitting by Fiebig & Güsten (1989) using single dish observations, there have been an increasing number of higher

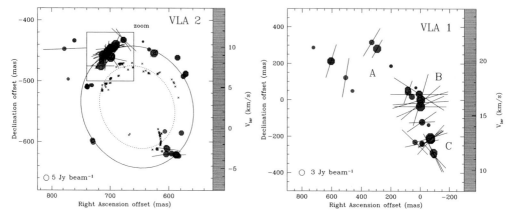

Figure 4. From Surcis *et al.* (2011). The H_2O masers of VLA2 (left) and VLA1 (right) in the massive star forming region W75N. The line segments denote the polarization fraction and direction and the maser features are scaled according to their flux. In the left panel, the ellipses indicate the expansion of the maser ring from previous observations (crosses, Torrelles *et al.* (2003))

spatial resolution circular polarization observations confirming the earlier results (e.g. Sarma *et al.* 2001). These observations typically reveal B-field strengths between 15 and 150 mG at densities of $n_{H_2} = 10^8 - 10^{11}$ cm^{-3}. In addition to the circular polarization, low levels of linear polarization (typically $\lesssim 2\%$) are also observed in star forming regions. While often structure in the B-field direction is detected, the observations show rapid changes of direction over small scales (see Fig. 4).

In addition to providing information on the magnetic field strength and structure, full maser polarization radiative transfer modeling can also provide important information on other physical quantities of the masing gas. In particular, the observations by Surcis *et al.* (2011) have revealed striking differences of intrinsic line-width and hence either turbulent widths or temperatures around two sources, VLA1 and VLA2, in the W75N massive star forming region. These sources are thought to represent two different evolutionary stages of proto-stars.

4.3. *Methanol masers*

The 6.7 and 12 GHz methanol masers are some of the most abundant masers in high-mass star forming regions. After the initial presentation of their polarization, the last few years have seen an increasing number of publications on polarimetric methanol maser studies (e.g. Vlemmings *et al.* 2006, 2008, 2011, Dodson 2008, Stack & Ellingsen 2011). In particular, 6.7 GHz methanol maser observations with MERLIN have shown that the magnetic field is likely regulating the infall on the circumstellar disk in the Cepheus A region (Vlemmings *et al.* 2010).The methanol maser polarimetry has now also been extended to higher frequency class-I masers (e.g. Sarma *et al.* 2009, and these proceedings). While as described above, the methanol circular polarization cannot yet be used to determine the magnetic field strength, the linear polarization has shown large scale magnetic field structures that are consistent with magnetic field observations at other frequencies (e.g. Surcis *et al.* 2009, and these proceedings).

4.4. *Summary*

The recent magnetic field observations, and in particular those of methanol, indicate a clear relation between the maser derived magnetic field direction and that of known outflows and/or circumstellar toroidal structures. Additionally, they often show a good agreement with dust polarization measurements at lower resolution. This indicates that maser do not, as occasionally thought, probe isolated pockets of magnetized gas, but rather are indeed good probes of the magnetic field in star forming regions.

The B-field measurements of both masers and non-maser observations as a function of number density seem to indicate that the field strength follows an approximate $B \propto n^{0.5}$ density scaling law over an enormous range of densities. This implies that the magnetic field remains partly coupled to the gas up to the highest number density. However, above $n \sim 10^8$ cm^{-3}, the shock excited H_2O maser are short-lived (with a typical lifetime $\tau_m \sim 10^8$ s) compared to the typical adiabatic diffusion timescale at the highest densities ($\tau_d \sim 10^9$ s), and in the non-masing gas of similar high densities, magnetic field strengths should be lower due to the adiabatic diffusion. Still, the maser B measurements strongly indicate a dynamical importance of magnetic fields during the high-mass star formation process, especially in shaping outflows and jets.

5. Further maser polarization studies

Besides the evolved stars and star forming regions, masers are found in several other types of sources, such as supernova remnants and megamaser galaxies. However, there have been few recent publications on polarization from these sources. However, work is progressing on the measurement of OH megamaser polarization such as presented by Robishaw *et al.* (2008). These authors measured the Zeeman splitting of OH megamaser emission at 1667 MHz from five ULIRGs using the Arecibo and Green Bank Telescope. They found line-of-sight magnetic field strengths ranging from ~ 0.5 to 18 mG, similar to those measured in Galactic OH masers. They infer that this suggests that the local process of massive star formation in ULIRGs occurs under similar conditions as in the Galaxy.

Finally, besides the maser species discussed above, hydrogen recombination line masers have also been shown to display polarization. This is described further in Thum *et al.* (these proceedings).

6. Future perspectives

The wealth of new maser polarization observations over the last few years has clearly demonstrated the relevance of masers in the study of Galactic as well as extra-galactic magnetic fields. Especially methanol maser polarization observations have shown their enormous potential. In the near future, maser polarization will be an important goal in the observations with the upgraded EMERLIN and EVLA and also the VLBI observations with the EVN and VLBA will make great strides in detailed imaging of circumstellar and protostellar magnetic fields. At the higher maser frequencies, ALMA will provide a further step when its polarimetric capabilities are offered (as shown by Pérez-Sánchez *et al.*, these proceedings). Additionally, SMA and ALMA high resolution dust polarization observations will also close the gap between the small scale maser magnetic field measurements and the very large scale single dish dust polarization observations.

Acknowledgements

WV acknowledges support by the Deutsche Forschungsgemeinschaft (DFG) through the Emmy Noether Research grant VL 61/3-1.

References

Amiri, N., Vlemmings, W. H. T., Kemball, A. J., & van Langevelde, H. J. 2012, *A&A*, 538, A136

Caswell, J. L., Kramer, B. H., & Reynolds, J. E. 2009, *MNRAS*, 398, 528

Caswell, J. L., Kramer, B. H., & Reynolds, J. E. 2011, *MNRAS*, 415, 3872

Cotton, W. D., Ragland, S., & Danchi, W. C. 2011, *ApJ*, 736, 96

Dinh-v-Trung 2009, *MNRAS*, 399, 1495

Dodson, R. 2008, *A&A*, 480, 767

Elitzur, M. 1996, *ApJ*, 457, 415

Etoka, S. & Diamond, P. J. 2010, *MNRAS*, 406, 2218

Fiebig, D. & Güsten, R. 1989, *A&A*, 214, 333

Fish, V. L. & Reid, M. J. 2006, *ApJS*, 164, 99

Goldreich, P., Keeley, D. A., & Kwan, J. Y. 1973, *ApJ*, 179, 111

Gómez, Y., Tafoya, D., Anglada, G., *et al.* 2009, *ApJ*, 695, 930

Gray, M. D. 2003, *MNRAS*, 343, L33

Herpin, F., Baudry, A., Thum, C., Morris, D., & Wiesemeyer, H. 2006, *A&A*, 450, 667

Jen, C. K. 1951, *Physical Review*, 81, 197

Leal-Ferreira, M. L., Vlemmings, W. H. T., Diamond, P. J., *et al.* 2012, arXiv:1201.3839

Pérez-Sánchez, A. F., Vlemmings, W. H. T., & Chapman, J. M. 2011, *MNRAS*, 418, 1402

Robishaw, T., Quataert, E., & Heiles, C. 2008, *ApJ*, 680, 981

Sarma, A. P., Troland, T. H., & Romney, J. D. 2001, *ApJ*, 554, L217

Sarma, A. P. & Momjian, E. 2009, *ApJ*, 705, L176

Stack, P. D., & Ellingsen, S. P. 2011, *PASA*, 28, 338

Surcis, G., Vlemmings, W. H. T., Dodson, R., & van Langevelde, H. J. 2009, *A&A*, 506, 757

Surcis, G., Vlemmings, W. H. T., Curiel, S., *et al.* 2011, *A&A*, 527, A48

Szymczak, M., Cohen, R. J., & Richards, A. M. S. 2001, *A&A*, 371, 1012

Szymczak, M. & Gérard, E. 2004, *A&A*, 423, 209

Torrelles, J. M., Patel, N. A., Anglada, G., *et al.* 2003, *ApJ*, 598, L115

Vlemmings, W. H. T., van Langevelde, H. J., & Diamond, P. J. 2005, *A&A*, 434, 1029

Vlemmings, W. H. T., Diamond, P. J., & Imai, H. 2006a, *Nature*, 440, 58

Vlemmings, W. H. T., Harvey-Smith, L., & Cohen, R. J. 2006c, *MNRAS*, 371, L26

Vlemmings, W. H. T. 2008, *A&A*, 484, 773

Vlemmings, W. H. T., Surcis, G., Torstensson, K. J. E., & van Langevelde, H. J. 2010, *MNRAS*, 404, 134

Vlemmings, W. H. T., Humphreys, E. M. L., & Franco-Hernández, R. 2011, *ApJ*, 728, 149

Vlemmings, W. H. T., Torres, R. M., & Dodson, R. 2011, *A&A*, 529, A95

Watson, W. D. 1994, *ApJ*, 424, L37

Wolak, P., Szymczak, M., & Gérard, E. 2012, *A&A*, 537, A5

Cosmic Masers - from OH to H$_0$
Proceedings IAU Symposium No. 287, 2012
R.S. Booth, E.M.L. Humphreys & W.H.T. Vlemmings, eds.

© International Astronomical Union 2012
doi:10.1017/S1743921312006618

Polarization of Class I methanol (CH$_3$OH) masers

A. P. Sarma

DePaul University, Chicago IL, USA
email: asarma@depaul.edu

Abstract. Magnetic fields are known to play an important role in several stages of the star formation process. Class I methanol (CH$_3$OH) masers offer the possibility of measuring the large-scale magnetic field in star forming regions at high angular resolution, due to connections between the large-scale magnetic field in the pre-shock regions to the observed magnetic field along the outflows in the post-shock regions where these masers are formed. The detection of the Zeeman effect in the 36 GHz and 44 GHz Class I methanol maser lines by Sarma and Momjian has opened an exciting new window into the study of the star formation process, but for the results to be interpreted correctly, the Zeeman splitting factor (z) for both these lines needs to be urgently measured by experiment. Ratios between the pre-shock and post-shock magnetic fields and densities lead to the conclusion that the value of z cannot be too different from 1 Hz mG^{-1}, unless the predicted densities at which 36 GHz and 44 GHz methanol masers are excited are drastically incorrect. Similarities between the detected fields in 36 GHz and 44 GHz Class I masers, and 6.7 GHz Class II masers, support the claim that these masers may be tracing the large-scale magnetic field or that the magnetic field remains the same during different evolutionary stages of the star formation process, provided such similarities are not just due to the assumption of a uniform nominal value for z, or result simply from selection effects due to orientation and/or the shock process. Given the exciting possibilities, a larger statistical sample of measurements in both the 36 GHz and 44 GHz lines is certainly needed.

Keywords. magnetic fields — masers — polarization — stars: formation — radio lines: ISM

1. Introduction

Magnetic fields in star forming regions present tremendous observational and theoretical challenges. Incorporating them into numerical models significantly increases the computational complexity. Observing them requires high sensitivity and, if they are to be of any use in understanding star formation processes, high angular resolution. Yet, they are believed to play such important roles in a number of stages in the star formation process that measuring them and understanding their influence is of paramount importance. As it stands, the role of magnetic fields in regulating the onset of star formation is still a matter of debate (e.g., Crutcher *et al.* 2009). It has become increasingly clear, though, that magnetic fields play a critical role in carrying angular momentum away from the protostar during collapse (McKee & Ostriker 2007, and references therein). Moreover, the outflows along which this takes place may be driven by dynamically enhanced magnetic fields in the protostellar disk (Banerjee & Pudritz 2006). In particular, the driving of outflows along magnetic field lines may be critical in allowing accretion to continue onto high mass protostars (Banerjee & Pudritz 2007).

The Zeeman effect remains the most direct method for measuring the magnetic field strength (Troland *et al.* 2008). Over the years, observations of the Zeeman effect in H I and OH thermal lines have revealed the strength of the magnetic field in the lower density envelopes of molecular clouds (e.g., Brogan & Troland 2001; Sarma *et al.* 2000). However, measuring fields in the dense gas nearer to the protostar is difficult to achieve with such lines. Thermal lines of CN hold promise (Falgarone *et al.* 2008), but must await the advent of high sensitivity and high angular resolution interferometers at their frequencies. On the other hand, interstellar masers, being compact and intense, offer a means of measuring the magnetic field in star forming regions at high angular resolution. For years, their effectiveness as probes of star forming regions was overshadowed by a perception that the specialized conditions in which such masers form necessarily prevented them from being linked to conditions on larger scales. However, recent discoveries indicate that rather than being a measure in isolated atypical fragments, the magnetic fields measured in masers are indeed linked to the larger scale magnetic field. Fish & Reid (2006) found a relative consistency in the magnetic fields measured in clusters of mainline (1665 and 1667 MHz) OH masers across a massive star forming region, and concluded that magnetic fields are ordered in massive star forming regions. More recently, Vlemmings *et al.* (2010) have determined from polarization observations of 6.7 GHz methanol masers toward Cepheus HW2 that the masers probe the large scale magnetic field.

It is in this context that the ability to measure the Zeeman effect in Class I methanol (CH_3OH) masers provides an important new tool. Historically, Class I methanol masers were categorized on the basis of their distance from observable indicators of star formation, whereas Class II methanol masers were known to be close to many of the acknowledged indicators of star formation (Menten 1991). Further study led to the conclusion that Class I methanol masers likely form in collisional shocked regions in protostellar outflows (Cragg *et al.* 1992; Sandell *et al.* 2005). Therefore, they offer us the potential to measure the magnetic field along the outflow. Of course, the effort to measure magnetic fields using these masers is still in its infancy, and this cherished goal will require sustained effort. Sarma & Momjian (2009) made the first measurement of the Zeeman effect in the 36 GHz Class I methanol maser line toward the star forming region M8E. Sarma & Momjian (2011) made the first measurement of the Zeeman effect in the 44 GHz Class I methanol maser line toward a star forming region in OMC-2. This contribution will discuss these discoveries, and what they tell us about the potential for learning about star forming regions by making polarization observations of Class I methanol masers.

2. Observations and Data Reduction

Observations of the $4_{-1} - 3_0$ E methanol maser emission line at 36 GHz toward M8E were carried out in the C-configuration of the Expanded Very Large Array (EVLA) of the NRAO† in two 2 hr sessions on 2009 July 9 and 25. Thirteen EVLA antennas equipped with the 27–40 GHz (Ka-band) receivers were used in these observations. Meanwhile, observations of the $7_0 - 6_1$ A^+ methanol maser emission line at 44 GHz were carried out using 22 antennas in the D-configuration of the EVLA in two 2 hr sessions on 2009 Oct 25 and Nov 25. Table 1 lists the observing parameters and other relevant data for these observations. Both the 36 GHz and 44 GHz data were correlated using the old VLA correlator, and in order to avoid the aliasing known to affect the lower 0.5 MHz of the bandwidth for EVLA data correlated with the old VLA correlator, the spectral line was centered in the second half of the 1.56 MHz wide band. For the 36 GHz observations toward M8E, the source 3C286 (J1331+3030) was used to set the absolute flux density scale, while the compact source J1733−1304 was used as an amplitude calibrator. For the 44 GHz observations toward OMC-2, the source 3C147 (J0542+4951) was used to

† The National Radio Astronomy Observatory (NRAO) is a facility of the National Science Foundation of the USA operated under cooperative agreement by Associated Universities, Inc.

Table 1. Parameters for EVLA Observations

Parameter	36 GHz Observations	44 GHz Observations
Observation Dates	2009 July 9 & 25	2009 Oct 25 & Nov 25
Configuration	C	D
R.A. of field center (J2000)	$18^{\mathrm{h}}04^{\mathrm{m}}53.3^{\mathrm{s}}$	$05^{\mathrm{h}}35^{\mathrm{m}}27.66^{\mathrm{s}}$
Decl. of field center (J2000)	$-24°26'42.0''$	$-05°09'39.6''$
Total Bandwidth	1.56 MHz	1.56 MHz
No. of channels	256	256
Channel Spacing	0.051 km s^{-1}	0.040 km s^{-1}
Total Observing Time	2 hr	4 hr
Rest Frequency	36.16929 GHz	44.069488 GHz
Velocity at band center[a]	13.7 km s^{-1}	13.2 km s^{-1}
Target source velocity	11.2 km s^{-1}	11.6 km s^{-1}
FWHM of synthesized beam	$1.76'' \times 0.58''$	$1.93'' \times 1.58''$
	P.A. $= -7.80°$	P.A. $= -10.40°$
Line rms noise[b]	18 mJy beam^{-1}	8 mJy beam^{-1}

Notes: [a] The line was centered in the second half of the 1.56 MHz band in order to avoid aliasing (see § 2). [b] The line rms noise was measured from the stokes I image cube using maser line free channels.

set the absolute flux density scale, while the compact source J0607−0834 was used as an amplitude calibrator.

The editing, calibration, Fourier transformation, deconvolution, and processing of the data were carried out using the Astronomical Image Processing System (AIPS) of the NRAO. After applying the amplitude gain corrections of J1733−1304 on the target source M8E, and J0607−0834 on the target source OMC-2 respectively, the spectral channel with the strongest maser emission signal in each of the two sets of data was split, then self-calibrated in both phase and amplitude in a succession of iterative cycles (e.g., Sarma *et al.* 2002). The final phase and amplitude solutions were then applied to the full spectral-line uv data set, and Stokes I and V image cubes were made with a synthesized beamwidth of $1.76'' \times 0.58''$ for M8E and $1.93'' \times 1.58''$ for OMC-2 respectively. Further processing of the data, including magnetic field estimates, was done using the MIRIAD software package.

3. Analysis

For cases in which the Zeeman splitting $\Delta\nu_z$ is much less than the line width $\Delta\nu$, the magnetic field can be obtained from the Stokes V spectrum, which exhibits a *scaled derivative* of the Stokes I spectrum (Heiles *et al.* 1993). Here, consistent with AIPS conventions, $I = (\mathrm{RCP}+\mathrm{LCP})/2$, and $V = (\mathrm{RCP}-\mathrm{LCP})/2$; RCP is right- and LCP is left-circular polarization incident on the antennas, where RCP has the standard radio definition of clockwise rotation of the electric vector when viewed along the direction of wave propagation. Since the observed V spectrum may also contain a scaled replica of the I spectrum itself, the Zeeman effect can be measured by fitting the Stokes V spectra in the least-squares sense to the equation

$$V = aI + \frac{b}{2}\frac{\mathrm{d}I}{\mathrm{d}\nu} \tag{3.1}$$

(Troland & Heiles 1982; Sault *et al.* 1990). The fit parameter a is usually the result of small calibration errors in RCP versus LCP, and is expected to be small. In both the 36 GHz and 44 GHz observations, a was of the order of 10^{-4} or less. While eq. (3.1) is strictly true only for thermal lines, numerical solutions of the equations of radiative transfer (e.g., Nedoluha & Watson 1992) have shown that it gives reasonable values for the magnetic fields in masers also. In eq. (3.1), the fit parameter $b = zB\cos\theta$, where z is the Zeeman splitting factor (Hz mG^{-1}), B is the magnetic field, and θ is the angle of the magnetic field to the line of sight (Crutcher $et\ al.$ 1993). For all cases in which $\Delta\nu_z \ll \Delta\nu$, the Zeeman effect reveals information only on the magnetic field along the line of sight, $B_{\mathrm{los}} = B\cos\theta$.

The value of the Zeeman splitting factor z is critical for determining B_{los} from the observations. Clearly, z for methanol masers is very small, because CH$_3$OH is a non-paramagnetic molecule. Unfortunately, there are no existing laboratory measurements for z at either 36 GHz or 44 GHz. Following the treatment of Vlemmings (2008) for the Zeeman splitting of 6.7 GHz methanol masers, Sarma & Momjian (2009) derived z for the 36 GHz CH$_3$OH line using the Landé g-factor based on laboratory measurements of 25 GHz methanol masers (Jen 1951). However, Vlemmings $et\ al.$ (2011) has since reported that such an extrapolation is likely to give a value of z that is in error by a factor of 2-10, depending on the methanol ladder (E or A). Therefore, it is best to refrain from quoting a value for B_{los} in this contribution, pending experimental measurement of the z factor. Instead, the values for zB_{los} will be stated, since they come directly from the observed data, and are not affected by any estimated value of z.

4. Results and Discussion

4.1. *Zeeman detection toward M8E at 36 GHz*

Figure 1 shows the Stokes I and V profiles in the 36 GHz Class I methanol maser line toward two positions in M8E; the two positions are to the northwest and southeast of the maser line peak. As described in § 3, the magnetic fields were determined by fitting the Stokes V spectra in the least-squares sense using equation (3.1). The values of the parameter b in eq. (3.1) obtained from this fit are $b = -53.2 \pm 6.0$ Hz for the northwest position, and $b = +34.4 \pm 5.9$ Hz for the southeast position. Since an experimentally measured value of z is not available, it is difficult to convert this result into a value for the magnetic field. Still, we can speculate that the true (experimentally determined) value of z for the 36 GHz line will not be too different from $\sim$1 Hz mG^{-1}, based on the discussion in § 4.3 below.

The line-of-sight magnetic field in M8E has opposite signs at the two positions for which Stokes I and V profiles are shown in Fig. 1; this is true irrespective of the eventually determined value of z. By convention, a negative value for B_{los} indicates a field pointing toward the observer. The observed change in the sign of B_{los} at these two positions, together with a slight asymmetry in the maser line profiles at each position, indicates that we are observing at least two masers that are very close in position and velocity. The masers are marginally resolved in these C-configuration observations, otherwise the opposite magnetic fields would sum to zero. EVLA B- or A-configuration observations will be necessary to fully resolve the maser components; Sarma & Momjian have an approved proposal for follow-up observations. The observed change in the sign of B_{los} occurs over a size scale of 0.9″, equal to 1300 AU (assuming the distance to M8E is 1.5 kpc). This may mean that the clumps where the 36 GHz maser is being excited come from two different regions where the field is truly different. Alternatively, it may mean that the field lines curve across the region in which the masers are being excited, so that the line-of-sight field traced by one maser is pointed toward us, whereas that traced by the other maser is pointed away from us.

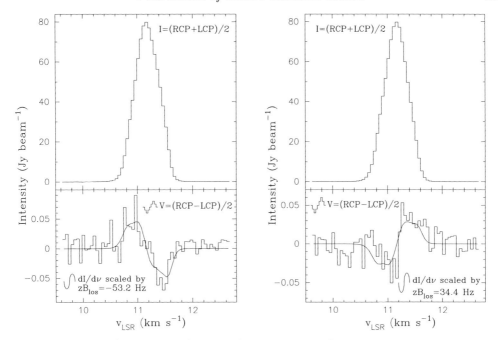

Figure 1. Stokes I (*top-histogram*) and V (*bottom-histogram*) profiles of the 36 GHz Class I methanol maser toward the northwest (*left panel*) and southeast (*right panel*) of the maser line peak in M8E. The curve superposed on V in each of the lower frames is the derivative of I scaled by a value of $zB_{los} = -53.2 \pm 6.0$ Hz in the left panel, and $zB_{los} = +34.4 \pm 5.9$ Hz in the right panel.

4.2. *Zeeman detection toward OMC-2 at 44 GHz*

Figure 2 shows the Stokes I and V profiles in the 44 GHz Class I methanol maser line toward OMC-2. Again, as described in § 3, the magnetic fields were determined by fitting the Stokes V spectra in the least-squares sense using equation (3.1). The value of the parameter b in eq. (3.1) obtained from this fit is equal to $b = 18.4 \pm 1.1$ Hz. As in the case for the 36 GHz line, it is difficult to derive a value for B_{los} without knowing the experimentally determined value for z. Once again, though, we can speculate that the true (experimentally determined) value of z for the 44 GHz line will not be too different from ~ 1 Hz mG^{-1}, based on the discussion in § 4.3 below.

4.3. *Magnetic Fields and Densities*

Class I methanol masers are known to be excited in collisional shocks along outflows in star forming regions. This appears to be the case in OMC-2, where Slysh & Kalenskii (2009) observed six 44 GHz methanol masers spread out along a line aligned at an angle approximately 30° east of north, and at larger scales there is a CO outflow aligned along the same direction (Takahashi *et al.* 2008). Cyganowski *et al.* (this conference) have also shown several excellent examples of Class I methanol masers lined up along outflows at comparable (and high) resolution. Since maser amplification is particularly efficient in directions approximately perpendicular to the shock propagation, the compression of an ordered magnetic field from this orientation would give the following relationship:

$$\frac{B_0}{\rho_0} = \frac{B_1}{\rho_1} \tag{4.1}$$

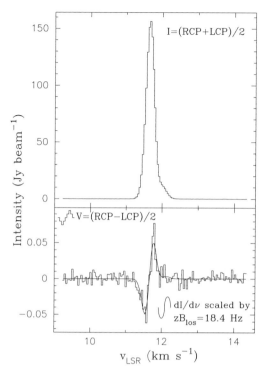

Figure 2. Stokes I (*top-histogram*) and V (*bottom-histogram*) profiles of the 44 GHz Class I methanol maser in OMC-2. The curve superposed on V in the lower frame is the derivative of I scaled by a value of $zB_{\mathrm{los}} = 18.4 \pm 1.1$ Hz.

where 0 and 1 refer to the preshock and postshock (maser) regions respectively, and ρ is the gas density (e.g., Sarma *et al.* 2008). The discussion below uses the molecular hydrogen number density n instead of ρ. If B_0 and n_0 are known, then the measured value of B_{los} can be used to find n_1, and compared to the density at which the 44 GHz methanol maser is known to be excited. Poidevin *et al.* (2010) used the Chandrasekhar-Fermi method to estimate the value of the magnetic field in this region from their 850 μm observations, and found it to be 0.13 mG; this is adopted as the value of B_0 in equation (4.1). The value $n_0 = 10^4$ cm^{-3} can be taken from the C^{18}O observations toward OMC-2 by Castets & Langer (1995). Finally, $B_1 = 2\,B_{\mathrm{los}}$ can be written based on statistical grounds (Crutcher 1999). While the current observations only give the value of zB_{los}, and B_{los} cannot be obtained from observations without knowing z (§ 3), a nominal value of $z = 1.0$ Hz mG^{-1} provides some interesting insights. Using these values in equation (4.1), one obtains $n_1 \sim 10^6$ cm^{-3} for the postshock number density. This is in excellent agreement with theoretical models that show the 44 GHz methanol maser action is maximized in regions with density $10^5 - 10^6$ cm^{-3} (Pratap *et al.* 2008, and references therein). This implies that when the value of z is eventually measured experimentally, it will likely not turn out to be too different from 1.0 Hz mG^{-1}, unless there is something drastically wrong with the theoretically predicted densities at which the 44 GHz maser is excited. Similar considerations apply to the value of z for the 36 GHz line.

4.4. *Additional Considerations and Future Directions*

While two examples, each at a different frequency, are far from the final word on a subject, there is no denying that the prospect of measuring the Zeeman effect in Class

I methanol masers opens up an exciting new window into the physics of star formation. Turning equation (4.1) around, one could say that knowing n_0 from observations and n_1 from theoretical models for methanol masers, the measured values of B_{los}, and hence B_1, would allow us to get a measure of the large-scale magnetic field B_0. This opens up the possibility of tracking the large scale magnetic field in star forming regions at high angular resolution by observing the Zeeman effect in Class I methanol masers. Moreover, the B_{los} values observed in the 36 GHz and 44 GHz lines (based on a nominal value of $z \sim 1$ Hz mG^{-1}) appear to be very similar to the B_{los} values observed in the 6.7 GHz Class II methanol maser line observed by Vlemmings (2008) and Vlemmings *et al.* (2011), who detected significant magnetic fields with the 100 m Effelsberg telescope in this line toward 44 sources. Since Class I and Class II methanol masers likely trace different spatial regions (Ellingsen 2005, and references therein), the likely similarities in the fields measured in these masers provide additional support for the claim that methanol masers may trace the larger scale magnetic fields in star forming regions. Another possibility might be that Class I masers occur in the very early stages of star formation (before the formation of an ultracompact H II region), and Class II masers occur later on in the evolutionary process. In that case, the similarity in B_{los} for these two classes may indicate that the magnetic field strength remains the same during the early stages of the star formation process. It is possible, however, that the similarities are simply due to the nominal choice of $z \sim 1$ Hz mG^{-1} for the 36 GHz and 44 GHz lines or, even if the value of z may really be similar for these two lines (which appears likely based on the discussion in § 4.3), that the similarities in measured magnetic fields may result simply from selection effects due to orientation and/or the shock process. Moving forward, of course, it is important to dwell on alternative possibilities, if for nothing else than to maintain a healthy dose of scientific skepticism until overwhelming examples point to the contrary. Ascribing the detections to a completely fake Zeeman pattern does not appear to be a possibility, especially given that the detections have been made at different frequencies (hence different receivers), and the Class II detections are even by another class of telescope. Moreover, detections taken on different days were imaged separately in order to verify that similar Stokes V patterns were obtained from different sets of observations. Next, if the experimentally measured value of z turns out to be a factor of 10 lower than the nominal value of 1 Hz mG^{-1}, the calculated magnetic fields would be too large, and some kind of non-Zeeman interpretation would have to be ascribed to the detected Stokes V profile. Such considerations certainly make the case for a larger statistical sample of measurements in both the 36 GHz and 44 GHz lines. Perhaps even more critical are experimental measurements of the Zeeman splitting factor z for both the 36 GHz and 44 GHz lines (and 6.7 GHz lines), in order to test possible correlations or anti-correlations between fields measured in Class I and II masers and at different frequencies within each of these types.

5. Conclusions

The detection of the Zeeman effect in the 36 GHz and 44 GHz Class I methanol (CH$_3$OH) maser lines opens a new window into the star formation process. Given the connections between pre-shock and post-shock magnetic fields and the densities in these regions, the magnetic fields detected in these lines could potentially be used to trace the large-scale magnetic field at high spatial resolution in star forming regions. At present, the Zeeman splitting factor z for both these lines has not been measured experimentally, and this complicates the interpretation of the 36 GHz and 44 GHz detections. The assumption of a nominal value of $z = 1$ Hz mG^{-1} for the 36 GHz and 44 GHz lines reveals that the magnetic fields near the protostar (as traced by Class II masers) may be similar to fields farther away along the outflow (as traced by Class I methanol masers), provided this similarity is not merely due to the adoption of a uniform nominal value for z, or at at deeper level, due to selection effects resulting from orientation and/or the shock process itself. However, if z is significantly different from this value, considerations of the ratio of magnetic fields to densities in pre-shock and post-shock regions indicates that models for the densities at which 36 GHz and 44 GHz methanol masers are excited would have to be significantly revised. All of this points to the urgent need for the experimental measurement of z and motivates a larger statistical sample of measurements in both the 36 GHz and 44 GHz Class I methanol maser lines.

References

Banerjee, R. & Pudritz, R. E. 2006, *ApJ*, 641, 949

Banerjee, R. & Pudritz, R. E. 2007, *ApJ*, 660, 479

Brogan, C. L. & Troland, T. H. 2001, *ApJ*, 560, 821

Castets, A. & Langer, W. D. 1995, *A&A*, 294, 835

Cragg, D. M., Johns, K. P., Godfrey, P. D., & Brown, R. D. 1992, *MNRAS*, 259, 203

Crutcher, R. M., Troland, T. H., Goodman, A. A., Heiles, C., Kazes, I., & Myers, P. C. 1993, *ApJ*, 407, 175

Crutcher, R. M. 1999, *ApJ*, 520, 706

Crutcher, R. M., Hakobian, N., & Troland, T. H. 2009, *ApJ*, 692, 844

Ellingsen, S. P. 2005, *MNRAS*, 359, 1498

Falgarone, E., Troland, T. H., Crutcher, R. M., & Paubert, G. 2008, *A&A*, 487, 247

Fish, V. L. & Reid, M. J. 2006, *ApJS*, 164, 99

Heiles, C., Goodman, A. A., McKee, C. F., & Zweibel, E. G. 1993, in *Protostars and Planets III*, ed. E. H. Levy & J. I. Lunine (Tucson: Univ. Arizona Press), 279

Jen, C. K. 1951, *Physical Review*, 81, 197

McKee, C. F. & Ostriker, E. C. 2007, *ARAA*, 45, 565

Menten, K. M. 1991, *ApJL*, 380, L75

Nedoluha, G. E. & Watson, W. D. 1992, *ApJ*, 384, 185

Poidevin, F., Bastien, P., & Matthews, B. C. 2010, *ApJ*, 716, 893

Pratap, P., Shute, P. A., Keane, T. C., Battersby, C., & Sterling, S. 2008, *AJ*, 135, 1718

Sandell, G., Goss, W. M., & Wright, M. 2005, *ApJ*, 621, 839

Sarma, A. P., Troland, T. H., Roberts, D. A., & Crutcher, R. M. 2000, *ApJ*, 533, 271

Sarma, A. P., Troland, T. H., Crutcher, R. M., & Roberts, D. A. 2002, *ApJ*, 580, 928

Sarma, A. P., Troland, T. H., Romney, J. D., & Huynh, T. H. 2008, *ApJ*, 674, 295

Sarma, A. P. & Momjian, E. 2009, *ApJL*, 705, L176

Sarma, A. P. & Momjian, E. 2011, *ApJL*, 730, L5

Sault, R. J., Killeen, N. E. B., Zmuidzinas, J., & Loushin, R. 1990, *ApJS*, 74, 437

Slysh, V. I. & Kalenskii, S. V. 2009, *Astronomy Reports*, 53, 519

Takahashi, S., Saito, M., Ohashi, N., Kusakabe, N., Takakuwa, S., Shimajiri, Y., Tamura, M., & Kawabe, R. 2008, *ApJ*, 688, 344

Troland, T. H. & Heiles, C. 1982, *ApJ*, 252, 179

Troland, T. H., Heiles, C., Sarma, A. P., Ferland, G. J., Crutcher, R. M., & Brogan, C. L. 2008, arXiv:0804.3396

Vlemmings, W. H. T. 2008, *A&A*, 484, 773

Vlemmings, W. H. T., Surcis, G., Torstensson, K. J. E., & van Langevelde, H. J. 2010, *MNRAS*, 404, 134

Vlemmings, W. H. T., Torres, R. M., & Dodson, R. 2011, *A&A*, 529, A95

Cosmic Masers - from OH to H_0
Proceedings IAU Symposium No. 287, 2012   © International Astronomical Union 2012
R.S. Booth, E.M.L. Humphreys & W.H.T. Vlemmings, eds.   doi:10.1017/S174392131200662X

Polarization of the Recombination Line Maser in MWC349

C. Thum[1], D. Morris[2], and H. Wiesemeyer[3]

[1] IRAM, Avenida Divina Pastora, 7,
Núcleo Central, E 18012 Granada, Spain
email: thum@iram.es

[2] IRAM, 300 rue de la Piscine,
Domaine Universitaire de Grenoble
38406 Saint Martin d'Hres, France
email: morris@iram.fr

[3] MPIfR, Auf dem Hügel 51
5300 Bonn, Germany
email: wiesemeyer@mpifr.mpi-bonn.de

Abstract. We present observations of the circular polarization of the recombination line maser in MWC 349. Six good quality H30α spectra were obtained during 2010 – 2011 which show that the Zeeman features are complex, time variable, and usually different for the blue- and red-shifted maser spikes. We propose that the magnetic field, located in the corona of the circumstellar disk, has toroidal and radial components. It is plausibly generated in a disk dynamo.

Keywords. stars: individual (MWC 349), stars: magnetic fields, stars: winds, outflows

1. Introduction

The optically inconspicuous peculiar emission line star MWC 349 is one of the brightest radio stars and a strong emitter in the mid infrared. Its spectral energy distribution (Fig. 1) has two components: *(i)* the emission from hot circumstellar dust peaking near $\lambda = 10\,\mu$m and *(ii)* the emission from an ionized wind which follows a $\nu^{0.7}$ power law from $\lambda \lesssim 350\,\mu$m to $\gtrsim 21$ cm. Radio recombination lines emitted by the wind at cm wavelengths were detected by Altenhoff *et al.* (1981). Observations at millimeter wavelengths reveiled that the recombination lines at wavelengths shorter than 3 mm are masing (Martín–Pintado *et al.* 1989). Subsequent investigations (Thum *et al.* 1998) showed that the amplification peaks near $\lambda = 300\,\mu$m and terminates at quantum number $n = 7$ at 19 μm.

The blue- and red-shifted maser spikes (Fig. 2) mark the radial velocities where maser amplification is strongest. These spikes are located symmetrically about the star in the plane of the circumstellar disk which is seen nearly edge–on (Martín–Pintado *et al.* 2011). These investigators measure the radial distance of the spikes from the star as 29 a.u. The strong magnetic field found by Thum & Morris (1999) from Zeeman features in the H30α spectrum is therefore located at this large distance from the star in the ionized corona of the circumstellar disk.

In the present contribution, we report new polarization observations of the H30α transition obtained with much improved instrumentation. The new data help to characterize the origin of the magnetic field and shed new light on the physical nature of the star.

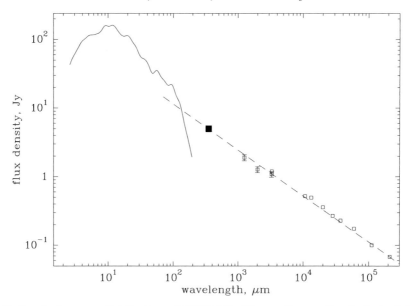

Figure 1. Spectral energy distribution of MWC 349. The continuous line is derived from ISO observations where the numerous emission lines have been subtracted. The dashed line is a fit to radio/mm/submm observations (Altenhoff *et al.* 1994; Tafoya *et al.* 2004). The 350 μm observation (filled square) was made by Weiss & Thum (unpubl.) at APEX.

2. Observations

Using the observing procedure XPOL at the IRAM 30m telescope (Thum *et al.* 2008) we monitored the H30α line at 231.9 GHz in all Stokes parameters during the period of April 2010 to January 2012. This transition was selected because its frequency is high enough for the maser to have strong amplification, while it is still easily accessible from the 30m telescope. Altogether, six spectra of sufficient quality were obtained. All spectra were observed with identical instrumental setup. The spectral backend was set to a resolution of 80 kHz (0.1 km s^{-1}).

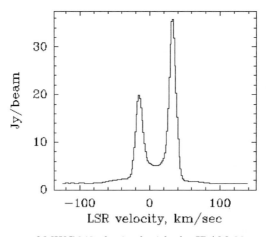

Figure 2. H30α spectrum of MWC 349 obtained with the IRAM 30m telescope at 231.9 GHz (Stokes *I*). The blue- ($v_{LSR} = -18$ km s^{-1}) and red-shifted ($v_{LSR} = +32$ km s^{-1}) maser spikes are superposed on a broad pedestal. The baseline offset of 1.7 Jy is due to free–free continuum emission of the stellar wind.

In this setup, the most important effects causing instrumental polarization are misalignment, phase errors, and the beam squint. Misalignment between the horizontal and vertical receiver channels of EMIR is small and stable (Carter *et al.* 2012). It causes a slighly elevation dependent instrumental Stokes V of the order of $V_i = 0.2\,\%$, as derived from observations of unpolarized point sources. Very similar values of V_i are obtained when subtracting from the V– spectra of MWC 349 a fraction of the Stokes I spectrum so that the net power on the V–spectrum is zero (positive and negative features are equally strong).

Errors of the phase between the two receiver channels transport power from Stokes U into V, and could thus cause false V signals. This effect can however be neglected, since the phase drift of XPOL is smaller than $1°$ per hour and the source is not linearly polarized. We also ignore beam squint which operates only on extended sources.

The observed V–spectra were thus corrected for the misalignment error. They were then normalized such that the negative velocity half ($v_{LSR} < v_s$, where $v_s = 8.2\;\mathrm{km\,s^{-1}}$ is the systemic velocity) of the spectra was devided by their peak power of the blue-shifted maser spike, and their positive velocity part by the peak power of the red-shifted spike. The resulting spectra are free of systematics down to a level of $\sim 0.1\%$. They are shown in Fig. 3, together with the original spectrum obtained by Thum & Morris (1999).

3. Results

The observed spectra clearly show spectral features similar to those already detected in 1999 by Thum & Morris. As argued by these authors, the observed features are best interpreted as due to the Zeeman effect. Although maser propagation effects can in principle generate Zeeman–like features (Nedoluha & Watson 1994), the special conditions needed for these effects to operate are not met in the case of the recombination line maser.

The time series of spectra (Fig. 3) clearly shows that the observed Zeeman features are time variable. Within the limits given by our rather incomplete sampling, we conclude that the typical time scale of variation is much less than a year and larger than a few days. The time scale of B–field variations is therefore similar to the time scale of total power variations of the H30α maser which was found to be ~ 1 month (Thum *et al.* 1992).

Further inspection of the spectra gathered in Fig. 3 shows that the blue- and red-shifted spectral features are mostly very different from the standard antisymmetric (S–shaped) Zeeman signature. Often, we observe rather symmetric (W–shaped) features. Whereas S–shaped features are produced by homogenous fields, W–shaped features require that the maser propagates in a medium where the velocity has gradient *and* where the magnetic field reverses its direction. Both conditions are easily met in the corona of a rotating disk. As a consequence of the edge–on viewing geometry, the observed line-of-sight component of the field translates into a toroidal disk field. The occurance of W–shaped features implies however that there are also radial components present.

The strength of the field is estimated from the approximately S–shaped features to be about 20–25 mG. This value refers however only to the line–of–sight component. The total field strength can only be higher. Such a field must be considered strong, both in comparison with fields reported for molecular cores (e.g. Myers & Goodman 1994) as well as compared with the thermal energy density of the plasma where the maser propagates. We estimate the plasma β–coefficient to be $\gtrsim 0.7$ which makes the observed field dynamically important.

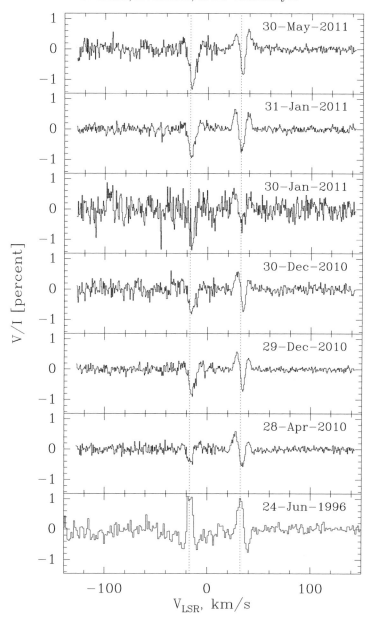

Figure 3. Time sequence of circular polarization spectra obtained with the IRAM 30m telescope. The transition observed is H30α at 231.9 GHz. Spectra were smoothed to a resolution of 0.4 km s^{-1}. The vertical dashed lines indicate the peak velocities of the blue- and red-shifted maser spikes.

4. Discussion

The observed polarization characteritics suggest that the magnetic field associated with MWC 349 is very complex. The field strength varies on the time scales of order one month, orders of magnitude faster than the disk rotation period ($\sim$ 35 years). The field probed by the blue- and red-shifted spikes which are located symmetrically about the star seem to vary independently from each other, both in strength and orientation.

We suggest on the basis of these characteristics that the field is generated locally in the dense disk corona by a dynamo process. Such a dynamo has been proposed by Tout & Pringle (1992).

References

Altenhoff, W. J., Strittmatter, P. A. & Wendker, H. J. 1981, *A&A*, 93, 48
Altenhoff, W. J., Thum, C., & Wendker, H. J. 1994, *A&A*, 281, 161
Carter, M., Lazareff, B., Maier, D., *et al.* 2012, *A&A*, 538, A89
Martín–Pintado, J., Bachiller, R., Thum, C., & Walmsley, C. M. 1989, *A&A*, 215, L13
Martín-Pintado, J., Thum, C., Planesas, P., & Báez-Rubio, A., 2011 *A&A*, 53, L15
Myers, P. C. & Goodman, A. A. 1988, *ApJ*, 326, L27
Nedoluha G. E. & Watson W. D. 1994, *ApJ*, 423, 394
Tafoya, D., Gómez, Y., & Rodríguez, L. F. 2004, *ApJ*, 610, 827
Thum C., Martín–Pintado J., & Bachiller R. 1992, *A&A*, 256, 507
Thum, C., Martín–Pintado, J., Quirrenbach, A., & Matthews, H. E. 1998, *A&A*, 333, L63 - L66
Thum, C. & Morris, D. 1999 *A&A*, 344, 923 – 929
Thum, C., Wiesemeyer, H., Paubert, G., Navarro, S., & Morris, D. 2008, *PASP*, 120, 777
Tout C. A. & Pringle J. E. 1992, *MNRAS*, 259, 604
White, R. L. & Becker, R. H. 1985, *ApJ*, 297, 677

Cosmic Masers - from OH to H_0
Proceedings IAU Symposium No. 287, 2012
R.S. Booth, E.M.L. Humphreys & W.H.T. Vlemmings, eds.

© International Astronomical Union 2012
doi:10.1017/S1743921312006631

VLBA SiO maser observations of the OH/IR star OH 44.8-2.3: magnetic field and morphology

N. Amiri[1], W. H. T. Vlemmings[2], A. J. Kemball[3] and H. J. van Langevelde[4]

[1]Sterrewacht Leiden, Leiden University, Niels Bohrweg 2, 2333 CA Leiden, The Netherlands
[2]Chalmers University of Technology, Onsala Space Observatory, SE-439 92, Onsala, Sweden
[3]Joint Institute for VLBI in Europe (JIVE), Postbus 2, 7990 AA Dwingeloo, The Netherlands
[4]Department of Astronomy and Institute for Advanced Computing Applications and Technologies/ NCSA, University of Illinois at Urbana-Champaign, 1002 W. Green Street, Urbana, IL 61801, USA

Abstract. We report on Very Long Baseline Array SiO maser observations of the OH/IR star OH 44.8 - 2.3. The observations show that the maser features form a ring located at a distance of 5.4 AU around the central star. The masers show high fractional linear polarization up to 100%. The polarization vectors are consistent with a dipole field morphology. Additionally, we report a tentative detection of circular polarization of 7% for the brightest maser feature. This indicates a magnetic field of 1.5 ± 0.3 G. The SiO masers and the 1612 MHz OH masers suggest a mildly preferred outflow direction in the circumstellar environment of this star. The observed polarization is consistent with magnetic field structures along the preferred outflow direction. This could indicate the possible role of the magnetic fields in shaping the circumstellar environment of this object.

Keywords. Stars, Magnetic Fields, Maser, Polarization

1. Introduction

SiO maser emission occurs in the inner circumstellar envelopes (CSEs) of asymptotic giant branch (AGB) stars and can be studied at high angular resolution using high resolution radio interferometers. Very long baseline interferometry (VLBI) observations of the masers in Mira variables have shown that the masers are confined to a region, sometimes ring-shaped, between the stellar photosphere and the dust formation zone (e.g. Cotton *et al.* 2008). Furthermore, VLBI observations revealed that the masers are significantly polarized with linear polarization fractions up to 100%. The circular polarization of the masers is in the range 3% to 5% (Kemball *et al.* 2009).

No information is available at high angular resolution for SiO masers in higher mass loss OH/IR stars. These objects have larger CSEs and much longer periods up to 2000 days than do Mira variables (Herman & Habing 1985). They are strong 1612 MHz OH maser emitters (Baud *et al.* 1979). The stars are surrounded by thick dust shells, which makes them optically obscured. Here, we report the SiO maser polarimetric observations of the OH/IR star OH44.8-2.3 with the VLBA. The observations enable us to obtain the spatial distribution of the SiO maser features in OH/IR stars for the first time. Additionally, our experiment probes the magnetic field strength and morphology in the SiO maser region of OH/IR stars and compares them with those of Mira variables.

2. Observations

We observed the v=1, $J = 1 \rightarrow 0$ SiO maser emission toward OH 44.8-2.3 on 6 July 2010 using the Very Long Baseline Array (VLBA) at 43 GHz. The observations were performed in dual circular polarization spectral line mode. The DiFX correlator was used with a bandwidth of 4 MHz and 1024 spectral channels, which results in 0.03 km s^{-1} spectral resolution.

Auxiliary Expanded Very Large Array (EVLA) observations were performed to measure the absolute electric vector polarization angle (EVPA) for the polarization calibrators. We observed the continuum sources J2253+1608 and J1751+0939 as transfer calibrators. The observations were performed in continuum mode and full polarization, using two 128 MHz spectral windows.

3. Results

3.1. *Total Intensity*

Fig. 1 shows the SiO maser emission map of OH 44.8-2.3 obtained from the VLBA observations. The emission is summed over all velocity channels. The maser features form a partial ring of 4.75 mas. The emission exhibits two opposite arcs and appears to be absent from the western and eastern side of the ring. Assuming a distance of 1.13 kpc obtained from the phase lag method (Van Langevelde *et al.* 1990), the ring radius corresponds to 5.4 AU. Feature 1 exhibits the largest flux density of 2.8 Jy. The ring pattern observed for the SiO emission indicates the tangential amplification of the masers.

3.2. *Linear Polarization*

The polarization morphology of the SiO maser emission is shown in Fig. 1. The stokes parameters (I, Q, and U) are summed over frequency. The polarized emission is plotted as vectors with a length proportional to the polarization intensity. The position angle of the vectors corresponds to the EVPA of the emission. The background contours show the total intensity. The observations reveal that the SiO maser features of OH 44.8-2.3 are highly linearly polarized with an average linear polarization fraction of 30%. The highest polarization fraction corresponds to 100% for feature 8.

The high fractional linear polarization observed for SiO masers could indicate the anisotropic pumping origin of the masers. It was shown that anisotropic background radiation from the central star generates anisotropic pumping, potentially producing a high degree of linear polarization in the SiO maser region (e.g. Watson 2002). However, the results from Nedoluha & Watson (1990) show that the polarization vectors still trace the direction of the magnetic field despite the possibility that the linear polarization develops from anisotropic pumping.

The observations show that the SiO masers of OH 44.8-2.3 are in a regime that $g\Omega \gg R > \Gamma$; where R, $g\Omega$, and Γ represent the stimulated emission rate, the Zeeman rate, and the collisional and radiative decay rate, respectively. In this regime the linear polarization vectors appear either parallel or perpendicular to the projected magnetic field, depending on the angle between the magnetic field direction and the line of sight (Goldreich *et al.* 1973). When taking the EVPA of $-50°$ for feature 8, which has the highest linear polarization fraction, this implies that the magnetic field direction is either parallel ($-50°$) or perpendicular ($40°$) to the linear polarization vectors. In either case, the EVPA vectors probably indicate a large-scale magnetic field in the SiO maser region of this star.

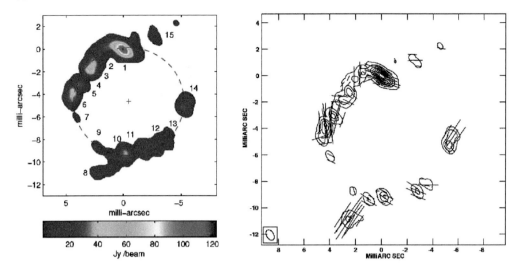

Figure 1. Left Panel: The VLBA map of v=1, $J = 1 \rightarrow 0$ SiO maser emission towards the OH/IR star OH 44.8-2.3. The features are color-coded according to the flux density (Jy / beam) integrated over all velocity channels. Right panel: Contour plot of the Stokes I image at levels [1, 2, 5, 10, 20, 40, 80, 100]% of the peak. Vectors are overlaid proportional to the linearly polarized intensity (on a scale 1 mas = 1.25 Jy beam^{-1}) and drawn at a position angle of the EVPA. All Stokes parameters (I,Q,U) are summed over velocity.

The linear polarization morphology is consistent with the dipole magnetic field morphology in the SiO maser region of this star. However, we cannot rule out toroidal or solar type field morphologies. Polarimetric observations of the OH and H_2O maser regions of the CSE of this star are required to clarify its magnetic field morphology.

3.3. *Circular Polarization*

We report a tentative detection of 7% for the circular polarization fraction for feature 1 in the modest spectral resolution dataset which has 128 spectral channels (Fig. 1). However, due to the increased noise in the high spectral resolution dataset (1024 spectral channels), we cannot confirm the detection. Further polarimetric VLBI observations of the SiO masers of this star with more integration time are needed to clarify this. The left panel in Fig. 2 shows the total intensity spectrum and the circular polarization profile of feature 1. The V spectrum is plotted after removing a scaled down version of the Stokes I spectrum, which is caused by the instrumental gain differences between the right circular polarization (RCP) and the left circular polarization (LCP) profiles.

We note that, based on polarization studies, we cannot distinguish between the Zeeman and non-Zeeman effects from the observations. Since the SiO masers are in the regime where $g\Omega \gg R > \Gamma$, the non-Zeeman effect introduced by Wiebe & Watson (1998) is applicable. According to this scenario the circular polarization can be generated if the magnetic field orientation changes along the direction of maser propagation. The circular polarization produced from this scenario is on average $\sim \frac{m_l^2}{4}$, where m_l indicates the linear polarization fraction. For the 7% linear polarization fraction measured for feature 1, the generated circular polarization it causes is 0.12%. This implies that the measured circular polarization for feature 1 is about six times higher than the estimated value from the non-Zeeman effect. Wiebe & Watson (1998) show that if the circular polarization is higher than the average of $\frac{m_l^2}{4}$, the circular polarization stems from other causes, probably the Zeeman effect. Therefore, it is likely that the circular polarization of this star comes from the Zeeman splitting. However, Wiebe & Watson (1998) explain that the average circular polarization in individual features can go up to 20%.

Assuming the Zeeman interpretation of the observed circular polarization, the magnetic field derived from circular polarization corresponds to the following equation (Kemball & Diamond 1997):

$$B = 3.2 \times m_c \times \Delta\nu_D \times \cos\theta, \qquad (3.1)$$

where m_c, $\Delta\nu_D$, and θ indicate the fractional circular polarization, the maser line width, and the angle between the magnetic field and line of sight, respectively. The full width half maximum line with for the Stokes I spectrum of feature 1 corresponds to $\sim$ 0.8 km s^{-1}. Using the preceding relation, a magnetic field of 1.8$\pm$0.5 G is derived for feature 1.

Alternatively, we use the cross-correlation method introduced by Modjaz *et al.* (2005) to measure the magnetic field due to the Zeeman splitting. In this method the right circular polarization (RCP) and the left circular polarization (LCP) spectra are cross-correlated to determine the velocity splitting. The magnetic field is determined by applying the Zeeman splitting coefficient for SiO masers. We measured a magnetic field of 1.5$\pm$0.3 G for feature 1.

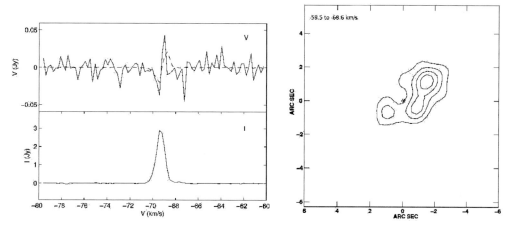

Figure 2. Left panel: Total power (I) and circular polarization (V) spectra of the brightest SiO maser feature of OH 44.8-2.3. The dashed line is the fit to the observed V-spectrum. The V spectrum is shown after removing the scaled down replica of Stokes I. Right panel: 1612 MHz OH maser red-shifted emission map of OH 44.8-2.3 summed over several velocity channels from -59.5 to -68.6 km s^{-1}. The red-shifted peak exhibits the highest flux density of 8.1 Jy. The contour levels are at 0.005, 0.011, 0.016, 0.022, and 0.027 Jy. The star symbol indicates the position of the peak in the red-shifted emission which probably indicates the position of the central star.

3.4. *OH maser observations of OH 44.8-2.3*

We found previous observations of the 1612 MHz OH masers of OH 44.8-2.3 using the Very Large Array (VLA) in the NRAO archive. The observations were performed, under project name 'AH127', in the A configuration, which gives a resolution of 1". The right panel in Fig. 2 displays a 1612 MHz OH maser map of this star. The emission is summed over several channels slightly red-shifted to the stellar velocity (from -59.5 to -68.6 km s^{-1}).

The 1612 MHz OH masers of this star indicate an elongated shell morphology in the direction where there is a gap in the SiO maser emission (Fig. 1; left panel). We note that the OH masers occur on much larger scale ($\sim$1471 AU) than the SiO masers (5.4 AU) around the star. It is therefore not obvious that both deviations from symmetry are related, but if they are, this indicates that there is a mechanism at work that can support the asymmetry on many scales.

Interestingly, the direction of the magnetic field is parallel or perpendicular to the location of the gaps in the SiO maser ring and the OH maser extent. This could indicate that there is a large-scale magnetic field that imposes a preferred direction on the outflow on scales that span two orders of magnitude. Since the timescales involved in forming the OH shell is much larger than that for SiO masers, one would then conclude that the magnetic field is important for imposing an asymmetric signature on the neutral outflow in the OH/IR phase. Furthermore, the high fractional linear polarization measured for the SiO masers of this star could potentially indicate the possible role of the magnetic field in shaping the circumstellar environment of this star.

4. Summary

We performed a pilot study to observe the SiO maser emission in OH/IR stars which have higher mass loss rates compared to Mira variables. Our observations indicate a ring morphology for the SiO maser region of the OH/IR star OH 44.8-2.3. The ring pattern is similar to what was observed previously for Mira variables. The SiO maser features exhibit high fractional linear polarization up to 100%. The polarization vectors are consistent with a dipole field morphology. Additionally, we report a tentative detection of circular polarization at 7% for the brightest maser feature, which corresponds to a magnetic field of 1.5 ± 0.3 G, assuming the Zeeman interpretation for the observed polarization.

In particular, we found that the observed polarization is consistent with magnetic field structures along the preferred outflow direction. This could potentially indicate a role of the magnetic field in shaping the circumstellar environment of this object, and that magnetic fields thus could be important for imposing asymmetries on many scales in the circumstellar environment of this star.

References

Baud, B., Habing, H. J., Matthews, H. E., & Winnberg, A., 1979, *A&A*, 35, 179
Cotton, W. D., Jaffe, W.m, Perrin, G., & Woillez, J., 2008, *A&A* , 477, 517
Goldreich Peter, Keeley, Douglas A., & Kwan, John Y., 1973, *ApJ*, 179, 111
Herman, J. & Habing, H. J., 1985, *A&A*, 59, 523
Kemball, A. J. & Diamond, P. J., 1997, *ApJ*, 481, 111
Kemball, Athol J., Diamond, Philip J., Gonidakis, Ioannis, Mitra, Modhurita, Yim, Kijeong, Pan, Kuo-Chuan, & Chiang, Hsin-Fang, 2009, *ApJ*, 698, 1721
Modjaz, Maryam, Moran, James M., Kondratko, Paul T., & Greenhill, Lincoln J., 2005, *ApJ*, 626, 104
Nedoluha, Gerald E. & Watson, William D., 1990, *ApJ*, 361, 53
Van Langevelde, H. J., van der Heiden, R., & van Schooneveld, C., 1990, *A&A*, 239, 193
Watson, W. D., 2002, IAUS206, 206, 464
Wiebe, D. S. & Watson, W. D., 1998, *ApJ*, 503, 71

Cosmic Masers - from OH to H_0
Proceedings IAU Symposium No. 287, 2012
R.S. Booth, E.M.L. Humphreys & W.H.T. Vlemmings, eds.

© International Astronomical Union 2012
doi:10.1017/S1743921312006643

Linear polarization of hydroxyl masers in circumstellar envelope outer regions

P. Wolak[1], M. Szymczak[1] and E. Gérard[2]

[1] Toruń Centre for Astronomy, Nicolaus Copernicus University, Gagarina 11,
87-100 Toruń, Poland
email: wolak@astro.uni.torun.pl

[2] GEPI, UMR 8111, Observatoire de Paris, 5 place J. Janssen, 92195 Meudon Cedex, France

Abstract. A recent polarimetric survey of OH masers in a large sample of AGB and post-AGB stars revealed widespread occurrence of polarized features. We made a statistical analysis of the polarization properties of this large data set. We discuss the alignment of polarization position angles between the extreme blue- and red-shifted parts of the 1612 MHz spectrum. The average polarization angle of OH masers from the opposite sides of the envelope agrees within 20° for 80% of the sources in the sample. For two objects monitored over ∼6 years the polarization position angle at 1612 MHz is constant within measurement uncertainties: this suggests a stable and a very regular structure of the circumstellar magnetic fields. Alternatively, this could indicate a galactic origin of the field which may be amplified by the stellar wind in the outermost parts of the envelopes.

Keywords. masers – polarization – circumstellar matter – magnetic field – stars: AGB and post-AGB

1. Introduction

Late-type stars showing high mass loss are known to harbor magnetic fields in their envelopes. The magnetic field strength and its structure have been determined by SiO, H_2O and OH maser observations (eg. Kemball & Diamond 1997, Szymczak *et al.* 1998, Vlemmings *et al.* 2005). As these species occur in different parts of the circumstellar envelope it is possible to recover a picture of the magnetic field throughout the entire envelope (Szymczak *et al.* 2001, Vlemmings *et al.* 2005, Vlemmings *et al.* 2011). In the inner regions of the envelope the field strengths are strong enough to dynamically influence the shaping of the stellar wind but it is still debated whether the magnetic field is important in forming asymmetric structures observed in planetary nebulae (Bains *et al.* 2003, Soker 2006).

Early studies of the OH 1612 MHz maser lines in AGB stars claimed a lack of polarized features. Improvements of spectral resolution up to $0.1\,\mathrm{km\,s^{-1}}$ enabled detection circularly polarized features in several objects (Zell & Fix 1991), whereas linear polarization was usually weak, if any (Olnon *et al.* 1980). Our recently published polarimetric survey of AGB and post-AGB stars revealed widespread occurrence of polarized OH maser emission (Wolak *et al.* 2012). More than 75% of the objects in the complete sample have polarized features. One of our findings is a very small difference in the position angle of linearly polarized emission from the extreme blue- and red-shifted parts of the 1612 MHz spectra. The standard model of circumstellar envelopes (Reid *et al.* 1977) predicts that the 1612 MHz outermost peaks come from two very compact regions on the opposite sides of the envelope. Indeed, early interferometric observations fully confirmed this model. For instance the 1612 MHz emission of ∼$0.5\,\mathrm{km\,s^{-1}}$ width from the near and

Table 1. Position angles of the linearly polarized emission for two stars. The $\Delta\chi_{1612\,\mathrm{B-R}}$ is the difference between the averaged polarization position angles for the extreme blue- and red-shifted parts of the 1612 MHz spectrum. The $\Delta\chi_{1667-1612\,\mathrm{R}}$ is the difference between the averaged polarization position angles of the red-shifted parts of the 1667 and 1612 MHz spectra. SD is the standard deviation and SDM is the standard deviation of the mean.

Name	$\Delta\chi$	Number of epochs	Mean	SD	SDM	Median
					(°)	
OH127.8+0.0	$\Delta\chi_{1612\,\mathrm{B-R}}$	181	1.8	8.1	0.6	3.1
	$\Delta\chi_{1667-1612\,\mathrm{R}}$		27.6	15.4	1.1	27.9
OH17.7-2.0	$\Delta\chi_{1612\,\mathrm{B-R}}$	172	2.1	3.9	0.3	1.7
	$\Delta\chi_{1667-1612\,\mathrm{R}}$		65.7	18.7	1.4	62.5

far caps of the shell of the archetypal OH/IR source OH127.8+0.0 remained unresolved with a 0.03" beam (Norris *et al.* 1984). This implies that the sizes of the extreme sides of the envelope are less than 0.8% of its diameter. Thus, high spectral resolution spectra of about 0.1 km s^{-1} obtained even with a 3.5'×19' beam reliably probe a local magnetic field because the depolarization effect is negligible. We discuss the properties of the linearly polarized emission by adding new data obtained in a monitoring project.

2. Observations

The observations were carried out with the Nançay Radio Telescope (NRT) from May 2003 to December 2008. The technical details of the NRT were described by van Driel *et al.* (1996). All four Stokes parameters were measured at 1612 and 1667 MHz simultaneously. The methods of observation, instrumental polarization calibrations and error budget are discussed elsewhere (Szymczak & Gérard 2004). The spectral resolution was $\sim$0.14 km s^{-1} or $\sim$0.07 km s^{-1}. The system directly provided three of the four Stokes parameters, namely I, Q and V, while the fourth parameter U was extracted by a horn rotation of 45°. The linearly polarized flux density $p = (Q^2 + U^2)^{0.5}$, fractional linear polarization, $m_\mathrm{L} = p/I$, and polarization position angle, $\chi = 0.5 \tan^{-1}(U/Q)$, were derived from the Stokes parameters. The properties of the star sample are given in Wolak *et al.* (2012). In the following we limit the discussion to the sources that have both extreme peaks at 1612 MHz brighter than 10σ, i.e. about 0.35 Jy.

3. Results

Linearly polarized features are detected in 59 and 22 sources at 1612 MHz and 1667 MHz, respectively. There are 29 1612 MHz sources with linearly polarized emission in both parts of the double peaked profile. Some examples of p spectra with the superimposed position angle of linear polarization are shown in Fig. 1.

At 1612 MHz most of the sources show only weak ($<$15°) variations of the polarization angle from channel to channel. The analysis presented in Wolak *et al.* (2012) is extended here by considering data from multi-epoch observations of 14 sources. The difference between the average polarization position angles for the four most blue- and red-shifted channels in the 1612 MHz p profile, $\Delta\chi_{1612\,\mathrm{B-R}}$, is calculated. Its value is lower than 20° for more than 80% (24/29) of the sources and is higher than 45° for two sources only. For the whole subsample the absolute average and median values of $\Delta\chi_{1612\,\mathrm{B-R}}$ are 16.2±2.9 and 13.9°, respectively. For a few sources observed at several epochs spanning 6−8 years, no systematic variations of $\Delta\chi_{1612\,\mathrm{B-R}}$ higher than 10° are detected.

Nine sources in the sample have linearly polarized emission at the two frequencies from the same usually red-shifted side of the shell. The scatter of χ angles of the OH mainline channels is on average 2.7 times greater than that of 1612 MHz lines. For seven

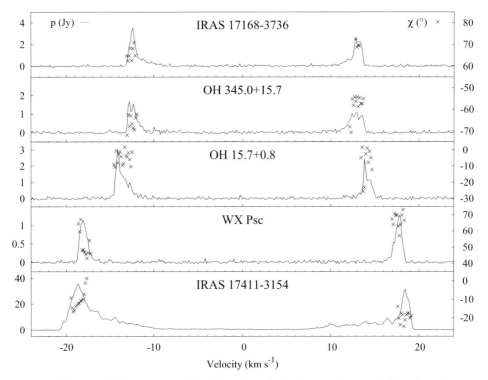

Figure 1. OH 1612 MHz spectra of the linearly polarized emission, p of the selected sources with a double peaked profiles. The polarization position angle (points) are superimposed (right ordinate) for the extreme outside channels. The velocity scale is shown with regard to the systemic velocity.

sources, the velocities of the extreme emission at both frequencies overlap within less than $0.1 \, \mathrm{km \, s^{-1}}$. In the majority of sources the difference in the average polarization angle for four neighbouring channels of the 1612 and 1667 MHz lines ranges from 36 to 80°. The mean difference for nine sources is 44 ± 8° and the median is 47°.

The Stokes I profiles at 1612 and 1667 MHz and the polarization position angle are shown superimposed for two sources only (Fig. 2) for the sake of conciseness. The times series of polarization position angles for the extreme blue- and red-shifted features, their differences at 1612 MHz, $\Delta\chi_{1612\,\mathrm{B-R}}$, and the differences between average polarization angles for the four red-shifted channels in the 1667 and 1612 MHz p profiles, $\Delta\chi_{1667-1612\,\mathrm{R}}$ are also shown. We note that $\Delta\chi_{1667-1612\,\mathrm{R}}$ is 2-4 times more scattered than the $\Delta\chi_{1612\,\mathrm{B-R}}$ (Tab. 1). For the sources OH127.8+0.0 and OH17.7-2.0 the averages $\Delta\chi_{1612\,\mathrm{B-R}}$ over $\sim$6 years are less than 1.8 and 2.1°, respectively (Tab.1). We note a gradual increase of this difference in the latter object which is known to experience secular changes. The scatter of χ angles within the OH mainline channels is on average 2.7 times greater than that within 1612 MHz lines. This may suggest that either the 1667 MHz masers come from more turbulent regions and/or the magnetic fields are less ordered.

4. Discussion

Small differences in the polarization angles at the near and far edges of the OH shells in our subsample suggest a regular magnetic field geometry in the 1612 MHz maser regions.

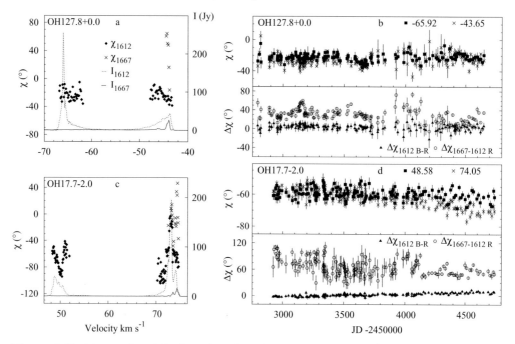

Figure 2. Position angles of the linearly polarized emission in two OH/IR stars. (a) The Stokes I spectra at 1612 MHz (dashed line) and 1667 MHz (solid line) of OH127.8+0.0 with superimposed polarization position angles, χ marked as diamonds and crosses for 1612 and 1667 MHz, respectively. (b) (*upper panel*) polarization position angle time series for the brightest extreme blue- and red-shifted four-channel features at 1612 MHz and (*lower panel*) the difference in polarization position angles of these features, $\Delta\chi_{1612B-R}$ (filled triangles) and the difference in polarization position angles between 1612 and 1667 MHz red-shifted features, $\Delta\chi_{1612-1667R}$ (open circles). (c) and (d) are the same as (a) and (b), respectively but for OH17.7-2.0.

However, it is hard to explain such a homogeneous magnetic field over OH shells of large diameters (e.g. Habing 1996). A plausible explanation for the low values of $\Delta\chi_{1612B-R}$ is that the origin of the field is intrinsic and the same field orientation is carried away (frozen in) by the stellar wind flowing towards the front and back of the envelope. Alternatively, the magnetic field could be of Galactic origin and be amplified and/or distorted by the stellar wind. The interstellar fields can pervade the outer $(10^{16-17}$ cm) regions of circumstellar envelopes and this hypothesis needs further investigation. High values of $\Delta\chi_{1612B-R}$ in a minority of sources suggest a deviation from globally ordered magnetic field caused by local outflows or outbursts. The latter possibility is likely for OH17.7−2.0 where $\Delta\chi_{1612B-R}$ is only 2.1° but the polarization angle of the eruptive feature near 73 km s^{-1} (Szymczak & Gérard 2005) is about $-15°$ and differs by about 64° from the mean value for the extreme blue- and red-shifted velocities. The polarization position angles at the extreme parts of the 1667 MHz spectrum are different and their scattering from channel to channel is much higher that those at 1612 MHz. This suggests that the 1667 MHz emission probes regions with less regular magnetic fields than those probed by the 1612 MHz line.

References

Bains, I., Gledhill, T. M., Yates, J. A., & Richards, A. M. S. 2003, *MNRAS*, 338, 287

Bowers, P. F. & Johnston, K. J. 1990, *ApJ*, 354, 676

Habing, H. J. 1996, *APPR*, 7, 97

Kemball, A. J. & Diamond, P. J. 1997, *ApJ*, 418, L111

Norris, R. P., Booth, R. S., Diamond, P. J., *et al.* 1984, *MNRAS*, 208, 435

Olnon, F. M., Winnberg, A., Matthews, H. E., & Schultz, G. V. 1980, *AAPS*, 42, 1190

Reid, M. J., Muhleman, D. O., Moran, J. M., *et al.* 1977, *ApJ*, 214, 60

Soker, N. 2006, *PASP* 118, 260S

Szymczak, M., Cohen, R. J., & Richards, A. M. S. 1998, *MNRAS*, 297, 1151

Szymczak, M., Cohen, R. J., & Richards, A. M. S. 2001, *A&A*, 371, 1012

Szymczak, M. & Gérard, E. 2004, *A&A*, 423, 209

Szymczak, M. & Gérard, E. 2005, *A&A*, 433, L29

van Driel W., Pezzani J., Gérard E. 1996, in High Sensitivity Radio Astronomy, ed. N. Jackson, & R. J. Davis (Cambridge Univ. Press), 229

Vlemmings, W. H. T., van Langevelde, H. J., & Diamond, P. J. 2005, *A&A*, 434, 1029

Vlemmings, W. H. T., Humphreys, E. M. L., & Franco-Hernández, R. 2011, *ApJ*, 728, 149

Wolak, P., Szymczak, M., & Gérard, E. 2012, *A&A*, 537, A5

Zell, P. J. & Fix, J. D. 1991, *ApJ*, 369, 506

Cosmic Masers - from OH to H_0
Proceedings IAU Symposium No. 287, 2012   © International Astronomical Union 2012
R.S. Booth, E.M.L. Humphreys & W.H.T. Vlemmings, eds.   doi:10.1017/S1743921312006655

Maser polarization with ALMA

Andrés F. Pérez-Sánchez[1] and Wouter Vlemmings[2]

[1]Argelander Institute for Astronomy, University of Bonn
D-53121 , Bonn, Germany
email: aperez@astro.uni-bonn.de

[2]Chalmers University of Technology, Onsala Space Observatory
SE-439 92 Onsala, Sweden.

Abstract. Once ALMA full polarization capabilities are offered, (sub-)mm polarization studies will enter a new era. It will become possible to perform detailed studies of polarized maser emission towards for example massive star forming regions and late-type stars such as (post-) Asymptotic Giant Branch stars and young Planetary Nebulae. In these environments, SiO, H_2O and HCN are molecules that can naturally generate polarized maser emission observable by ALMA. The maser polarization can then be used to derive the strength and morphology of the magnetic field in the masing regions. However, in order to derive, in particular, the magnetic field orientation from maser linear polarization, a number of conditions involving the rate of stimulated emission, molecular state decay and Zeeman splitting need to be satisfied. In this work, we discuss these conditions for the maser transitions in the ALMA frequency range and highlight the optimum transitions to further our understanding of star formation and evolved star magnetic fields.

Keywords. masers, polarization, stars: magnetic fields, stars: AGB and post-AGB, stars: formation.

1. Introduction

Polarized maser emission has been detected towards Star Forming Regions (SFR) and expanding Circumstellar Envelopes (CSE) of late-type stars like (post-) Asymptotic Giant Branch (AGB) stars and young Planetary Nebulae (PNe). Both single dish and interferometric observations have revealed that Silicon monoxide (SiO), Water (H_2O), Hydrogen cyanide (HCN), Hydroxyl (OH) and Methanol (CH_3OH), among others; can naturally generate polarized maser emission in such enviroments (e.g. Vlemmings et al 2006). According to the Zeeman interpretation, the character of the polarization depends strongly on the ratio between the Zeeman frequency ($g\Omega$), the rate of stimulated emission (R) and the rate of the decay of the molecular state (Γ). The Zeeman splitting induced by a magnetic field depends on the molecule's shell structure. Since SiO, H_2O and HCN are non-paramagnetic, closed-shell molecules; $g\Omega$ is expected to be less than the intrinsic line breadth $\Delta\omega$. In this case, the fractional polarization (p_L) is typically of order a few percent. To explain the high levels of fractional linear polarization detected in some cases (specifically SiO masers, e.g. Amiri et al 2012, Vlemmings *et al.* 2011) other physical processes that can produce polarized maser radiation need to be considered. Anisotropic pumping can, for example, enhance the fractional linear polarized maser radiation that is produced due to a magnetic field permeating the masing region, towards SFR or CSE of late-type stars (Vlemmings *et al.* 2011, Nedoluha & Watson 1990).

The different conditions (temperature and gas density) required by each molecular specie to generate polarized maser emission, allow us to constrain the magnetic field properties, i.e., field strength and/or direction, in different regions around proto-stars or

Table 1. In the ALMA frequency range a number of SiO maser transitions, of up to $J = 7 - 6$ for vibrational-excited states of up to $\nu = 3$, could be detected. Here we list some of the most relevant ones detected mostly towards CSE of late-type stars, including the ALMA band where the different lines could be observed.

	SiO	^{29}SiO	^{30}SiO
	GHz; Band	GHz; Band	GHz; Band
$\nu = 1$	43.12 (J=1-0); B1 86.24 (J=2-1); B2 301.81 (J=7-6); B7	127.74 (J=3-2); B4 170.32 (J=4-3); B5 253.70 (J=6-5); B6	168.32 (J=4-3); B5
$\nu = 2$	128.45 (J=3-2); B4 171.27 (J=4-3); B5 214.08 (J=5-4); B6	255.47 (J=6-5); B6	167.16 (J=4-3); B5
$\nu = 3$	127.55 (J=3-2); B4 170.07 (J=4-3); B5 212.58 (J=5-4); B6		

in CSEs. Interferometric observations at radio wavelengths have become a useful tool to perform a deeper study of polarized maser emission towards SFR and late-type stars. The new generation of observatories and instruments enable the study of maser radiation from higher vibrational-excited rotational transition with better sensitivity and resolution. Nowadays, The Atacama Large (sub-) Millimeter Array (ALMA) is starting to perform early science observations. Gradually all of its capabilities including polarimetry will be available for full science observations. In the ALMA frequency range a number of SiO, H_2O and HCN maser transitions belonging to vibrational-excited levels of up to $\nu = 3$ can be observed (Table 1 and 2). Maser polarization theory (e.g. Western & Watson 1984) predict fractional linear polarization levels of up to 100% for $J = 1 - 0$ rotational transitions of diatomic molecules. Such polarization levels can be reached considering the presence of a magnetic field of a few Gauss in, for example, the SiO masing region in the CSE of late-type stars. But in the case of higher rotational transitions (i.e. $J = 2 - 1$, $J = 3 - 2$, etc.) theory also predicts that the fractional linear polarization should decrease as the angular momentum number of the involved state increases. Using Submillimeter Array (SMA) observations, Vlemmings *et al.* (2011) have measured fractional linear polarization of the $\nu = 1, J = 5 - 4$ SiO masers of VX Sgr of up to $\sim 40\%$. In the Zeeman interpretation frame, such high fractional polarization levels need very strong magnetic fields in the masing region in order to be generated, which results in an unlikely strong magnetic field in the surface of the stars. However, the fractional linear polarization level can increase with the angular momentum number if anisotropic pumping is considered. Thus, the problem turns into determining if the fractional linear polarization measured for those molecular states still can be used to trace the magnetic field structure in the masing region. In order to determine that, it is necessary to evaluate the ratios between the maser parameters R, $g\Omega$ and Γ for each single rotational transition detected with angular momentum higher than $J = 1$. The main goal of this work is to determine if the polarized maser radiation produced by molecular transitions of SiO, H_2O and HCN in the ALMA frequency range, could generate suitable levels of linear fractional polarization to study the magnetic field structure towards SFR and CSE of (post-) AGB stars. To do this, we have run numerical models adapted from Nedoluha & Watson (1992) in order to calculate the fractional linear polarization level that could be generated by the

Table 2. A few of both the H_2O and HCN maser lines that will be observed in the ALMA frequency range.

	H_2O				HCN	
J_{K_a,K_c}	GHz; Band	Source type			GHz; Band	Source type
3_{30}-2_{20}	183.31; B5	SFR, O-CSE		J=1-0	88.6; B2	C-CSE
5_{15}-4_{22}	325.15; B7	SFR, O-CSE		$\nu_2 = 1^{1c}$		
6_{43}-5_{50}	439.15; B8	SFR, O-CSE		J= 2-1	177.2; B5	C-CSE
				J= 3-2	267.2; B6	C-CSE
				J= 4-3	354.5; B7	C-CSE
$\nu_2 = 2$ 1_{10}-1_{01}	658.01; B9	O-CSE				

interaction of the molecular states of SiO, H_2O and HCN in the ALMA frequency range, with a magnetic field in the masing region.

2. Overview

Low frequency observations of maser emission from astrophysical sources have been shown to be excellent tools to study the magnetic field structure towards SFR and late-type stars. As mentioned above, the polarization characteristics of the maser emission strongly depend on the radiative conditions of the region where the maser radiation is being generate (saturated or unsaturated). Saturated maser is defined as the radiative regime where the stimulated emission rate R overcomes the decay rate, Γ. In this case, the growth of the possible polarization modes is determined by the population of the molecular states which can interact with a particular mode of polarization (Nedoluha & Watson 1990). Maser polarized emission can be produced in both the unsaturated and the saturated regimes. Goldreich *et al.* (1973) first identified two regimes in the saturated frame where, in the presence of a magnetic field, polarized radiation could be generated: a) The strong magnetic field regime, $g\Omega \gg R$; b) The intermediate strength magnetic field regime, $(g\Omega)^2/\Gamma \gg R \gg g\Omega$, for angular molecular transition J=1-0. Watson and co-workers (Western & Watson 1984, Deguchi & Watson 1990, Nedoluha & Watson 1990) did extend this treatment solving the radiative transfer equations for polarized maser radiation as a function of the emergent intensity R/Γ, taking into account molecular transitions of up to J=3-2. Nedoluha & Watson (1990) have shown that p_L decreases when the rotational transition involved has molecular levels with higher angular momentum, and varies as a function of the angle (θ) between the magnetic field lines ($\vec{B}$) and the direction of propagation of the maser radiation ($\vec{k}$). When the magnetic field is perpendicular to the propagation direction of the maser radiation, the fractional linear polarization reaches its maximum value, decreasing to its minimum value when θ comes close to zero.

The angle that the vector of polarization makes with the $\vec{k} \cdot \vec{B}$ plane, ϕ, changes as a function of the ratio $g\Omega/R$ (Nedoluha & Watson 1990). For a fixed value of θ, the vector of polarization should be either parallel or perpendicular to the kB plane if the Zeeman frequency overcomes the stimulated emission rate, $g\Omega > R$. In the other hand, for larger values of R, but still in the intermediate strength magnetic field regime, the vector of polarization is already rotated away from the kB plane, ie, if $R > g\Omega$ the vector of polarization is neither parallel nor perpedicular to the kB plane. Therefore, as long as the condition $g\Omega > R$ is satisfied, the information about the morphology of the magnetic field in the masing region could be extracted from the polarization vector.

Observationally, fractional linear polarization of up to 100% for J=1-0 SiO maser transitions (Amiri *et al.* 2012) and values over 33% for SiO molecular transitions involving higher angular momentum states (e.g. Vlemmings *et al.* 2011) have been detected. Anisotropic pumping seems to be a plausible explanation for such high p_L values (Nedoluha & Watson 1990). Contrary to the effect of a magnetic field alone in the masing region, the linear polarization can increase with the angular momentum of the involved states due to the anisotropic pumping. But even considering the linear polarization has been enhanced by anisotropic pumping, if $g\Omega > R$, the magnetic field structure still can be traced directly from the polarization vector.

3. Results

The goal of this work is to determine if the different maser transitions in the ALMA frequency range can generate reliable levels of fractional linear polarization in the strong magnetic field regime. In order to evaluate whether R satisfied the condition $g\Omega \gg R$, we have used a radiative transfer code adapted from Nedoluha & Watson (1992). Since R cannot be determined directly from the observations, it has been calculated using:

$$R \approx \frac{AkT_b\Delta\Omega}{4\pi\hbar\omega},$$

where A is the Einstein coefficient of the involved transition, k the boltzmann constant, T_b the brightness temperature and $\Delta\Omega$ the beaming solid angle for the maser radiation. From the observations, it is possible to estimate the value of T_b, but it is not straightforward to determine the value of $\Delta\Omega$.

Here, we present partial numerical results of our models (Perez-Sanchez & Vlemmings 2012, in prep). In Fig 1 we present the results for the 325 GHz, $5_{15} - 4_{22}$ H_2O maser transition, as well as for the 215 GHz, $\nu = 1, J = 5 - 4$ SiO maser line.

4. Conclusions

From our models for H_2O and SiO, it seems likely that polarized maser emission lines in the ALMA frequency range can be used in order of determining the magnetic field structure towards late-type stars and SFR. In Figure 1 it is clear that, in the case of SiO, suitable levels of fractional linear polarization, even in the case of $\theta > 75$ degrees, could be observed in the strong magnetic field regime $g\Omega > R$. In our models, we have used typical magnetic field strength reported in the literature for both the H_2O (50mG) and the SiO (1 G) masing regions (Nedoluha & Watson 1990, Nedoluha & Watson 1992). Thus, still in the presence of anisotropic pumping, the polarization vector will trace the magnetic field, being either parallel or perpendicular to the direction of magnetic field in the plane of the sky. This agrees pretty well with what Vlemmings *et al.* (2011) suggested about tracing the magnetic field around VX sgr using the vector of polarization of the $\nu = 1, J = 5 - 4$ maser line. The uncertainties or our results are on the beaming angle of the maser radiation, which cannot be derived from observations. In the case of water, it seems that up to 10% of fractional linear polarization could be detected having $g\Omega > R$. A complete analysis, including other maser lines of up to $\nu = 3$ in the case of SiO, is going to be submitted soon (Perez-Sanchez & Vlemmings 2012, in prep).

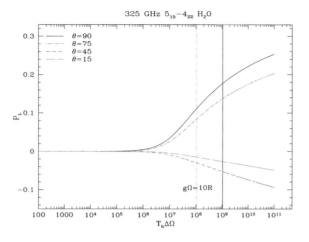

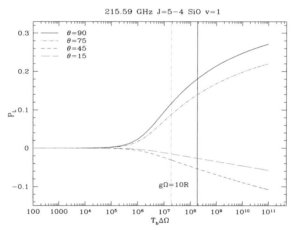

Figure 1. Fractional linear polarization as a function of $T_b\Delta\Omega$ for two representative lines of H_2O and SiO masers in the ALMA frequency range, for four different values of θ (90, 75, 45, 15 degrees). The gray line points out the value of $T_b\Delta\Omega$ where polarization should be reliable observed for each line, whereas the black one shows the limit value where $g\Omega = R$. For water we have considered a magnetic field strength of 50mG and decay rate of $\Gamma = 1s^{-1}$. Taking into account a beaming solid angle of $\Delta\Omega = 10^{-2}$, water masers with brigthness temperature of a few $\times 10^{10}$ K can generate at least 10% of fractional linear polarization in the presence of a magnetic field perpendicular to the direction of propagation of the maser radiation. In the case of SiO, considering $|B| = 1G$ and $\Gamma = 5s^{-2}$, and once again taking $\Delta\Omega = 10^{-2}$, it seems likely to observe about 10% of fractional linear polarization due to the action of a magnetic field, still satisfying the inequality $g\Omega > R$.

References

Amiri, N., Vlemmings, W. H. T., Kemball, A. J., & van Langevelde, H. J. 2012, *A&A*, 538, A136

Deguchi, S. & Watson, W. D. 1990, *ApJ*, 354, 649

Goldreich, P., Keeley, D. A., & Kwan, J. Y. 1973, *ApJ*, 179, 111

Nedoluha, G. E. & Watson, W. D. 1990, *ApJ*, 354, 660

Nedoluha, G. E. & Watson, W. D. 1992, *ApJ*, 384, 185

Perez-Sanchez, A. F. & Vlemmings, W. 2012, in preparation

Vlemmings, W. H. T., Diamond, P. J., & Imai, H. 2006, *Nature*, 440, 58

Vlemmings, W. H. T., Humphreys, E. M. L., & Franco-Hernández, R. 2011, *ApJ*, 728, 149

Western, L. R. & Watson, W. D. 1984, *ApJ*, 285, 158

Cosmic Masers - from OH to H$_0$
Proceedings IAU Symposium No. 287, 2012
R.S. Booth, E.M.L. Humphreys & W.H.T. Vlemmings, eds.

© International Astronomical Union 2012
doi:10.1017/S1743921312006667

High resolution magnetic field measurements in high-mass star-forming regions using masers

Gabriele Surcis[1], Wouter H. T. Vlemmings[2], Huib J. van Langevelde[1,3] and Busaba Hutawarakorn Kramer[4]

[1]Joint Institute for VLBI in Europe
Postbus 2, 7990AA, Dwingeloo, the Netherlands
email: surcis@jive.nl

[2]Chalmers University of Technology, Onsala Space Observatory
SE-439 92 Onsala, Sweden
email: wouter.vlemmings@chalmers.se

[3]Sterrewacht Leiden, Leiden University
Postbus 9513, 2300RA Leiden, the Netherlands
email: langevelde@jive.nl

[4]Max-Planck Institut für Radioastronomie
Auf dem Hügel 69, 53121 Bonn, Germany
email: bkramer@mpifr-bonn.mpg.de

Abstract. The bright and narrow spectral line emission of masers is ideal for measuring the Zeeman-splitting as well as for determining the orientation of magnetic fields in 3-dimensions around massive protostars. Recently, polarization observations at milliarcsecond resolution of 6.7-GHz CH$_3$OH masers have uniquely been able to resolve the morphology of magnetic fields close to massive protostars. The observations reveal that the magnetic fields are along outflows and/or on the surfaces of circumstellar tori. Here we present three different examples selected from a total number of 7 massive star-forming regions that were investigated at 6.7-GHz with the EVN in the last years.

Keywords. Stars:formation, masers:methanol, polarization

1. Introduction

Three different scenarios have been proposed to explain the formation of high-mass stars. In one of these scenarios, *Core accretion* (McKee & Tan 2003), massive stars form through gravitational collapse, which involves disc-assisted accretion to overcome radiation pressure. This scenario is similar to the favored picture of low-mass star-formation, in which magnetic fields are thought to play an important role by removing excess angular momentum, thereby allowing accretion to continue onto the star. However, the role of magnetic fields during the protostellar phase of high-mass star-formation is still a debated topic. In particular, it is still unclear how magnetic fields influence the formation and dynamics of discs and outflows. Most current information on magnetic fields close to high-mass protostars comes from polarized maser emissions, which allow us to investigate the magnetic field on small scales (10s–1000s AU) by using interferometers, such as the European VLBI Network (EVN) and the Very Long Baseline Array (VLBA).

In this contribution we summarize the results obtained by observing the full polarized emission of 6.7-GHz CH$_3$OH towards the massive protostars W75N–VLA 1, NGC7538-IRS 1, and W51–e2. For W75N–VLA 1 and NGC7538-IRS 1 we also observed the polarized emission of 22-GHz H$_2$O masers. The linear polarization and total intensity of

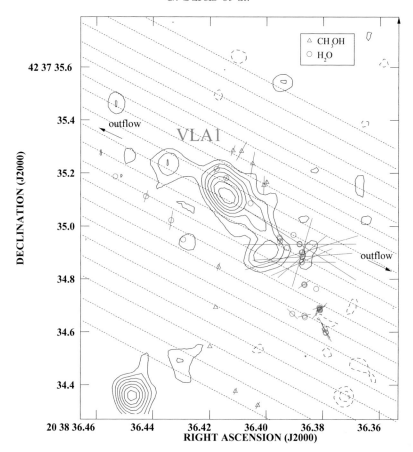

Figure 1. Modified version of Fig.3 of Surcis *et al.* (2011b) of the CH$_3$OH (triangles) and H$_2$O (circles) masers of W75N-VLA 1. The contours are the 1.3 cm continuum emission observed with the VLA. The linear polarization vectors are also reported (20 mas correspond to a linear polarization fraction of 1%). The dashed lines indicate the large-scale direction of the magnetic field.

22-GHz H$_2$O and 6.7-GHz CH$_3$OH masers were analysed by using a full radiative transfer method code based on the models of Nedoluha & Watson (1992). The output of the code are the emerging brightness temperature ($T_b\Delta\Omega$) and the intrinsic thermal linewidth (ΔV_i) from which the θ angle between the magnetic field orientation and the maser propagation direction can be estimated (e.g., Surcis *et al.* 2011a). From the circularly polarized emission of the masers we were able to measure the Zeeman-splitting of the two maser transitions, but only in the case of H$_2$O masers this led to the magnetic field strength, as the value of the Landé g-factor appropriate for the 6.7-GHz CH$_3$OH masers is still unknown (Vlemmings *et al.* 2011).

2. W75N-VLA 1

VLA 1 is a massive protostar located in the active high-mass star-forming region W75N(B) (Torrelles *et al.* 1997), which is at a distance of 1.30±0.07 kpc (Rygl *et al.* 2012). In the region two compact H II regions were also detected, named VLA 2 and VLA 3, which are though to be in an earlier evolutionary stage than VLA 1 (Torrelles *et al.* 1997). A large-scale high-velocity outflow, with an extension greater than 3 pc and

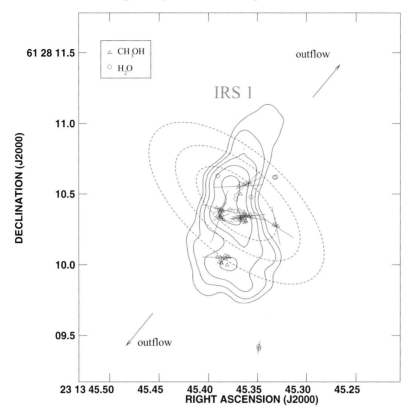

Figure 2. Modified version of Fig.1 of Surcis *et al.* (2011a) of the CH_3OH (triangles) and H_2O (circles) masers of NGC7538-IRS1. The contours are the 2 cm continuum emission observed with the VLA. The linear polarization vectors are also reported (60 mas correspond to a linear polarization fraction of 1%). The dashed ellipses indicate the direction of the magnetic field.

a total molecular mass greater than 255 $M_\odot$, was also detected from W75N(B) (e.g., Shepherd *et al.* 2003). VLA 1 was proposed to be the powering source of the outflow (e.g., Torrelles *et al.* 1997).

For the first time we were able to compare the orientation of the magnetic field determined from the polarized emission of two different maser species, i.e. CH_3OH and H_2O masers, that sample different physical characteristics. Both maser species are linearly distributed along the radio jet of VLA 1 (Fig. 1). The CH_3OH masers indicate a magnetic field orientation (NE-SW) at a position angle of about 73°, while the field in the H_2O maser region has an average position angle of $\sim 71^o$. The magnetic field derived from both maser species is thus almost aligned with the outflow, which has a position angle of 66° (Hunter *et al.* 1994), suggesting that VLA 1 might indeed be the powering source of the outflow. We measured a magnetic field strength from the circularly polarized emission of H_2O masers of about 670 mG. For more details see Surcis *et al.* (2009) and Surcis *et al.* (2011a).

3. NGC7538-IRS1

IRS 1 is the brightest source of the complex massive star-forming region NGC7538, which is located at a distance of 2.65 kpc in the Perseus arm of the Galaxy (Moscadelli *et al.* 2009). The central star of IRS 1 has been suggested to be an O6 star of about 30 $M_\odot$

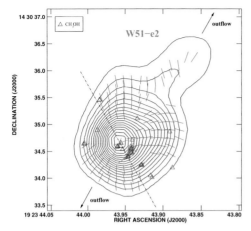

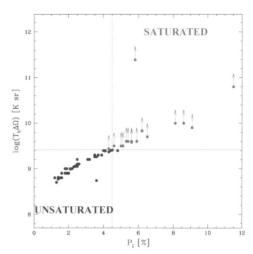

Figure 3. Modified version of Fig. 5a of Tang *et al.* (2009) of the dust polarization of W51-e2. The magnetic field (red segments) detected with the SMA (angular resolution $0''.7$ that corresponds to ~ 4000 AU) is superimposed on the 870 μm continuum contour map of W51–e2. The green segments mark the direction of the magnetic fields as derived from the CH$_3$OH masers (angular resolution $0''.001$ corresponding to ~ 5 AU). The dot-dashed black line indicates the direction of the ionized accreting flow by Keto & Klaassen (2008).

Figure 4. The emerging brightness temperatures ($T_b \Delta \Omega$) as function of the linear polarization fraction (P_l) of all CH$_3$OH masers detected so far with the EVN. The red arrows indicate that the $T_b \Delta \Omega$ values obtained from the radiative transfer method code are lower limits. The red full line is the limit of $T_b \Delta \Omega$ from which the CH$_3$OH masers are considered saturated, and the dotted line shows the lower P_l for saturated masers.

(e.g. Sandell *et al.* 2009). A molecular bipolar outflow (PA = 140°), with a velocity of about 250 km s^{-1} and a mass of 82.8 M$_\odot$, and a molecular torus (angular size ~ 2 arcsec, PA = 50°) have also been detected towards IRS 1 (Qiu *et al.* 2011, Klaassen *et al.* 2009).

We detected both CH$_3$OH and H$_2$O masers around IRS 1 but no linearly polarized emission from the H$_2$O masers close to the protostar was found (Fig. 2). Comparing the velocities of the CH$_3$OH masers with the velocities of the large-scale torus we suggest that the masers are tracing the interface between the infall and the large-scale torus. The H$_2$O masers are instead associated with the blue-shifted part of the outflow. Analysing the orientation of the linear polarization vectors of the CH$_3$OH masers and taking into account the sign of the Zeeman-splitting measurements (positive = magnetic field points away from the observer, negative = towards the observer) we determine that the magnetic field is situated on the two surfaces of the torus with a counterclockwise direction on the top surface. For more details see Surcis *et al.* (2011b).

4. W51-e2

W51–e2 is one of the brightest molecular cores located in the eastern edge of the luminous star-forming region W51 at a distance of $5.41^{+0.31}_{-0.28}$ kpc (Sato *et al.* 2010). Keto & Klaassen (2008) showed evidence both for infalling, or accreting, gas with a possible rotation around W51–e2 and for a bipolar outflow (PA$\approx 150^o$). The dust polarization emission at 870 μm revealed a hourglass-like morphology for the inferred magnetic field (red segments in Fig. 3) near the collapsing core of W51–e2 (Tang *et al.* 2009). The magnetic field morphology (green segments in Fig. 3) determined from the linearly polarized emission of masers, which have been detected close to the 870 μm continuum peak, is consistent with the hourglass morphology. This indicates that the CH_3OH masers are able to probe the large-scale magnetic fields close to the protostars.

5. Linear polarization fraction and saturation state of CH_3OH masers

We detected a total number of 213 CH_3OH masers towards all the 7 massive star-forming regions observed with the EVN. We were able to model 72 masers by using the adapted full radiative transfer method code for CH_3OH masers, which is based on the models of Nedoluha & Watson (1992). When the code is applied to a saturated maser, it gives a lower limit for $T_b\Delta\Omega$ and an upper limit for ΔV_i. Since the masers are unsaturated when $R/\Gamma < 1$ and because the stimulated emission rate is $R \simeq Ak_BT_b\Delta\Omega/4\pi h\nu$, we can estimate an upper limit for $T_b\Delta\Omega$ below which the masers can be considered unsaturated. For CH_3OH masers ($A = 2 \times 10^9 s^{-1}$) the limit is $T_b\Delta\Omega < 2.6 \times 10^9$ K sr. In Fig. 4 we report $T_b\Delta\Omega$ as function of the linear polarization fraction (P_l). From the plot we see that all the 6.7-GHz CH_3OH masers with $P_l \lesssim 4.5\%$ are unsaturated.

6. Conclusion

In conclusion we have found that the magnetic field in massive star-forming regions plays a role as important as in the formation of low-mass stars. All our results are in agreement with the theoretical simulations of Banerjee & Pudritz (2007), who demonstrated that the formation of an early outflow driven by the magnetic field is necessary in order to form the observed disc-outflow systems. From our measurements it seems that the magnetic fields are oriented along the outflow with an hourglass morphology at large-scales and when we observe closer to the protostar the magnetic field appear to be on the surfaces of a torus/disc structure from which the matter accreates onto the protostar along the magnetic field lines.

References

Banerjee, R. & Pudritz, R. E. 2007, *ApJ*, 660, 479
Hunter, T. R., Taylor, G. B., Felli. M., & Tofani, G. 1994, *A&A*, 284, 215
Keto, E. & Klaassen, P. 2008, *ApJ*, 678, L109
Klaassen, P. D., Wilson, C. D., Keto, E. R., & Zhang, Q. 2009, *ApJ*, 703, 1308
McKee, C. F. & Tan, J. C. 2003, *ApJ*, 585, 850
Moscadelli, L., Reid, M. J., Menten, K. M., Brunthaler, A., Zheng, X. W., Xu, Y. *et al.* 2009, *ApJ*, 693, 406
Nedoluha, G. E. & Watson, W. D. 1992, *ApJ*, 384, 185
Qiu, Keping, Zhang, Qizhou, & Menten, Karl M. 2011, *ApJ*, 728, 6
Rygl, K. L. J., Brunthaler, A., Sanna, A., Menten, K. M., Reid, M. J., van Langevelde, H. J., Honma, M., Torstesson, K. J. E., & Fujisawa, K. 2012, *A&A*, arXiv1111.7023R
Sandell, G. Goss, W. M., Wright, M., & Corder, S. 2009, *ApJ*, 699, L31
Sato, M., Reid, M. J., Brunthaler, A., & Menten, K. M. 2010, *ApJ*, 720, 1055
Shepherd, D. S., Testi, L., & Stark, D. P. 2003, *ApJ*, 584, 882
Surcis, G., Vlemmings, W. H. T., Dodson, R., & van Langevelde, H. J. 2009, *A&A*, 506, 757
Surcis, G., Vlemmings, Curiel, S., Hutawarakorn Kramer, B., Torrelles, J. M., & Sarma, P. 2011a, *A&A*, 527, A48
Surcis, G., Vlemmings, W. H. T., Torres, R. M., van Langevelde, H. J., & Hutawarakorn Kramer, B. 2011b, *A&A*, 533, A47
Tang, Y.-W., Ho, P. T. P., Koch, P. M., Girart, J. M., Lai, S.-P., & Rao, R. 2009, *ApJ*, 700, 251
Torrelles, J. M., Gómez, J. F., Rodríguez, L. F., Ho, P. T. P., Curiel, S., & Vazquez, R. 1997, *ApJ*, 489, 744
Vlemmings, W. H. T., Torres, R. M., & Dodson, R. 2011, *A&A*, 529, 95

Cosmic Masers - from OH to H_0
Proceedings IAU Symposium No. 287, 2012
R.S. Booth, E.M.L. Humphreys & W.H.T. Vlemmings, eds.
© International Astronomical Union 2012
doi:10.1017/S1743921312006679

The magnetic field of IRAS 16293-2422 as traced by shock-induced H_2O masers

Felipe O. Alves[1], Wouter H. T. Vlemmings[2], Josep M. Girart[3] and José M. Torrelles[4]

[1] Argelander-Institut für Astronomie, University of Bonn,
Auf dem Hügel 71, D-53121, Bonn, Germany
email: falves@astro.uni-bonn.de

[2] Chalmers University of Technology, Onsala Space Observatory,
SE-439 92 Onsala, Sweden
email: vlemmings@chalmers.se

[3] Institut de Ciències de l'Espai (IEEC-CSIC), Campus UAB, Facultat de Ciències,
C5 par 2ª, 08193 Bellaterra, Catalunya, Spain
email: girart@ice.cat

[4] Institut de Ciències de l'Espai (CSIC)-UB/IEEC, Universitat de Barcelona,
Martí i Franquès 1, E-08028 Barcelona, Spain
email: torrelles@ieec.cat

Abstract. H_2O masers are important magnetic field tracers in very high density gas. We show one of the first magnetic field determinations at such high density in a low-mass protostar: IRAS 16293-2422. We used the Very Large Array (VLA) to carry out spectro-polarimetric observations of the 22 GHz Zeeman emission of H_2O masers. A blend of at least three maser features can be inferred from our data. They are excited in zones of compressed gas produced by shocks between the outflows ejected by this source and the ambient gas. The post-shock particle density is in the range $1 - 3 \times 10^9$ cm^{-3}, and the line-of-sight component of the magnetic field is estimated as ~ 113 mG. The outflow dynamics is likely magnetically dominated.

Keywords. stars: formation, masers, polarization, ISM: magnetic fields, ISM: individual: IRAS 16293-2422

1. Introduction

Water masers are unique because they are found in a variey of astrophysical environments, including star-forming regions at distinct mass ranges. The most commonly observed water maser line is the $(6_{16} - 5_{23})$ transition at 22 GHz, an excellent probe of molecular gas at very high volume densities ($n \sim 10^{8-10}$ cm^{-3}, Elitzur *et al.* 1989). Spectro-polarimetry of water masers is a powerful tool to study magnetic fields at such high densities, since the strength of the maser line emission is such that the small splitting due to the Zeeman effect can be detected. This allows the line-of-sight (LOS) component of the magnetic field to be determined. If the linear polarization is also measured, then the full 3D magnetic field configuration can be derived.

Very Long Baseline Interferometry (VLBI) of water masers reveals the magnetic field properties at very small (subarcseconds) spatial scales, resolving its structure in the dense molecular material around circumstellar envelopes of young stars (Vlemmings *et al.* 2006). Moreover, this technique overcomes the depolarization issue, when a decrease in the polarization degree is observed toward high extinction media due to unresolved fields or roundness of dust grains (Goodman *et al.* 1995, Lazarian *et al.* 1997). Therefore, multi-wavelength polarimetry allows for tracking the magnetic field morphology from the

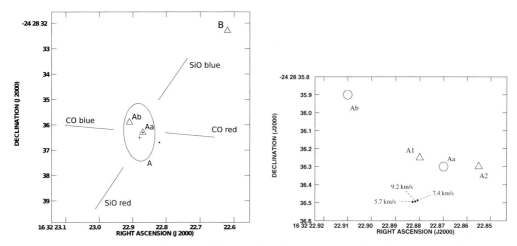

Figure 1. *Left panel*: Distribution of dust and molecular material in I16293. The plus signal indicates the position of our VLA water maser data. The ellipse is the deconvolved size of the source A as derived by Rao *et al.* (2009). Triangles denote the position of the submm condensations observed by Chandler *et al.* (2005). Stars denote the VLBI water maser detections of Imai *et al.* (2007) and straight lines denote the direction of the quadrupolar outflow. *Right panel*: Possible independent maser features (stars) as derived by our gaussian fit. The numeric labels are the maser velocities with respect to the Local Standard of Rest (LSR). The 3.7 cm sources $A1$ and $A2$ (triangles) and the submm sources Aa and Ab (circles) are also shown (Chandler *et al.* 2005).

diffuse gas of molecular clouds to dense cores. In this case, the magnetic field strength is expected to increase as a function of the volume density, as observationally shown by Fiebig & Guesten (1989).

In this work, we report spectro-polarimetric Very Large Array (VLA) observations of H_2O masers toward IRAS 16293-2422 (hereafter, I16293). This object is a well studied Class 0 low-mass binary protostar (separation $\sim$ 600 AU), with a very rich chemistry (Jørgensen *et al.* 2011). The brightest component (source A) has a very strong submillimeter (submm) flux and is resolved into two condensations: Aa and Ab (Chandler *et al.* 2005). A quadrupolar outflow configuration is also observed in SiO and CO molecular emission.

H_2O maser emission in I16293 has been well monitored and shows strong (> 200 Jy) and stable emission over a few weeks but highly variable over months (Claussen *et al.* 1996, Furuya *et al.* 2003). The observed features are detected only toward source A and have proper motions of $\sim$ 5.6 mas/yr (Imai *et al.* 2007). The maser spots are mainly associated with the outflows around the source and its circumstellar disk.

2. Observations and Results

The observations were performed with the VLA in extended A-configuration, on 2007 June 25th and 27th. Each track lasted $\sim$ 5.5 hours. A total of 27 antennas were used, 10 of them already retrofitted with the new system, resulting in a combined VLA/EVLA (Extended VLA) observation. We used the K band receivers tuned at the frequency of the water maser ($6_{16} - 5_{23}$) rotational transition ($\mu_0 = 22.23508$ GHz), and the full polarization capability of the correlator. The spectral setup is 10.5 km s^{-1} wide, with a spectral resolution of 0.08 km s^{-1}, and it covers the brightest maser features observed around 7 km s^{-1}. Data reduction was done with the Astronomical Image Processing

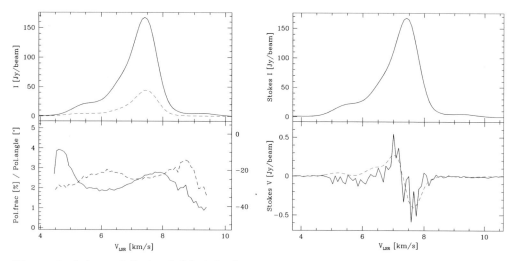

Figure 2. *Left panel*: Stokes I (black line) and linear polarization intensity (multiplied by a factor of 10, red dashed line) spectra of water maser emission (upper panel). The spectra of polarization fraction (black line, left scale) and position angle (dashed line, right scale) are also shown (lower panel). *Right panel*: Stokes I (upper panel) and Stokes V (lower panel) spectra of the water maser emission. The red dashed line is the scaled derivative of total power I.

Table 1. H_2O maser components in IRAS 16294-2422.

V_{LSR} (km s^{-1})	I_{peak}^a (Jy beam^{-1})	α (2000) (h m s)	δ (2000) (o $\prime$ $\prime\prime$)
5.7	23	16 32 22.8830	-24 28 36.495
7.4	168	16 32 22.8808	-24 28 36.487
9.2	5	16 32 22.8818	-24 28 36.493

Notes: ^{a} Equatorial coordinates derived with the JMFIT task of AIPS.

Software package (AIPS). Imaging of Stokes parameters I, Q and U were generated with a quasi-uniform weighting. The resulting synthesized beam is $0.14'' \times 0.08''$.

The water masers detected in our observations are associated with source A with a projected distance of ~ 30 AU from the dust condensation Aa (Fig. 1, left panel). The maser line peaks at 7.4 km s^{-1} with a flux density of 170 Jy beam^{-1} and a rms noise of 8 mJy beam^{-1} and 23 mJy beam^{-1} for edge channels and channels containing emission, respectively. The Stokes I spectrum has a non-Gaussian line profile, and at least three unresolved components can be inferred from the spectrum (Fig. 2, upper panels). A two-dimensional Gaussian profile fit on those components provides a mean spatial separation of 22 milliarcseconds. The equatorial coordinates and the peak intensity of each component is shown in Table 1. The three features are linearly distributed in a E-W direction (Fig. 1, right panel) and have the same LSR velocity as the SE outflow lobes (Rao *et al.* 2009). In addition to the submm sources, two centimeter (cm) sources ($A1$ and $A2$) were observed in the same region and might be associated with a radio jet and to a protostar (Chandler *et al.* 2005). This reinforces that our maser features are excited by shocks in the dense gas at the surroundings of condensation Aa.

Polarized emission: The spectrum of linearly polarized intensity is very similar to the Stokes I line profile except for the flux scale, which is weaker by a factor of ~ 30. (Fig. 2, left-upper panel). The measured polarization fraction is $2.5 \pm 0.2\%$, and the position angle of the polarization vectors (θ, counted from North to East) is -23°. Both quantities are very stable across the maser (Fig. 2, left-lower panel).

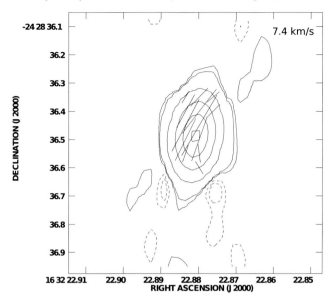

Figure 3. Distribution of H_2O linear polarization vectors in the brightest emission channel. Contours are -50, -30, 30, 50, 500, 3×10^3, 1×10^4, $2 \times 10^4 \times 8$ mJy beam^{-1}. Only polarization vectors whose $P > 1\%$ are plotted.

The line profile of the circular polarization (Stokes V) has the characteristic S-shape (Fig. 2, right-lower panel). It is proportional to the first derivative of the Stokes I spectrum and the LOS component of the magnetic field. The fraction of circular polarization, calculated as $(V_{max} - V_{min})/I_{max}$, is 0.45% for the brightest component. The remaining maser featuers show only residual Zeeman profiles with amplitudes at the rms level.

3. The B-field in I16293

From the Zeeman splitting formalism, the magnetic field strength can be correlated to the fraction of circular polarization P_V by (Fiebig & Guesten, 1989)

$$B \cos \theta = \frac{P_V \times \Delta V_I}{2 \times A_{F-F'}}, \qquad (3.1)$$

where ΔV_I is the *FWHM* of the total power spectrum and $A_{F-F'}$ is a coefficient which depends on the maser rotational levels F and F', the intrisic thermal linewidth $\Delta \nu_{th}$ and the maser saturation degree. The angle θ is the angle between the magnetic field and the maser propagation direction. Assuming that $A_{F-F'}$ ranges from 0.012 to 0.018 in order to be consistent with models and observations (Nedoluha & Watson 1992; Vlemmings *et al.* 2002), and a linewidth of 0.75 km s^{-1}, which is more realistic for resolved maser lines, the field strength is estimated to range between -94 and -141 mG, pointing towards the observer. Given that the linear polarization fraction is about 3%, this maser is likely unsaturated and the field strength determination is a fair approximation.

Fig. 3 shows the linear polarization vectors at the brightest channel. The direction of the plane-of-sky (POS) component of the magnetic field can be either perpendicular or parallel to the polarization vectors. However, since that our line is a blend of features spatially unresolved, we are unable to solve for this degeneracy, which depends on the maser saturation level and θ for each feature. Nevertheless, we can at least claim that the POS field topology is quite ordered in both cases, i. e., it would remain ordered when rotated by 90°.

4. Shock properties

The observed maser emission is excited in zones of compressed gas produced by shocks between the outflows and the ambient gas. In an ionized medium, shocks compress the gas and the magnetic field by a factor m_A, where $m_A = v_S/v_A$. While m_A is the so called Afvénic Mach number, v_S and v_A are the shock velocity and the Alfvén speed, respectively. Given that $v_A \propto B_0(n)^{-1/2}$ (where B_0 is the preshock magnetic field), the preshock density can be estimated by (Kaufman & Neufeld 1996)

$$n_0 = 1.6 \times 10^6 \left(\frac{B_s}{\mathrm{mG}}\right)^2 \left(\frac{v_S}{\mathrm{kms}^{-1}}\right)^2 \mathrm{cm}^{-3}, \tag{4.1}$$

where B_s is the magnetic field in the shocked region, calculated in §3. Using the shock velocities modeled by Ristorcelli *et al.*(2005), the preshock density is estimated as $\sim 1 \times 10^8$ cm^{-3}, consistent with typical preshock densities where water masers will be eventually pumped. Moreover, for such densities, the magnetic pressure ($B_s^2/8\pi$) is similar or higher than the shock ram pressure ($\rho_0 v_S$), and the outflow evolution is magnetically controlled.

Assuming flux-freezing of the well stablished magnetic field morphology reported by Rao *et al.* (2009), who estimated a POS field strength of 4.5 mG at densities of 5×10^7 cm^{-3} for I16293, the compression of this ordered field implies that

$$\frac{B_{Rao}}{n_{Rao}} = \frac{B_s}{n_s} \rightarrow n_s = 1.3 \times 10^9 \mathrm{cm}^{-3}, \tag{4.2}$$

which is expected for effective maser pumping.

5. Conclusions

We report spectro-polarimetric VLA observations of water masers toward I16293. The strong, but unresolved spatially emission is excited in zones of compressed gas where a mean LOS magnetic field strength of 113 mG is estimated. The postshock densities are consistent with the physical conditions for H$_2$O maser pumping, and the dynamical evolution of the outflow is likely regulated by the magnetic field.

References

Chandler, C. J., Brogan, C. L., Shirley, Y. L., & Loinard, L. 2005, *ApJ*, 632, 371
Claussen, M. J., Wilking, B. A., Benson, P. J., *et al.* 1996, *ApJS*, 106, 111
Elitzur, M., Hollenbach, D. J., & McKee, C. F. 1989, *ApJ*, 346, 983
Fiebig, D. & Guesten, R. 1989, *A&A*, 214, 333
Furuya, R. S., Kitamura, Y., Wootten, A., Claussen, M. J., & Kawabe, R. 2003, *ApJS*, 144, 71
Goodman, A. A., Jones, T. J., Lada, E. A., & Myers, P. C. 1995, *ApJ*, 448, 748
Imai, H., Nakashima, K., Bushimata, T., *et al.* 2007, *PASJ*, 59, 1107
Jørgensen, J. K., Bourke, T. L., Nguyen Luong, Q., & Takakuwa, S. 2011, *A&A*, 534, A100
Kaufman, M. J. & Neufeld, D. A. 1996, *ApJ*, 456, 250
Lazarian, A., Goodman, A. A., & Myers, P. C. 1997, *ApJ*, 490, 273
Nedoluha, G. E. & Watson, W. D. 1992, *ApJ*, 384, 185
Rao, R., Girart, J. M., Marrone, D. P., Lai, S.-P., & Schnee, S. 2009, *ApJ*, 707, 921
Ristorcelli, I., Falgarone, E., Schöier, F., *et al.* 2005, in: D. C. Lis, G. A. Blake & E. Herbst (eds.), *Astrochemistry: Recent Successes and Current Challenges*, Proc. IAU Symposium No. 231, p. 227
Vlemmings, W. H. T., Diamond, P. J., & van Langevelde, H. J. 2002, *A&A*, 394, 589
Vlemmings, W. H. T., Diamond, P. J., van Langevelde, H. J., & Torrelles, J. M. 2006, *A&A*, 448, 597

Cosmic Masers - from OH to H$_0$
Proceedings IAU Symposium No. 287, 2012
R.S. Booth, E.M.L. Humphreys & W.H.T. Vlemmings, eds.

© International Astronomical Union 2012
doi:10.1017/S1743921312006680

Water Maser Emission Around Low/Intermediate Mass Evolved Stars

M. L. Leal-Ferreira[1], W. H. T. Vlemmings[2], P. J. Diamond[3], A. Kemball[4], N. Amiri[5,6] and J.-F. Desmurs[7]

[1] Argelander Institute für Astronomie, Uni-Bonn, Auf dem Hügel 71, 53121 Bonn, Germany
email: ferreira@astro.uni-bonn.de

[2] OSO (Sweden), [3] JBCA (UK), [4] University of Illinois (USA), [5] JIVE (Netherlands), [6] Leiden Observatory (Netherlands), [7] OAN (Spain)

Abstract. We present results of Very Long Baseline Array (VLBA) polarimetric 22 GHz H$_2$O maser observations of a number of low/intermediate mass evolved stars. We observed 3 Miras (Ap Lyn, IK Tau and IRC+60370), 1 semi-regular variable (RT Vir) and 1 pPN (OH231.8+4.2). Circular polarization is detected in the H$_2$O maser region of OH231.8+4.2 and we infer a magnetic field of $|B_{||}| = \sim$45 mG. This implies an extrapolated magnetic field of $\sim$2.5 G on the surface of the central star. The preliminary results on RT Vir and IRC+60370 also indicate the first detection of weak H$_2$O maser linear polarization.

Keywords. masers, magnetic fields, polarization, stars: AGB and post-AGB

1. Introduction

During the transition from an AGB star to a planetary nebula (PN), most low and intermediate mass stars lose their spherical symmetry. Magnetic fields are one of the candidates that can play a role in shaping asymmetrical PNe. However, magnetic field observations around evolved stars are still rare. We observed 22.235080 GHz H$_2$O masers around 5 evolved stars with the goal of measuring linear and circular polarization. The results may allow us to infer the magnetic fields properties around these 5 stars.

2. Results

2.1. OH231.8+4.2 / Rotten Egg Nebula / Calabash Nebula (Leal-Ferreira et al. 2012)

We detected 30 H$_2$0 masers around OH231.8+4.2. With respect to the central star position, 20 masers are located on the north (NReg), and 10 on the south (SReg). We compared our detections with those of Desmurs *et al.* (2007), and found that the offset between the mean position of the detections of Desmurs *et al.* (2007) and ours is 14.4 mas. Taking a distance d = 1540 pc (Choi *et al.* 2012) and $i = 36°$ (Kastner *et al.* 1992; Shure *et al.* 1995), the separation velocity between the masers in the 2 regions is 21 ± 11 km/s. The masers appear to be moving in the direction of the nebula jet, albeit at much lower velocity. This could indicate that they arise in a turbulent material entrained by the jet.

We found linear polarization for 3 features (all in SReg; Fig. 1). The high scatter between the linear polarization vectors can be caused by turbulence or, in case of a toroidal field, it could represent the tangent points of the field lines. We also found circular polarization on the 2 strongest masers; one in each region (Fig. 1). From these results, we inferred $|B_{||}|_{NReg} = 44 \pm 7$ mG and $|B_{||}|_{SReg} = -29\pm21$ mG. Although the morphology of the field is still not determined, the strength of the field on the surface of the star (with a typical radius of 1 AU) is $\sim$2.5 G if we assume a toroidal magnetic field (B $\propto 1/r$).

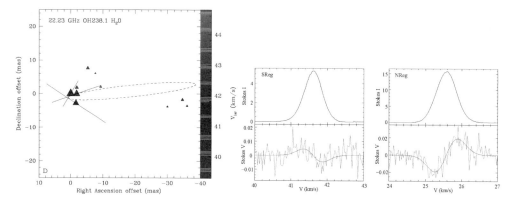

Figure 1. Left: Maser features observed in the region (SReg) around the Southern jet of OH231.8+4.2. The sizes of the triangles are scaled by their fluxes, and the colors follow the velocity scale. The vectors indicate the linear polarization directions, and the detached ellipse is a potential field morphology. Middle and right: I (top) and V (bottom) spectra of the brightest features we detected in the SReg (middle) and in the NReg (right) of OH231.8+4.2. The S-shape solid line is the best fit derivative of the total power spectrum.

2.2. *RT Vir, AP Lyn, IK Tau and IRC+60370*

We did not detect any maser emission around AP Lyn. Several masers were found around RT Vir, IK Tau and IRC+60370. Our preliminary results show that linear polarization between $\sim0.5\%$ and $\sim0.8\%$ is present around RT Vir and IRC+60370 (Fig 2). This is less than typically found in star forming regions and consistent with the previously found upper limits on the linear polarization in the H_2O maser envelopes of Mira and supergiant stars (Vlemmings *et al.* 2005).

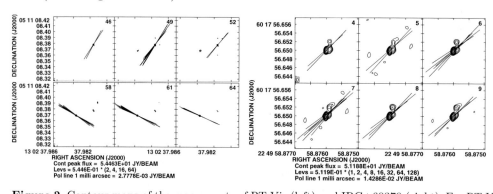

Figure 2. Contour maps of the maser spots of RT Vir (left) and IRC+60370 (right). For RT Vir we show one in every three channels in which linear polarization is present. For IRC+60370 we show 6 consecutive channels. The vectors indicate the linear polarization direction.

References

Choi Y. K., Brunthaler A., Menten K. M. & Reid M. J. 2012, this proceedings
Desmurs J. -F., Alcolea J., Bujarrabal V., Sánchez Contreras C., *et al.* 2007, *A&A*, 468, 189
Kastner J. H., Weintraub D. A., Zuckerman B., Becklin E. E., *et al.* 1992, *ApJ*, 398, 552
Leal-Ferreira M. L., Vlemmings W. H. T., Diamond P. J., *et al.* 2012, *A&A* (accepted)
Shure M., Sellgren K., Jones T. J. & Klebe D. 1995, *AJ*, 109, 721
Vlemmings, W. H. T., van Langevelde, H. J., & Diamond, P. J. 2005, *A&A*, 434, 1029

Cosmic Masers - from OH to H$_0$
Proceedings IAU Symposium No. 287, 2012
R.S. Booth, E.M.L. Humphreys & W.H.T. Vlemmings, eds.

© International Astronomical Union 2012
doi:10.1017/S1743921312006692

Observational tests of SiO maser polarisation models

L. L. Richter[1], A. J. Kemball[1,2] and J. L. Jonas[1]

[1] Physics and Electronics Department, Rhodes University, Grahamstown, 6140, South Africa

[2] Department of Astronomy, University of Illinois at Urbana-Champaign, 1002 W. Green Street, Urbana, IL 61801, USA

Abstract. SiO masers are often observed in the near-circumstellar envelope of late-type evolved stars. The polarisation of the masers can be used as a probe of the magnetic field in this region, subject to maser polarisation radiative transfer model. Two main maser polarisation models have been developed for the weak Zeeman splitting case applicable to circumstellar SiO masers. Observational tests aimed at discriminating between these models were performed at maser component level, using VLBA observations of v=1 J=1-0, v=2 J=1-0 and v=1 J=2-1 SiO masers towards the high-luminosity source VY CMa.

Keywords. stars: late-type, polarization, masers

The magnetic fields in the near-circumstellar envelopes of late-type, evolved stars can potentially be derived from observations of polarised Silicon Monoxide masers. This requires a theory of maser polarisation transport for the weak-splitting Zeeman regime appropriate to circumstellar SiO masers.

Two main weak-splitting maser polarisation theories have been developed, by Elitzur (2002) (and references therein) and Watson (2009) (and references therein). The theories conflict in several key respects, and can provide orders of magnitude different magnetic field estimates from the same observed levels of SiO maser circular polarisation. This work describes observational tests aimed at discriminating between the two theories.

The observations were performed with the VLBA (Very Large Baseline Array) in March 2007, towards the supergiant star VY CMa, chosen for it's high SiO luminosity. The SiO transitions v=1 J=1-0, v=2 J=1-0 and v=1 J=2-1 were observed. The data were reduced and imaged in AIPS, using methods described in Kemball & Richter (2011).

The total intensity images of each line are plotted in Figure 1. The images were aligned through spatial cross-correlation of the zero moment maps, to an estimated accuracy of less than 0.05 mas. Three observational tests of weak-splitting maser polarisation theory are described below. Two of the tests were performed using the six maser features shown in the figure, which overlap in multiple transitions.

Comparison of m_l at J=1-0 and J=2-1

The Elitzur model predicts similar levels of m_l in v=1 J=1-0 and v=1 J=2-1. The Watson model predicts larger m_l in the J=1-0 transition, when the level of saturation of the two transitions is similar.

The percentage m_l was compared in the overlapping maser features shown in Figure 1. There is no consistent ordinal relationship between the measured m_l in J=1-0 and J=2-1. However, the results are more consistent with the Elitzur model prediction of similar m_l in the two lines.

Comparison of m_c at J=1-0 and J=2-1

Under the Elitzur model, circular polarisation is caused by standard Zeeman splitting, so the v=1 J=1-0 m_c will be double that of v=1 J=2-1. Under the Watson model, circular

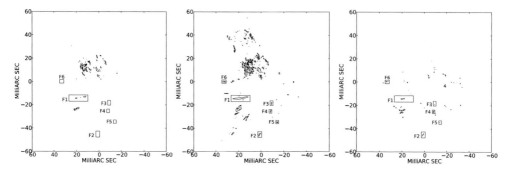

Figure 1. Contour plots of v=2 J=1-0 (left), v=1 J=1-0 (middle) and v=1 J=2-1 (right) SiO maser emission towards VY CMa. The contours are plotted at a level of 5%, 2% and 2% of the image maxima of 12.15 Jy/beam, 22.15 Jy/beam and 46.96 Jy/beam respectively.

polarisation caused by standard Zeeman splitting may be increased by a factor of up to a few due to saturation, and non-Zeeman circular polarisation can arise which is not significantly dependent on the J level of the transition.

The m_c comparison could only be performed for feature F1, at a single channel. There the m_c values were: $4.61 \pm 0.30\%$ for v=1 J=1-0, and $4.31 \pm 2.06\%$ for v=1 J=2-1. Due to the large uncertainties, standard Zeeman circular polarisation cannot be ruled out by this result.

Comparison of m_c and m_l

The non-Zeeman circular polarisation of the Watson model is created by rotation of linear polarisation, and may consequently be correlated with the linear polarisation (Watson 2009). If many maser features show m_c much greater than the average value of $m_l^2/4$, then the circular polarisation is unlikely to be caused by non-Zeeman effects (Wiebe & Watson 1998).

In the current observations, 14 maser features show statistically significant m_c and m_l. There is no correlation between m_c and m_l for these features. For all 14 features, m_c is greater than $m_l^2/4$. This is strong evidence against the non-Zeeman circular polarisation mechanism of the Watson model (Wiebe & Watson 1998).

These component-level observational tests of the SiO maser polarisation theory are an important improvement on previous single-dish implementations of the tests, which are affected by spatial blending of the many maser components.

Overall, the tests are not supportive of Watson model non-Zeeman circular polarisation. The most definitive test of Zeeman circular polarisation is the v=1 J=1-0 and J=2-1 circular polarisation comparison, which provided inconclusive results in these observations. Further observations are suggested to provide additional data for these tests.

References

Elitzur, M. 2002, in: V. Migenes & M. J. Reid (eds.), *IAU Symposium 206* (Astronomical Society of the Pacific), p. 452

Kemball, A. J. & Richter, L. L. 2011, *A&A*, 533, A26

Watson, W. D. 2009, in: A. Esquivel, J. Franco, G. Garcia-Segura, E. de Gouveia Dal Pino, A. Lazarian, S. Lizano & A. Raga (eds.), *Revista Mexicana de Astronomia y Astrofísica Conference Series Vol. 36* (Instituto de Astronomia), p. 113

Wiebe, D. S. & Watson, W. D. 1998, *ApJ* (Letters), 503, L71

T3a:

ar formation: maser variability *Chair: Philip Diamond*

Cosmic Masers - from OH to H_0
Proceedings IAU Symposium No. 287, 2012   © International Astronomical Union 2012
R.S. Booth, E.M.L. Humphreys & W.H.T. Vlemmings, eds.   doi:10.1017/S1743921312006709

Variability of Class II methanol masers in massive star forming regions

Sharmila Goedhart[1,2], Mike Gaylard[2], and Johan van der Walt[3]

[1]SKA SA, Third Floor, The Park, Park Rd, Pinelands 7405, South Africa

[2]Hartebeesthoek Radio Astronomy Observatory, PO Box 443, Krugersdorp 1740, South Africa

[3]Centre for Space Research, North-West University, Private Bag X6001, Potchefstroom 2520, South Africa

Abstract. Class II methanol masers are known to be tracers of an early phase of massive star formation. The 6.7- and 12.2-GHz methanol maser transitions can show a significant amount of variability, including periodic variations. Studying maser variability can lead to important insights into conditions in the maser environment but first the maser time-series need to be characterised. The results of long-term monitoring of 8 regularly-varying sources will be presented and methods of period-search discussed.

Keywords. masers, radio lines: ISM, stars: formation

1. Introduction

Class II methanol masers are known to be tracers of the earliest stages of massive starformation (Breen *et al.*, 2010). In many cases the maser is the only readily detectable emission in the region of the massive young star. Masers are extremely sensitive to all changes in their environment. This would include local conditions in the masing gas volume as well as the incoming radiation field. Class II methanol masers are believed to be primarily pumped by infrared radiation (Cragg, Sobolev & Godfrey, 2005). Thus studying variability in methanol masers could lead to insights into changing conditions in massive star forming regions. Some methanol masers are known to be highly variable (e.g. Caswell, Vaile & Ellingsen 1995, MacLeod & Gaylard 1996).

The Hartebeesthoek Radio Astronomy Observatory has been monitoring a selection of methanol masers since 1992. The source G351.78-0.54 has now been monitored for 19 years. In 1999, Goedhart, Gaylard & van der Walt (2004) started an intensive program to monitor 54 methanol maser sources over four years. The sample showed a range of behaviour, including the very surprising result of periodicity in six of the sources. Monitoring continues on sources of interest. The time-series on G351.78-0.54 and the periodic sources will be discussed.

2. Observations

The observations were done using the 26-m telescope at the Hartebeesthoek Radio Astronomy Observatory. The observations from September 1992 to April 2003 were of left-circular polarisation using a 256-channel spectrometer. Observations resumed in September 2003 with the cryogenic receiver upgraded from single to dual polarisation, a 2x1024 channel spectrometer and a new telescope control system. The telescope surface was replaced by solid panels, with the final alignment taking place in September 2004. Amplitude calibrations were based on continuum monitoring of Virgo A, Hydra A and 3C123. Pointing corrections were calculated by observing at half a beamwidth

S. Goedhart *et al.*

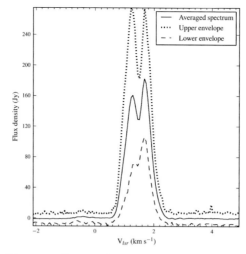

Figure 1. Range of variation of G351.78-0.54.

offset to the north, south, east and west of the source position. The calibrations were
checked against the relatively quiescent source G351.42+0.64. The sources were generally
observed once a week, with daily observations when a source was seen to be flaring.

3. Time-series analysis

The most variable features in a maser spectrum can be found visually by plotting
the upper and lower envelopes of all of the spectra. This is calculated by finding the
maximum and minimum value of the time-series in each spectrum. An example of such
a plot is shown in Figure 1, which shows the range of variation in G351.78-0.54 for the
period January 1999 to April 2003. One can select channels of interest from these spectra
for further analysis.

Determining whether an astronomical time-series showing regular variations is periodic
can be challenging. The samples are generally unevenly spaced, making it impossible to
use a standard Fast Fourier Transform. A number of methods have been developed for
period searches of unevenly sampled time-series. The most popular of these are phase
dispersion minimisation or epoch-folding, the Discrete Fourier Transform (DFT) and the
Lomb-Scargle periodogram. Additional challenges in the methanol data are long-term
trends, non-periodic outbursts, flares of varying amplitudes and profiles that change
with each repeat.

Epoch-folding using the test statistic developed by Davies (1990) is effective for short
time-series, even when there are large gaps in the data. However, the data need to be de-
trended, which could potentially introduce a bias to the results. Various DFT algorithms
were tried (Deeming, 1975; Scargle, 1982; Kurtz, 1985, Lenz & Breger 2005), but these
had limited success since the variations were generally non-sinusoidal. The Lomb-Scargle
periodiogram (Press & Rybicki 1989) has proven to be the most effective in determining
the underlying periods in the maser time-series. Long-term trends manifest as a very
low-frequency sinusoid, which can be subtracted from the data before calculating a new
periodogram. This also gives the potential to verify multiple periods in the data.

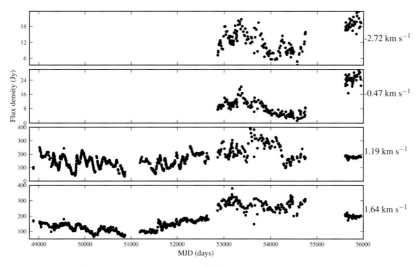

Figure 2. Time-series of G351.78-0.54 at 6.7 GHz.

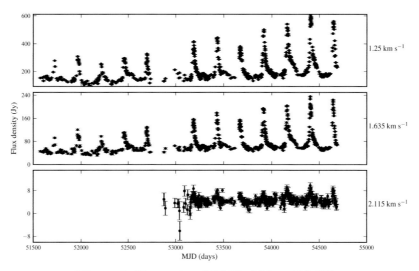

Figure 3. Time-series of G9.62+0.20 at 12.2 GHz.

4. Results

The range of variation and time-series of the dominant spectral features in G351.78-0.54 are shown in Figures 1 and 2. This source does not exhibit periodic behaviour but is highly variable. The most interesting characteristic is a time delay across the velocity features in the left-hand peak. It has been suggested by Macleod & Gaylard (1996) that this could be caused by an outflow passing behind the masers. LBA observations show that the masers do indeed show a strong position-velocity gradient (Goedhart *et al.* in prep).

The time-series of the most variable features in G9.62+0.20 at 12.2- and 6.7- GHz are shown in Figures 3 and 4, respectively. The flares have a period of 244 days. The

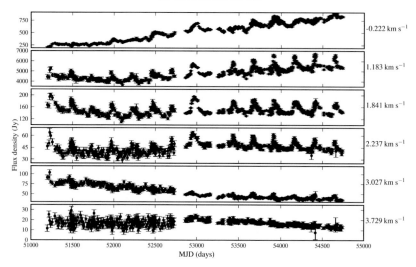

Figure 4. Time-series of G9.62+0.20 at 6.7 GHz.

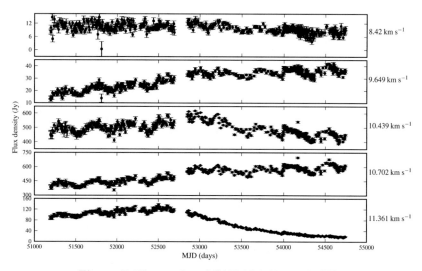

Figure 5. Time-series of G188.95-0.89 at 6.7 GHz.

12.2 GHz masers have been showing progressively stronger flares. Correlated flares are seen at 6.7 GHz, but they are not as strong. The baseline intensity between flares has remained relatively constant at 12.2 GHz, while some features at 6.7 GHz have shown steady increases (-0.22 km/s) or decay (3.027 km/s).

G188.95+0.89 (Figure 5) showed low-amplitude but clear sinusoidal variations in all its spectral features up to mid-2003. The period of 394 days can still be seen in the time-series but the cycle profile has changed to show a sharp rise and slow decay. The feature at 11.361 km/s has been showing a steady decay since 2003.

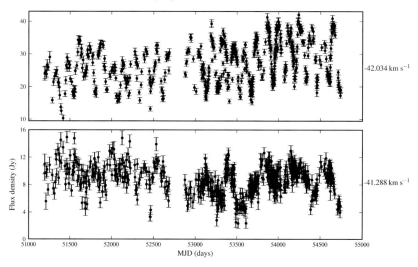

Figure 6. Time-series of G338.93-0.06 at 6.7 GHz.

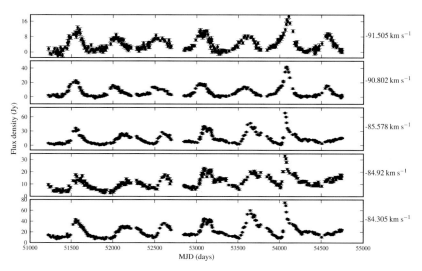

Figure 7. Time-series of G331.13-0.24 at 6.7 GHz.

G338.93-0.06 (Figure 6) shows a clearly defined period of 132.5 days in the feature at -42.034 km/s, but does not show correlated variability in the other feature at -41.288 km/s. The time-series is characterised by sharply defined minima.

The time series for the peak channels at 6.7 GHz in G331.13-0.24 is shown in Figure 7. All of the features show a periodic signature of 502 days. The group at -87 to -83.5 km/s show flares which start at regular intervals, but the duration of these flares varies, as well as the shape of the flares. Figure 8 shows the result of folding the time-series modulo 502.7 days. A time delay of $\sim$70 days is apparent between the two velocity groups. The second group shows a well-defined minimum just before the start of the flares.

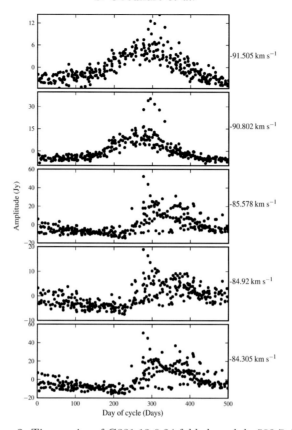

Figure 8. Time-series of G331.13-0.24 folded modulo 502.7 days.

G12.89+0.49 (Figure 9 has the shortest observed period of 29.4 days and shows a well-defined minimum just before the start of a flare, with flare amplitude and peak time varying from cycle to cycle.

G339.62-0.12 (Figure 10) and G328.24-0.55 (Figure 11) show very similar behaviour and similar periods of 202 and 222 days, respectively. Not all of the peaks show periodicity. The features at -37.178 and -35.729 km/s in G339.62-0.12 show additional, non-periodic flares, which presumably indicate localised changes in those particular maser features.

5. Discussion and conclusion

A wide range of periods and cycle profiles are seen. While the light curves do not appear to be strictly periodic, it is likely that there is an underlying periodic trigger, as indicated by the minima of the light curves which appear to be the most stable feature in the cycle profiles. Changes in cycle profile from flare to flare can be explained by varying conditions in the volume of masing gas (eg. changes in maser path length due to turbulence).

In van der Walt (2011) it was shown that the light curves for G9.62+0.20 and G188.95+ 0.89 can be explained by a simple colliding wind binary model, and the decay of the 11.361 km/s feature in G188.95+0.89 could be due to the recombination of the ionizing

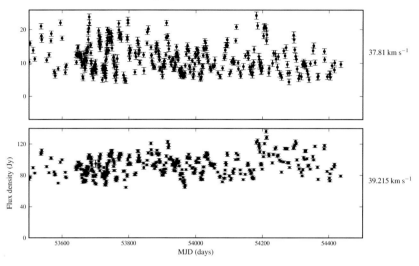

Figure 9. Time-series of G12.89 at 6.7 GHz. A subset of the full time-series is shown here for clarity.

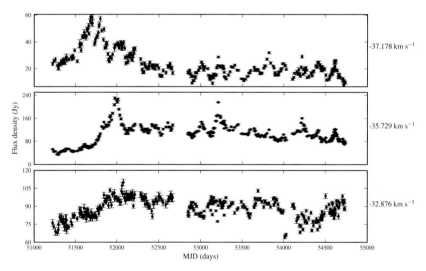

Figure 10. Time-series of G339.62-0.12 at 6.7 GHz.

gas in the background. Different light-curves can be reproduced by changing the orbital parameters of the binary system. Recently Szymczak *et al.*(2011) discovered a new periodic methanol maser source, G22.357+0.66 which has a period of 179 days and shows a similar flare profile to G9.62+0.20. Araya *et al.*(2010) detected another periodic source, G37.55+0.20, which exhibits correlated quasi-periodic flares in methanol and formaldehyde. In this case, the period of the flares appears to be getting shorter. They propose that the flares are caused by periodic accretion of circumbinary disk material.

Much more work needs to be done before the implications of periodic and quasi-periodic maser variability are understood. Observationally, further searches for periodic sources to increase the sample size will be helpful in characterising the properties of these sources.

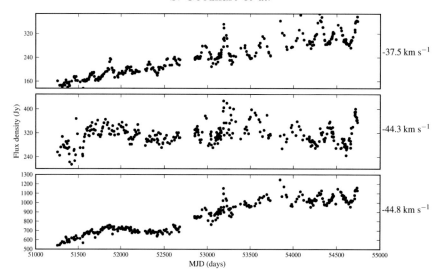

Figure 11. Time-series of G328.24-0.55 at 6.7 GHz.

High resolution, multi-wavelength maps are needed to understand the relation between the masers and other objects such as HII regions and outflows. Simultaneous monitoring of other maser lines will also be useful in understanding whether the root cause of the maser variability comes from the pump or background radiation. Detailed modelling will be necessary to understand the effects of radiative transfer in different morphologies and maser models with time-dependent components would have to be developed.

References

Araya, E. D., Hofner, P., Goss, W. M., Kurtz, S., Richards, A. M. S., Linz, H., Olmi, L., & Sewio, M., 2010, *ApJ*, 717, L133

Breen, S. L., Ellingsen, S. E., Caswell, J. L., & Lewis, B. E., 2010, *MNRAS*, 401, 2219

Caswell, J. L., Vaile, R. A., & Ellingsen, S. P., 1995, *Proc. Astron. Soc. Aust.*, 12, 37

Cragg, D. M., Sobolev, A. M., & Godfrey, P. D., 2005, *MNRAS*, 360, 533

Davies, S. R., 1990, *MNRAS*, 244, 93

Deeming, T. J., 1975, *Astrophy. & Sp. Sc.*, 36, 137

Goedhart, S., Gaylard, M. J., & van der Walt, D. J., 2004, *MNRAS*, 355, 553

Kurtz, D. W., 1985, *MNRAS*, 213, 773

Lenz, P. & Breger, M., 2005, *Communications in Asteroseismology*, 146, 53

MacLeod, G. C. & Gaylard, M. J., 1996, *MNRAS*, 280, 868

Press, W. H. & Rybicki, G. B., 1989, *ApJ*, 338, 277

Scargle, J. D., 1982, *ApJ*, 263, 835

Szymczak, M., Wolak, P., Bartkiewicz, A., & van Langevelde, H. J., 2011, *A&A*, 531, L3

van der Walt, D. J., 2011, *AJ*, 141, 152

Cosmic Masers - from OH to H$_0$
Proceedings IAU Symposium No. 287, 2012
R.S. Booth, E.M.L. Humphreys & W.H.T. Vlemmings, eds.

© International Astronomical Union 2012
doi:10.1017/S1743921312006710

Binary systems: implications for outflows & periodicities relevant to masers

Nishant K. Singh[1,2] and Avinash A. Deshpande[1]

[1]Raman Research Institute, C. V. Raman Avenue, Sadashivanagar, Bangalore 560080, India
emails: nishant@rri.res.in, desh@rri.res.in

[2]Joint Astronomy Programme, Indian Institute of Science, Bangalore 560 012, India

Abstract. Bipolar molecular outflows have been observed and studied extensively in the past, but some recent observations of periodic variations in maser intensity pose new challenges. Even quasi-periodic maser flares have been observed and reported in the literature. Motivated by these data, we have tried to study situations in binary systems with specific attention to the two observed features, i.e., the bipolar flows and the variabilities in the maser intensity. We have studied the evolution of spherically symmetric wind from one of the bodies in the binary system, in the plane of the binary. Our approach includes the analytical study of rotating flows with numerical computation of streamlines of fluid particles using PLUTO code. We present the results of our findings assuming simple configurations, and discuss the implications.

Keywords. masers, radio lines: general, (stars:) binaries: general, stars: winds, outflows

1. Introduction

Bipolar outflows are ubiquitous in nature and presumably thought to be associated with star forming regions in molecular clouds. Since the discovery of these bipolar outflows (Snell *et al.* (1980)) various attempts have been made to understand the physical nature of these phenomena. Such outflows are generally believed to occur around young stellar objects (YSOs) and are thought to have intimate relationship with the process of star formation and early stage of stellar evolution (see reviews Snell (1983), Lada (1985), Bachiller (1996)). Unanimous view about the process of star formation and its early evolution is yet to emerge which in turn, inevitably, makes it difficult to have a clear understanding of the nature of these bipolar outflows. Observations of periodic/quasi-periodic variations in maser intensity from such regions add further complications (Goedhart, Gaylard & van der Walt (2004), Goedhart *et al.* (2005), Goedhart *et al.* (2009), van der Walt, Goedhart & Gaylard (2009), Szymczak *et al.* (2011), Araya *et al.* (2010)).

Periodicities observed in the maser light curves are one of the most challenging, poorly understood, features. In the present work, we have tried to demonstrate that a binary system, consisting of a star and another gravitating object, could be a potential candidate to explain *together* both of the following observed features of the maser sources, the bipolarity and the periodicity in the intensity variations.

2. The Model and Simulations

Consider a *binary system* consisting of two bodies, S and P, which are rotating around their common center of mass, O, in a plane. Let $\overline{X}\,\overline{Y}\,\overline{Z}$ be the *inertial* (fixed) coordinate frame in which the two bodies lie in the $\overline{X}\,\overline{Y}$−plane with angular velocity vector in $\overline{Z}$−direction. Let $X\,Y\,Z$ be the *rotating* (corotating) coordinate frame which rotates with an angular velocity same as that of the two bodies in the binary system and therefore

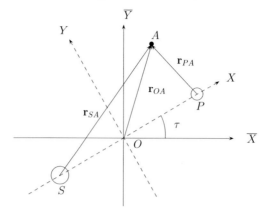

Figure 1. The inertial $(\overline{XY})$ and rotating (XY) coordinate frames are shown in the plane of a binary. The center of mass of the two bodies labelled as S and P is the origin of both the coordinate frames and is denoted by O.

both the bodies appear to be at rest in this frame. The origins of both the coordinate frames coincide and are taken to be at the center of mass, O, of the binary system. The units of various quantities are chosen such that the properties of the system depend only on a single parameter. Let the total mass $(\mathcal{M})$ of the primaries $(S$ and $P)$ be the unit of mass; the distance between them $(\mathcal{D})$ be the unit of distance; and the unit of time be chosen in such a way that the angular speed of the primaries, denoted by Ω, be unity. Let ξ be the mass of P, thus mass of S is $(1 - \xi)$. In the rotating reference frame, with positive X in the direction of the body P, the coordinates of P and S will be $(1 - \xi, 0)$ and $(-\xi, 0)$ respectively. If $(X, Y, 0)$ be the coordinate of an arbitrary point A, then from Figure 1,

$$r_{SA} = |\mathbf{r}_{SA}| = \sqrt{(X + \xi)^2 + Y^2} \; ; \quad r_{PA} = |\mathbf{r}_{PA}| = \sqrt{(X - (1 - \xi))^2 + Y^2} \qquad (2.1)$$

The gravitational potential in the corotating frame at the point A may be written as,

$$\Phi = -\frac{(1 - \xi)}{r_{SA}} - \frac{\xi}{r_{PA}} \qquad (2.2)$$

Let us assume that one of the bodies, say S, has a spherically symmetric wind very near to its upper atmosphere, whereas the other body, P, is interacting only gravitationally. If $\mathbf{v}(\mathbf{X}, \tau)$ be the fluid velocity of the wind in the rotating frame then we may write the Euler equations in rotating frame for steady flow, with p and ρ as the fluid pressure and density respectively, as,

$$(\mathbf{v} \cdot \boldsymbol{\nabla}) \mathbf{v} = -\frac{\boldsymbol{\nabla} p}{\rho} - \boldsymbol{\nabla} \Phi - \hat{\boldsymbol{\Omega}} \times \left(\hat{\boldsymbol{\Omega}} \times \mathbf{X} \right) - 2 \hat{\boldsymbol{\Omega}} \times \mathbf{v} \qquad (2.3)$$

where $(\mathbf{X}, \tau) \equiv (X, Y, Z, \tau)$ and $\hat{\boldsymbol{\Omega}}$ $(= \hat{e}_Z$, which is the unit vector along $Z \equiv \overline{Z})$ is the angular velocity of the corotating frame relative to the inertial frame.

It can be shown that,

$$(\mathbf{v} \cdot \boldsymbol{\nabla}) \mathcal{B} = 0 \qquad (2.4)$$

where

$$\mathcal{B} = \left(\frac{1}{2} v^2 + \int \frac{dp}{\rho} + \Phi_{\text{eff}} \right) \; ; \quad \Phi_{\text{eff}} = \Phi - \frac{1}{2} |\hat{\boldsymbol{\Omega}} \times \mathbf{X}|^2 \qquad (2.5)$$

Therefore the quantity, $\mathcal{B}$, is constant along a particular streamline for steady flows, although it could be a different constant for different streamlines. Noting the fact that the particle paths and streamlines are the same for steady flows, we can see that $\mathcal{B}$ remains the same for a particular fluid element as it moves along a particular streamline. Thus, for a particular streamline, we may write

$$\mathcal{B} = \frac{1}{2}v^2 + \left(\frac{\gamma}{\gamma - 1}\right) RT + \Phi_{\text{eff}} = \text{constant} = C \qquad (2.6)$$

where we have used adiabatic equation of state and note that the term $\int dp/\rho$ appearing in equation 2.5 may be replaced by specific enthalpy (w) for isentropic evolution of fluid element. γ is ratio of specific heats at constant pressure and constant volume, R is the gas constant and T is the temperature. As the terms $v^2/2$ and $\{\gamma/(\gamma - 1)\} RT$ in equation 2.6 cannot be negative, we infer from equation 2.6 that the motion of a fluid element, and hence the corresponding streamline, is restricted to the region where $\Phi_{\text{eff}} < C$.

2.1. *Numerical simulations using PLUTO code*

To study how the spherically symmetric wind from the body S flows in the presence of another gravitating body P in a binary system, we use PLUTO code. The details of the code may be found in Mignone *et al.* (2007) (and references therein, and code at http://plutocode.ph.unito.it/). We use the hydrodynamic module of the code and solve the equations in three-dimensional cartesian geometry. We adapt the code in the corotating frame of the binary in which the two bodies, S and P, appear to be at rest, by adding the necessary body-forces, namely, coriolis and centrifugal, to the equation of motion. To understand the evolution of the wind in the plane of the binary, we start with a high-pressure circularly symmetric region at the location of S and the fluid pressure outside this region being very small, the fluid particles experience force which is pointed radially outward from the location of S. We have performed simulations for various initial conditions (by chosing different values for high/low pressure/density regions) and for different values of the parameter ξ. Also, the flow need not be steady. Our simulations are done in two ways: (a) we set the initial conditions and study the flow. In this case the matter eventually flows out of the computational domain as there is no supply of matter from the location of the body S, and the code stops; (b) Having set the initial conditions we may supply the matter at some arbitrary time intervals. In this case, as the matter is not completely depleted out of the computational domain, the code runs for longer time. How quickly the matter is depleted out of the domain depends on the initial condition. Thus, we may study the average properties of the wind flow (i.e. density, pressure etc.) by plotting density/pressure maps at different times.

2.2. *Results from PLUTO simulations and discussion*

We choose to present the result by demonstrating the evolution of density maps with time as seen in the corotating frame which is the rest frame of the binary system. This choice of plotting the density map seems relevant, as ultimately we will be interested in knowing the distribution of matter in space and its evolution in time to identify the regions which could potentially be *maser emitting spots*, particularly due to relatively high concentration of matter. From various panels in Figure 2, we see that the isotropy of the wind is broken very near to the binary system, as desired for density modulation seen for a fixed line-of-sight of an inertial observer. Although it is known that the orbits of a test-body are chaotic in the binary system, the problem known as *the restricted three body problem* (Poincaré (1890)), it is remarkable to note that the third test-body being replaced by the wind (continuous matter) evolves in a similar fashion for various initial conditions and for different values of ξ as seen from simulations (e. g., the average property of the wind, say, density, seems to evolve in a definitive way). Figure 2 is the

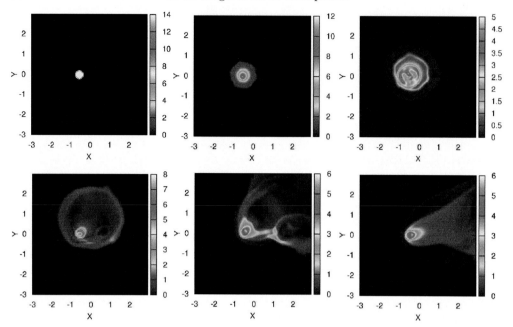

Figure 2. Snap-shots of density maps, as seen from the corotating frame (for $\xi = 0.5$ and with arbitrary supply of matter at S), where the bodies S and P are at $(-0.5, 0)$ and $(0.5, 0)$, respectively. Time increases from upper-left (initial time) to lower-right.

result of one of many simulations performed to study this problem. The anisotropy of the wind may be understood to be due to the shapes of the isocontours of the effective potential (Φ_{eff}) which has five Lagrange-points near to the binary, and these isocontours tend to become Keplerian far away from the binary system. Hence one may expect that the outflowing matter, which need not escape, settles into Keplerian orbits depending on their initial velocities. These results prompt us to imagine that on an average, much of the matter spirals outward with certain pitch-angle, which depends on the intial conditions, and tends to settle in Keplerian orbits far away from the binary system, forming a torus skirting the spiral pattern.

2.3. *Results of variability (light-curve) simulation*

We simulate the situation discussed at the end of last subsection and present the results by showing light-curves of maser intensities from different locations as seen along a fixed sight-line of an inertial observer. As it should, the modulations in the maser intensity will depend on the inclination of the binary system with respect to the sky-plane, and so we show our results for an arbitrarily chosen inclination angle. Columns of negligible gradients in the sight-line velocity component with adequate concentration of matter are the most preferred sites of maser emission.

3. Summary and Conclusions

We have investigated the flow of the wind from one of the bodies in a binary system, and tried to understand the plausible mechanisms for modulations in the maser intensity. Anisotropies seen in the rotating frame will have the desired character of bipolar flows and also will appear to be periodic (with binary-period) to an inertial observer for relevant sight-lines. Further, in our model the potential maser spots in the sky-plane do *not* move and the *minima* in intensity variation cycle repeat at regular intervals of the orbital

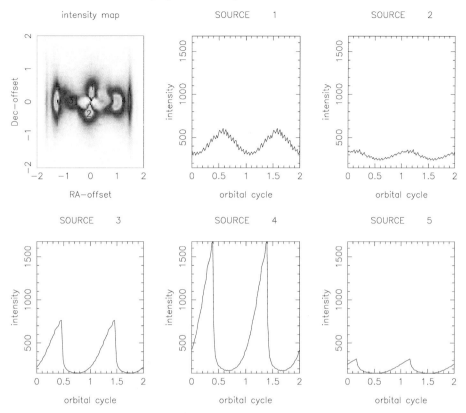

Figure 3. Light-curves (shown as line plots corresponding to the locations marked in the color panel, and over two orbital cycles) that an inertial observer would see, as monitored from our simulation of maser emission from flows in binary systems.

period. These characteristics are naturally produced in our model wherein the variabilities observed in the maser intensity is due to the density modulation resulting from the flow in the binary as seen from the simulations.

Acknowledgments: We thank the IAU, Roy Booth and other symposium organizers for providing the financial support to one of us (NKS).

References

Araya, E. D., Hofner, P., Goss, W. M., Kurtz, S., Richards, A. M. S., Linz, H., Olmi, L., & Sewilo, M., 2010, *ApJ*, **717**, L133
Bachiller, R., 1996, *ARA&A*, **34**, 115
Goedhart, S., Gaylard, M. J., & van der Walt, D. J., 2004, *MNRAS*, **355**, 553
Goedhart, S., Minier, V., Gaylard, M. J., & van der Walt, D. J., 2005, *MNRAS*, **356**, 839
Goedhart, S., Langa, M. C., Gaylard, M. J., & van der Walt, D. J., 2009, *MNRAS*, **398**, 995
Lada, C. J., 1985, *ARA&A*, **23**, 267
Mignone, A., Bodo, G., Massaglia, S., Matsakos, T., Tesileanu, O., Zanni, C., & Ferrari, A., 2007, *ApJS*, **170**, 228
Poincaré, H., 1890, *Acta Math.*, **13**, 1-270
Snell, R. L., Loren, R. B., & Plambeck, R. L., 1980, *ApJ*, **239**, L17
Snell, R. L., 1983, *RMAA*, **7**, 79
Szymczak, M., Wolak, P., Bartkiewicz, A., & van Langevelde, H. J., 2011, *A&A*, **531**, L3
van der Walt, D. J., Goedhart, S., & Gaylard, M. J., 2009, *MNRAS*, **398**, 961

Cosmic Masers - from OH to H_0
Proceedings IAU Symposium No. 287, 2012
R.S. Booth, E.M.L. Humphreys & W.H.T. Vlemmings, eds.

© International Astronomical Union 2012
doi:10.1017/S1743921312006722

Intermittent maser flare around the high-mass young stellar object G353.273+0.641

Kazuhito Motogi[1], Kazuo Sorai[1,2], Kenta Fujisawa[3,4], Koichiro Sugiyama[3] and Mareki Honma[5,6]

[1] Department of Cosmosciences, Graduate School of Science, Hokkaido University,
N10 W8, Sapporo 060-0810, Japan
email: motogi@astro1.sci.hokudai.ac.jp

[2] Department of Physics, Faculty of Science, Hokkaido University,
N10 W8, Sapporo 060-0810, Japan

[3] Department of Physics, Faculty of Science, Yamaguchi University,
Yoshida 1677-1, Yamaguchi-city, Yamaguchi 753-8512 2, Japan

[4] The Research Institute of Time Studies, Yamaguchi University,
Yoshida 1677-1, Yamaguchi-city, Yamaguchi 753-8511, Japan

[5] Department of Astronomical Science, The Graduate University for Advanced Studies,
2-21-1 Osawa, Mitaka, Tokyo 181-8588, Japan

[6] Mizusawa VLBI Observatory, National Astronomical Observatory of Japan,
2-12 Hoshi-ga-oka, Mizusawa-ku, Oshu, Iwate 023-0861, Japan

Abstract. The water maser site associated with G353.273+0.641 is classified as a dominant blueshifted H_2O maser, which shows an extremely wide velocity range ($\pm$ 100 km s^{-1}) with almost all flux concentrated in the highly blueshifted emission. The previous study has proposed that this peculiar H_2O maser site is excited by a pole-on jet from high mass protostellar object. We report on the monitoring of 22-GHz H_2O maser emission from G353.273+0.641 with the VLBI Exploration of Radio Astrometry (VERA) and the Tomakamai 11-m radio telescope. Our VLBI imaging has shown that all maser features are distributed within a very small area of 200×200 au^2, in spite of the wide velocity range (> 100 km s^{-1}). The light curve obtained by weekly single-dish monitoring shows notably intermittent variation. We have detected three maser flares during three years. Frequent VLBI monitoring has revealed that these flare activities have been accompanied by a significant change of the maser alignments. We have also detected synchronized linear acceleration (-5 km s^{-1}yr^{-1}) of two isolated velocity components, suggesting a lower-limit momentum rate of 10^{-3} M$_\odot$ km s^{-1}yr^{-1} for the maser acceleration. All our results support the previously proposed pole-on jet scenario, and finally, a radio jet itself has been detected in our follow-up ATCA observation. If highly intermittent maser flares directly reflect episodic jet-launchings, G353.273+0.641 and similar dominant blueshifted water maser sources can be suitable targets for a time-resolved study of high mass protostellar jet.

Keywords. masers, stars: early-type, stars: formation, ISM: jets and outflows

---

1. Introduction

G353.273+0.641 (hereafter G353) is a strong 22 GHz H_2O maser site in the southern high mass star-forming region NGC6357 (Skellis *et al.* 1984). The source distance is 1.7 kpc from the sun (Neckel 1978). Multi-epoch ATCA observation has been reported in Caswell & Phillips (2008) (hereafter CP08). Class II CH$_3$OH maser emission ($J_k = 5_1$–6_0 A$^+$) at 6.668519 GHz is also associated with this source (CP08), and hence, the host young stellar object (YSO) is identified as a high mass YSO (e.g., Minier *et al.* 2003). CP08 has also suggested that G353 is still in the pre-ultra compact (UC) H II

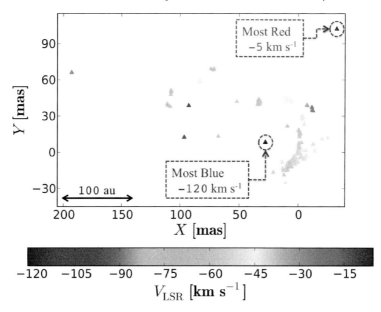

Figure 1. Spatial distribution of detected maser spots. Here, each triangle indicates detected maser spot and its color represents a line of sight velocity. The coordinate origin is $\alpha_{2000} = 17^{\mathrm{h}}26^{\mathrm{m}}01^{\mathrm{s}}.5883, \delta_{2000} = -34°15'14''.905$.

region phase, i.e., high mass protostellar phase, based on the absence of any detectable OH masers (e.g., Caswell 1997; Breen *et al.* 2010).

G353 has been classified as a dominant blue-shifted H_2O maser in CP08. That is, almost all flux is concentrated in the blue-shifted emission, despite the very broad velocity range of ± 100 km s^{-1} with respect to the systemic velocity of -5 km s^{-1}. They argued that this type of masers can be caused by well-collimated jet aligned close to the line of sight. There are a few maser sources which show similar blue-shift dominance (e.g., Caswell 2004; CP08; Caswell & Breen 2010). Some of them suggest acceleration of outflowing materials in the jet. The statistical analysis in Caswell & Breen (2010) indicates that such a blue-shift dominance is a characteristic of H_2O masers at the earliest evolutionary stage of star-formation.

The exact relation between jet activity and variability of the maser is still unclear, but, once they appear, H_2O masers are an excellent tool to survey dynamic jet activities in small scale, since its bright emission allows us easy and frequent monitoring even with a small size radio telescope.

2. Long-term monitoring of the H_2O maser

Our VLBI and single dish monitoring of G353 using VERA (VLBI Exploration of Radio Astrometry) and the Hokkaido University Tomakomai 11-m radio telescope, which has been started in 2008 November and is still ongoing, has shown intermittent flare activities (see Motogi *et al.* 2011 in details). Figure 1 presents overall distribution of maser spots. We have detected only the blueshifted side of the entire maser emission reported in CP08. The most blueshifted and redshifted components are separated by only 100 mas (200 au) along the SE-NW direction.

Figure 2 shows the light-curve of the main velocity components (-53 ± 7 km s^{-1}). The observed variation is notably intermittent and there are three significant maser flares. Motogi *et al.* (2011) has revealed that these flares are accompanied by spatial varia-

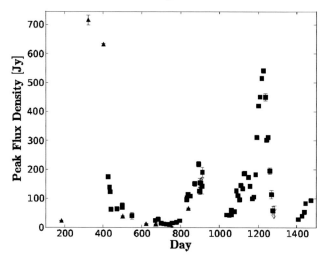

Figure 2. The light curve of the maser components at the line of sight velocity of -53 ± 7 km s^{-1}. The x-axis show relative days from the first day of 2008. The squares and triangles show single dish and VLBI data points, respectively. Follow-up single-dish data, which show the 3rd flare, are added to the dataset presented in Motogi *et al.* (2011)

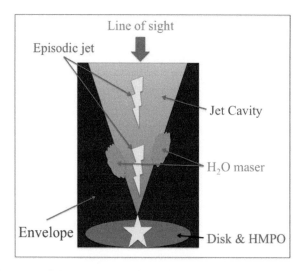

Figure 3. Schematic view of the "pole-on jet" model in G353.

tions, which can be explained by episodic shock propagation along the SE-NW direction. The characteristic velocity, which is estimated from the spatial scale of maser distribution ($\sim$100 au) divided by the time interval of three times flares ($\sim$1yr), is $\sim$500 km s^{-1}. This is rather larger than the typical three-dimensional velocity of this maser site ($\sim$100 km s^{-1}), and more like that of a radio jet seen in several HMPOs (Curiel *et al.* 2006; Martí, Rodrígues & Reipurth 1998). Motogi *et al.* (2011) has also measured the linear acceleration (5 km s^{-1} yr^{-1}) of two distinct maser clusters (see Motogi *et al.* 2011 in detail). The lower-limit momentum rate required for the acceleration is $\sim 1.1 \times 10^{-3}$ M$_\odot$ km s^{-1} yr^{-1}. This value is consistent with that of the outflows driven by high mass YSOs (Arce *et al.* 2007).

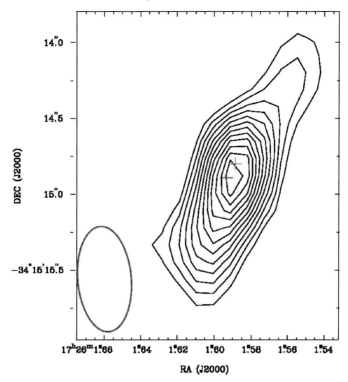

Figure 4. ATCA contour image of 22 GHz radio continuum emission from G353. Contour levels are 105 to 545 μJy beam^{-1} in steps of 40 μJy beam^{-1}. The lowest level corresponds to 3 σ noise. The synthesized beam (0$''$.69 $\times$ 0$''$.36) is shown in the lower left corner. The most red and blue maser components are shown by red and blue cross, respectively. Relative positional error between masers and continuum is less than 0$''$.1.

3. Follow-up jet survey

Since all maser properties can be consistently explained by the "pole-on jet" model (see figure 3), we have finally performed direct survey of a radio jet itself using the Australia Telescope Compact Array (ATCA). Our ATCA radio continuum observation was made at January 13 2012 in the 6A configuration. We observed two 2 GHz bands centered on 18 and 22 GHz simultaneously with the Compact Array Broadband Backend (CABB; Wilson *et al.* 2011). The observed hour-angle range is 10 hour and total on-source time is 6 hour because of fast switching (the cycle time of 3 min). The 1-σ noise level is 19 and 35 μJy beam^{-1} for 18 and 22 GHz, respectively.

As a result, we have successfully detected a weak and optically thick thermal radio jet (Motogi *et al.* 2012, in prep). Figure 4 shows ATCA contour image of 22 GHz continuum emission. The spectral index estimated from the total flux of 18 and 22 GHz continuum is about +1.5 (see table 1) and is consistent with that of a typical radio jet from an HMPO (e.g., Hofner *et al.* 2007). The jet is clearly elongated along the NW — SE direction, indicating that the jet has a finite inclination angle. This direction is consistent with the direction of the velocity gradient and shock propagation shown by our maser monitoring. This fact strongly suggests physical association between the maser and jet activities.

Table 1. Properties of the detected radio jet

Frequency (GHz)	Peak Flux (mJy beam^{-1})	Total Flux (mJy)	Spectral Index
18	0.54 ± 0.02	0.76 ± 0.07	$+1.5$
22	0.59 ± 0.04	1.02 ± 0.11	

4. Future Works

Our studies have validated the "pole-on jet" model for the dominant blueshifted water masers, and we are now planning further follow-up studies in order to reveal exact relation between H_2O maser and jet activities. Simultaneous proper motion measurements of the masers are, especially, a direct way to examine their relation. Careful comparison between the maser light-curve and propagation of the jet is also valuable, because such a direct correlation between a maser and thermal emission in the time-domain have not yet been reported in a star-forming region. It will be a good example of what expected in extremely high-resolution studies of a high mass protostellar system in the ALMA era.

Furthermore, if highly intermittent variation in the maser-light curve actually traces episodic jet-launchings, then a G353-type H_2O maser source can be a suitable target for a time-resolved study of high mass protostellar jet. Single-dish based, dense maser monitoring can give us not only a unique statistical dataset about jet-variation, but also a chance to survey any time variation in the innermost region of an accretion disk along jet-launching activities.

5. Acknowledgements

This work was financially supported by the Grant-in-Aid for the Japan Society for the Promotion of Scinence Fellows (KM).

References

Arce, H. G., *et al.* 2007, in Reipurth, B., Jewitt, D., & Keil, K., eds, *Protostars and Planets V.* Univ. of Arizona Press, Tucson, p. 245
Breen, S. L., Caswell J. L., Ellingsen, S. P., & Phillips, C. J. 2010, *MNRAS*, 406, 1487
Caswell J. L. 1997, *MNRAS*, 209, 203
Caswell J. L. 2004, *MNRAS*, 351, 279
Caswell J. L. & Phillips C. J. 2008, *MNRAS*, 386, 1521 (CP08)
Caswell J. L. & Breen S. L. 2010, *MNRAS*, 407, 2599
Curiel S., *et al.* 2006, *ApJ*, 638, 878
Hofner, P., Cesaroni, R., Olmi, L., Rodrígues, L. F., Martí, J. & Araya, E. 2007, *A&A*, 465, 197
Martí, J., Rodrígues, L. F., & Reipurth, B. 1998, *ApJ*, 502, 337
Minier, V., Ellingsen, S. P., Norris, R. P. & Booth, R. S. 2003, *A&A*, 403, 1095
Motogi, K., *et al.* 2011, *MNRAS*, 417, 238
Neckel, T. 1978, *A&A*, 69, 51
Sakellis, S., Taylor, M. I., Taylor, K. N. R., Vaile, K. A., & Han, T. D. 1984, *PASP*, 96, 543
Wilson, W. E., *et al.* 2011, *MNRAS*, 416, 832

Cosmic Masers - from OH to H_0
Proceedings IAU Symposium No. 287, 2012
R.S. Booth, E.M.L. Humphreys & W.H.T. Vlemmings, eds.

© International Astronomical Union 2012
doi:10.1017/S1743921312006734

VERA Observations of the H_2O Maser Burst in Orion KL

Tomoya Hirota[1,2], Masato Tsuboi[3], Kenta Fujisawa[4], Mareki Honma[1,2], Noriyuki Kawaguchi[2,5], Mi Kyoung Kim[1], Hideyuki Kobayashi[1,2], Hiroshi Imai[6], Toshihiro Omodaka[6], Katsunori, M. Shibata[1,2], Tomomi Shimoikura[7], and Yoshinori Yonekura[8]

[1] National Astronomical Observatory of Japan, Mitaka, Tokyo 181-8588, Japan
email: `tomoya.hirota@nao.ac.jp`

[2] Department of Astronomical Sciences, The Graduate University for Advanced Studies (SOKENDAI), Mitaka, Tokyo 181-8588, Japan

[3] Institute of Space and Astronautical Science, Japan Aerospace Exploration Agency, Sagamihara, Kanagawa 229-8510, Japan

[4] Faculty of Science, Yamaguchi University, Yamaguchi, Yamaguchi 753-8512, Japan

[5] National Astronomical Observatory of Japan, Oshu, Iwate 023-0861, Japan

[6] Graduate School of Science and Engineering, Kagoshima University, Kagoshima, Kagoshima 890-0065, Japan

[7] Department of Astronomy and Earth Sciences, Tokyo Gakugei University, Koganei, Tokyo 184-8501, Japan

[8] Center for Astronomy, Ibaraki University, Mito, Ibaraki 310-8512, Japan

Abstract. In 2011 February, a burst of the 22 GHz H_2O maser in Orion KL was reported. In order to identify the bursting maser features, we have been carrying out observations of the 22 GHz H_2O maser in Orion KL with VERA, a Japanese VLBI network dedicated for astrometry. The bursting maser turns out to consist of two spatially different features at 7.58 and 6.95 km s^{-1}. We determine their absolute positions and find that they are coincident with the shocked molecular gas called the Orion Compact Ridge. We tentatively detect the absolute proper motions of the bursting features toward the southwest direction, perpendicular to the elongation of the maser features. It is most likely that the outflow from the radio source I or another young stellar object interacting with Compact Ridge is a possible origin of the H_2O maser burst. We will also carry out observations with ALMA in the cycle 0 period to monitor the submillimeter H_2O maser lines in the Orion Compact Ridge region. These follow-up observations will provide novel information on the physical and chemical properties of the mastering region.

Keywords. masers, ISM: individual (Orion KL), ISM: jets and outflows, radio lines: ISM

1. Introduction

An enormous outburst of the 22 GHz H_2O maser in Orion KL (D=420 pc; Hirota *et al.* 2007, Kim *et al.* 2008) is one of the most enigmatic phenomena in terms of cosmic masers. The first H_2O maser burst in Orion KL was discovered in 1979 and the active phase continued until 1985 with several flare-up events (Garay *et al.* 1989). The second burst was detected from 1997 to 1999 and monitoring observations were carried out with single-dish telescopes and VLBI (Omodaka *et al.* 1999, Shimoikura *et al.* 2005). The maximum flux density reached up to an order of 10^6 Jy in both burst phases. Although either a circumstellar disk or a jet/outflow associated with a young stellar object (YSO)

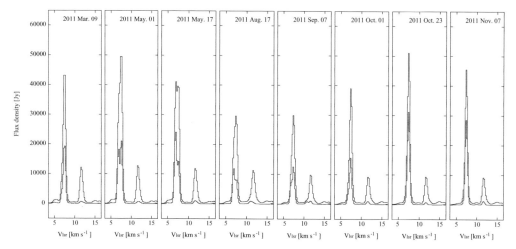

Figure 1. Observed spectra of the H$_2$O maser in Orion KL. Bold and thin solid lines represent the scalar-averaged cross power spectra and total power spectra, respectively.

is proposed as a possible origin of the maser burst, the origin is still unclear because of the lack of observational evidences other than the H$_2$O maser itself.

The third-time burst of the H$_2$O maser in Orion KL has started in February 2011 (Tolmachev 2011, Gaylard 2012). The maser burst appears to be periodic with an interval of 13 years (1985, 1998, and 2011). If the current burst event is the same as those of previous ones, the H$_2$O maser is expected to flare up to 10^6 Jy again, providing a rare opportunity to investigate the nature of this burst event. With this in mind, we have started astrometric observations of the bursting H$_2$O maser in Orion KL with VERA (VLBI Exploration of Radio Astrometry), a Japanese VLBI network developed for astrometry. Further details are described in Hirota *et al.* (2011).

2. Observations

Observations of the H$_2$O maser (6_{16}-5_{23}, 22235.080 MHz) in Orion KL were carried out on 2011 March 09, May 01, and May 17 with VERA and are still ongoing. The maximum baseline length was 2270 km and the uniform-weighted synthesized beam size (FWMH) was 1.7 mas×0.9 mas with a position angle of 143 degrees on average. We employed the dual beam observation mode, in which Orion KL and an ICRF source J054138.0-054149 were observed simultaneously. The data were recorded onto magnetic tapes at a rate of 1024 Mbps, providing a total bandwidth of 256 MHz. One IF channel with 16 MHz bandwidth was assigned to Orion KL and the remaining 15 IF channels were assigned to ICRF J054138.0-054149. For the H$_2$O maser lines, the spectral resolution was set to be 15.625 kHz, corresponding to a velocity resolution of 0.21 km s^{-1}. A bright continuum source, ICRF J053056.4+133155, was observed every 80 minutes for bandpass and delay calibration. Calibration and imaging were performed using the NRAO Astronomical Image Processing System (AIPS) software package.

3. Results

The H$_2$O maser burst is detected at the peak LSR velocity of 7.58 km s^{-1} in our first epoch of observation on 2011 March 09, as shown in Figure 1. The total flux density of $(4.4\pm0.3)\times10^4$ Jy is three orders of magnitude larger than that in the quiescent phase in

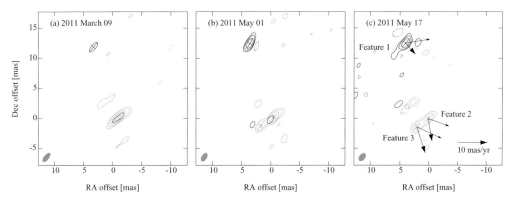

Figure 2. Phase-referenced images of the bursting maser features. The delay tracking center position $(0, 0)$ is $\alpha=05^h35^m14^s.1255$ and $\delta=-05°22'36''.475$ (J2000). Gray and black contours represent the 6.95 km s^{-1} and 7.58 km s^{-1} features, respectively. The contour levels are -1600, 1600, 3200, 6400, and 12800 Jy beam^{-1}. The bold and thin arrows in panel (c) represent the absolute proper motion vectors and those with respect to source I, respectively.

2006 (e.g. Hirota *et al.* 2007). Due to the lack of short baselines in VERA, the correlated flux, $(2.2\pm0.2)\times10^4$ Jy, recovers 50% of the total flux density. On 2011 May 01, another new velocity component at 6.95 km s^{-1} appears while the flux density of the 7.58 km s^{-1} component gradually decreases. Two weaks later, the total flux density of the 6.95 km s^{-1} component becomes comparable with that of the 7.58 km s^{-1} component on 2011 May 17. These two velocity components are consistent with those detected during the first burst event (Garay *et al.* 1989) while they are slightly shifted toward lower velocity compared to that reported for the second event, 7.64 km s^{-1}(Shimoikura *et al.* 2005).

Phase-referenced images of the bursting H_2O maser features are shown in Figure 2. The maser features show elongated structure along the northwest-southeast direction. It is consistent with that observed with VLBA during the previous burst in 1997-1999 (Shimoikura *et al.* 2005). In the first epoch of observation in March 2011, the 7.58 km s^{-1} feature shows single-peaked structure while it splits into double peaks in May 2011. In addition, another spatially distinct feature at the LSR velocity of 6.95 km s^{-1} appears at 12 mas north. The variation of the spatial structure is consistent with that seen in the H_2O maser spectra shown in Figure 1.

The absolute position of the target maser source is measured with respect to the extragalactic position reference source J054138.0-054149 with the dual-beam astrometry observations with VERA. Compared with the absolute positions reported for the previous burst events (e.g. Greenhill *et al.* 1998), the current position of the bursting maser is shifted by ~200 mas. Although this could be due to the much better astrometric accuracy of VERA (~1 mas, see Hirota *et al.* 2007) than that previously achieved with VLA in A-configuration. The positions of the bursting masers are thus almost consistent with each other. We have also measured the absolute proper motions of three bursting maser features as shown in Figure 2. Taking into account the annual parallax of Orion KL of 2.39 mas (Kim *et al.* 2008), we obtain the absolute proper motions of 4-9 mas yr^{-1} or 8-18 km s^{-1} toward southwest direction.

Judging from the similarities in peak LSR velocity, position, and structure in all of the bust events, they could be attributed to the common origin. However, it is unlikely that the bulk of the bursting gas clump would be identical throughout every burst events because the typical life time of the H_2O masers in Orion KL is much shorter than that of the bursting phase. In fact, we can see significant time variation of velocity and spatial

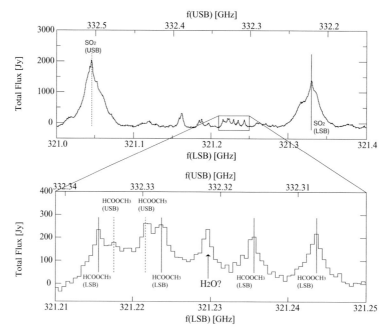

Figure 3. A single-dish 321 GHz spectrum toward Orion KL taken with the Atacama Submillimeter Telescope Experiment (ASTE) 10 m telescope. Note that the spectrum was taken with the double side-band (DSB) receiver.

structure of the bursting maser features even within the present monitoring observations for only two months from March to May 2011.

4. Discussion

Here we discuss a possible powering source of the bursting maser features. One of the most plausible interpretations is related to the outflow from a YSO in the Orion KL region as proposed by Garay *et al.* (1989). It has long been established that the H_2O masers in Orion KL are excited by interaction with outflow and ambient dense gas (Genzel *et al.* 1981, Liu *et al.* 2002). The most plausible powering source is the radio source I, which drives a northeast-southwest low-velocity outflow (Plambeck *et al.* 2009). The bursting maser features are located at 8" or 3400 AU southwest from source I, and are coincident with the interacting region named as Orion Compact Ridge. The elongation of the bursting maser features, which is perpendicular to the direction to source I, would suggest a shocked layer between the low-velocity outflow and the Compact Ridge. The magnitude of the proper motions of the bursting maser features are 16-20 km s^{-1} pointing toward the west to southwest direction when we subtract the absolute proper motion of source I (6.3 mas yr^{-1}, -4.2 mas yr^{-1}) recently reported by Goddi *et al.* (2011). Interestingly, they are roughly consistent with the low-velocity outflow from source I (Genzel *et al.* 1981). Thus, the H_2O maser burst may occur at a different part of the shocked layer when the episodic or possibly 13 year-periodic outflow from source I interacts with Compact Ridge.

Nevertheless, it is still unclear why only the $\sim$8 km s^{-1} components show such anomalous amplification phenomena episodically or with a possible 13 year-periodicity. It may imply a special condition to stimulate such maser burst. One of the possibilities is an existence of another pre-existing YSO in Compact Ridge interacting with the powerful outflow from source I (Garay *et al.* 1989). In fact, there are number of closer infrared, radio and X-ray sources around the bursting maser features than source I (see discussion in Hirota *et al.* 2011 and Favre *et al.* 2011). Because of the lack of high resolution observations resolving these continuum sources as well as their velocity structures, the powering source of the bursting maser features is still debatable. Higher-resolution observations with ALMA of the dust continuum and thermal molecular line emissions will be helpful to reveal physical properties of the bursting maser features and its origin. In addition, observational studies on submillimeter H_2O maser lines (see Figure 3) in parallel to the monitoring with VLBI of the 22GHz maser lines will also provide information on pumping mechanism of the bursting H_2O maser features. Follow-up observations will be carried out in the ALMA early sciences (cycle 0).

Acknowledgements

This work is supported in part by The Graduate University for Advanced Studies (Sokendai), and has made use of the SIMBAD database, operated at CDS, Strasbourg, France. TH is partly supported by Grant-in-Aid from The Ministry of Education, Culture, Sports, Science and Technology of Japan (No. 21224002).

References

Favre, C. *et al.* 2011, *A&A*, 532, A32
Garay, G., Moran, J. M., & Haschick, A. D. 1989, *ApJ*, 338, 244
Gaylard, M. J. 2012, in: R. Booth, L. Humphries, & W. Vlemmings (eds.), *Cosmic Masers – from OH to H0*, Proc. IAU Symposium No. 287 (Cambridge: Cambridge University Press), in press
Genzel, R., Reid, M. J., Moran, J. M., & Downes, D. 1981, *ApJ*, 244, 884
Goddi, C., Humphreys, E. M. L., Greenhill, L. J., Chandler, C. J., & Matthews, L. D. 2011, *ApJ*, 728, 15
Greenhill, L. J., Gwinn, C. R., Schwartz, C., Moran, J. M., & Diamond, P. J. 1998, *Nature*, 396, 650
Liu, S. -Y., Girart, J. M., Remijan, A., & Snyder, L. E. 2002, *ApJ*, 576, 255
Hirota, T. *et al.* 2007, *PASJ*, 59, 897
Hirota, T. *et al.* 2011, *ApJL*, 739, L59
Kim, M. K. *et al.* 2008, *PASJ*, 60, 991
Omodaka, T. *et al.* 1999, *PASJ*, 51, 333
Plambeck, R. L. *et al.* 2009, *ApJ*, 704, L25
Shimoikura, T., Kobayashi, H., Omodaka, T., Diamond, P. J., Matveyenko, L. I., & Fujisawa, K. 2005, *ApJ*, 634, 459
Tolmachev, A. 2011, *ATel*, 3177

Cosmic Masers - from OH to H_0
Proceedings IAU Symposium No. 287, 2012
R.S. Booth, E.M.L. Humphreys & W.H.T. Vlemmings, eds.

© International Astronomical Union 2012
doi:10.1017/S1743921312006746

The variability of cosmic methanol masers in massive star-forming regions

Jabulani P. Maswanganye and Michael J. Gaylard

HartRAO, P O BOX 443, Krugersdorp 1740, South Africa
email: jabulani@hartrao.ac.za, mike@hartrao.ac.za

Abstract. The methanol masers associated with G35.20-1.74 were monitored at 12178 MHz for four years and 6668 MHz for five years using the 26m Hartebeesthoek telescope. This source showed irregular variability and a single large flare event during the monitoring window.

Keywords. masers, radio lines: ISM, ISM: molecules, stars: formation.

1. Introduction and Observations

Class II methanol masers (MMII) are associated with high mass star-forming regions and often appear to be projected on ultra-compact ionised hydrogen (UCHII) regions. The 6668 MHz $J_k = 5_1 - 6_0$ A^+ and 12178 MHz $J_k = 2_0 - 3_{-1}$ E are the brightest class II methanol masers. Hartebeesthoek Radio Astronomy Observatory (HartRAO) monitors samples of methanol masers in the two transitions in a quest to understand the source of the observed variability. G35.20-1.74 (W48, IRAS 18592+0108) is one such maser source, which is associated with a small UCHII region (Wood & Churchwell, 1989)

The monitoring was carried out using the 26m telescope. Both the 6668 MHz and 12178 MHz receivers provide dual left- and right- circular polarisation outputs. Spectra were obtained at half-power beamwidth offsets to measure and correct for pointing errors. Spectra were calibrated using Ott *et al.* (1994) scale with 3C123 and 3C218 as the primary calibrators.

2. Results and discussion

The spectra and time series for G35.20-1.74 at 6668 MHz and 12178 MHz are shown in Figures 1 and 2 respectively. The variations appear to occur randomly, but are correlated (i) in many of the peaks in each transition and (ii) in peaks at similar velocities in the two transitions. The large flare at about MJD 54172 is notable. The 42.663 km.s^{-1} time series at 12178 MHz and 42.780 km.s^{-1} at 6668 MHz are atypical in showing flare onset changing to a sharp decrease to a short duration minimum. The 46.205 km.s^{-1} time series at 6668 MHz shows an exponential decay with e-folding time of 361 days. Similar behaviour was noted in a maser peak in G188.95+0.89 by van der Walt (2011).

The maser spots in this source were modelled as lying in a rotating Keplerian disk by Minier *et al.* (2000). Referring to the maser spot map in their Fig. 3, the separation of the extreme North and South maser spots in the map is 1300 AU = 7.5 light days (also interpreted as the disk diameter) and 3.2 days for West-East separation. The maximum of the large flare occurred earliest in masers at velocities around 41 km.s^{-1} (Southern spot group) and 44.5 to 45.3 km.s^{-1} (Northern group); then after 12 - 15 days at 42.3 - 42.8 km.s^{-1} (West central group), then after another ~15 days at 43.5 - 44 km.s^{-1} (East central group).

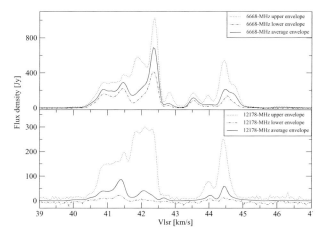

Figure 1. Methanol spectra for G35.20-1.74 at 6668 MHz and 12178 MHz.

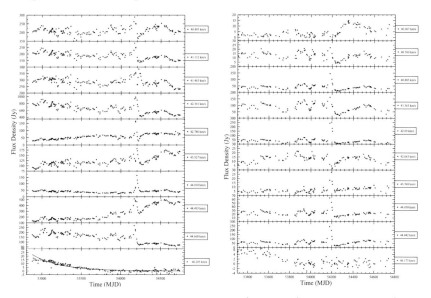

Figure 2. Time series for G35.20-1.74 at 6668 MHz (left panel) and 12178 MHz (right panel).

By contrast, the correlated random variations before the large flare, e.g. from MJD 53800 to 54100, show little or no time delay in the different time series. This suggests the flare and the other random variations have different origins. The correlated random variations may represent a response to infrared or radio continuum emission changes in the central source, which will produce small time delays, given the extent of masing spots. The much larger delays seen in the large flare could represent a response to a rotating or precessing jet.

References

Minier, V., Booth, R. S., & Conway, J. E. 2000, *A&A*, 362, 1093
Ott, M., Witzel, A., Quirrenbach, A., Krichbaum, T. P., Standke, K. J., Schalinski, C. J., & Hummel, C. A. 1994, *A&A*, 284, 331
van der Walt, D. J. 2011, *A&A*, 141, 152
Wood, D. O. S. & Churchwell, E. 1989, *ApJS*, 69, 831

Cosmic Masers - from OH to H_0
Proceedings IAU Symposium No. 287, 2012
R.S. Booth, E.M.L. Humphreys & W.H.T. Vlemmings, eds.

© International Astronomical Union 2012
doi:10.1017/S1743921312006758

Chasing the flare in Orion KL: Observations of the 22 GHz H_2O Masers at HartRAO

Sunelle Otto and Michael J. Gaylard

Hartebeesthoek Radio Astronomy Observatory,
PO Box 443, Krugersdorp, 1740, South Africa
email: sunelle@hartrao.ac.za, mike@hartrao.ac.za

Abstract. The work here represents the summary of observations made with the Hartebeesthoek Radio Astronomy Observatory (HartRAO) 26 m telescope of the water vapour (H_2O) Masers in the Orion KL source region for a period of 8 months during the flare of 2011. The observation setup, calibration method together with the resulting maser time series are discussed.

Keywords. masers, ISM: individual (Orion KL), radio lines: ISM

1. Background

The source of interest discussed here is the H_2O maser in the Orion KL (Kleinman-Low) nebula source region (Reid *et al.* 1981). Flares in this region from the 1.35 cm wavelength water maser emission line occurred during the period of 1979–1985, with a second flare in 1998 (Tolmachev 2000). The previous flares were very intense, increasing in flux density from around hundreds to millions of Janskys. A new flare was reported in 2011 (Tolmachev 2011) and it was decided to use this to test the performance of the new 22 GHz receiver on the 26m Hartebeesthoek telescope.

2. Observation setup

The 26 m radio telescope at HartRAO was used with the experimental 1.3 cm receiver to make spectral line observations of the water masers in Orion KL at a centre frequency of 22235.12 MHz and 8 MHz bandwidth. Radio continuum observations at the same centre frequency but with 32 MHz bandwidth were made of Jupiter over the same period to assist in calibration. The observations were made from the time period of 12 March 2011 to 22 November 2011, at 05 32 46.6 and (1950) −05 24 27.2. The antenna half power beamwidth at this frequency is 0.035 degrees, with the spectrometer velocity resolution being 0.105 km.s^{-1} per channel and the centre velocity 10 km.s^{-1}.

3. Data reduction

Jupiter was used as the calibrator source as it has a constant brightness temperature of 136 K at 22 GHz (Gibson *et al.* 2005, Hill *et al.* 2009). Drift scans were done on this source and the resulting observed antenna temperatures were corrected for various factors including pointing, angular size, focus offset and atmospheric absorption corrections. The maser spectral line observations were done using the position switching method. To determine the pointing correction, spectra were obtained at the half power points of the beam in the North, South, East and West directions. Bad data were removed; these were due to bad weather, high system temperature or large errors in pointing. Corrections for atmospheric absorption were then made using water vapour measured by GPS after which the data were converted to flux densities using the results from the Jupiter calibration.

4. Results

The average total intensity spectrum (lcp and rcp added) of the reduced 2011 maser data is given by Fig. 1, together with HartRAO data from 2007 overlain. It is clear that some of the maser peaks are flaring. The highest peak is at $V_{lsr} = 7.893$ km s^{-1} (A) reaching an average of over 80,000 Jy total intensity, with the second and third brightest peaks at $V_{lsr} = 7.367$ km s^{-1} (B) and V_{lsr} 12.107 km s^{-1} (C) respectively. This can be compared with Hirota *et al.* (2011) who found two main maser features flared during the period of March to May 2011, at 7.58 km s^{-1} and 6.95 km s^{-1}.

Time series plots of the maser data (lcp) were made for the main feature velocities as well as for other weaker peaks are shown in Fig. 2. Some of the features show an increase in flux density stretched over a long period of time as seen at $V_{lsr} = -30.102$ km s^{-1} and peak B, with $V_{lsr} = -6.642$ km s^{-1} having a shorter burst in intensity and an average intensity of ~50 Jy at other times.

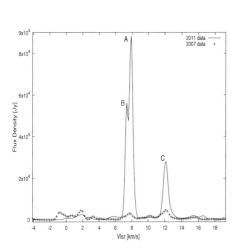

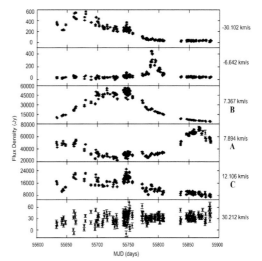

Figure 1. Average total intensity spectrum of the H$_2$O Masers in Orion KL from 2011 with 2007 data overlain.

Figure 2. Light curves (lcp) of the H$_2$O Masers in Orion KL at the peak and other weaker velocities.

Altough flaring clearly occured during this period of 2011, the flux density has not yet increased to values comparable to that of the previous flares. Continued observations will reveal more of the behaviour of these water masers.

References

Gibson, J., Welch, W. J., & De Pater, I. 2005, *Icarus*, 173, 439
Hill, R. S., Weiland, J. L., & Odegard, N. 2009, *ApJ*, 180, 246
Hirota, T., Tsuboi, M., & Fujisawa, K. 2011, *ApJ*, 180, 246
Reid, M. J. & Moran, J. M 1981, *ARAA*, 19, 231
Tolmachev, A. M. 2000, *Astron. Lett.*, 26, 34
Tolmachev, A. 2011, *ATel*, 3177

Cosmic Masers - from OH to H_0
Proceedings IAU Symposium No. 287, 2012
R.S. Booth, E.M.L. Humphreys & W.H.T. Vlemmings, eds.

© International Astronomical Union 2012
doi:10.1017/S174392131200676X

On the Methanol masers in G9.62+0.20E: Preliminary colliding-wind binary (CWB) calculations

S. P. van den Heever[1], D. J. van der Walt[1], J. M. Pittard[2] and M. G. Hoare[2]

[1]Centre for Space Research, North west University, Potchefstroom, South Africa
[2]Dept. of Astronomy & Astrophysics, Leeds University, Leeds, England

Abstract. A comparison between the observed periodic flaring of methanol maser sources in the star forming region G9.62+0.20E and the continuum emission from parts of a background HII region is made. Using a colliding wind binary (CWB) model preliminary calculations show that the CWB model results fit the maser light curves very well.

Keywords. masers, hydrodynamics

1. Introduction

Class II methanol masers are exclusively associated with massive star forming regions. At present about seven class II methanol masers show periodic or highly regular flaring behavior. Recently van der Walt *et al.* (2009) and van der Walt (2011) proposed that periodic methanol masers might be due to changes in the background free-free emission associated with a CWB system. Although the toy model of van der Walt (2011) is able to reproduce the observed maser light curves quite well, some untested assumptions have to be checked. The two most important assumptions are:

- The shocked gas cools adiabatically and the shock luminosity L_{shock} scales as $1/r$ where r is the distance between the stars.
- The flux of ionizing radiation emerging from the shocked gas is sufficient to produce the ionization required to explain the observed variation in the masers.

As a start, before some more elaborate numerical calculations we first investigated the validity of the above mentioned assumptions and present some very early results here. For the ionizing source, a B0 type star was assumed with $L_\star = 38.5 ergs^{-1}$ cm^{-2}, M = 20 $M_\odot$ and $\dot{M} = 10^{-6} M_\odot yr^{-1}$ (see Sternberg *et al.* 2003). For possible post-shock temperatures we assumed $T_{shock} = 10^6$ K and 5×10^6 K, and calculated the corresponding emission spectra with Cloudy. The luminosity of the shocked gas was taken as $\log(L_{shock}) = 35, 35.5, 36,$ and 36.5. Figure 1(left) shows the combined spectra of the central star and the hot shocked gas for the two temperatures and $\log(L_{shock}) = 35.5$. Figure 1(right) compares the resulting electron density distributions at the ionization front (based on G9.62+0.20E) for $T_{shock} = 10^6$ K and for different shock luminosities as indicated in the figure.

Inspection of Figure 1(right) shows that for $\log L_{shock} < 35$ there is practically no effect, while for $\log L_{shock} = 35.5$ and $\log L_{shock} = 36$ there is a marked increase in the electron density. In fact the required change in electron density to explain the observed changes in 12.2 GHz masers in G9.62+0.20E can be obtained with $L_{shock} = 10^{35.5}$.

A 2D numerical hydrodynamic code (see e.g. Pittard & Stevens 1997)) was used to investigate whether $L_{shock} \simeq 10^{35.5} erg\, s^{-1}$ can be produced and what r dependence

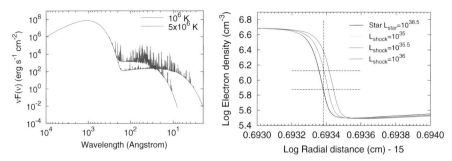

Figure 1. (left) Combined stellar and hot shocked gas spectra for temperatures as indicated in the graph. (right) Electron density distribution in the ionization front for various shock luminosities. Horizontal lines show required change in electron density.

follows from reasonable assumptions about mass loss rate and terminal wind speed. These calculations showed that most probably the shocks cool radiatively resulting in an $L_{shock} \propto r^{-2}$ dependence. It was also found that the requirement that $\log L_{shock} \simeq 35.5$ can be met with reasonable assumptions on the mass loss rates and wind speeds.

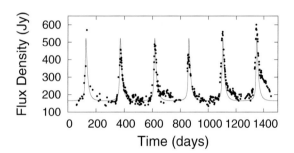

Figure 2. Comparison between the simulated results and the observed 12.2 GHz maser light curves.

An $L_{shock} \propto r^{-2}$ dependence was used in the toy model of van der Walt (2011). Figure 2 compares the observed time series for the 12.2 GHz maser in G9.62 + 0.20E with the model results. The $L_{shock} \propto r^{-2}$ dependence requires the eccentricity of the orbit to be 0.81 compared to 0.9 for $L_{shock} \propto r^{-1}$. Otherwise the model seems to reproduce the observed time series very well.

From an energetics point of view it seems that the shock is sufficient to reproduce the required energy. From the results the 12.2 GHz masers in G9.62+0.20E can be explained by a radiative shock ($L_{shock} \propto r^{-2}$) rather than adiabatic ($L_{shock} \propto r^{-1}$).

References

Goedhart, S., Gaylard, M., & van der Walt, D. J. 2003, *MNRAS*, 333, L33–L36
van der Walt, D. J., Goedhart, S., & Gaylard, M. 2009, *MNRAS*, 398, 961–970
van der Walt, D. J. 2011, *AJ*, 141, 152
Stevens, I. R., Blondin, J. M., & Pollock, A. M. T. 1992, *APJ*, 386, 265–287
Sternberg, A., Hoffmann, T. L., & Pauldrach, A. W. A. 2003, *ApJ*, 599, 1333–1343
Pittard J. M. & Stevens I. R. 1997, *MNRAS*, 292, 298

T3b:

Star formation masers *Chair: Crystal Brogan*

Cosmic Masers - from OH to H$_0$
Proceedings IAU Symposium No. 287, 2012
R.S. Booth, E.M.L. Humphreys & W.H.T. Vlemmings, eds.

© International Astronomical Union 2012
doi:10.1017/S1743921312006771

Masers in star forming regions

Anna Bartkiewicz[1] and Huib Jan van Langevelde[2,3]

[1] Toruń Centre for Astronomy, Nicolaus Copernicus University,
Gagarina 11, 87-100 Toruń, Poland, email: `annan@astro.uni.torun.pl`

[2] Joint Institute for VLBI in Europe,
Postbus 2, 7990 AA Dwingeloo, The Netherlands, email: `langevelde@jive.nl`

[3] Sterrewacht Leiden, Leiden University,
Postbus 9513, 2300 RA Leiden, The Netherlands

Abstract. Maser emission plays an important role as a tool in star formation studies. It is widely used for deriving kinematics, as well as the physical conditions of different structures, hidden in the dense environment very close to the young stars, for example associated with the onset of jets and outflows. We will summarize here the recent observational and theoretical progress on this topic since the last maser symposium: the IAU Symposium 242 in Alice Springs.

Keywords. masers – stars: formation – stars: early-type – radio lines: ISM – ISM: molecules – ISM: jets and outflows – ISM: kinematics and dynamic

1. Introduction

Cosmic masers are known as a unique tool in star-formation studies and are one of the first observed signposts of high-mass star formation, particularly the hydroxyl (OH), water (H_2O) and methanol (CH_3OH) masers, that are common and intense. This is demonstrated by the many results presented in this volume.

At the last IAU Symposium 242 in Alice Springs, Fish (2007) summarized the relation between masers and star-formation in the following way *"maser observations are at the vanguard of star formation research: yesterdays observations can be explained by complementary data and theory today, and todays observations lay the groundwork for the breakthroughs that will be achieved in the context of tomorrow."* In this review we will summarize some of the achievements and discoveries in the area of star-formation masers presented in the literature since the Australian Symposium. It is important to evaluate what "yesterday's tomorrow" has unveiled in the area of star formation and identify the possible "today's tomorrow" breakthroughs.

2. Population studies

It was relatively well established from earlier surveys of masers in our Galaxy that massive star-forming regions can be associated with OH, H_2O and Class II CH_3OH masers (e.g., Caswell *et al.* 1995, Szymczak *et al.* 2002). However, there is still a great need for verifying what the relation is between specific stages and classes of star formation and different masers. For this, complete and ever more sensitive surveys with better astrometric precision are most valuable. For example, since our previous meeting Green *et al.* (2009) discovered that high-mass star formation (HMSF) is present in both the far and near 3 kpc arms through 49 detections of 6.7 GHz methanol masers. The Red MSX Sources (RMS) based survey by Urquhart *et al.* (2009) investigated the statistical correlation of water masers with early-stages of massive star-formation. They found similar detection rates for UC H II regions and MYSOs, suggesting that the conditions needed for maser activity are equally likely in these two stages of star formation.

Nowadays, more instruments have become available, especially to focus systematically on different maser transitions like the higher excited methanol masers from both Class I (collisional excitation) and II (radiative excitation) methanol, as well as silicon monoxide (SiO), ammonia (NH_3) and formaldehyde (H_2CO) masers. Details are presented by e.g. Kurtz, Kalenskii, Voronkov, Sjouwerman, Wootten, Booth, Kim, Pestalozzi, Brogan and their collaborators in these proceedings.

In addition, yesterday's key-questions, which were deemed essential in order to use masers for studying the physics of star formation, are still not fully answered:

- Is there an evolutionary sequence based on maser occurrence?
- Are Class I and Class II methanol masers associated?
- Where, when and how exactly do masers arise?
- What physical conditions are needed to produce the maser(s)?

These questions require systematic studies of a large number of sources, possibly at high angular resolution, as well as observations of specific sources using multi-wavelength observations, in order to converge and refine our hypotheses. For example, single-dish studies suggested at first that the methanol Class I and II are coincident, but later interferometric images showed they are not co-spatial on arcsecond scales, even though they may be driven by the same YSO (Cyganowski *et al.* 2009).

2.1. *Methanol masers, the most widespread masers in HMSFRs*

Methanol masers have been widely studied, particularly after the discovery of the widespread, bright 6.7 GHz transition (Menten 1991). Many transitions have been found to emit maser emission from both the A and E types and they have been classified into Class I and Class II, which are collisionally and radiatively excited, respectively (Cragg *et al.* 1992). In general, the Class I (e.g., 36 and 44 GHz lines) are likely associated with outflows (lying further from the central objects) while the Class II (e.g., 6.7 and 12.2 GHz transitions) often coincide with hot molecular cores, UC H II regions, OH masers and near-IR sources.

The most common Class II masers are 6.7 and 12.2 GHz lines that according to the models (e.g., Cragg *et al.* 2005) should co-propagate quite often, as confirmed by observations (e.g., Breen *et al.* 2011). Both masers are strongly associated with HMSFRs and enable us to probe the dense environments where stars are being born. A large sample of 113 sources with known 6.7 GHz masers and 1.2-mm dust clumps was searched at 12.2 GHz by Breen *et al.* (2010b). These authors found out that when the 6.7 GHz emission is more luminous, then the evolutionary stage of the central object tends to be more advanced, while also the 12.2 GHz is often associated with more evolved regions. Based on this an evolutionary sequence for masers associated with massive star formation regions was proposed, which is consistent with conclusions of e.g. the survey of Class I by Pratap *et al.* (2008). However, some discussion continues to refine this relation, e.g. Fontani *et al.* (2010) and Chen *et al.* (2011).

2.1.1. *Class I and Class II methanol masers*

One of the key research areas over the past few years has been on relatively rare methanol masers. Voronkov *et al.* (2010b) found two new detections of 9.9 GHz Class I masers. To date we know of 5 masers at 9.9 GHz (additionally from Slysh *et al.* 1993, Voronkov *et al.* 2006, 2011). The detection rate is likely so low because of the strong dependence of the maser brightness on the physical conditions (Sobolev *et al.* 2005).

This maser is believed to trace shocks caused by different phenomena (e.g., expanding H II regions, outflows). A particularly interesting case is G331.13−00.24, which shows periodic variability at 6.7 GHz with a period of 500 days (Goedhart *et al.* 2004). There is an obvious urgency to verify whether the variations of both lines correlate, pointing to a common origin of the seed radiation and providing an estimate of the physical conditions for that. For more details see Voronkov *et al.* (these proceedings).

There are now also first arcsecond-resolution images of the 36 GHz methanol masers in HMSFRs thanks to the upgrades of both ATCA and EVLA, e.g. in M8E (Sarma & Momjian 2009), Sgr A (Sjouwerman *et al.* 2010b), G309.38−0.13 (Voronkov *et al.* 2010a) and DR21 (Fish *et al.* 2011b). In the latter case it was found that surprisingly the Class I 36 GHz and 229 GHz masers appear in close proximity (also in velocity) with the Class II 6.7 GHz maser, while the 44 GHz Class I masers is absent. According to the model by Voronkov *et al.* (2005) such cases require an intermixed environment of dust and gas at a lowish temperature of ≈60 K.

One may wonder whether we have come closer to answering the question *"when do Class I masers appear?"*. Chen *et al.* (2011) searched 192 EGOs (the candidates associated with ongoing outflows) for 95 GHz methanol masers, resulting in a 55 per cent detection rate. These detections are likely associated with the redder GLIMPSE point-source colors. There are two possible explanations, either the Class I objects are associated with lower stellar masses or they are associated with more than one evolutionary phase during high-mass star formation, apparently contradicting the most straightforward schemes (e.g. Breen *et al.*, 2010b). Marseille *et al.* (2010) compared the physical conditions by observing several molecular tracers in both weak and bright mid-IR emitting massive dense cores. The methanol Class I maser at 84.5 GHz was found to be strongly anti-correlated with the 12 μm source brightness, leading to an interpretation that these represent more embedded mid-IR sources with a spherically symmetric distribution of the envelope material.

Ellingsen *et al.* (2011) searched for rare and weak methanol masers at 37.7, 38.3, 38.5 GHz Class II methanol masers towards 70 HMSFRs. They detected 13 at 37.7 GHz and 3 at 38.3/5 GHz and found that the 37.7 GHz masers are associated with the most luminous 6.7 and 12.2 GHz masers, likely representing a short (of 1000–4000 years) period in an advanced stage of the evolution. Therefore, the 37.7 GHz methanol masers may be called the *horsemen of the apocalypse* for the Class II methanol maser phase.

2.1.2. *The morphology of 6.7 GHz masers*

More sensitive VLBI surveys have led to the discovery of more complex 6.7 GHz maser structures, including several that show a ring-like morphology (Bartkiewicz *et al.* 2005, 2009). Kinematics of the maser spots revealed that outflow/infall dominates over the possible Keplerian rotation in a disc. A similar morphology with a similar kinematic signature was found in the well-known HMSFR Cep A, where, due to additional constraints on the orientation, the radial motions are more likely resulting from infall (Torstensson *et al.* 2011a, Sugiyama *et al.* 2008). Moreover, it seems that the magnetic field plays a role in shaping this morphology (Vlemmings *et al.* 2010). Such ring-like characteristics were also seen in water masers associated with slightly more advanced stages, where masers were likely tracing an accretion disc or its remnant (Motogi *et al.* 2011a). Torstensson *et al.* (2011b) analysed some of these ring-like maser sources using thermal emission at arcsec scale and found that mostly the distribution of the methanol gas peaks at the maser position with the larger scale gas showing a modest outflow velocity. They argued that the methanol gas has a single origin in these sources, possibly associated with an accretion shock. ALMA resolution is necessary for probing the regions of interest at size scales of 1000 AU.

For all of these studies it is important to remember that the VLBI technique resolves out some of the emission. Pandian *et al.* (2011) noted that more complex morphologies and often larger structures become apparent when using shorter baseline interferometers (EVLA, MERLIN) compared to VLBI, analyzing a study of 72 sources from the Arecibo Methanol Masers Galactic Survey. Thus, the 29% detection of the methanol rings of Bartkiewicz *et al.* (2009) may be biased by observational effects. Similarly, Cyganowski *et al.* (2009) also noted that shorter baselines observations resulted in more complex and extended emission for two targets from the EVN sample of Bartkiewicz *et al.* (2009). On the other hand, comparing Pandian *et al.* (2011) and Bartkiewicz *et al.* (2009) results, we note that in three out five cases the emission is very similar on EVN and MERLIN images, but this does not include any of the ring sources.

3. Gas kinematics through proper motion studies of masers

There has also been significant progress on maser proper motion studies at the milli-arcsecond scale. These result from multi-epoch observations that often have two simultaneous objectives: proper motions of the maser features in order to derive the kinematics of the gas in the direct environment of the YSOs and accurate direct distances by means of detecting the parallax. The parallax measurements are summarized and presented in this volume by e.g., Reid, Honma *et al.*, Nagayama *et al.*, Choi *et al.* (these proceedings).

Here we focus on the dynamical studies. In a series of papers, Moscadelli *et al.* (e.g., 2007, 2011a) demonstrated the power of multi-epoch VLBI for tracing the 3D kinematics close to an YSO towards nine well-studied HMSFRs. Combining observations of 22 GHz water and 6.7 GHz methanol masers within a time-span of a few years they detected various motions such as outflow, rotation, infall, all happening in the direct environments of these YSOs. They pointed out that these velocity gradients on milli-arcsecond scales still reflect large-scale (100-1000 AU) motions (Moscadelli *et al.* 2011b). In some cases such studies can constrain the YSO position and mass. For example, Goddi *et al.* (2011) directly measured these different phenomena going on within 400 AU from the high-mass protostar AFGL 5142; the gas infall is traced by the 3D velocities of the methanol masers, while a slow, massive, collimated, bipolar outflow is detectable through the water masers. Very detailed dynamics were registered by Torrelles *et al.* (2011) who used multi-epoch data of water masers towards Cep A HW2, noticing morphological changes at scales of 70 AU in a time-span of 5 years. They also argued that the R5 expanding bubble structure has been dissipating in the circumstellar medium and that a slow, wide-angle outflow at the scale of 1000 AU co-exists with the well-known high-velocity jets.

In addition to these methanol and water maser observations, there are unique data from SiO masers, although they are quite rare around YSOs. Matthews *et al.* (2010) observed the Orion Source I at both 43.1 and 42.8 GHz transitions, resulting in a most detailed view of the inner 20-100 AU of a MYSO. The SiO masers lie in an X-shaped structure, with clearly separated blue- and redshifted emission, while bridges of intermediate-velocity emission connect both sides. They proposed that these masers are related to a wide-angle bipolar wind emanating from a rotating, edge-on disc. This provides direct evidence of the formation of a MYSO via disc-mediated accretion. Other examples and more explanations can be found in contributions by e.g. Sanna *et al.*, Goddi *et al.*, Sawada-Satoh *et al.*, Sugiyama *et al.* (these proceedings).

4. Physical conditions for maser emission

In order to answer the key question *where, when and how exactly do masers arise?*, we must probe the physical conditions in which they form. Such studies concern multiple maser transitions and studying the masing regions at a wide range of wavelengths. Both surveys of a large number of sources, as well as detailed individual source observations are needed to complete the scenario of maser formation. A very good example is the result obtained by Cyganowski *et al.* (2008) that Class I and II methanol masers coincide with so-called extended green objects (EGOs) which are indicators of outflows and are a promising starting point for identifying MYSOs (Cyganowski, these proceedings).

Breen *et al.* (2010a) investigated the OH/H_2O/CH_3OH relation for a large sample of HMSFRs and noticed a closer similarity of the velocities of OH and methanol masers than of either of these species compared to the water maser peak velocity. In spite of the different pumping schemes of water and methanol masers, they both show a similar, 80% detection rate association with OH sources. It also has been found by comparing high-luminosity masers with low-luminosity ones that the high brightness ones are related to lower NH3(1,1) excitation temperatures, smaller densities, but three times larger column densities. Moreover, the high-luminosity sources are associated with 10 times more massive molecular cores, larger outflows and their internal motions are more pronounced (Wu *et al.* 2010). Interestingly, Pandian *et al.* (2010) showed that the continuum of the counterparts of 6.7 GHz methanol masers is consistent with rapidly accreting massive YSOs (>0.001 $M_\odot$ yr^{-1}) by constraining their SEDs. Only a minority of the sample, 30%, coincided with H II regions that are usually ultra- or hyper compact. The latter was also confirmed by Sánchez-Monge *et al.* (2011) and Sewilo *et al.* (2011). Indeed the majority of 6.7 GHz masers seems to appear before the H II stage of MYSOs, as was suggested by earlier studies, e.g., Walsh *et al.* (1998). Alternatively, we may still not have been able to reach the proper sensitivity for such conclusions.

Studies of specific sources in multiple maser transitions and their counterparts in other tracers are of special value. In the well known ON 1 source OH transitions at 1.612, 1.665, 1.667, 1.720, 6.031 and 6.035 GHz lie in a similar region as 6.7 GHz methanol masers. Green *et al.* (2007) concluded that they possibly trace a shock front in the form of a torus/ring around the YSO. That interpretation is also supported by polarization angles and velocity gradients. In the HMSFR NGC 7538 IRS 1 new masers at 12.2 GHz were found (Moscadelli *et al.* 2009) and in addition, 23.1 GHz Class II methanol masers were accurately registered (Galván-Madrid *et al.* 2010). They appear closely associated with 4.8 GHz H_2CO masers, indicating that the conditions must be similar for both of these relatively rare masers. It is possible that they are excited by the free-free emission from an H II region. However, surprisingly, they are not accompanied by any 6.7 or 12.2 GHz methanol masers.

Although we are collecting more and more information, the origin of maser structures in high-mass star formation is still not clear. A long-term question *what do linearly distributed methanol masers trace, an edge-on disc or an outflow?* is still open. De Buizer *et al.* (2009) showed that orientations of SiO outflows were not consistent with the methanol masers delineating a disk orientation. Moreover, for the methanol rings the proposed morphology could generally not be confirmed by infrared high resolution imaging (De Buizer *et al.*, these proceedings). Beuther *et al.* (2009), using NH_3 as a tracer towards methanol Class II masers found that if Keplerian accretion disks exist, they should be confined to regions smaller than 1000 AU. Therefore, ALMA-resolution observations are really needed in order to reach the relevant scales in the direct environment of MYSOs.

5. Masers as a signpost of star formation

Masers are readily usable as a diagnostic in complex SFRs, for example as indicators that star formation has begun. Purcell et al. (2009) investigated the NGC 3576 region and verified the evolutionary status of the various molecular components. Water masers were found towards the NH_3 emission peaks, lying in the arms of the filament. In the HMSFR G19.61$-$0.23 water masers trace the outflow/jet associated with the most massive core, SMA 1, also traced by $H^{13}CO^+$ emission (Furuya et al. 2011). The massive cold dense core G333.125–0.562 showed water and methanol masers, as well as SiO thermal emission, but remained undetected at wavelengths shorter than 70 μm (Lo et al. 2007). Moreover, 44 GHz methanol masers coincide with presumably masing 23 GHz NH_3 emission in the EGO G35.03+0.35 (Brogan et al. 2011). The latter project is based on simultaneous observations of continuum emission and a comprehensive set of lines, something that has become possible with the Expanded Very Large Array (EVLA) and should contribute significantly to our understanding of star formation (Brogan et al., this volume).

6. Variability of masers

The 6.7 and 12.2 GHz methanol masers have been monitored and unexpectedly periodic variations were discovered from some masers in HMSFRs (e.g. Goedhart et al. 2003). Such studies are possible with single-dish observations and often require long-term commitments. Monitoring can provide important clues about which phenomena are responsible for the variability, but also about the more general physical conditions in the masing regions or the background radiation field.

Recently, Goedhart et al. (2009) summarized nine years of monitoring of G12.89+0.49 at the 6.7 and 12.2 GHz transitions and suggested that the stability of the period is best explained by assuming an underlying binary system. In G9.62+0.20E three methanol lines at 6.7, 12.2 and 107 GHz showed flaring (van der Walt et al. 2009) and a colliding-wind binary (CWB) scenario is found to explain periodicity through variations in the seed photon flux and/or the pumping radiation field (van der Walt 2011). Follow-up studies were required in order to provide more details about the source; VLBI imaging revealed the maser distribution (Goedhart et al. 2005) and multi-epoch observations enabled the direct estimation of its distance of 5.2±0.6 kpc via trigonometric parallax (Sanna et al. 2009). Recently, Szymczak et al. (2011) discovered a similar case of variability in G22.357+0.066 that can also be explained by changes in the background free-free emission. A period of 179 days was derived from single dish monitoring. The time delays seen between maser features can be combined with the VLBI imaging to construct the 3D structure of the maser region. Another example is G33.641$-$0.228 where the 6.7 GHz methanol bursts originate from a region of 70 AU (Fujisawa et al. 2012). The authors interpret this as coming from an impulsive energy release like a stellar flare. By monitoring many different objects we may also find more newly appearing masers as was the case with 6.7 GHz emission in IRAS 22506+5944 (Wu et al. 2009).

Variability was also detected for other maser transitions. In G353.273+0.641 intermittent 22 GHz maser flare activity appeared to be accompanied by structural changes, likely indicating that the excitation is linked to an episodic radio jet (Motogi et al. 2011b). Lekht et al. (2012) presented a catalog of 22 GHz H_2O spectra monitored over 30 years towards G34.3+0.15 (aka W44C). They detected a long-term variability with an average period of 14 years and two series of flares that are likely associated with some cyclic activity of the protostar in the UC H II. 1720 MHz OH masers towards W75N also showed flaring, possibly related to the very dense molecular material that is excited and slowly accelerated by the outflow (Fish et al. 2011a). A surprising event was registered in

IRAS18566+0408 by Araya *et al.* (2010): the 4.8 GHz H_2CO maser showed flaring and a period correlated with the 6.7 GHz methanol outbursts. Both regions are separated spatially, so both phenomena likely indicate variations in the infrared radiation field, maybe related to periodic accretion events.

7. Masers in low and intermediate-mass protostars

Because low and intermediate-mass stars are more common and evolve more slowly, they should in principle be easier to study, at closer distances, maybe less embedded and in less confusing environments. It should therefore be possible to study the star-formation process in more detail. However, masers are not so common in these kinds of objects and appear possibly in different stages. Bae *et al.* (2011) found that only 9 and 6% of a sample of 180 intermediate-mass YSOs showed 22 GHz water and 44 GHz methanol maser emission, respectively. Water is likely related to the inner parts of outflows and can be highly variable, while methanol is possibly associated with the interfaces of the outflows with the ambient dense gas. The detection rates of both masers rapidly decreases as the central (proto)stars evolve and the excitations of the two masers appear closely related. The most embedded (Class 0-like) intermediate-mass YSOs known to date are all associated with water masers (e.g., Sánchez-Monge *et al.* 2010; de Oliveira Alves, these proceedings). However, an intriguing object, IRAS 00117+6412 was found with water masers in MM2, where no outflow seems present (Palau *et al.* 2010). In order to understand this case, more observations with the best sensitivity and resolution are needed.

The 44 and 36 GHz Class I methanol masers rarely appear in lower-mass YSOs. Kalenskii *et al.* (2010a) found only four 44 GHz and one 36 GHz maser, while no emission was detected towards the remaining 39 outflows. They also noticed the masers have lower luminosity compared to those in HMSFRs. Imaging of L1157 indicates that the 44 GHz maser may form in thin layers of turbulent post-shock gas or in collapsing clumps (Kalenskii *et al.* 2010b; Kalenskii *et al.* these proceedings).

8. HMSFRs in the Galactic Centre and beyond the Galaxy

Recent searches towards Sgr A revealed that the 36 GHz Class I methanol masers correlate with NH_3(3,3) density peaks and outline regions of cloud-cloud collisions, maybe just before the onset of local massive star formation (Sjouwerman *et al.* 2010a). The 44 GHz masers correlate with the 36 GHz locations, but less with the OH masers at 1720 MHz (Pihlstöm *et al.* 2011a), which are associated with the interaction between the supernova remnant Sgr A East and the interstellar medium (Pihlstöm *et al.* 2011b). One particularly interesting group of 44 GHz masers was found that does not overlap with any 36 GHz emission. These masers may signal the presence of a hotter and denser environment than the material swept up from the shock, maybe related to advanced star formation (Pihlstöm *et al.* 2011a).

Kilomasers are masers beyond our Galaxy, with luminosities comparable to the brightest galactic maser, which either amplify a background AGN or originate from star-forming regions. These Galactic analog H_2O masers have become a great tool in studies of young super-star cluster formation with high angular resolution. They are detected in a few nearby galaxies only (e.g., Castangia *et al.* 2008). Brogan *et al.* (2010) found that water masers in the Antennae Galaxies are associated with star-formation, as they show kinematic and spatial agreement with massive and dense CO molecular clouds. The various early stages of star formation in the components of the Antennae Galaxies was confirmed by Ueda *et al.* (2012).

The Large and Small Magellanic Clouds were also searched for maser emission. Green *et al.* (2008) detected four 6.7 GHz methanol (of which one is new) and two 6.035 GHz OH (of which one is new) masers in the LMC. Both transitions indicate much more modest maser populations compared to the Milky Way, likely originating from lower oxygen and carbon abundances. Moreover, Ellingsen *et al.* (2010) found the first 12.2 GHz methanol maser towards the LMC. They also detected 22 GHz water and 6.7 GHz methanol masers that are associated with more luminous and redder YSOs. The first 6.7 GHz methanol spectrum towards Andromeda Galaxy (M31) was presented by Sjouwerman *et al.* (2010b). More on kilo- and also mega-masers is presented by e.g. Tarchi (these proceedings).

9. Magnetic field

Masers are a particularly unique tool for studying the magnetic field studies in HMS-FRs (Vlemmings, these proceedings). This is an important subject in star formation, as the magnetic field could be a dominant force in the process by supporting the molecular cloud against gravitational collapse, regulating the accretion and shaping the outflows as has been argued for Cepheus A (Vlemmings *et al.* 2006). We have already mentioned the work by Green *et al.*(2007) who found a shock front in the form of a torus/ring around the YSO in ON 1. That scenario is also supported by the measured polarization angles of the masers. A magnetic field strength of a few mG was detected through the Zeeman splitting of OH and methanol masers. Linearly and circularly polarized emission of 22 GHz water masers was used for measuring the orientation and strength of magnetic fields in W75N. The magnetic fields around the young massive protostar VLA2 are found to be well ordered around an expanding gas shell (Surcis *et al.* 2011b). In NGC 7538–IRS 1 the water masers did not show significant Zeeman splitting while the 6.7 GHz methanol masers indicated a possible range of magnetic field strength of 50 mG $< |B_\parallel| < 500$ mG, depending on the value of Zeeman-splitting coefficient. These masers likely are related to the outflow and the interface between infall and the large-scale torus, respectively (Surcis *et al.* 2011a).

10. Summary

It is clear that many new results and discoveries have been obtained in our attempts to understand star formation and the associated masers. Are we closer to answer the key-questions? What is the current state of masers in SFRs?

We note, that:

• The time sequence for masers seems to take shape and is confirmed by (most) new methanol transition measurements, however, some issues still exist,

• Convergence can be seen on the issue where water, methanol (both Classes), OH and SiO arise, but testing the hypotheses with high-resolution observations and at other wavelengths is critical,

• One may hope that in synergy with ALMA the role of masers to study small scale dynamics will be strengthened,

   • Monitoring programs are starting to give important clues about co-evolving binaries,
   • There should be more focus given to low-mass stars,
   • More work on models to get accurate physical conditions is needed.

We started the review with the popular statement that *masers are an unique tool in star-formation studies* and in the end we are more convinced that they really are. In order to advance the use of masers and solve the detailed questions we *just* need better instruments and of course the patience to work on data to obtain more and more interesting and even surprising results. We all can wait (and work) with curiosity for the further discoveries that will be presented in the next maser symposium.

Acknowledgements

AB acknowledges support by the Polish Ministry of Science and Higher Education through grant N N203 386937.

References

Araya, E. D., Hofner, P., Goss, W. M., *et al.* 2010, *ApJ*, 717, L133
Bae, J. H., Kim, K. T., Youn, S. Y., *et al.* 2011, *ApJSS*, 196, 21
Bartkiewicz, A., Szymczak, M., & van Langevelde, H. J. 2005, *A&A*, 442, L61
Bartkiewicz, A., Szymczak, M., van Langevelde, H. J., Richards, A. M. S., & Pihlström, Y. M. 2009, *A&A*, 502, 155
Beuther, H., Walsh, A. J., & Longmore, S. N. 2009, *ApJSS*, 184, 366
Breen, S. L., Caswell, J. L., Ellingsen, S. P., & Phillips, C. J. 2010a, *MNRAS*, 406, 1487
Breen, S. L., Ellingsen, S. P., Caswell, J. L., & Lewis, B. E. 2010b, *MNRAS*, 401, 2219
Breen, S. L., Ellingsen, S. P., Caswell, J. L., *et al.* 2011, *ApJ*, 733, 80
Brogan, C. L., Johnson, K., & Darling, J. 2010, *ApJ*, 716, L51
Brogan, C. L., Hunter, T. R., Cyganowski, C. J., *et al.* 2011, *ApJ*, 739, L16
Castangia, P., Tarchi, A., Henkel, C., & Menten, K. M. 2008, *A&A*, 479, 111
Caswell, J. L., Vaile, R. A., Ellingsen, S. P., Whiteoak, J. B., & Norris, R. P. 1995, *MNRAS*, 272, 96
Chen, X., Ellingsen, S. P., Shen, Z. Q., *et al.* 2011, *ApJS*, 196, 1, 9
Cragg, D. M., Johns, K. P., Godfrey, P. D., & Brown, R. D. 1992, *MNRAS*, 259, 203
Cragg D. M., Sobolev, A. M., & Godfrey, P. D. 2005, *MNRAS*, 360, 533
Cyganowski, C. J., Whitney, B. A., Holden, E., *et al.* 2008, *AJ*, 136, 2391
Cyganowski, C. J., Brogan, C. L., Hunter, T. R., & Churchwell, E. 2009, *AJ*, 702, 1615
De Buizer, J. M., Redman, R. O., Longmore, S. N., Caswell, J., & Feldman, P. A. 2009, *A&A*, 493, 127
Ellingsen, S. P., Breen, S. L., Caswell, J. L., Quinn, L. J., & Fuller, G. A. 2010, *MNRAS*, 404, 779
Ellingsen, S. P., Breen, S. L., Sobolev, A. M., *et al.* 2011, *ApJ*, 742, 109
Fish, V. L. 2007, *Masers and star formation. Proceedings of the International Astronomical Union*, 3, 71
Fish, V. L., Gray, M., Goss, W. M., & Richards, A. M. S. 2011a, *MNRAS*, 417, 555
Fish, V. L., Muehlbrad, T. C., Pratap, P., *et al.* 2011b, *ApJ*, 729, 14
Fontani, F., Cesaroni, R., & Furuya, R. S. 2010, *A&A*, 517, A56
Fujisawa, K., Sugiyama, K., Aoki, N., *et al.* 2012, *PASJ*, in print, *arXiv:1109.2429*
Furuya, R. S., Cesaroni, R., & Shinnaga, H. 2011, *A&A*, 525, A72
Galván-Madrid, R., Montes, G., Ramírez, E. A., Kurtz, S., Araya, E., & Hofner, P. 2010, *ApJ*, 713, 423
Goddi, C., Moscadelli, L., & Sanna, A. 2011, *A&A*, 535, 8
Goedhart, S., Gaylard, M. J., & van der Walt, D. J. 2003, *MNRAS*, 339, 33
Goedhart, S., Gaylard, M. J., & van der Walt, D. J. 2004, *MNRAS*, 355, 553
Goedhart, S., Minier, V., Gaylard, M. J., & van der Walt, D. J. 2005, *MNRAS*, 356, 839
Goedhart, S., Langa, M. C., Gaylard, M. J., & van der Walt, D. J. 2009, *MNRAS*, 398, 995
Green, J. A., Richards, A. M. S., Vlemmings, W. H. T., Diamond, P., & Cohen, R. J. 2007, *MNRAS*, 382, 770
Green, J. A., Caswell, J. L., Fuller, G. A., *et al.* 2008, *MNRAS*, 385, 948
Green, J. A., McClure-Griffiths, N. M., Caswell, J. L., *et al.* 2009, *ApJ*, 696, L156
Kalenskii, S. V., Johansson, L. E. B., Bergman, P., *et al.* 2010a, *MNRAS*, 405, 613
Kalenskii, S. V., Kurtz, S., Slysh, V. I., *et al.* 2010b, *AR*, 54, 10, 932
Lekht, E. E., Pashchenko, M. I., & Rudnitskii, G. M. 2012, *AR*, 56, 1, 45
Lo, N., Cunningham, M., Bains, I., Burton, M. G., & Garay, G. 2007, *MNRAS*, 381, L30

Marseille, M. G., van der Tak, F. F. S., Herpin, F., & Jacq, T. 2010, *A&A*, 522, A40
Matthews, L. D., Greenhill, L. J., Goddi, C., *et al.* 2010, *ApJ*, 708, 80
Menten, K. M. 1991, *ApJ*, 380, 75
Moscadelli, L., Goddi, C., Cesaroni, R., Beltrán, M. T., & Furuya, R. S. 2007, *A&A*, 472, 867
Moscadelli, L., Reid, M. J., Menten, K. M., *et al.* 2009, *ApJ*, 693, 406
Moscadelli, L., Cesaroni, R., Rioja, M. J., Dodson, R., & Reid, M. J. 2011a, *A&A*, 526, A66
Moscadelli, L., Sanna, A., & Goddi, C. 2011b, *A&A*, 536, A38
Motogi, K., Sorai, K., Habe, A., Honma, M., Kobayashi, H., & Sato, K. 2011a, *PASJ*, 63, 31
Motogi, K., Sorai, K., Honma, M., *et al.* 2011b, *MNRAS*, 417, 238
Palau, A., Sánchez-Monge, Á., Busquet, G., *et al.* 2010, *A&A*, 510, A5
Pandian, J. D., Momjian, E., Xu, Y., Menten, K. M., & Goldsmith, P. F. 2010, *A&A*, 522, A8
Pandian, J. D., Momjian, E., Xu, Y., Menten, K. M., & Goldsmith, P. F. 2011, *ApJ*, 730, 55
Pihlström, Y. M., Sjouwerman, L. O., & Fish, V. L. 2011, *ApJ*, 739, L21
Pihlström, Y. M., Sjouwerman, L. O., & Mesler, R. A. 2011, *ApJ*, 740, 66
Pratap, P., Shute, P. A., Keane, T. C., Battersby, C., & Sterling, S. 2008, *AJ*, 135, 1718
Purcell, C. R., Minier, V., Longmore, S. N., *et al.* 2009, *A&A*, 504, 139
Sánchez-Monge, Á., Pandian, J. D., & Kurtz, S. 2011, *ApJ*, 739, L9
Sanna, A., Reid, M. J., Moscadelli, L., *et al.* 2009, *ApJ*, 706, 464
Sarma, A. P. & Momjian, E. 2009, *ApJ*, 705, 176
Sewilo, M., Churchwell, E., Kurtz, S., Goss, M. W., & Hofner, P. 2011, *ApJSS*, 194, 44
Sjouwerman, L. O., Pihlström, Y. M., & Fish, V. L. 2010a, *ApJ*, 710, L111
Sjouwerman, L. O., Murray, C. E., Pihlström, Y. M., Fish, V. L., & Araya, E. D. 2010b, *ApJ*, 724, L158
Slysh, V. I., Kalenskij, S. V., & Val'tts, I. E. 1993, *ApJ*, 413, 133
Sobolev, A. M., Ostrovskii, A. B., Kirsanova, M. S., *et al.* 2005, *IAUS 227*, 174
Sugiyama, K., Fujisawa, K., Doi, A., *et al.* 2008, *PASJ*, 60,23
Surcis, G., Vlemmings, W. H. T., Torres, R. M., van Langevelde, H. J., & Hutawarakorn Kramer, B. 2011a, *A&A*, 533, A47
Surcis, G., Vlemmings, W. H. T., Curiel, S., *et al.* 2011b, *A&A*, 527, A48
Szymczak, M., Kus, A. J., Hrynek, G., Kepa, A., & Pazderski, E. 2002, *A&A*, 392, 277
Szymczak, M., Wolak, P., Bartkiewicz, A., & van Langevelde, H. J. 2011, *A&A*, 531, L3
Torrelles, J. M., Patel, N. A., Curiel, S., Estalella, R., Gómez, J. F., *et al.* 2011, *MNRAS*, 410, 627
Torstensson, K. J. E., van Langevelde, H. J., Vlemmings, W. H. T., & Bourke, S. 2011a, *A&A*, 526, 38
Torstensson, K. J. E., van der Tak, F. F. S., van Langevelde, H. J., Kristensen, L. E., & Vlemmings, W. H. T. 2011b, *A&A*, 529, 32
Ueda, J., Iono, D., Petitpas, G., *et al.* 2012, *ApJ*, 745, 65
Urquhart, J. S., Hoare, M. G., Lumsden, S. L., Oudmaijer, R. D., Moore, T. J. T., Brook, P. R., Mottram, J. C., Davies, B., & Stead, J. J. 2009, *A&A*, 507, 795
van der Walt, D. J., Goedhart, S., & Gaylard, M. J. 2009, *MNRAS*, 398, 961
van der Walt, D. J. 2011, *AJ*, 141, 152
Vlemmings, W. H. T., Diamond, P. J., van Langevelde, H. J., & Torrelles, J. M. 2006, *A&A*, 448, 597
Vlemmings, W. H. T., Surcis, G., Torstensson, K. J.,E., & van Langevelde, H. J. 2010, *MNRAS*, 404, 134
Voronkov, M. A., Sobolev, A. M., Ellingsen, S. P., & Ostrovskii, A. B. 2005, *MNRAS*, 362, 995
Voronkov, M. A., Brooks, K. J., Sobolev, A. M., *et al.* 2006, *MNRAS*, 373, 411
Voronkov, M. A., Caswell, J. L., Britton, T. R., *et al.* 2010a, *MNRAS*, 408, 133
Voronkov, M. A., Caswell, J. L., Ellingsen, S. P., & Sobolev, A. M. 2010b, *MNRAS*, 405, 2471
Voronkov, M. A., Walsh, A. J., Caswell, J. L., *et al.* 2011, *MNRAS*, 413, 2339
Walsh, A. J., Burton, M. G., Hyland, A. R., & Robinson, G. 1998, *MNRAS*, 301, 640
Wu, Y. W., Xu, Y., Yang, J., & Jing-Jing, L. 2009, *RAA*, 9, 12, 1343
Wu, Y. W., Xu, Y., Pandian, J. D., *et al.* 2010, *ApJ*, 720, 392

Cosmic Masers - from OH to H₀
Proceedings IAU Symposium No. 287, 2012
R.S. Booth, E.M.L. Humphreys & W.H.T. Vlemmings, eds.

© International Astronomical Union 2012
doi:10.1017/S1743921312006783

Masers in GLIMPSE Extended Green Objects (EGOs)

**Claudia J. Cyganowski[1], Crystal L. Brogan[2], Todd R. Hunter[2],
Ed Churchwell[3], Jin Koda[4], Erik Rosolowsky[5], Sarah Towers[4],
Barb Whitney[3], and Qizhou Zhang[6]**

[1] Harvard-Smithsonian Center for Astrophysics,
Cambridge, MA 02138 USA
NSF Astronomy and Astrophysics Postdoctoral Fellow
email: `ccyganowski@cfa.harvard.edu`

[2] National Radio Astronomy Observatory, Charlottesville, VA 22902 USA

[3] Department of Astronomy, University of Wisconsin-Madison, Madison, WI, 53706, USA

[4] Department of Physics and Astronomy, Stony Brook University, Stony Brook, NY 11794, USA

[5] Department of Physics and Astronomy, University of British Columbia, Okanagan, Kelowna
BC, Canada

[6] Harvard-Smithsonian Center for Astrophysics, Cambridge, MA 02138 USA

Abstract. Large-scale *Spitzer* surveys of the Galactic plane have yielded a new diagnostic
for massive young stellar objects (MYSOs) that are actively accreting and driving outflows:
extended emission in the IRAC 4.5 μm band, believed to trace shocked molecular gas. Maser
studies of these extended 4.5 μm sources (called EGOs, Extended Green Objects, for the common
coding of 3-color IRAC images) have been and remain crucial for understanding the nature of
EGOs. High detection rates in VLA CH$_3$OH maser surveys provided the first proof that EGOs
were indeed MYSOs driving outflows; our recent Nobeyama 45-m survey of northern EGOs
shows that the majority are associated with H$_2$O masers. Maser studies of EGOs also provide
important constraints for the longstanding goal of a maser evolutionary sequence for MYSOs,
particularly in combination with high resolution (sub)mm data. New SMA results show that
Class I methanol masers can be excited by both young (hot core) and evolved (ultracompact
HII region) sources within the same massive star-forming region.

Keywords. infrared: ISM– infrared: stars– ISM: jets and outflows– ISM: molecules– masers–
radio continuum: ISM– stars: formation– techniques: interferometric

1. Introduction

What are GLIMPSE Extended Green Objects (EGOs)? GLIMPSE–the Galactic Legacy
Infrared Mid-Plane Survey Extraordinaire (Churchwell *et al.* 2009)–is a *Spitzer Space
Telescope* survey of the Galactic Plane in the four bands of the Infrared Array Camera
(IRAC; Fazio *et al.* 2004): 3.6, 4.5, 5.8, and 8.0 μm. The broad IRAC bands include emission features from a range of interstellar species (e.g. Fig. 1 of Reach *et al.* 2006). All of
the IRAC bands include H$_2$ lines; the 4.5 μm band also includes the CO (v = 1-0) band-
head and Brα. Notably, 4.5 μm is also the only IRAC band to *lack* polycyclic aromatic
hydrocarbon (PAH) emission features. This combination of characteristics means that
in massive star-forming regions, shock-excited gas in (proto)stellar molecular outflows
can stand out as morphologically distinct, extended 4.5 μm emission. Thus, the ability
to search for extended 4.5 μm emission in GLIMPSE images presented an exciting opportunity to compile a new sample of candidate massive young stellar objects (MYSOs)

127

with *active* outflows, independent of previous CO imaging surveys based largely on IRAS point sources (e.g. Zhang *et al.* 2001). Because the 4.5 μm band is commonly coded as green in three-color IRAC images (RGB: 8.0, 4.5, 3.6 μm), these extended 4.5 μm sources are known as Extended Green Objects (EGOs; Cyganowski *et al.* 2008).

Cyganowski *et al.* (2008) cataloged over 300 EGOs in the GLIMPSE-I survey area ($10° \leqslant |l| \leqslant 65°$, $|b| \leqslant 1°$). Based on their mid-infrared (MIR) colors and association with infrared dark clouds (IRDCs), Cyganowski *et al.* (2008) suggested that EGOs were specifically *massive* YSOs driving active outflows.

2. The Nature of EGOs (as revealed by CH$_3$OH masers)

The first step, after the identification of EGOs as a class of objects by Cyganowski *et al.* (2008), was to test whether GLIMPSE EGOs were in fact massive YSOs driving outflows. To do this, Cyganowski *et al.* (2009) conducted sensitive, high-angular resolution searches for two diagnostic types of CH$_3$OH masers towards a sample of EGOs with the Very Large Array (VLA)†: 6.7 GHz Class II CH$_3$OH masers, associated exclusively with *massive* YSOs (e.g. Minier *et al.* 2003) and 44 GHz Class I CH$_3$OH masers, associated with molecular outflows and outflow-cloud interfaces (e.g. Kurtz *et al.* 2004). An initial sample of 28 EGOs was selected to (1) cover a range of MIR properties, including morphology, angular extent of 4.5 μm emission, and the presence of 8 and/or 24 μm counterparts, and (2) be visible from the northern hemisphere. Due to technical problems, however, there is a bias in the final survey sample: only 19 sources with (strong) 6.7 GHz masers near the phase center were reobserved at 6.7 GHz, and only these 19 sources were observed at 44 GHz (see Cyganowski *et al.* 2009 for details). As a result, the Cyganowski *et al.* (2009) sample is, in essence, a 6.7 GHz CH$_3$OH maser-selected EGO subsample.

The detection rates for both 6.7 GHz Class II and 44 GHz Class I CH$_3$OH masers were extremely high: >64% (of the original 28 sources) and $\sim$ 90% (of the 19 observed sources), respectively. For 6.7 GHz CH$_3$OH masers, this detection rate was nearly twice that towards other MYSO samples (see also Cyganowski *et al.* 2009). The spatial distribution and kinematics of the two types of CH$_3$OH masers are strikingly different. The 6.7 GHz masers are centrally concentrated and usually coincident with 24 μm emission, while the 44 GHz masers are spatially distributed, often over tens of arcseconds, and coincident with 4.5 μm emission. To complement the maser surveys and provide information about the thermal gas emission, Cyganowski *et al.* (2009) used the James Clerk Maxwell Telescope (JCMT) ‡ to observe HCO$^+$(3-2), H^{13}CO$^+$(3-2), thermal CH$_3$OH ($5_{2,3}$-$4_{1,3}$), and SiO(5-4) emission. The velocities of the 44 GHz masers cluster near the systemic velocity (as measured from the dense gas tracers observed with the JCMT), consistent with their excitation at interfaces between outflows and the surrounding molecular gas. In contrast, the velocities of the 6.7 GHz masers exhibit every possible permutation with respect to the thermal gas v$_{LSR}$: different sources provide examples of 6.7 GHz masers at and near the systemic velocity, only redshifted, only blueshifted, and both red and blueshifted but *not* at the systemic velocity. This diversity suggests that the 6.7 GHz masers observed towards this EGO sample do not all arise in any single physical/dynamical structure.

The JCMT survey also provided further evidence, in addition to the Class I masers, that the target EGOs were associated with outflows. For all sources, the HCO$^+$ spectra

---

† The National Radio Astronomy Observatory operates the VLA and is a facility of the National Science Foundation operated under agreement by the Associated Universities, Inc.

‡ The JCMT is operated by The Joint Astronomy Centre on behalf of the Science and Technology Facilities Council of the United Kingdom, the Netherlands Organisation for Scientific Research, and the National Research Council of Canada.

showed broad line wings, and the detection rate for thermal SiO emission was very high: 90% (9 of a subset of 10 EGOs). Gas-phase SiO abundance is enhanced for only $\sim 10^4$ years after a shock (e.g. Pineau de Forets *et al.* 1997), so the SiO emission indicates that the outflows are being actively driven, as has been observationally confirmed in a survey comparing low-mass Class 0 and I sources (Gibb *et al.* 2004). High-resolution mm-λ followup observations have confirmed the presence of bipolar molecular outflows in EGOs. Submillimeter Array (SMA)¶ imaging of the EGOs G11.92−0.61 and G19.01−0.03 reveals high-velocity, well-collimated outflows traced by ^{12}CO(2-1) emission (Cyganowski *et al.* 2011a). The CO outflows are coincident with the extended 4.5 μm emission in these sources, and many of the 44 GHz Class I CH$_3$OH masers coincide with or trace the edges of the high-velocity CO outflow lobes.

In sum, the initial CH$_3$OH maser studies of EGOs provided strong evidence that the survey targets were young MYSOs with active outflows, and so presumably ongoing accretion. High resolution observations of direct tracers of molecular outflows and hot cores have confirmed this picture for EGOs observed to date (Cyganowski *et al.* 2011a), and further high-resolution mm-λ followup is ongoing.

It is worth emphasizing that extended 4.5 μm emission acts as an effective way to find MYSOs *in GLIMPSE*, and that this is to an important extent a function of the shallowness of the GLIMPSE-I survey. Outflows in nearby low-mass star-forming regions exhibit extended 4.5 μm emission in *Spitzer* images (e.g. Noriega-Crespo *et al.* 2004, Velusamy *et al.* 2007). However, the extended 4.5 μm emission from such low-mass outflows is too faint to be detected in GLIMPSE (see also discussion in Cyganowski *et al.* 2008). In deeper *Spitzer* Galactic Plane surveys (such as the GLIMPSE-360 survey of the outer Galaxy, and the ongoing Deep GLIMPSE), extended 4.5 μm emission will pick out outflows, including those from low- and intermediate- mass YSOs.

Millimeter Methanol Masers. The SMA observations of G11.92−0.61 and G19.01−0.03 also revealed probable CH$_3$OH maser emission in the 229.759 GHz Class I transition (Cyganowski *et al.* 2011a). The observed 229.759 GHz emission is spectrally narrow, and spatially and spectrally coincident with 44 GHz masers. While the angular resolution of the SMA data is not sufficient to definitively establish masing based on brightness temperature, the 229.759/230.027 line ratios indicate nonthermal emission. Our ongoing SMA observations of EGOs indicate that 229.759 GHz Class I masers may be common towards MYSO outflows. If so, such masers offer potential for self-calibrating very high-resolution 1.3 mm observations of MYSOs, e.g. with ALMA.

Evolutionary State. Constraining the evolutionary state of EGOs is crucial for placing maser studies of EGOs in the broader context of maser studies of massive star-forming regions. In particular, the presence of an ultracompact (UC) HII region signals a comparatively late stage of massive star formation, when a young O/B star has already ionized its environment. While it is important to remember that high resolution (sub)millimeter studies of UC HIIs typically reveal additional objects in a range of evolutionary stages (Hunter *et al.* 2008; Brogan *et al.* 2008; Hunter *et al.* 2004), the absence/presence of UC HIIs has been used in the past as a key divide between younger/older MYSO samples. To search for UC HIIs, shallow 44 GHz (7 mm) continuum data were obtained simultaneously during the 44 GHz maser survey. The *non*detection rate was 95%, but the sensitivity was sufficient to rule out only bright UC HII regions (Cyganowski *et al.* 2009). To better constrain the cm-λ emission of EGOs, Cyganowski *et al.* (2011b) carried out deep VLA

¶ The Submillimeter Array is a joint project between the Smithsonian Astrophysical Observatory and the Academia Sinica Institute of Astronomy and Astrophysics and is funded by the Smithsonian Institution and the Academia Sinica.

3.6 and 1.3 cm continuum observations of a sample of 14 EGOs associated with 6.7 GHz CH_3OH masers, 44 GHz CH_3OH masers, or both. The surveys had angular resolution of $\sim 1''$ and $\sigma \sim 30\ \mu$Jy beam^{-1} (3.6 cm) and 250 μJy beam^{-1} (1.3 cm). The *non*detection rate of these VLA surveys was 57% (8/14). The cm-λ EGO counterparts that were detected were generally weak ($\lesssim 1$ mJy at 3.6 cm), and not detected at 1.3 cm due to the higher noise. Only 2 EGOs are associated with ultracompact or compact HII regions; both show cm-λ multiplicity, with morphological evidence that a less evolved source may be driving the 4.5 μm outflow. One EGO is detected only at 1.3 cm; comparison with 1.4 mm data suggests that the 1.3 cm emission is free-free from an optically thick hypercompact (HC) HII region (Cyganowski *et al.* 2011b). This EGO (G11.92−0.61) in fact hosts a protocluster of 3 compact mm continuum cores (Cyganowski *et al.* 2011a). The 1.3 cm source is associated with a 1.4 mm continuum core, hot core line emission, 24 μm emission, and H_2O and 6.7 GHz CH_3OH masers.

3. Masers in EGOs: Insights for masers in massive star forming regions

In addition to constraining the nature of EGOs themselves, the wealth of multiwavelength data now available for many EGOs allows us to examine cross-correlations between different maser types and other star formation indicators (see also Brogan *et al.* 2011 and in this volume).

Class I CH_3OH Masers as Evolutionary Indicators? The EGO G18.67+0.03 is particularly interesting in the context of proposed maser evolutionary sequences, which posit that the youngest MYSOs exhibit only Class I CH_3OH maser emission (e.g. Ellingsen 2006, Ellingsen *et al.* 2007, Breen *et al.* 2010). Our VLA CH_3OH maser surveys revealed maser emission associated with 3 MIR sources in this region: two (including the EGO) have both 6.7 GHz Class II and 44 GHz Class I CH_3OH masers, while the third MIR source has only 44 GHz Class I masers (Cyganowski *et al.* 2009). The Class I maser-only MIR source has a cm-λ continuum counterpart, with properties consistent with a UC HII region (Cyganowski *et al.* 2011b). None of the other sources in the field have cm-λ continuum emission in our deep VLA surveys. The association of 44 GHz CH_3OH masers with a UC HII region in G18.67+0.03 is consistent with recent suggestions by Voronkov *et al.* (2010) that Class I masers may be excited by shocks driven by expanding HII regions. However, younger sources (such as hot cores) are often found in close proximity to UC HII regions (e.g. Hunter *et al.* 2006; Cyganowski *et al.* 2007 and references therein); therefore, additional multiwavelength information was needed. We conducted SMA 1.3 mm observations to determine the evolutionary states of the G18.67+0.03 maser sources based on their molecular line emission.

Our SMA observations reveal hot core line emission towards both Class II+Class I maser sources (Cyganowski *et al.* in prep.). In contrast, emission from only a few common species is detected towards the UC HII region/Class I maser-only source. All three maser sources are associated with molecular outflows traced by ^{13}CO (2-1) emission. However, the UC HII region outflow is *not* detected in SiO(5-4). As described above, SiO emission provides a discriminant between active (SiO present) and "relic" or "fossil" outflows (SiO absent). Thus, the lack of SiO emission is strong additional evidence that no younger YSO is present near the Class I-only maser source, and that these Class I masers are associated with the comparatively evolved UC HII region.

H_2O maser (and NH_3) Survey of Northern EGOs. Most maser studies of EGOs to date have focused on CH_3OH masers. For the broader picture of masers in high-mass star forming regions, H_2O masers in EGOs are an important missing piece of the puzzle. We

have recently completed a Nobeyama 45-m survey of all 94 northern EGOs ($\delta \gtrsim -20°$) for H_2O maser and $NH_3(1,1)$, (2,2), and (3,3) emission (Cyganowski *et al.*, in prep.). The median rms of the H_2O maser survey is ~0.11 Jy. The goals of this survey were (1) to evaluate the significance of the MIR categories from the Cyganowski *et al.* (2008) EGO catalog, (2) to compare the H_2O maser properties of EGO subsamples (for example, those associated with Class I/II CH_3OH masers), and (3) to compare H_2O maser properties with clump physical properties from other tracers (T_{kin}, density, etc.).

The overall H_2O maser detection rate in our Nobeyama survey is ~68%. Cyganowski *et al.* (2008) classified EGOs as "likely" or "possible" outflow candidates based on MIR morphology; they also tabulated whether each EGO was or was not associated with an IRDC in the GLIMPSE images. The H_2O maser detection rate is somewhat higher towards EGOs that are "likely" outflow candidates, and roughly comparable towards EGOs that are and are not associated with IRDCs. We find little evidence of statistically significant differences in the H_2O maser properties (for example, the isotropic maser luminosity) of various EGO subsamples.

Recent studies have found correlations between the isotropic H_2O maser luminosity and the properties of the driving source or surrounding clump. Urquhart *et al.* (2011) found a positive correlation between H_2O maser luminosity and bolometric luminosity for both HII regions and MYSOs from the Red *MSX* Source (RMS) sample, while Breen & Ellingsen (2011) report an anticorrelation between H_2O maser luminosity and the H_2 number density of the associated clump. Breen & Ellingsen (2011) attribute this anticorrelation to an evolutionary effect, but caution that the clump densities were calculated assuming a single temperature for all clumps. Our Nobeyama survey and the 1.1 mm Bolocam Galactic Plane Survey (BGPS; Aguirre *et al.* 2011, Rosolowsky *et al.* 2010) together provide the data necessary to test this evolutionary interpretation and explore connections between H_2O maser and clump properties: H_2O maser spectra, T_{kin} measurements from NH_3, and clump properties from 1.1 mm dust continuum emission. Combining these datasets, we find no correlation between H_2O maser luminosity and clump density. There is a weak correlation between H_2O maser luminosity and clump temperature (T_{kin}). This is consistent with the correlation of maser luminosity with bolometric luminosity found by Urquhart *et al.* (2011): the more luminous the central source, the more it will heat the surrounding clump.

4. Conclusions

In sum, maser studies have been instrumental in showing that GLIMPSE EGOs are associated with MYSOs driving active, massive outflows. Sensitive VLA surveys yielded exceptionally high detection rates for CH_3OH Class I and II masers. In a somewhat less sensitive ($\sigma \sim 0.11$ Jy) Nobeyama 45-m H_2O maser survey, the detection rate is slightly higher than in similar searches towards other MYSO samples. By and large, cm continuum emission towards EGOs is weak, with most harboring an earlier phase of massive star formation than UC HII regions. Our SMA case study of G18.67+0.03 shows that Class I CH_3OH masers can be excited by both young (hot core) and older (UC HII) sources in the same massive star-forming region, indicating that simple evolutionary cartoons are probably not realistic. In our EGO H_2O maser survey, we see no trends in H_2O maser luminosity with clump density, association with IRDCs, or with CH_3OH maser type (Class I or II or both). There is a weak trend in H_2O maser luminosity with clump temperature, consistent with the L_{H_2O} vs. L_{bol} correlation.

Acknowledgements

Support for this work was provided by NSF grant AST-0808119. C.J.C. was partially supported during this work by a National Science Foundation Graduate Research Fellowship, and is currently supported by an NSF Astronomy and Astrophysics Postdoctoral Fellowship under award AST-1003134.

References

Aguirre, J. E., Ginsburg, A. G., Dunham, M. K., *et al.* 2011, *ApJS*, 192, 4
Breen, S. L., Ellingsen, S. P., Caswell, J. L., & Lewis, B. E. 2010, *MNRAS*, 401, 2219
Breen, S. L. & Ellingsen, S. P. 2011, *MNRAS*, 416, 178
Brogan, C. L., Hunter, T. R., Cyganowski, C. J., *et al.* 2011, *ApJ Letters*, 739, L16
Brogan, C. L., Hunter, T. R., Indebetouw, R., *et al.* 2008, *Ap&SS*, 313, 53
Churchwell, E., Babler, B. L., Meade, M. R., *et al.* 2009, *PASP*, 121, 213
Cyganowski, C. J., Brogan, C. L., & Hunter, T. R. 2007, *AJ*, 134, 346
Cyganowski, C. J., Whitney, B. A., Holden, E., *et al.* 2008, *AJ*, 136, 2391
Cyganowski, C. J., Brogan, C. L., Hunter, T. R., & Churchwell, E. 2009, *ApJ*, 702, 1615
Cyganowski, C. J., Brogan, C. L., Hunter, T. R., Churchwell, E., & Zhang, Q. 2011a, *ApJ*, 729, 124
Cyganowski, C. J., Brogan, C. L., Hunter, T. R., & Churchwell, E. 2011b, *ApJ*, 743, 56
Ellingsen, S. P. 2006, *ApJ*, 638, 241
Ellingsen, S. P., Voronkov, M. A., Cragg, D. M., *et al.* 2007, in: J. M. Chapman & W. A. Baan (eds.), *Astrophysical Masers & their Environments*, Proc. IAU Symposium No. 242 (Cambridge, UK: Cambridge University Press), p. 213
Fazio, G. G., Hora, J. L., Allen, L. E., *et al.* 2004, *ApJS*, 154, 10
Gibb, A. G., Richer, J. S., Chandler, C. J., & Davis, C. J. 2004, *ApJ*, 603, 198
Hunter, T. R., Brogan, C. L., Indebetouw, R., & Cyganowski, C. J. 2008, *ApJ*, 680, 1271
Hunter, T. R., Brogan, C. L., Megeath, S. T., *et al.* 2006, *ApJ*, 649, 888
Hunter, T. R., Zhang, Q., & Sridharan, T. K. 2004, *ApJ*, 606, 929
Kurtz, S., Hofner, P., & Álvarez, C. V. 2004, *ApJS*, 155, 149
Minier, V., Ellingsen, S. P., Norris, R. P., & Booth, R. S. 2003, *A&A*, 403, 1095
Noriega-Crespo, A., Morris, P., Marleau, F. R., *et al.* 2004, *ApJS*, 154, 352
Pineau des Forets, G., Flower, D. R., & Chieze, J.-P. 1997, in: B. Reipurth & C. Bertout (eds.), *Herbig-Haro Flows and the Birth of Low-Mass Stars* Proc. IAU Symposium No. 182 (Boston: Kluwer Academic Publishers), p. 199
Reach, W. T., Rho, J., Tappe, A., *et al.* 2006, *AJ*, 131, 1479
Rosolowsky, E., Dunham, M. K., Ginsburg, A., *et al.* 2010, *ApJS*, 188, 123
Urquhart, J. S., Morgan, L. K., Figura, C. C., *et al.* 2011, *MNRAS*, 418, 1689
Velusamy, T., Langer, W. D., & Marsh, K. A. 2007, *ApJ Letters*, 668, L159
Voronkov, M. A., Caswell, J. L., Ellingsen, S. P., & Sobolev, A. M. 2010, *MNRAS*, 405, 2471
Zhang, Q., Hunter, T. R., Brand, J., *et al.* 2001, *ApJ Letters*, 552, L167

Cosmic Masers - from OH to H$_0$
Proceedings IAU Symposium No. 287, 2012
R.S. Booth, E.M.L. Humphreys & W.H.T. Vlemmings, eds.

© International Astronomical Union 2012
doi:10.1017/S1743921312006795

44 GHz Methanol Maser Surveys

S. E. Kurtz

Centro de Radioastronomía y Astrofísica, Universidad Nacional Autonoma de México
(Morelia, Michoacán, México)
email: s.kurtz@crya.unam.mx

Abstract. Class I 44 GHz methanol masers are not as well-known, as common, or as bright as their more famous Class II cousins at 6.7 and 12.2 GHz. Nevertheless, the 44 GHz masers are commonly found in high-mass star forming regions. At times they appear to trace dynamically important phenomena; at other times they show no obvious link to the star formation process. Here, we summarize the major observational efforts to date, including both dedicated surveys and collateral observations. The principal results are presented, some that were expected, and others that were unexpected.

Keywords. masers, surveys, stars:formation

1. The Early Days

The first detection of maser emission in the 44 GHz methanol line was reported in the literature by Morimoto *et al.* (1985). Emission was found in four galactic sources, all of them regions hosting high-mass star formation. It wasn't immediately clear if the 44 GHz line (the $7_0 - 6_1$ A$^+$ transition) was a class I or class II maser. In fact, determining to which maser class it belonged was one of the chief motivations of the early surveys. The first survey, made by Haschick, Menten & Baan (1990) with the Haystack 37-m telescope, targeted 50 galactic star formation regions. Their 50% detection rate not only identified many new masers, but confirmed the 44 GHz transition as belonging to class I, along with the 25, 36, 84, and 95 GHz lines. Until it was largely taken over by the U.S. Air Force, the Haystack telescope continued to be productive for methanol maser surveys, most notably that of Pratap *et al.* (2008).

Additional northern hemisphere single-dish surveys were reported by Bachiller *et al.* (1990) and Kalenskii *et al.* (1992). Both of these surveys used the Yebes 14-m telescope. The former targeted 124 regions of known water maser emission and yielded a 13% detection rate; all of the maser detections were associated with compact HII regions. The latter targeted 137 cold IRAS sources and had a frigid detection rate of 2%, detecting only 3 masers.

A southern single-dish survey was reported by Slysh *et al.* (1994) who observed with the Parkes 64-m telescope. Their sample of 250 objects consisted mostly of HII regions and sources known to present maser emission in some other molecular species; they had a 22% detection rate, discovering 25 masers.

These early single-dish surveys produced a sample of about 110 known 44 GHz masers and clearly established this transition as being class I, and arising in massive star formation regions. All four of these surveys, and several later interferometric surveys, have been nicely cataloged by Val'tts & Larionov (2007).

2. Follow-up Interferometric and Single-Dish Surveys

The single-dish surveys laid the crucial groundwork for establishing the nature of 44 GHz masers, but of course they lacked the angular resolution needed for more detailed studies. Interferometric surveys became possible when the VLA was outfitted with 7 mm receivers in the mid-1990s.

Kogan & Slysh (1998) published positions for masers in eight well-known star formation regions that had been identified in the early single-dish surveys. They noted the appearance of 44 GHz masers in regions where class II methanol masers had been reported, suggesting that some of the more empirical aspects of the early class I/II definitions might need revision.

A similar conclusion was reached in a second VLA survey, reported by Kurtz, Hofner & Vargas (2004). They observed 44 galactic star formation regions with an 84% detection rate. The heterogeneous nature of their sample limits the significance of this detection rate; in particular, about one third of their targets were selected from the detections reported by Bachiller *et al.* (1990).

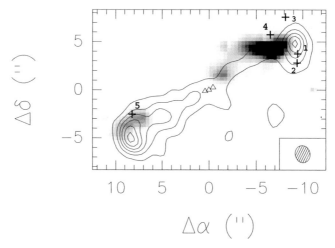

Figure 1. The IRAS 20126+4104 molecular outflow. The greyscale shows H_2 emission (Cesaroni *et al.* 1997) while the contours show SiO emission (Cesaroni *et al.* 1999). The 44 GHz methanol masers reported by Kurtz *et al.* (2004), shown as crosses, clearly coincide with the edges of the outflow lobes, where the molecular outflow interacts with the ambient medium, as was speculated by Plambeck & Menten (1990).

In some cases the interferometric surveys simply confirmed what was already known from the earlier, single-dish surveys. For example, Kurtz *et al.* (2004) found evidence for thermal methanol emission in some sources, but such emission was already well-known from the Haschick *et al.* (1990) survey. Interferometric observations, however, did permit the location of this thermal emission with respect to other star formation tracers. For example, Araya *et al.* (2008) used thermal methanol emission to trace the hot molecular core in G31.41+0.31, which is spatially offset from both the ultracompact HII region and the masers. An additional example of affirming earlier results was the discovery of the close association of methanol masers with the outflow lobes of IRAS 20126+4104. Such an association was anticipated by Plambeck & Menten (1990) who originally proposed that class I masers arise in molecular gas shocked by outflows (see Fig. 1).

Kurtz *et al.* (2004) also found evidence for the maser–outflow association in DR21, locating two distinct lobes of maser emission, with a velocity separation of about

5 km s^{-1}. The intriguing aspect of this outflow is that the masers themselves seem to define the outflow lobes. Their finding was followed up by more sensitive maser and centimeter continuum observations reported by Araya *et al.* (2009), who found four maser arcs (two blue-shifted and two red-shifted) and speculated that there were two distinct outflow events, each one producing a pair of maser arcs (see Fig. 2).

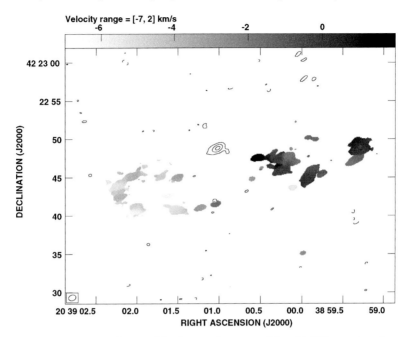

Figure 2. The DR21 region shows a bipolar outflow traced by 44 GHz methanol masers. The red lobe (toward the west) and the blue lobe (toward the east) have a velocity difference of about 5 km s^{-1}. The grey scale shows the first moment map while contours show the 7 mm continuum emission. The figure is adapted from Araya *et al.* (2009).

In other cases, the interferometric surveys produced unexpected results. For example, Kurtz *et al.* (2004) found a median projected separation of 0.2 pc between the 44 GHz masers and the compact HII regions in their sample (see Fig. 3). This was substantially closer than the nominal 1 pc typically quoted for class I masers.

3. Methanol Masers in Low-Mass Star Forming Regions

Until recently, methanol masers were thought to arise *only* in high-mass star formation regions — unlike the far more ubiquitous water masers, which occur in both low- and high-mass star forming regions (and indeed, in evolved objects as well). Although class II masers still appear to be unique to high-mass star formation regions, recent surveys by Kalenskii *et al.* (2006, 2010a) have identified several 44 GHz maser candidates. Their two single-dish surveys, made with the Onsala 20-m telescope, targeted nearby, chemically rich outflow regions. Although 39 regions had no emission at the 3–5 Jy level, four regions did show 44 GHz emission: NGC1333 I2, NGC1333 I4A, HH25 MMS, and L1157. So far only the L1157 region has been observed interferometrically (Kalenskii *et al.* 2010b); their VLA observations confirm the maser nature of the emission (see Fig. 4).

Kalenskii *et al.* (2010b) developed a model in which the maser emission arises in a collapsing clump of molecular gas. Although the model qualitatively explains the L1157

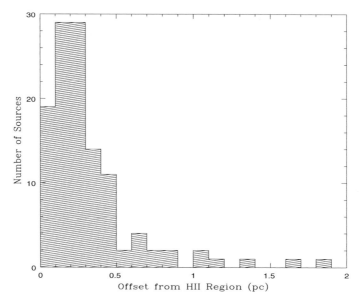

Figure 3. Histogram of the projected distance between the 44 GHz maser position and the geometric center of nearby ionized gas emission, from Kurtz *et al.* (2004). The median projected distance is 0.2 pc — substantially less than the nominal distance of 1 pc normally used in the class I maser definition.

maser, they are hesitant to suggest it as a general explanation for low-mass methanol masers because none of the other sources present a double-line, red-asymmetry profile (see Kalenskii *et al.* this volume). Based on their single-dish survey they argue that a minimum methanol column density of 10^{14} cm^{-2} is required for maser emission. One result of their work is to break the strangle-hold of high-mass star forming regions on class I masers. In a sense it is a loss to no longer have class I masers as a *unique* indicator of high-mass regions. Hopefully there will be some compensation by having nearby masers in less-complicated environments (compared to high-mass star forming regions) thus facilitating their study. In any case, the upper limit of a 10% detection rate implies that although class I masers my not be unique to high-mass star forming regions, they certainly are not a common phenomenon in low-mass star forming regions.

4. More Recent Single-Dish Surveys

Two recent single-dish surveys of the 44 GHz methanol maser line are particularly noteworthy. One of these, Bae *et al.* (2011), a simultaneous water plus methanol survey of 180 young stellar objects, made with a 22-meter dish of the Korean VLBI network, is described in detail elsewhere in this volume by K. T. Kim.

The other survey, Fontani *et al.* (2010), observed a sample of about 300 young stellar objects (YSOs) in the 6.7 GHz (class II) methanol line with the Effelsberg telescope, and a randomly selected sub-sample of 88 objects in the 44 and 95 GHz (class I) lines with the Nobeyama 45-m. A statistical comparison of their data for the different lines suggests that the 95 GHz maser is intrinsically fainter, and that the class I masers are spread over a larger sky area than the class II masers. More significantly, as we will discuss below, they find a higher 44 GHz maser detection rate in older objects (48% ±8%), and a lower detection rate in younger objects (17% ± 5%), suggesting that class I masers preferentially arise in *older* regions.

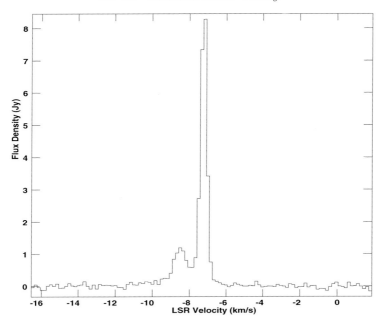

Figure 4. Spectrum of two of the 44 GHz masers in the low-mass star formation region L1157, reported by Kalenskii *et al.* (2010b). The stronger peak corresponds to source M1a while the weaker peak is M1b. Note that both the velocity and the brightness axes in their paper were incorrectly labeled. The correct units for the brightness scale are Jy beam^{-1}, giving brightness temperatures in excess of 2000 K for the brighter maser component.

5. Two New Interferometric Surveys

Two well-known and well-studied samples of high mass protostellar objects (HMPOs) and/or YSOs have emerged in the past 10 years: the samples of Molinari *et al.* (1996) and of Sridharan *et al.* (2002). The Molinari sample consists of 163 IRAS sources, half with colors of cold clouds, half with colors of ultracompact (UC) HII regions. The sample has been the subject of many additional studies, including a search for molecular outflows by Zhang *et al.* (2005) that identified 35 outflows in a sub-sample of 69 objects.

A second sample, also of 69 objects, was identified by Sridharan *et al.* (2002). Their selection criteria included colors indicative of UC HII regions but with 6 cm flux densities lower than 25 mJy, and the presence of molecular gas. Thus, like the Molinari sample, the Sridharan sample identifies young (possibly proto-) stellar objects.

We have completed a 44 GHz maser survey of the Molinari/Zhang sample and are mid-way through observations of the Sridharan sample. Both surveys are made with the VLA, with angular resolutions of about 2″ and 3σ detection limits of about 0.15 Jy. The Molinari/Zhang sample will soon be submitted (Gómez-Ruíz *et al.* in prep.) and here we mention several results from that survey.

As in the Kurtz *et al.* (2004) survey, some findings were no surprise. For example, Fig. 5 shows a histogram of the relative velocity between the masers and the systemic velocity of the region. The narrow velocity range found in this survey is consistent with earlier findings that class I masers usually occur at velocities very close to systemic.

An unexpected result of this survey is that the masers appear to be slightly closer to the ionized gas (when present) than they are to the IRAS positions. The trend is not particularly pronounced, but on average the masers appear to be about 0.1 pc closer to the ionized gas than to the embedded infrared sources. The IRAS positional uncertainty

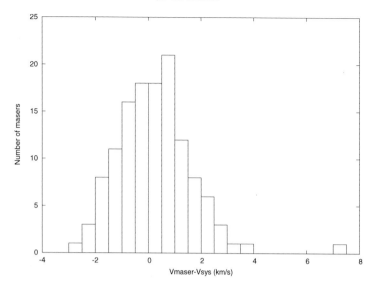

Figure 5. Histogram of the relative velocity between the maser component and the systemic velocity of the region. The latter is traced by ammonia velocities, as reported by Molinari *et al.* (1996). Apart from one outlier, the distribution is fairly narrow, and approximately centered on zero.

is of the same order as the statistical difference in the separations, so it isn't clear how robust the result is; a comparison with Spitzer images would definitely be worthwhile.

It's reasonable to expect that the physical conditions might be different closer to the UCHII regions; for example, the dust is likely to be hotter, and the grain size distribution is likely modified. But changes in the dust would most likely affect the radiation field, and class I masers are thought to be collisionally — not radiatively — pumped.

One way to understand how the difference in distances might arise is shown in Fig. 6. If the masers arise close to an outflow interaction region, then their distance to ionized emission *or* the dust emission could be rather meaningless; whichever of these two components is closer to the outflow would show smaller offsets from the maser, but without implying any physical connection between them. As noted by Hoare *et al.* (2007), emission from ionized gas and from warm dust are generally not spatially correlated in massive star forming regions. Because the two components generally do not coincide, some difference in separation is expected.

A distinct surprise in the Gómez-Ruíz survey was the maser detection rate when the sample was divided by source type. In particular, the 40% *overall* detection rate of the survey changes when the sample is separated by those sources with UC HII regions (presumably older) and those without UC HII regions (presumably younger or HMPO). The detection rates are 59% for the former and 35% for the latter. The same trend is seen by Fontani *et al.* (2010) who, as mentioned above, report a 48% ± 8% detection rate for the "high" sources and 17% ± 5% for the "low" sources.

These trends are in sharp contrast to the phenomenally high 90% detection rate for 44 GHz masers in EGOs (Extended Green Objects) reported by Cyganowski *et al.* (2009). EGOs are thought to be in a fairly young evolutionary state. And indeed, deep centimeter continuum observations reported by Cyganowski *et al.* (2011) found compact/ultracompact HII regions in only 2 of 14 sources observed, The lack of free-free emission strongly indicates that EGOs are young objects, typically in a pre-UCHII

Figure 6. A Spitzer 3-color image (3.6, 4.5 and 8.0 μm) of M10 (AFGL 5142) with the IRAS error ellipse shown and the 44 GHz masers indicated with crosses. Image from Gómez-Ruíz (in prep.)

region phase. But the two surveys mentioned above find *lower* detection rates in younger objects — certainly nothing approaching 90%.

The strikingly high detection rate of 44 GHz masers in EGOs probably has more to do with EGOs being good tracers of outflows than with the overall evolutionary state of region. The association of Class I masers with outflows is fairly clear, but the relation of 44 GHz masers to the overall evolutionary state of the region is still rather murky. Moreover, as the Gómez-Ruíz survey shows, although the presence of an EGO is an exceedingly good predictor of the presence of a 44 GHz maser, the inverse relationship does not hold. To wit, the majority of regions where they detect 44 GHz masers are *not* regions that host EGOs.

6. Conclusions

Numerous surveys have been made of the 44 GHz methanol maser, using both single-dish telescopes and interferometers. The more recent (and extensive) surveys have confirmed a number of behaviors that were seen in the early (and generally less extensive) surveys. Examples include the low number of velocity components in the 44 GHz class I spectra, the close agreement between maser velocities and ambient gas systemic velocities, and the fact that the maser positions generally do not coincide with infrared and centimeter continuum emission.

Nevertheless, the newer surveys have also produced some surprises, One of these is that the 44 GHz class I maser *can* appear in low-mass star formation regions. Additionally, the closer (but still offset) maser positions with respect to infrared/centimeter sources was an unexpected contribution of the interferometric surveys. Perhaps most significant is the finding of the phenomenally high detection rate of these masers toward EGOs.

The latter point is still somewhat mysterious, because of the finding in several surveys that detection rates increase with the apparently increasing age of host star formation region. Reconciling these detection rates with the evolutionary picture of methanol masers in star forming regions will require further study.

References

Araya, E., Hofner, P., Kurtz, S., Olmi, L., & Linz, H. 2008, *ApJ* 675, 420

Araya, E., Kurtz, S., Hofner, P., & Linz, H. 2009, *ApJ* 698, 1321

Bachiller, R., Gómez-González, J, Barcia, A., & Menten, K. M. 1990, *A&A* 240, 116

Bae, J.-H., Kim, K.-T., Youn, S.-Y., Kim, W.-J., Byun, D.-Y., Kang, H., & Oh, C. S. 2011, *ApJS* 196, 21

Cesaroni, R., Felli, M., Testi, L., Walmsley, C. M., & Olmi, L. 1997, *A&A*, 325, 725

Cesaroni, R., Felli, M., Jenness, T., Neri, R., Olmi, L., Robberto, M., Testi, L., & Walmsley, C. M. 1999, *A&A* 345, 949

Cyganowski, C. J., Brogan, C. L., Hunter, T. R., & Churchwell, E. 2009, *ApJ* 702, 1615

Cyganowski, C. J., Brogan, C. L., Hunter, T. R., & Churchwell, E. 2011, *ApJ* 743, 56

Fontani, F., Cesaroni, R., & Furuya, R. S. 2010, *A&A* 517, 56

Haschick, A. D., Menten, K. M., & Baan, W. A. 1990, *ApJ* 354, 556

Hoare, M. G., Kurtz., S. E., Lizano, S., Keto, E., & Hofner, P. 2007, in: B. Reipurth, D. Jewitt, & K. Keil (eds.), *Protostars and Planets V*, (Tucson: University of Arizona Press), p. 181

Kalenskii, S. V., Bachiller, R., Berulis, I. I., Val'tts, I. E., Gómez-González, J., Martin-Pintado, J., Rodríguez-Franco, A., & Slysh, V. I. 1992, *SvA* 36, 517

Kalenskii, S. V., Promyslov, V. G., Slysh, V. I., Bergman, P., & Winnberg, A. 2006, *Astron. Rep.*, 50, 289

Kalenskii, S. V., Johansson, L. E. B., Bergman, P., Kurtz, S., Hofner, P., Walmsley, C. M., & Slysh, V. I. 2010a, *MNRAS*, 405, 613

Kalenskii, S. V., Kurtz, S., Slysh, V. I., Hofner, P., Walmsley, C. M., Johansson, L. E. B., & Bergman, P. 2010b, *Astron. Rep.*, 54, 932

Kogan, L. & Slysh, V. 1998, *ApJ* 497, 800

Kurtz, S., Hofner, P., & Vargas-Álvarez, C. 2004, *ApJS* 155, 149

Molinari, S., Brand, J., Cesaroni, R., & Palla, F. 1996 *A&A* 308, 573

Morimoto, M., Kanzawa, T., & Ohishi, M. 1985 *ApJ* (Letters) 288, L11

Plambeck, R. L. & Menten, K. M. 1990, *ApJ*, 364, 555

Pratap, P., Shute, P. A., Keane, T. C., & Battersby, C., Sterling S. 2008, *AJ* 135, 1718

Sridharan, T. K., Beuther, H., & Schilke, P. 2002, *ApJ* 566, 931

Slysh, V. I., Kalenskii, S. V., Val'tts, I. E., & Otrupcek, R. 1994, *Mon. Not. Royal Astron. Soc.* 268, 464

Val'tts, I. E. & Larionov, G. M. 2007 *Astron. Rep.* 51, 519

Zhang, Q., Hunter, T. R., Brand, J., Sridharan, T. K., Cesaroni, R., Molinari, S., Wang, J., & Kramer, M. 2005, *ApJ* 625, 864

Cosmic Masers - from OH to H$_0$
Proceedings IAU Symposium No. 287, 2012
R.S. Booth, E.M.L. Humphreys & W.H.T. Vlemmings, eds.

© International Astronomical Union 2012
doi:10.1017/S1743921312006801

A Highly-collimated Water Maser Bipolar Outflow in the Cepheus A HW3d Massive Protostellar Object

James O. Chibueze[1,2], Hiroshi Imai[1], Daniel Tafoya[1], Toshihiro Omodaka[1], Osamu Kameya[3], Tomoya Hirota[3], Sze-Ning Chong[1], and José M. Torrelles[4]

[1] Department of Physics and Astronomy, Graduate School of Science and Engineering, Kagoshima University, 1-21-35 Korimoto, Kagoshima 890-0065, Japan
email: james@milkyway.sci.kagoshima-u.ac.jp

[2] Department of Physics and Astronomy, Faculty of Physical Sciences, University of Nigeria, Carver Building, 1 University Road, Nsukka, Nigeria

[3] Mizusawa VLBI Observatory, National Astronomical Observatory of Japan, 2-21-1 Osawa, Mitaka, Tokyo 181-8588, Japan

[4] Instituto de Ciencias del Espacio (CSIC)-UB/IEEC, Facultat de Física, Universitat de Barcelona, Martí i Franquès 1, E-08028 Barcelona, Spain

Abstract. We report the results of multi-epoch very long baseline interferometry (VLBI) water (H$_2$O) maser observations carried out with the VLBI Exploration of Radio Astrometry (VERA) toward the HW3d object within the Cepheus A star-forming region. We measured proper motions of 30 water maser features, tracing a compact bipolar outflow. This outflow is highly collimated, extending through $\sim$400 mas (290 AU), and having a typical proper motion velocity of $\sim$6 mas yr^{-1} ($\sim$21 km s^{-1}). The dynamical timescale of the outflow was estimated to be $\sim$100 years, showing that the outflow is tracing a very early star-formation phase. Our results provide strong support that the HW3d object harbors an internal massive protostar, as previous observations suggested. In addition, we have analyzed Very Large Array (VLA) archive 1.3 cm continuum data of the 1995 and 2006 epochs obtained towards Cepheus A. These results indicate possible different protostars around HW3d and/or strong variability in its radio continuum emission.

Keywords. ISM: individual objects (Cepheus A), ISM: jets and outflows, masers: H$_2$O, stars: formation

1. Introduction

Complex environments due to the close proximity of many young massive stellar objects characterize the formation of massive stars. Sites of massive star formation are usually large distances away from the Earth ($\sim$few kpc), and their relatively short formation timescale of $\sim$10^5 yr underscores the difficulty in understanding their formation and evolution (Zinnecker & Yorke 2007). There are a number of competing concepts on how high-mass stars form. Among them are the concept of low (or medium) mass star mergers proposed by Bonnell *et al.* (1998); competitive accretion in protostellar cluster environment proposed by Bonnell & Bate (2006); and the concept of gravitational collapse/high core accretion rate (Yorke & Sonnhalter 2002).

Previous results of water (H$_2$O) maser studies in high-mass star-forming regions show that H$_2$O maser spatio-kinematics can trace star-formation activities at different evolutionary phases. Cepheus A (Cep A) at a distance of $\sim$700 pc (Dzib *et al.* 2011) is one of the closest massive star-forming regions, second only to Orion KL. Cep A HW2 is the most studied object of all the 16 radio continuum sources reported by Hughes & Wouter-loot (1984). It is possibly the most evolved of all the objects in the region, harboring a massive star exhibiting a YSO-disk-jet system features, with the presence of a strong magnetic field and a wide-angle outflow traced by H_2O masers (see reports by Rodríguez *et al.* 1994; Patel *et al.* 2005; Curiel *et al.* 2006; Jiménez-Serra *et al.* 2007; Torrelles *et al.* 2007, 2011; Vlemmings *et al.* 2010).

In this proceedings, we direct our attention to the HW3d object, located ~3" south-east of HW2. Judging from the positive spectral index of HW3d, the presence of H_2O masers and OH masers (Hughes *et al.* 1995; Cohen *et al.* 1984), Garay *et al.* (1996) suggested that HW3d may be harboring a YSO rather than being excited externally by shocks.

2. VERA & VLA Observations of Cepheus A

From 2006 May 13 to 2007 August 31, we conducted 9-epoch spectral observations toward Cep A H_2O masers at ~22 GHz using the VLBI Exploration of Radio Astrometry (VERA). J2005+7752, BL Lac, J2015+3710 were also observed for the calibration of the H_2O maser data. Using the dual-beam system of VERA, we were able to simultaneously observe a position reference source, J2303+6405, 2.19° away from Cep A for astrometric purposes. In the AIPS data reduction, we executed the self-calibration procedures using a bright maser feature in order to detect weak maser emission. We also conducted an inverse phase referencing procedure to detect the weak continuum emission from the position reference source, after which astrometry was possible in order to determine the absolute positions of the H_2O masers features (see Chibueze *et al.* 2012). From our detected maser features in HW3d, we traced 30 relative proper motions associated with the object.

To explore possible variation in the HW3d continuum emission, we retrieved Very Large Array (VLA) 1.3 cm archival data of 1995 and 2006. Standard calibration procedures were applied in reducing the data. A close look at the two obtained 1.3 cm continuum emission maps shows some interesting variabilities (see Figure 1).

3. Results

VLA Results

Figure 1 shows the superimposed maps of the 1995 (red contours) and the 2006 (white contours) VLA 1.3 cm continuum emissions from the HW2, HW3d and HW3c. The insert therein shows the 2006 continuum while the crosses represent the positions of the continuum peaks in 1995. A couple of variations can be clearly seen in the HW3d object. First, the 1995 HW3d compact structure has evolved into an elongated structure. Second, there is a shift of ~200 mas in the position of the HW3d continuum peak between the epochs. Assuming that there is only one YSO in HW3d, we estimated its proper motion to be ~65 km s^{-1} using the 200 mas change is its continuum peak position. We also found a variation in the total flux density of HW3d (~9 mJy in 1995, and ~6.9 mJy in 2006).

To explain the observed changes in HW3d, one may think of three possible scenarios:
1. A single YSO having a proper motion of ~65 km s^{-1} (This is likely the most improbable considering the high velocity of the object);
2. The presence of multiple sources in HW3d;
3. Internal proper motions of clumps in a jet with flux density variation.

VERA Results

From the detected maser features in HW3d we measured, for the first time, 30 H_2O maser relative proper motions, by adopting a reference position derived using the method

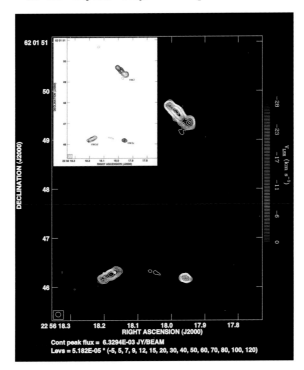

Figure 1. VLA 1.3 cm continuum emission map of Cep A with the H_2O masers of HW3d. The white and red contours represent the 2006 and 1995 continuum respectively. Insert: 2006 map showing a cross indicating the position of the HW3d continuum peak in 1995. HW2, HW3d and HW3c are labeled appropriately (Figure adapted from Chibueze *et al.* 2012).

used in Torrelles *et al.* (2001). The proper motions show a 400-mas linear structure which is aligned with the direction of elongation of the VLA 2006 1.3 cm continuum emission. It is also bipolar in nature, thus implying that the H_2O maser relative proper motions are tracing a bipolar outflow in HW3d (see Figure 2). The opening angle of the outflow derived from the deconvolved image of the continuum emission is $\sim30°$, showing the bipolar outflow to be highly collimated. The C-cluster of H_2O maser proper motions showed high velocity dispersion (~20 km s^{-1}), which is suggestive of the fact that the outflow exciting source may be located in or near it (see Figure 2).

4. Analyses and Discussion

Position & Velocity Variance/Covariance Matrix Analyses (PVCM & VVCM)
PVCM/VVCM, first introduced by Bloemhof (1993), is good tool for extracting the spatial and kinematic essentials from the maser positions and proper motions. It involves the diagonalization of the matrices built from the maser position and velocity (proper motion) information using

$$\sigma_{ij} = \frac{1}{N-1}\sum_{n=1}^{N}(v_{i,n} - \bar{v}_i)(v_{j,n} - \bar{v}_j), \tag{4.1}$$

where i and j denote the two dimensional space axes in the case of position, or three orthogonal space axes in the case of velocity. n is the n-th maser feature in the collection summing up to N $(= 30)$. The bar indicates averaging over the maser features.

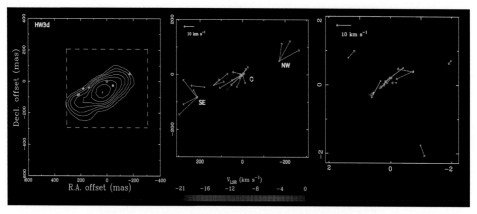

Figure 2. left: Distribution of the H_2O masers detected in our VERA observations superposed on the HW3d 1.3 cm continuum map obtained with the VLA in 2006. middle: The proper motions of these maser features. The position, length, and direction of an arrow indicate the position, speed, and direction of the maser feature motion on the sky, respectively. The continuous lines represent proper motions traced in 3 or more epochs while the dashed lines represent those traced in 2 epochs only. The color codes represent the maser LSR velocities. right: Zoom of the C-cluster showing high velocity dispersion within a ~ 4 mas area (Figure adapted from Chibueze *et al.* 2012).

In the 2-D PVCM, the maximum eigenvalue, ($\psi_{max} = 12697.3$ mas^2; at position angle, PA, 108.5°) is ~ 400 times the minimum one, indicating high collimation in the maser position dispersion. In the case of the 3-D VVCM, ψ_{max} (PA 123°), ψ_{mid}, and ψ_{min} are 111.4, 34.9. and 11.6 km^2 s^{-2}. The largest eigenvalue is 3 times the second largest eigenvalue, which shows a collimation factor of 3 in the velocity dispersion. There is correspondence between the PA of ψ_{max} in both PVCM and VVCM.

Modeling the H_2O Maser Spatio-kinematics

We have also made an expanding flow model of the H_2O maser proper motions based on a least-squares method (see Imai *et al.* 2011; Chibueze *et al.* 2012). We have assumed a single source scenario, located near the turbulent C-cluster, and that all the masers are moving away from the exciting source position with expansion velocities, $V_{exp}(i)$ (expected to be positive). Figure 3 shows the plot of the obtained expansion velocities

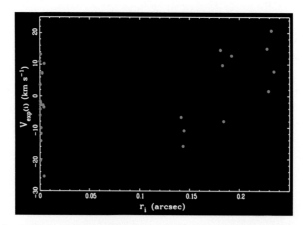

Figure 3. Expansion velocity distribution derived from the expansion flow model fitting of the maser proper motions in HW3d (Figure adapted from Chibueze *et al.* 2012).

versus the offset positions from the possible position of the exciting source. Evidently, the northwest and southeast (NW, SE) clusters fit properly into the model, but the proper motions of the cluster located around (+135 mas, - 39 mas) did not fit into the expansion flow model. Near the originating center (20 mas, -17 mas), there are also proper motions with negative $V_{exp}(i)$ which may be indication of infall, but the high infall velocities seen in the result may be due to another object in close proximity.

5. Summary/Conclusion

Based on our results and analyses we conclude the following:

1. We detected outflow activity in HW3d, indicating that this source is internally excited by a young massive stellar object.

2. The bipolar motions of the H_2O masers indicate the presence of a YSO-Jet system in HW3d.

3. The shift in position and flux variation of the continuum emission in HW3d suggests the possibility of multiplicity in the region or internal proper motions of clumps in the jet.

4. This multiplicity and/or proper motions of clumps in the jet can be tested in the future with SMA and EVLA observations.

References

Bloemhof, E. E. 1993, *ApJ*, 406, L75
Bonnell, I. A., Bate, M. R., & Zinnecker, H., 1998, *MNRAS*, 298, 93
Bonnell, I. A. & Bate, M. R., 2006, *MNRAS*, 370, 488
Chibueze, J. O., Imai, H., Tafoya, D., *et al.* 2012, *ApJ*, in press
Cohen, R. J., Rowland, P. R., & Blair, M. M., 1984, *MNRAS*, 210, 425
Curiel, S., Ho, P. T. P., Patel, N. A., *et al.* 2006, *ApJ*, 638, 878
Dzib, S., Loinard, L., Rodríguez, L. F., Mioduszewski, A. J., & Torres, R. M., 2011, *ApJ*, 733,71
Garay, G., Ramírez, S., Rodríguez, L. F., Curiel, S., & Torrelles, J. M., 1996, *ApJ*, 459,193
Hughes, V. A. & Wouterloot, J. G. A., 1984, *ApJ*, 276, 204
Hughes, V. A., Cohen, R. J., & Garrington, S., 1995, *MNRAS*, 272, 469
Imai, H., Omi, R., Kurayama, T., *et al.* 2011, *PASJ*, 63, 1293
Jiménez-Serra, I., Martín-Pintado, J., Rodríguez-Franco, A., *et al.* 2007, *ApJ*, 661, L187
Patel, N. A., Curiel, S., Sridharan, *et al.* 2005, *Nature*, 437, 109
Rodríguez, L. F., Garay, G., Curiel, S., *et al.* 1994, *ApJ*, 430, L65
Torrelles, J. M., Patel, N. A., Gómez, J. F., *et al.* 2001b, *ApJ*, 560, 853
Torrelles, J. M., Patel, N. A., Curiel, S., *et al.* 2007, *ApJ*, 666, L37
Torrelles, J. M., Patel, N. A., Curiel, S., *et al.* 2011, *MNRAS*, 410, 627
Vlemmings, W. H., T., Surcis, G., Torstensson, K. J. E., & van Langevelde, H., J., 2010, *MNRAS*, 404, 134
Yorke, H. W. & Sonnhalter, C., 2002, *ApJ*, 569, 846
Zinnecker, H. & Yorke, H. W., 2007, *ARA&A*, 45, 881

Cosmic Masers - from OH to H_0
Proceedings IAU Symposium No. 287, 2012
R.S. Booth, E.M.L. Humphreys & W.H.T. Vlemmings, eds.

© International Astronomical Union 2012
doi:10.1017/S1743921312006813

Methanol masers and millimetre lines: a common origin in protostellar envelopes

Karl J. E. Torstensson[1,2], Huib Jan van Langevelde[1,2], Floris F. S. van der Tak[3,4], Wouter H. T. Vlemmings[5], Lars E. Kristensen[2], Stephen Bourke[1] and Anna Bartkiewicz[6]

[1]Joint Institute of VLBI in Europe,
PO Box 2, NL-7990 AA Dwingeloo, The Netherlands
email: langevelde@jive.nl

[2]Leiden Observatory, Leiden University,
PO Box 9513, NL-2300 RA Leiden, The Netherlands
email: kalle@strw.leidenuniv.nl

[3]SRON Netherlands Institute for Space Research,
Landleven 12, NL-9747 AD Groningen, The Netherlands
email: vdtak@sron.nl

[4]Kapteyn Astronomical Institute
University of Groningen, The Netherlands

[5]Chalmers University of Technology, Department of Earth and Space Science,
SE-412 96 Gothenburg, Sweden
email: wouter.vlemmings@chalmers.se

[6]Toruń Centre for Astronomy, Nicolaus Copernicus University,
Gagarina 11, 87-100 Toruń, Poland
email: annan@astro.uni.torun.pl

Abstract. To understand the origin of the CH_3OH maser emission, we map the distribution and excitation of the thermal CH_3OH emission in a sample of 14 relatively nearby (<6 kpc) high-mass star forming regions that are identified through 6.7 GHz maser emission. The images are velocity-resolved and allow us to study the kinematics of the regions. Further, rotation diagrams are created to derive rotation temperatures and column densities of the large scale molecular gas. The effects of optical depth and subthermal excitation are studied with population diagrams. For eight of the sources in our sample the thermal CH_3OH emission is compact and confined to a region <0.4 pc and with a central peak close (<0.03 pc) to the position of the CH_3OH maser emission. Four sources have more extended thermal CH_3OH emission without a clear peak, and for the remaining two sources, the emission is too weak to map. The compact sources have linear velocity gradients along the semi-major axis of the emission of $0.3 - 13$ km s^{-1} pc^{-1}. The rotation diagram analysis shows that in general the highest rotation temperature is found close to the maser position. The confined and centrally peaked CH_3OH emission in the compact sources indicates a single source for the CH_3OH gas and the velocity fields show signs of outflow in all but one of the sources. The high detection rate of the torsionally excited $v_t = 1$ line and signs of high-K lines at the maser position indicate radiative pumping, though the general lack of measurable beam dilution effects may mean that the masing gas is not sampled well and originates in a very small region.

Keywords. Masers, stars: formation, ISM: jets and outflows, ISM: kinematics and dynamics, ISM: molecules, submillimeter

1. Introduction

Though high-mass star formation has been extensively studied, the particulars of the early evolution is still widely debated and several theories have been proposed (Zinnecker & Yorke 2007). The currently favoured scenario is that of monolithic collapse or core accretion, a scaled up version of lower-mass star formation. In this scenario the accretion occurs via a disk with associated outflows. However, because of the short life time and rapid evolution, massive stars are few and far between. Furthermore, they form in a clustered environment in the densest part of the molecular cloud where they are heavily obscured. These conditions lead to several observational challenges, and high angular resolution observations at wavelengths that can penetrate the dense molecular clouds are therefore needed.

Since its discovery, the 6.7 GHz methanol maser has been associated with the early stages of high-mass star formation (Menten 1991). Initial studies found the maser emission to be associated with ultra-compact (UC) HII regions (Phillips, Norris, Ellingsen, *et al.* 1998; Walsh, Burton, Hyland, *et al.* 1998). Follow-up studies have since shown that in general the maser emission is associated with an earlier evolutionary stage (Beuther, Walsh, Schilke, *et al.* 2002; Walsh, Macdonald, Alvey, *et al.* 2003). This scenario is supported by large studies in which most maser sources are associated with mm cores rather than UCHII regions. Moreover, searches for methanol maser emission towards low-mass star forming regions have shown the Class II masers to be exclusively associated with high-mass star formation (Minier, Ellingsen, Norris, *et al.* 2003; Xu, Li, Hachisuka, *et al.* 2008).

Detailed high-resolution VLBI studies of the methanol masers have led to several different classifications based on their morphology. It appears that the masers occur in outflows, disks, and possibly other physical structures. The extent of the masers appears to be between a few hundred and a couple of thousand AU, an observation common to all sources. A recent VLBI study of 30 methanol maser sources based on a blind survey found 30% of the masers to have an elliptical distribution (Bartkiewicz, Szymczak, van Langevelde, *et al.* 2009). A similar elliptical distribution has been observed in the nearby source Cepheus A HW2, for which we propose that the maser emission occurs in a shock interface in the equatorial region of the protostellar object (Torstensson, van Langevelde, Vlemmings, *et al.* 2011a). This geometry is supported by the magnetic field being aligned along the outflow axis, regulating the infall of the accreting material (Vlemmings, Surcis, Torstensson, *et al.* 2010).

To understand the origin of the CH_3OH maser emission, we map the distribution and excitation of the thermal CH_3OH emission in a sample of 14 relatively nearby (<6 kpc) high-mass star forming regions that are identified through 6.7 GHz maser emission.

2. Observations and data reduction

We have used the HARP-B instrument on the JCMT† to map the thermal CH_3OH emission in the $J = 7_K \rightarrow 6_K$ at 338 GHz towards 14 high-mass star-forming regions associated with 6.7 GHz methanol maser emission. The observations were performed in a jiggle-chop mode (harp5) to create $2' \times 2'$ maps with a pixel spacing of $6''$. This ensures proper Nyquist sampling of the $14''$ JCMT beam at 338 GHz. During the observations regular pointings were done on calibrators and we estimate an absolute pointing accuracy

† The James Clerk Maxwell Telescope is operated by the Joint Astronomy Centre on behalf of the Science and Technology Facilities Council of the United Kingdom, the Netherlands Organisation for Scientific Research, and the National Research Council of Canada.

of ~1″. The 1 GHz bandwidth covers a total of 25 lines of both E and A type CH₃OH
with upper energy levels between 65 K and 260 K, see Fig. 1 for a sample spectrum. In
our analysis we have adopted a main beam efficiency of 0.6 (Buckle, Hills, Smith, *et al.*
2009) and estimate a calibration uncertainty of 20%.

The initial data inspection and re-gridding of the data was done with the Starlink
package using Gaia/SPLAT after which the data were converted to GILDAS/CLASS
format and the remaining data reduction and analysis was performed in CLASS. To
increase the signal-to-noise of the spectra the data were smoothed to a velocity resolution
of 0.87 km s⁻¹, after which a linear baseline was fitted to the emission free regions of each
spectra and subtracted. The analysis was then performed on a pixel by pixel basis in
which the strongest unblended line in the spectra (-1E) was fitted with a Gaussian. The
velocity of the −1E line was then used to calculate the velocities of the other 24 methanol
lines. Around each calculated velocity the moments were calculated within a window size
of two times the line width of the fitted −1E line. Due to line blending we have also used
the first and second moment of the lines to help determine whether a line is real.

3. Results

The CH₃OH −1E is the strongest unblended line in the spectra, and is used to map the
integrated line flux, central velocity, and line width. To analyse the excitation of the large-
scale methanol gas, we have performed a rotational diagram analysis of each position, for
which at least three lines are detected. Furthermore, a population diagram analysis was
used at the position of the methanol maser emission. The population diagram analysis
takes into account the optical depth of the lines and the source size, both of which
can affect the measured line intensity (Torstensson, van der Tak, van Langevelde, *et al.*
2011b).

Four of the sources have extended and complex thermal CH₃OH emission and are part
of larger star-forming regions. In two of the sources the methanol emission was too weak
to be mapped. The remaining eight sources all have compact thermal CH₃OH emission
and will be the focus of this paper.

The thermal CH₃OH emission for the the eight compact sources is confined to a region
with a major-axis $a < 0.4$ pc, and centrally peaked close to the maser position ($r <
0.03$ pc), Fig. 2. Moreover, all but one of the eight sources have a linear velocity gradient
of a few km s⁻¹ pc⁻¹ across the source.

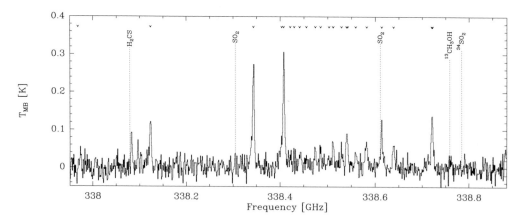

Figure 1. Sample spectrum of the source G23.389+00.185. The tick marks at the top indicate
the CH₃OH $7_K \to 6_K$ lines, also indicated are other identified species.

The rotational analysis indicates moderate temperatures (30–50 K) in the more extended gas, and elevated rotation temperatures ($\sim$100 K) at the maser position. Contrary to what is found in the source Cepheus A HW2 (Torstensson, van der Tak, van Langevelde, *et al.* 2011b) we only detect one single gas component in the other sources. Furthermore, despite signs of high-K in some of the sources we do not find any beam dilution in the population diagram analysis and the optical depth of the low-K lines is at most moderate. Still, for half of the 14 sources in our sample we detect the torsionally excited $v_t = 1$ line at the maser position.

4. Discussion

We have mapped the large scale distribution and excitation of the thermal methanol gas towards a sample of 14 high-mass star-forming regions associated with 6.7 GHz methanol maser emission. Eight of the sources have compact thermal CH_3OH emission with a central peak close to the maser position. The linear velocity gradient across the sources lines up with other outflows tracers and the rotational diagram analysis points to a single highly excited source at the centre. We interpret this result as the large-scale methanol gas being entrained in an outflow with a common driving source, and that it can be traced back to the protostellar object, as in the case of Cepheus A HW2 (Torstensson, van Langevelde, Vlemmings, *et al.* 2011a).

Because the 6.7 GHz methanol maser emission typically has an extent of a few hundred to a couple of thousands of AU and is located close to the protostellar object we expect the highly excited gas associated with the maser emission to have a similar extent (Bartkiewicz, Szymczak, van Langevelde, *et al.* 2009). We therefore performed a population diagram analysis of the gas at the maser location to allow for source size and optical depth effects. However, the general lack of beam dilution effects that we find leads us to believe that we are not sampling the highly excited gas with these observations. Rather we are only sampling the more extended gas traced by the low-K lines.

An independent indication of highly excited CH_3OH gas is the detection of the -1A $v_t = 1$ line at 337.97 GHz in half of the sources in our sample. Because collisional pumping alone is very inefficient at populating the torsionally excited state, the $v_t = 1$ line is an indicator of infrared pumping at work (Leurini, Schilke, Wyrowski, *et al.* 2007). However, due to the low signal-to-noise of the only torsionally excited line in our spectra we are unable constrain the radiative excitation and break the degeneracy between density and temperature. The sources for which the $v_t = 1$ line is detected also show signs of the

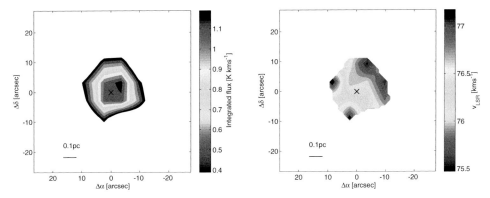

Figure 2. Integrated intensity (left) and central velocity (right) of the CH_3OH $7_{-1} \to 6_{-1}$ E line for the source G23.389+00.185. The black cross marks the methanol maser position.

high-K lines, supporting the argument of highly excited gas. Although the high-K lines are weak, we have a considerable detection rate of the $v_t = 1$ line. The high detection rate of the $v_t = 1$ line could be an effect of our sample selection, because we have intentionally targeted bright 6.7 GHz methanol maser sites, which are thought to be pumped by infrared radiation. For both the $v_t = 1$ line and the high-K lines beam dilution effects are likely important. There does however not seem to be any correlation between the distance of the source or the luminosity and the occurrence of the $v_t = 1$ line which one might expect. Neither are the sources with the $v_t = 1$ line detection those with the largest integrated flux of the $-1E$ line. Instead, the $v_t = 1$ line seems to be correlated with the sources for which we find a non-negligible optical depth in the low-K lines in the population diagram analysis. Higher resolution interferometric observations with beam sizes similar to the extent of the maser regions (and more accurate distance estimates) would greatly help to constrain the excitation of the highly excited gas (Beuther, Zhang, Hunter, *et al.* 2007).

The methanol maser emission can be described as having an elliptical distribution for three of the compact sources (Cepheus A HW2, G23.207−00.377, and G23.389+00.185) (Torstensson, van Langevelde, Vlemmings, *et al.* 2011a; Bartkiewicz, Szymczak, van Langevelde, *et al.* 2009). Assuming that the maser emission arises in the equatorial region of the protostellar object, a velocity model with a radial and a rotational velocity component can be fitted to the maser velocity field. In all three sources the radial velocity component is the dominant one. With the orientation of the larger scale thermal methanol outflow we can constrain the geometry of the systems and find that the radial velocity is that of infall rather than outflow.

All in all, our results are in agreement with the monolithic collapse scenario, in which accretion proceeds through a disk with associated outflows.

References

Bartkiewicz, A., Szymczak, M., van Langevelde, H. J., Richards, A. M. S., & Pihlström, Y. M. 2009, *A&A*, 502, 155

Beuther, H., Walsh, A., Schilke, P., Sridharan, T. K., Menten, K. M., & Wyrowski, F. 2002, *A&A*, 390, 289

Beuther, H., Zhang, Q., Hunter, T. R., Sridharan, T. K., & Bergin, E. A. 2007, *A&A*, 473, 493

Buckle, J. V., Hills, R. E., Smith, H., *et al.* 2009, *MNRAS*, 399, 1026

Leurini, S., Schilke, P., Wyrowski, F., & Menten, K. M. 2007, *A&A*, 466, 215

Menten, K. M. 1991, *ApJ*, 380, L75

Minier, V., Ellingsen, S. P., Norris, R. P., & Booth, R. S. 2003, *A&A*, 403, 1095

Phillips, C. J., Norris, R. P., Ellingsen, S. P., & McCulloch, P. M 1998, *MNRAS*, 300, 1131

Torstensson, K. J. E., van Langevelde, H. J., Vlemmings, W. H. T., & Bourke, S. 2011, *A&A*, 526, A38

Torstensson, K. J. E., van der Tak, F. F. S., van Langevelde, H. J., Kristensen, L. E., & Vlemmings, W. H. T. 2011, *A&A*, 529, A32

Vlemmings, W. H. T., Surcis, G., Torstensson, K. J. E., & van Langevelde, H. J. 2010, *MNRAS*, 404, 134

Walsh, A. J., Burton, M. G., Hyland, A. R., & Robinson, G. 1998, *MNRAS*, 301, 640

Walsh, A. J., Macdonald, G. H., Alvey, N. D. S., Burton, M. G., & Lee, J.-K. 2003, *A&A*, 410, 597

Xu, Y., Li, J. J., Hachisuka, K., Pandian, J. D., Menten, K. M., & Henkel, C. 2008, *A&A*, 485, 729

Zinnecker, H. & Yorke, H. W. 2007, *ARA&A*, 45, 481

Cosmic Masers - from OH to H$_0$
Proceedings IAU Symposium No. 287, 2012
R.S. Booth, E.M.L. Humphreys & W.H.T. Vlemmings, eds.

© International Astronomical Union 2012
doi:10.1017/S1743921312006825

The infrared environment of methanol maser rings at high spatial resolution

James M. De Buizer[1], Anna Bartkiewicz[2] and Marian Szymczak[2]

[1]SOFIA-USRA, NASA Ames Research Center, Moffett Field, CA 94035, USA
email: jdebuizer@sofia.usra.edu

[2]Toruń Centre for Astronomy, Nicolas Copernicus University, Gagarina 11, 87-100, Toruń,
Poland

Abstract. The recent discovery of methanol maser emission coming from ring-like distributions has led to the plausible hypothesis that they may be tracing circumstellar disks around forming high mass stars. In this article we discuss the distribution of circumstellar material around such young and massive accreting (proto)stars, and what infrared emission geometries would be expected for different disk/outflow orientations. For four targets we then compare the expected infrared geometries (as inferred from the properties of the maser rings) with actual high spatial resolution near-infrared and mid-infrared images. We find that the observed infrared emission geometries are not consistent with the masers residing in circumstellar disks.

Keywords. Masers, stars: formation, stars: early-type, circumstellar matter, infrared: ISM, instrumentation: high angular resolution

1. Infrared Observations of Young Stellar Objects

The first infrared observations to resolve disks around stars were of debris disks (e.g., Telesco *et al.* 1988, Jayawardhana *et al.* 1998). Debris disks are circumstellar disks that occur around stars with more evolved, (post)planet building disks. Since one can directly view a debris disk in its infrared dust emission it appears as an elongated ellipsoidal structures whose "flatness" is a function of the viewing angle of the disk. However, at earlier stages of star formation, such circumstellar disks are likely to be actively accreting on to their parent stars, and are thus much more massive and dense. In fact, accretion disks around very young stars began to be found, not by their direct emission, but by their scattered and reflected emission off of their upper and lower disk surfaces. Originally discovered in the optical (e.g., McCaughrean & O'Dell 1996), these "silhouette" disks demonstrated that if a disk is large and dense enough, it could be optically thick even at infrared wavelengths (e.g., Cotera *et al.* 2001). These sources are typified by a "dark lane" demarcating the disk itself, between two "lobes" of scattered/reradiated emission demarcating the upper and lower flared disk surfaces.

However, all of these observations were of disks around low-mass stars. The first claims of infrared detections of circumstellar disks around young massive stars (Stecklum & Kaufl 1998, De Buizer *et al.* 2000) were made, in part, on the assumption that the disk would appear as an elongated structure in its infrared dust emission, similar to what was seen with debris disks around low-mass stars. However, higher resolution follow-up observations (De Buizer *et al.* 2002) proved that these were not disks, and it was soon after realized that disks accreting onto young massive (proto)stars would be optically thick in the infrared as well, and would be even more dense and massive than those accreting onto low-mass protostars. Furthermore, in order to have a large enough reservoirs of material to accrete to such high masses, massive stars must form in the densest parts of giant molecular clouds, which are extremely obscured environments. Moreover, the earliest stages of massive star formation are deeply self-embedded; the (proto)stars are surrounded by massive and dense accretion envelopes, as well as the aforementioned circumstellar disks, which are believed to be large, thick, and flared.

Due to the high amount of obscuration, one would have no hope of observing these stages of massive star formation at even mid-infrared wavelengths if not for their bipolar outflows. Disk accretion is accompanied by outflow, and it is this outflow that punctures holes through the obscuring material surrounding a forming massive star. The outflow axis is oriented perpendicular to the disk, and the outflow starts out narrow and colli- mated. At such early stages of formation, we may only detect the presence of the massive young stellar object at infrared wavelengths if we are lucky enough to have a line of sight looking down the outflow axis. We could then see into the envelope, and view the scat- tered/reprocessed dust emission off of the cavity walls, and if the angle is just right, we may see down into the central disk or (proto)star. Such a chance alignment is rare, and it is in part due to this geometrical effect that detecting massive young stars in the infrared is so difficult. However, over time the young stellar object evolves and the outflow angle widens (Shu & Adams 1987). With the widened outflow, comes a large range of angles that allow for a higher probability of detection in the infrared. At some point the angle is so wide, that the distinction between what is the outflow cavity surface and what is the surface of the flared disk is blurred.

This notional sequence of events has been supported by several studies in the recent decade. For instance, observations of the earliest stages of massive star formation with collimated outflow cavities emitting brightly in their mid-infrared continuum emission were first identified by De Buizer (2006). The first claim of a candidate infrared "sil- houette disk" around a massive star was made by Chini et al. (2004) demonstrating the later stages where the outflow has widened considerably (though whether the mass of the central object is high mass is the subject of debate, e.g. Sako et al. 2005). In addition to these observations, radiative transfer models of Alvarez et al. (2004) and Zhang & Tan (2011) also produce these geometries; the first showing the results of the earliest stages of collimated outflow cavities, and the second for more open-angled outflows.

2. Infrared Imaging Tests of the Disk Hypothesis of Maser Rings

One phenomenon that seems to be related only to massive star formation is the appear- ance of methanol maser emission. Milliarcsecond-scale very long baseline interferometry (VLBI) observations of massive star-forming regions with 6.7 GHz methanol maser emis- sion resolve the discrete locations of the emission (known as "maser spots") and show a wide range of spot distributions. The maser spots can be distributed randomly without any regularity, however often they group into seemingly coherent structures like lines or arcs (e.g., Norris et al. 1993). We have recently completed a survey of 31 sources at 6.7 GHz using European VLBI Network (EVN) (Bartkiewicz et al. 2009). In addition to the curved and complex morphologies observed in other samples, we have discovered for the first time nine sources (29 % of the sample) with ring-like distributions with typical sizes having major axes of 0.2-0.3″. Though not apparent in the radio data, these ring- like structures strongly suggest the existence of a central stellar object and lead to the obvious question: *Are methanol maser rings tracing circumstellar disks around massive protostars?*

Because methanol masers are believed to be pumped by mid-infrared photons from dust (Cragg, Sobolev, & Godfrey 2002), if the masers are arising from gas in a dusty circumstellar disk, they would have to trace the inner few hundred AU radius at most, given any reasonable heating argument†. Furthermore, the presence of a circumstellar disk around a massive star is expected to be short-lived, owing to the generally more rapid evolution of massive stars at all phases (Povich & Whitney 2010) and the very caustic nature of the near stellar environment of a massive star. As such, it is believed that once accretion ends, the disk would summarily dissipate (i.e. there would be no analogous debris disk phase for massive stars). This means that if the methanol masers are within circumstellar disks, then they must trace a rather young stage of the massive star formation process where the disk is an active accretion disk and the massive young stellar object will be very self-embedded, and thus should have an appearance in the infrared as discussed in the previous section.

Consequently, there are expected spatial and morphological relationships that can be tested between what one sees in the infrared when looking toward methanol masers rings under the assumption that the rings are in disks and that they are associated with an early and obscured accretion phase of massive star formation:

Scenario 1: The more circular the maser ring, the more face-on the disk, meaning one would be looking right down the outflow axis into the outflow cavity. Therefore, the infrared emission should be unresolved or circularly symmetric with the peak of emission coincident with the maser ring center;

Scenario 2: Slightly elliptical rings would indicate a disk orientation close, but not quite face-on. In this case we would expect to either only see the blue-shifted outflow cavity, or just a hint of the red-shifted cavity (depending on disk/envelope extinction, outflow opening angle, and disk inclination). In this case, the maser ring center will be slightly offset from the infrared emission center in a direction given by the outflow axis.

Scenario 3: The more highly elliptical the ring, the closer to edge-on will be the disk. If detectable in the infrared, we would expect for moderately to highly elliptical maser rings to see something more like a silhouette disk in the infrared, where the maser ring would lie between two infrared bright sources (the outflow cavities), in the "dark lane" of the optically thick disk. Of course, if the source is extremely obscured and/or very close to edge-on, we may not expect any infrared emission to be detected at all.

3. The Experimental Design

Inspection of mid-infrared data from the *Spitzer* IRAC maps, GLIMPSE and MIPS-GAL, revealed that all of our sources with ring-like morphologies coincide with unresolved mid-infrared sources within one pixel in a GLIMPSE map (1.2″). However, the resolutions of the *Spitzer* data (2.3″ at 8 μm, 7″ at 24 μm) and 2MASS data (2.5″) are relatively poor. Therefore, we decided to obtain the highest spatial resolution near-infrared and mid-infrared imaging available to allow resolution of details on par with the scale of the maser rings (∼300 mas).

Using the Gemini 8-m telescopes, we obtained data with NIRI/Altair, an adaptive optics near-infrared instrument that can achieve resolutions better than ∼150 mas at 2 μm. The mid-infrared observations were made with T-ReCS, employing a method which fully characterized the system point spread function accurately enough to allow the imaging data to be reliably deconvolved, achieving spatial resolutions of ∼150 mas at 8 μm and ∼250 mas at 18 μm.

---

† Remember however, the flaring of the disk and the dense accretion envelope prevent one from viewing the mid-infrared emission coming directly from the disk for most, if not all, viewing angles.

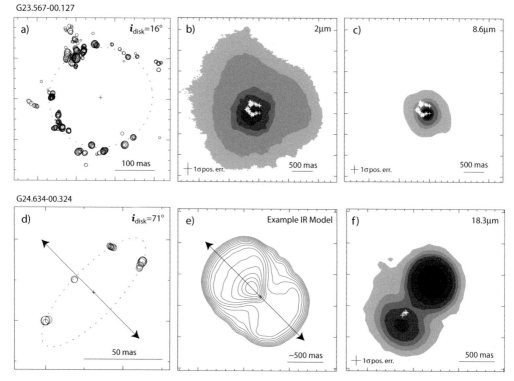

Figure 1. Results for G23.567-00.127 (a-c) and G24.634-00.324 (d-f). Methanol masers are shown as black open circles in panels a and d, and as white crosses in the other panels. Inferred disk inclinations from the maser rings are given in panels a and d (from fits by Bartkiewicz *et al.* 2009), with d also showing the angle (as projected on the sky) of the axis of the disk rotation/outflow expected (for panel a this axis is straight out of the page). The 1σ positional uncertainty between the maser positions and the infrared emission are given in the relevant panels. Panel e is not a fit to any of our data, but an example model of infrared emission from a massive young stellar object taken from Zhang & Tan (2011) for comparison.

Since it was of the highest importance that we know precisely where the maser rings are with respect to the dust emission in the infrared, we employed Gemini Observatory's high accuracy astrometric technique for our mid-infrared observations, which yields an absolute astrometry accuracy of $1\sigma = 90$ mas. For the near-infrared images, since the fields containing the science targets also contained near infrared stars from the 2MASS Point Source Catalog, we were able to use their measured positions to accurately define absolute astrometry of the NIRI images. A χ^2-minimization technique was used to register the 2MASS fields to the NIRI fields. This yielded an $1\sigma < 150$ mas error in the absolute astrometry of the near infrared images, where the actual errors are different for each source and dependent on the number of 2MASS stars on the NIRI field and their astrometric accuracy. Since we are comparing the offsets of infrared emission from the methanol masers, it should be stated that the absolute astrometrical error for the methanol masers are on the order of a few mas owing to the phase-referencing technique that was employed (Bartkiewicz *et al.* 2009). This means that when comparing the maser and infrared positions, the infrared astrometric errors quoted here for each source dominate in all cases.

4. The Results

Of the four methanol maser rings we observed in the infrared, none convincingly displayed the characteristics expected if the maser rings are indeed tracing circumstellar disks. We show the results for two of these in Figure 1. In the case of G23.567-00.127, the maser ring (Fig. 1a) is nearly circular, indicating that we should be looking almost straight down the outflow cavity (i.e., Scenario 1). This would imply that the infrared emission should be circularly symmetric with the infrared peak at the center of the maser ring. We see that the infrared emission peaks at 2 and 8.6 μm (Fig. 1b and c) do indeed lie near the maser ring center to within the positional errors, however the 2 μm morphology is fan-shaped as one would expect for the reflected light from a outflow cavity with a much more inclined geometry. In the case of G24.634-00.324, the maser ring is more highly elliptical (Fig. 1d), indicating that we should be seeing a more edge-on disk geometry, and perhaps a silhouette-like infrared morphology where the masers reside in the "dark lane" of the infrared emission (i.e., Scenario 3). Figure 1e shows the infrared emission geometry we would expected to see (adapted from the infrared models of Zhang & Tan 2011). We do indeed see this type of morphology in the infrared at 18.3 μm (Fig. 1f), as well as 8.6 μm (not shown), and the masers are coincident to within the positional errors with the dark lane. However the angle of the "silhouette disk" is almost 90° from the expected rotation axis as given by the maser ring (demonstrated by the arrow in Fig. 1d and e).

The other two sources in our sample have similar problems, leading to the conclusion that the *infrared emission from these sources does not seem to support the scenario where methanol maser rings trace circumstellar disks around young massive stars.*

Acknowledgements

AB and MS acknowledge support by the Polish Ministry of Science and Higher Education through grant N N203 386937.

References

Alvarez, C., Hoare, M., & Lucas, P. 2004, *A&A*, 419, 203
Bartkiewicz, A., Szymczak, M., van Langevelde, H. J., Richards, A. M. S., & Pihlström, Y. M. 2009, *A&A*, 502, 155
Chini, R., Hoffmeister, V., Kimeswenger, S., et al. 2004, *Nature*, 429, 155
Cotera, A. S., Whitney, B. A., Young, E., et al. 2001, *ApJ*, 556, 958
Cragg, D. M., Sobolev, A. M., & Godfrey, P. D. 2002, *MNRAS*, 331, 521
De Buizer, J. M. 2006, *ApJL*, 642, L57
De Buizer, J. M., Piña, R. K., & Telesco, C. M. 2000, *ApJS*, 130, 437
De Buizer, J. M., Walsh, A. J., Piña, R. K., Phillips, C. J., & Telesco, C. M. 2002, *ApJ*, 564, 327
Jayawardhana, R., Fisher, R. S., Hartmann, L., et al. 1998, *ApJL*, 503, L79
McCaughrean, M. J. & O'Dell, C. R. 1996, *AJ*, 111, 1977
Norris, R. P., Whiteoak, J. B., Caswell, J. L., Wieringa, M. H., & Gough, R. G. 1993, *ApJ*, 412, 222
Povich, M. S. & Whitney, B. A. 2010, *ApJl*, 714, L285
Sako, S., Yamashita, T., Kataza, H., et al. 2005, *Nature*, 434, 995
Shu, F. H. & Adams, F. C. 1987, *IAUS*, 122, 7
Stecklum, B. & Kaufl, H. 1998, *ESO Press Release*, PR 08/98
Telesco, C. M., Decher, R., Becklin, E. E., & Wolstencroft, R. D. 1988, *Nature*, 335, 51
Zhang, Y. & Tan, J. C. 2011, *ApJ*, 733, 55

Cosmic Masers - from OH to H$_0$
Proceedings IAU Symposium No. 287, 2012
R.S. Booth, E.M.L. Humphreys & W.H.T. Vlemmings, eds.

© International Astronomical Union 2012
doi:10.1017/S1743921312006837

Masers as evolutionary tracers of high-mass star formation

Shari L. Breen[1] and Simon P. Ellingsen[2]

[1]CSIRO Astronomy and Space Science, Australia Telescope National Facility, PO Box 76,
Epping NSW 1710, Australia
email: Shari.Breen@csiro.au

[2]School of Mathematics and Physics, University of Tasmania, GPO Box 37, Hobart, Tasmania
7000, Australia

Abstract. Determining an evolutionary clock for high-mass star formation is an important step towards realising a unified theory of star formation, as it will enable qualitative studies of the associated high-mass stars to be executed. We have carried out detailed studies of a large number of sources suspected of undergoing high-mass star formation and have found that common maser transitions offer the best opportunity to determine an evolutionary scheme for these objects. We have investigated the relative evolutionary phases of massive star formation associated with the presence or absence of combinations of water, methanol and main-line hydroxyl masers. The locations of the different maser species have been compared with the positions of 1.2 mm dust clumps, radio continuum, GLIMPSE point sources and Extended Green Objects. Comparison between the characteristics of coincident sources has revealed strong evidence for an evolutionary sequence for the different maser species in high-mass star formation regions. We present our proposed sequence for the presence of the common maser species associated with young high-mass stars and highlight recent advances. We discuss future investigations that will be made in this area by comparing data from the Methanol Multibeam (MMB) Survey with chemical clocks from the Millimetre Astronomy Legacy Team 90 GHz (MALT90) Survey.

Keywords. masers – ISM: molecules – stars: formation

1. Introduction

The process through which high-mass stars form is one of the hottest topics in modern astrophysics, with implications for fields as diverse as galactic evolution and the epoch of reionization. Masers are one of the best, if not *the best* signpost of young high-mass star formation regions. They are relatively common, intense and because they arise at centimetre wavelengths, are not affected by the high extinction that plagues observations in other wavelength ranges. To date progress has been slow towards the overall goal of utilising masers as tools to study star formation, however, the pace of advancement has recently accelerated. The proliferation of complementary high-resolution observations of star formation regions at millimetre through mid-infrared wavelengths means that this trend is likely to continue.

One of the difficulties in understanding the process through which high mass stars form is the lack of good sequential signposts in identifying different evolutionary stages of star formation, especially during the early stages while the young stellar objects are still embedded in their natal molecular clouds. Some types of masers are very common in high-mass star formation regions, for example, the 22 GHz water, 6.7 and 12.2 GHz methanol and 1.6 GHz hydroxyl transitions, while others are much more rare, such as the 37.7 GHz and 107 GHz methanol masers. Masers are created under a very specific set of physical conditions (e.g. Cragg *et al.* 2005) so those transitions which are common and strong, likely trace conditions that arise often and persist, while those that are rare likely trace rare or short-lived phases in the evolution of these regions.

Historically, attempts to construct a sequential timeline of the different common maser species in high-mass star formation regions have uncovered mixed results, primarily due to being based on heavily biased samples or small numbers of special sources. Furthermore, the observations often had poor spatial resolution and/or poor sensitivity, inducing confusion amongst sources with small angular separations, or failing to detect weak maser emission. There are some exceptions such as the early work by Forster & Caswell (1989) who showed that water masers appear prior to hydroxyl masers. Other works such as those by Walsh *et al.* (1998), Garay & Lizano (1999), Ellingsen (2006) and Fontani *et al.* (2010) have provided further strong evidence that masers make excellent evolutionary probes, but their samples are plagued to varying degrees by strong selection biases which limits the meaningfulness of their results.

At the last IAU maser conference in Alice Springs, Australia, Ellingsen *et al.* (2007) presented a 'straw man' evolutionary sequence for masers in high-mass star formation regions using a combination of new results and previously established facts from the literature. Here we outline advances towards establishing a robust evolutionary timeline for high-mass star formation regions that have been made since then.

2. Advances towards an accurate maser evolutionary timeline

Recently a great deal of progress has been made toward a robust maser evolutionary timeline, owing to the abundance of new high-sensitivity, high-resolution, large maser datasets together with complementary data. Chief amongst these important advances is the Methanol Multibeam survey (described in Section 2.1 and in more detail in Green *et al.* (2009)). Other contributors include 12.2 GHz methanol maser observations towards the MMB sources (Breen *et al.* 2010a, 2011, 2012) and new water maser catalogues with accurately determined positions (Breen *et al.* 2010b; Breen & Ellingsen 2011).

2.1. *The Methanol Multibeam Survey (MMB)*

The MMB is a complete survey for 6.7 GHz methanol masers within the Galactic plane (Green *et al.* 2009). These masers are especially useful as they exclusively trace sites of high-mass star formation (e.g. Minier *et al.* 2003) and trace the systemic velocities (e.g. Szymczak *et al.* 2007) of the regions that they are associated with, making them excellent tools for investigating not only the kinematics and the physical conditions of the regions that they are tracing, but also aspects like Galactic structure (Green *et al.* 2011).

The southern hemisphere component of the survey has been completed using the Parkes radio telescope. The location of Parkes allowed about 60% of the Galactic plane to be surveyed, from a longitude of 186°, through the Galactic centre, to a longitude of 60°, with a latitude extent of ±2°. The survey was completed in a scanning mode with a purpose built seven beam receiver which dramatically reduced the time required to complete the

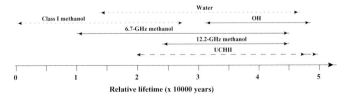

Figure 1. Proposed evolutionary timeline for the common maser species found towards high-mass star formation regions (Breen *et al.* 2010).

observations. During the course of the survey ~1000 methanol masers (3σ sensitivity of 0.7 Jy) were detected, of which ~40% are new detections, pinpointing sites of high-mass star formation. All detected sources without previously delivered accurate positions were followed up with an interferometer, chiefly the Australia Telescope Compact Array, resulting in positional accuracies of at least 0.4 arcsec for all detected masers.

The MMB maser catalogues in the longitude range: 186° (through the Galactic centre) to 20° have now been published and include the accurate source positions (Green *et al.* 2012; Caswell *et al.* 2011; Caswell *et al.* 2010; Green *et al.* 2010).

2.2. *Constructing an improved maser timeline*

The first quantitative timeline (see Fig. 1) for the common maser species in high-mass star formation was created by Breen *et al.* (2010). Using new, large samples of methanol masers at 6.7 (from the MMB survey) and 12.2 GHz, water masers and OH masers (Caswell 1998), together with mid-infrared (GLIMPSE), 1.2 mm dust continuum (Hill *et al.* 2005) and cm radio continuum data (Walsh *et al.* 1998), Breen *et al.* (2010, 2011, 2012) showed, through statistical analysis of their data, strong evidence that it is not only the presence or absence of the different maser species that can indicate the evolutionary stage of the high-mass star formation region that they are associated with, but that the properties of those masers can give even finer evolutionary details. Most notably, they show that both the luminosity and velocity range of detected 6.7 GHz and 12.2 GHz methanol and water maser emission increases as the star forming region evolves. The left panel of Fig. 2 shows that the 6.7 and 12.2 GHz methanol masers associated with OH masers (known to be associated with a later phase of evolution) tend to have higher luminosities. This is a notion supported by comparisons made by Wu *et al.* (2010) who showed that the most luminous methanol masers were associated with ammonia profiles indicative of more evolved objects than the methanol masers with lower luminosities.

Subsequent work by Ellingsen *et al.* (2011) used the results of Breen *et al.* (2010, 2011) to show that the presence of rare 37.7 GHz methanol masers may signal the end of the methanol maser phase. The right panel of Fig. 2 shows these 37.7 GHz methanol masers are associated with only the most luminous 6.7 and 12.2 GHz methanol masers, which combined with the rarity of these objects is consistent with them being a short lived phase towards the end of the 6.7 and 12.2 GHz methanol maser lifetime.

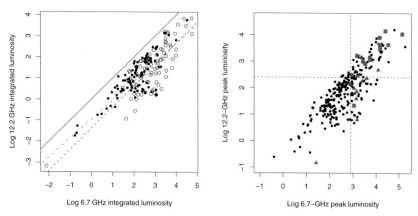

Figure 2. Left: Log 12.2 GHz versus log 6.7 GHz integrated luminosities (Breen *et al.* 2012). Methanol masers associated with OH masers are shown by unfilled circles and those without are shown by dots. **Right:** Luminosity of 12.2 GHz versus luminosity of the 6.7 GHz methanol masers (black dots). Sources that also exhibit emission from 37.7 GHz methanol masers are shown by purple squares (Ellingsen *et al.* 2011).

2.2.1. *The evolutionary stage traced by water and class I methanol masers*

As noted in Breen *et al.* (2010), the relative lifetime of water and class I methanol masers could only be estimated from the maser samples at the time. While some progress has been made towards understanding the evolutionary stage (or stages) that these transitions are tracing, such as the tendency of some class I methanol masers to be associated with a later evolution phase (e.g. Voronkov *et al.* 2010), much larger samples of sources need to be investigated before we can refine the timeline shown in Fig. 1.

3. Comparisons with Chemical Clocks

Non-masing molecular transitions have also proved to be useful as signposts of the evolutionary stage of high-mass stars (e.g. Longmore *et al.* 2007; Lee *et al.* 2004). Therefore an independent test of our proposed maser evolutionary scenario can be achieved by comparing evolutionary stage estimates derived from maser observations with those deduced from observations of certain molecules (or chemical clocks).

3.1. *The MALT90 Survey*

The Millimetre Astronomy Legacy Team 90 GHz (MALT90) Survey is a large project designed to characterise the physical and chemical evolution of dense cores (see Foster *et al.* (2011) for a description of the pilot survey). The survey will map more than 2000 cores with the Mopra radio telescope in 16 molecular lines near 90 GHz. At this frequency the spectrum is rich in diagnostic lines, spanning a range of excitation energies and densities, revealing distinct physical conditions and different stages in the chemical evolution of each core. The targets were selected from the 870 μm ATLASGAL survey (Schuller *et al.* 2009) of dust continuum emission and cover a wide range of evolutionary states, from pre-stellar cores, to protostellar cores, and finally, cores with HII regions. MALT90 data from the first two observing seasons (1135 cores observed in 2010 and 2011) have been publicly released and can be downloaded from the Australia Telescope Online Archive.

3.2. *Comparing the MMB to MALT90*

MALT90 provides the perfect dataset to test the maser evolutionary timeline. About 25% of the MALT90 cores observed are coincident with MMB sources, providing a large enough sample for meaningful analysis of the evolutionary stage associated with the masers, and also how this stage fits into the broader picture of star formation. Preliminary comparisons between these data shows that the methanol masers are overwhelmingly associated with MALT90 sources classified as 'protostellar'. Fig. 3 shows a typical example of MALT90 data towards a region associated with a methanol maser, exhibiting strong N_2H^+, but no HCO^+ or HCN. It is such obvious chemical signatures as these that can be used as a 'chemical clock' and test the current maser evolutionary timeline.

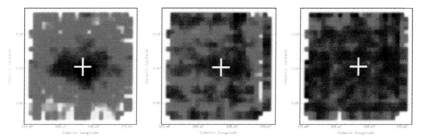

Figure 3. MALT90 maps of N_2H^+, HCO^+ and HCN for one source associated with a MMB methanol maser (white cross).

4. Conclusions

Masers provide the best opportunity for a robust and detailed timeline for the evolution of high-mass star formation. We have shown that not only the presence or absence of different maser species can give an indication of the evolutionary stage of the associated object, but that maser properties such as luminosity and velocity range can offer finer evolutionary detail. Investigations of large samples of water and class I methanol masers will allow their position on the current maser evolutionary timeline to be more accurately estimated. Comparisons with molecular data from the MALT90 survey will allow the maser evolutionary timeline to be independently confirmed. Preliminary results show that the 6.7 GHz methanol masers are associated with MALT90 sources with similar evolutionary stage classifications and chemical properties - showing great promise for future detailed investigations.

5. Acknowledgements

We acknowledge the contributions made to this work by the MMB and MALT90 survey teams, and in particular: James Caswell, James Green, Gary Fuller, Maxim Voronkov, Jill Rathborne, James Jackson and Jonathan Foster. The Australia Telescope Compact Array, Parkes and Mopra telescopes are part of the Australia Telescope National Facility which is funded by the Commonwealth of Australia for operation as a National Facility managed by CSIRO.

References

Breen, S. L., Ellingsen, S. P., Caswell, J. L., & Lewis, B. E., 2010a, *MNRAS*, 401, 2219
Breen, S. L., Ellingsen, S. P., Caswell, J. L., & Phillips, C. J., 2010b, *MNRAS*, 733, 406, 1487
Breen, S. L., Ellingsen, S. P., Caswell, J. L., Green, J. A., et al., 2011, *ApJ*, 733, 80
Breen, S. L. & Ellingsen, S. P., 2011, *MNRAS*, 416, 178
Breen, S. L., Ellingsen, S. P., Caswell, J. L., Green, J. A., et al., 2012, *MNRAS*, 421, 2511
Caswell, J. L., Fuller, G. A., Green, J. A., Avison, A., et al., 2010, *MNRAS*, 404, 1029
Caswell, J. L., Fuller, G. A., Green, J. A., Avison, A., et al., 2011, *MNRAS*, 417, 1964
Cragg, D. M., Sobolev, A. M., & Godfrey, P. D., 2005, *MNRAS*, 360, 533
Ellingsen, S. P., 2006, *ApJ*, 638, 241
Ellingsen S. P., et al., 2007, in Chapman J. M., Baan W. A., eds., Proc. IAU Symp., 242, Astrophysical Masers and their Environments. Cambridge Univ. Press, Cambridge, p. 213
Ellingsen, S. P., Breen, S. L., Sobolev, A. M., Voronkov, M. A., et al., 2011, *ApJ*, 742, 109
Fontani, F., Cesaroni, R., & Furuya, R. S., 2010, *A&A*, 517, 56
Forster, J. R. & Caswell, J. L., 1989, *A&A*, 213, 339
Foster, J. B., Jackson, J. M., Barnes, P. J., & Barris, E., 2011, *ApJS*, 197, 25
Garay, G. & Lizano, S., 1999, *PASP*, 111, 1049
Green, J. A., Caswell, J. L., Fuller, G. A., Avison, A. et al., 2009, *MNRAS*, 392, 783
Green, J. A., Caswell, J. L., Fuller, G. A., Avison, A. et al., 2010, *MNRAS*, 409, 913
Green, J. A., Caswell, J. L., McClure-Griffiths, N., Avison, A. et al., 2011, *ApJ*, 733, 27
Green, J. A., Caswell, J. L., Fuller, G. A., Avison, A. et al., 2012, *MNRAS*, 420, 3108
Lee, J.-E., Bergin, E. A., & Evanse, N. J., 2004, *ApJ*, 617, 360
Longmore S. N., et al., 2007, *MNRAS*, 379, 535
Minier, V., Ellingsen, S. P., Norris, R. P., & Booth, R. S., 2003, *A&A*, 403, 1095
Schuller, F., Menten, K., Contreras, K. M., & Wyrowski, F., 2009, *A&A*, 504, 415
Szymczak, M., Bartkiewicz, A., & Richards, A. M. S., 2007, *ApJ*, 706, 1609
Voronkov, M. A., Caswell, J. L., Ellingsen, S. P., & Sobolev, A. M., 2010, *MNRAS*, 405, 2471
Walsh, A. J., Burton, M. G., Hyland, A. R., & Robinson, G., 1998, *MNRAS*, 301, 640
Wu, Y. W., Xu, Y., Pandian, J. D., Yang, J., et al., 2010, *ApJ*, 720, 392

Cosmic Masers - from OH to H$_0$
Proceedings IAU Symposium No. 287, 2012
R.S. Booth, E.M.L. Humphreys & W.H.T. Vlemmings, eds.

© International Astronomical Union 2012
doi:10.1017/S1743921312006849

Class I methanol masers in low-mass star formation regions

S. V. Kalenskii[1], V. I. Slysh[1], L. E. B. Johansson[2], P. Bergman[2], S. Kurtz[3], P. Hofner[4], and C. M. Walmsley[5]

[1] Astro Space Center, Lebedev Physical Institute, 84/32 Profsoyuznaya st., Moscow, 117997, Russia
email: kalensky@asc.rssi.ru

[2] Onsala Space Observatory, Chalmers University of Technology, 439 92 Onsala, Sweden
email: pbergman@chalmers.se

[3] Centro de Radioastronomía y Astrofísica, Universidad Nacional Autonoma de México
(Morelia, Michoacán, México)
email: s.kurtz@crya.unam.mx

[4] Physics Department, New Mexico Tech., 801 Leroy Pl., Socorro, NM 87801, and National
Radio Astronomy Observatory, Socorro, NM 87801, USA
email: hofner_p@yahoo.com

[5] Osservatorio Astrofisico di Arcetri, Largo E. Fermi 5,1-50125 Firenze, Italy
email: walmsley@arcetri.astro.it

Abstract. Four Class I maser sources were detected at 44, 84, and 95 GHz toward chemically rich outflows in the regions of low-mass star formation NGC 1333I4A, NGC 1333I2A, HH25, and L1157. One more maser was found at 36 GHz toward a similar outflow, NGC 2023. Flux densities of the newly detected masers are no more than 18 Jy, being much lower than those of strong masers in regions of high-mass star formation. The brightness temperatures of the strongest peaks in NGC 1333I4A, HH25, and L1157 at 44 GHz are higher than 2000 K, whereas that of the peak in NGC 1333I2A is only 176 K. However, a rotational diagram analysis showed that the latter source is also a maser. The main properties of the newly detected masers are similar to those of Class I methanol masers in regions of massive star formation. The former masers are likely to be an extension of the latter maser population toward low luminosities of both the masers and the corresponding YSOs.

Keywords. masers, ISM: jets and outflows, ISM: molecules

1. Introduction

In spite of a number of observations and theoretical works, the nature of Class I methanol masers is still unknown. This is partly because until recently these masers have been observed only in regions of massive star formation, which are typically distant (2–3 kpc from the Sun or farther) and highly obscured at optical and even NIR wavelengths. In addition, high mass stars usually form in clusters. These properties make it difficult to resolve maser spots and to associate masers with other objects in these regions. In contrast, regions of low-mass star formation are much more widespread and many of them are only 200–300 pc from the Sun; they are less heavily obscured than regions of high-mass star formation, and there are many isolated low-mass protostars. Therefore, the study of masers in these regions might be more straightforward compared to that of high-mass regions, and hence, the detection of Class I masers there might have a strong impact on maser exploration.

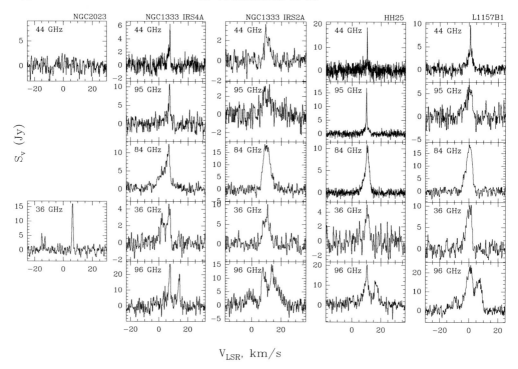

V_{LSR}, km/s

Figure 1. Spectra of the newly detected masers.

Bearing this in mind, we undertook a search for Class I methanol masers in regions of low-mass star formation. Since the most common viewpoint is that these masers arise in postshock gas in the wings of bipolar outflows (Plambeck & Menten, 1990; Chen *et al.* 2009) our source list was composed of these objects. The naive expectation is to find methanol masers towards bright thermal sources of methanol; therefore the basis of our source list consists of "chemically rich outflows", where methanol abundance is significantly enhanced relative to that in quiescent gas. Because methanol enhancement has been detected in young, well-collimated outflows from Class 0 and I sources, we included several such objects in our list regardless of whether methanol enhancement had been previously found there. A subsample of our list consisted of YSOs with known outflows and/or H_2O masers located in Bok globules. Like other objects from our list, these YSOs are typically isolated objects of low or intermediate mass, located in nearby (<500 pc) small and relatively simple molecular clouds. In total, our source list consisted of 37 regions which harbor 46 known outflows driven by Class 0 and I low-mass protostars, taken from the literature. All of them were observed in the $7_0 - 6_1 A^+$ transition at 44 GHz, where the strongest Class I masers have been found so far. In addition to the $7_0 - 6_1 A^+$ transition, most sources were observed in other Class I maser lines, namely, in the $4_{-1} - 3_0 E$ line at 36 GHz, in the $5_{-1} - 4_0 E$ line at 84 GHz, and in the $8_0 - 7_1 A^+$ line at 95 GHz, as well as the "purely thermal" $2_K - 1_K$ lines at 96 GHz.

2. Observations and results

Single-dish observations. The single-dish observations are described in detail by Kalenskii *et al.* (2006, 2010a). They were carried out with the 20-m radio telescope of the Onsala Space Observatory (OSO) during several observing sessions in 2004–2011. As a result, we detected maser candidates at 44 GHz towards NGC 1333I2A, NGC 1333I4A, HH 25

Table 1. Parameters of maser sources determined by the VLA observations.

Source	R.A. (J2000)	Dec. (J2000)	Major axis ($''$)	Minor axis ($''$)	T_{BR} (K)	V_{lsr} (km s^{-1})
NGC 1333I2A M1	03 29 00.802	31 14 21.32	2.2	1.4	170	10.9
NGC 1333I2A M2	03 29 01.422	31 14 18.80	2.7	2.0	35	8.8
NGC 1333I4A	03 29 10.829	31 13 18.68	1.0	0.4	2400	6.9
HH25 M1	05 46 08.071	-00 14 05.66	2.7	0.0	inf	9.6
HH25 M2	05 46 07.967	-00 14 02.42	0.7	0.0	inf	9.6
L1157 M1	20 39 10.033	68 01 42.20	0.4	0.2	53000	0.8
L1157 M2	20 39 09.465	68 01 15.59	1.9	0.7	470	1.7

and L1157. Toward NGC 1333I4A and HH 25, narrow features were also found at 95 and 84 GHz. In addition, a narrow line was detected at 36 GHz toward the blue lobe of an extremely high-velocity outflow in the vicinity of the bright reflection nebula NGC 2023. The source spectra are shown in Fig. 1.

VLA/EVLA observations. To check whether the newly detected sources are really masers we observed them with the NRAO† VLA/EVLA array in the D configuration, which provides an angular resolution about 1.″5 at 44 GHz. L1157 was observed with the VLA on March 17, 2007; the other sources were observed with the EVLA on August 08, 2010. The data were reduced using the NRAO Astronomical Image Processing System (AIPS) package. The source parameters are presented in Table 1.

3. Are the newly detected sources really masers?

The small sizes and high brightness temperatures at 44 GHz indicate that the newly detected sources are masers. The exceptions are NGC 2023, which was not found at 44 GHz, and NGC 1333I2A, with a line brightness temperature of only 170 K (Table 1). The nature of the 36 GHz line in the blue lobe of the bipolar outflow in NGC 2023 is unclear. On the one hand, the line is fairly narrow, and offset measurements showed that the source is compact at least with respect to the 105-arcsec Onsala beam. These properties suggest that the source is a maser. This assumption has further support in the fact that the line LSR velocity, $\approx$6.5 km s^{-1}, is less than the systemic velocity of about 10 km s^{-1}. On the other hand, the line has no counterpart at 44 GHz, which is more typical for thermal emission. Note, however, that there are known masers at 36 GHz without 44-GHz counterparts; in particular, no 44 GHz emission was found at the velocity of a fairly strong 36-GHz maser detected 3′ north of DR21(OH) by Pratap *et al.* (2008). Therefore, we tentatively conclude that the narrow line in NGC 2023 is a maser.

The fairly low brightness temperature and finite sizes (Table 1) of NGC 1333I2A (M1 and M2) suggest that they are thermal sources. However, a rotational diagram analysis (Kalenskii *et al.*, in prep.) shows that they are low-gain masers or a cluster of weak masers.

4. Properties of the new masers

Association with chemically rich outflows. New masers were found towards the lobes of outflows in NGC 1333I4A, NGC 1333I2A, NGC 2023, HH25, and L1157. These outflows are known to be chemically rich outflows with enhanced methanol abundances.

† The National Radio Astronomy Observatory is operated by Associated Universities, Inc., under contract with the National Science Foundation.

Comparison of the VLA maps with high-resolution maps of thermal methanol and other molecules shows that the masers coincide with chemically rich gas clumps, where the abundances of methanol and other molecules are enhanced (e.g., Gibb & Davis (1988); Bachiller *et al.* (1998); Benedettini *et al.* (2007)). In L1157, the masers are located in gas clumps, which, according to chemical modeling of Viti *et al.* (2004), probably pre-existed the outflow.

LSR velocities and intensities. Comparison of the maser LSR velocities with those of thermal methanol lines, observed in the same directions, show that these velocities coincide within 0.5 km/s. This coincidence occurs even when the LSR velocities of some other molecular lines toward the maser positions are significantly different. The LSR velocities of both maser and thermal methanol lines are usually close to the systemic velocities. An exception is the 36-GHz maser in the EHV outflow NGC 2023. Its radial velocity is less than the systemic velocity by about 3.5 km $^{-1}$. Note that Voronkov (this volume) has detected a high-velocity Class I maser just at 36 GHz.

Maser intensities. The new masers are weaker than the bright masers typical in regions of massive star formation. However, they obey the same relationship between the maser and YSO luminosities as reported by Bae *et al.* (2011) for masers in regions of high- and intermediate-mass star formation, thus extending this relationship toward low luminosities (Kalenskii *et al.*, in prep).

Variability. Several sessions of repeated observations of NGC 1333I4A, HH25, and L1157 at 44 GHz were performed in 2008–2011. No notable variations were found. Slight changes in line intensities can be attributed to poor signal-to-noise ratios and calibration uncertainties. However, further monitoring of these sources is desirable in order to search for flares similar to that which occurred in DR 21(OH).

To summarize, the main properties of the newly detected masers are similar to those of Class I methanol masers in regions of massive star formation. The former masers are likely to be an extension of the latter maser population toward lower luminosities of both the masers and the corresponding YSOs.

5. Maser models

The fact that the maser LSR velocities coincide with the systemic velocities allows us to conclude that the masers appear in dense clumps of gas, probably pre-existing the outflows. However, the exact nature of the masers remains unknown. Sobolev *et al.* (1998) suggested that compact maser spots arise in extended, turbulent clumps because in a turbulent velocity field the coherence lengths along some directions are larger than the mean coherence length, resulting in a random increase of the optical depth absolute values along certain sight lines in a clump. According to Kalenskii *et al.* (2010b) such a model can easily explain the observed brightness of the maser lines, but within the framework of this model it is difficult to explain why *single peaks* dominate the maser emission in the L1157 clumps. However, natural additional assumptions, such as the existence of shocks or centrally condensed clumps, makes it possible to explain the observational data.

An examination of the maser spectra in L1157 may lead to another interpretation of our results. Both 44 GHz masers detected in this source have double line profiles. It is known that a double thermal line with a "blue asymmetry" may be a signature of collapse (Zhou (1996)). Contrary to this, the masers in L1157 exhibit a "red asymmetry". However, just such an asymmetry is what one would expect for Class I masers arising in a collapsing clump. This model is discussed in more detail by Kalenskii *et al.* (2010b). Note that this model, if correct, is specific for the masers in L1157; no other maser in our sample exhibits a double line profile.

Acknowledgements

The work was financially supported by RFBR (grants No. 04-02-17547, 07-02-00248, and 10-02-00147-a), and Federal National Scientific and Educational Program (project number 16.740.11.0155). P.H. acknowledges partial support from NSF grant AST 0908901. S.Kurtz acknowledges support from UNAM DGAPA grant IN101310. The Onsala Space Observatory is the Swedish National Facility for Radio Astronomy and is operated by Chalmers University of Technology, Göteborg, Sweden, with financial support from the Swedish Research Council and the Swedish Board for Technical Development.

References

Bachiller, R., Codella, C., Colomer, F., Liechti, S., & Walmsley, C. M. 1998, *A&A*, 335, 266

Bae, J.-H., Kim, K.-T., Youn, S.-Y., Kim, W.-J., Byun, D.-Y., Kang, H., & Oh, C. S., 2011, *ApJS* 196, 21

Benedettini, M., Viti, S., Codella, C., Bachiller, R., Gueth, F., Beltran, M. T., Dutrey, A., & Guilloteau, S. 2007, *Mon. Not. R. Astron. Soc.*, 381, 1127

Chen, X., Ellingsen, S. P., & Shen, Z.-Q. 2009, *Mon. Not. R. Astron. Soc.*, 396, 1603

Gibb, A. & Davis, C. J. 1998, *Mon. Not. R. Astron. Soc.*, 298, 644

Kalenskii, S. V., Promyslov, V. G., Slysh, V. I., Bergman, P., & Winnberg, A. 2006, *Astron. Rep.*, 50, 289

Kalenskii, S. V., Johansson, L. E. B., Bergman, P., Kurtz, S., Hofner, P., Walmsley, C. M., & Slysh, V. I., 2010, *Mon. Not. R. Astron. Soc.*, 405, 613

Kalenskii, S. V., Kurtz, S., Slysh, V. I., Hofner, P., Walmsley, C. M., Johansson, L. E. B., & Bergman, P. 2010, *Astron. Rep.*, 54, 932

Plambeck, R. L. & Menten, K. M. 1990, *ApJ*, 364, 555

Pratap, P., Shute, P. A., Keane, T. C., Battersby, C., & Sterling S. 2008, *AJ*, 135, 1718

Sobolev, A. M., Wallin, B. K., & Watson, W. D. 1998, *ApJ*, 498, 763

Zhou S. 1996, in van Dishoeck E. F., ed, Proc. IAU Symp. 178, Molecules in Astrophysics: Probes and Processes, Kluwer, Dordrecht, p. 195

Viti, S., Codella, C., Benedettini, M., & Bachiller, R. 2004, *Mon. Not. R. Astron. Soc.*, 350, 1029

Cosmic Masers - from OH to H_0
Proceedings IAU Symposium No. 287, 2012
R.S. Booth, E.M.L. Humphreys & W.H.T. Vlemmings, eds.

© International Astronomical Union 2012
doi:10.1017/S1743921312006850

Dynamical detection of a magnetocentrifugal wind driven by a 20 M⊙ YSO

L. J. Greenhill[1], C. Goddi[2], C. J. Chandler[3], E. M. L. Humphreys[2], and L. D. Matthews[4]

[1] Harvard-Smithsonian CfA, 60 Garden St, Cambridge, MA, USA
email: greenhill@cfa.harvard.edu

[2] ESO, Karl-Schwarzschild-Strasse 2, 85748 Garching, Germany
email: cgoddi@eso.org, ehumphre@eso.org

[3] NRAO, PO Box O, Socorro, NM, USA
email: cchandle@nrao.edu

[4] MIT Haystack Observatory, Off Route 40, Westford, MA, USA
email: lmatthew@haystack.mit.edu

Abstract. We have tracked the proper motions of ground-state λ7mm SiO maser emission excited by radio Source I in the Orion BN/KL region. Based on dynamical arguments, Source I is believed to be a hard 20 M⊙ binary. The SiO masers trace a linear bipolar outflow (NE-SW) 100 to 1000 AU from the binary. The median 3D velocity is 18 $km\,s^{-1}$. An overlying distribution of 1.3 cm H_2O masers betrays similar characteristics. The outflow is aligned with the rotation axis of an edge-on disk and wide angle flow known inside 100 AU. Gas dynamics and emission morphology traced by masers around Source I provide dynamical evidence of a magnetocentrifugal disk-wind around this massive YSO, notably a measured gradient in line-of-sight velocity perpendicular to the flow axis, in the same direction as the disk rotation and with comparable speed. The linearity of the flow, despite the high proper motion of Source I and the proximity of dense gas associated with the Orion Hot Core, is also more readily explained for a magnetized flow. The extended arcs of ground-state maser emission bracketing Source I are a striking feature, in particular since dust formation occurs at smaller radii. We propose that the arcs mark two C-type shocks at the transition radius to super-Alfvénic flow.

Keywords. ISM: individual (Orion KL) — ISM: jets and outflows — ISM: Kinematics and dynamics — ISM: molecules — masers — stars: formation

1. Orion BN/KL

The predominant outflow in Orion BN/KL ($\sim$420 pc distant) is an uncollimated expansion of up to 300 $km\,s^{-1}$, traced by shocked fingers of H_2 and by high-velocity CO. It is not directly tied to any of the known YSOs. Rather, the apparent origin is proximate (in time and position) with the closest approach of three YSOs (radio Source I, infrared source n, and the BN object) inferred from their diverging proper motions (Gómez *et al.* 2008, Goddi *et al.* 2011a, Bally *et al.* 2011). With regard to this inferred dynamical history, BN/KL may indeed be unusual, but elements are familiar, specifically a disk-outflow model of Source I . The gas structure and dynamics have been resolved in studies of SiO and H_2O masers (Goddi *et al.* 2009, and references therein). In these studies, the position-velocity structure of the SiO masers, resolved with very long baseline interferometry (VLBI), has been shown to outline a disk and bipolar wide-angle outflow at radii of 10-100 AU (Greenhill *et al.* 2004, Matthews *et al.* 2010, Kim *et al.* 2008), where proper motions clearly separate outflow and rotating disk components (Matthews *et al.* 2010).

Table 1. Observing Log

Date (yymmdd)	Tracer	Array	Beam (mas) (°)	RMS[a] (mJy)
1999.08.28	SiO v=0, 1	A+	61×43 @ $+44°$	10-40
2002.03.31	SiO v=0, 1	A+	55×45 @ $-31°$	6-20
2006.04.15	SiO v=0, 1	A+	53×39 @ $-3.8°$	2.5-8.5
2009.01.12	SiO v=0, 1	A	53×43 @ $+3.6°$	3-20
2001.01.23	$H_2O\ 6_{16} - 5_{23}$	A+	113×48 @ $+30°$	> 6

[a] For a $1.3\,\mathrm{km\,s^{-1}}$ channel. Images were dynamic range limited.

We describe the angular distributions and evolution of ground-state $\lambda 7\,\mathrm{mm}$ SiO and $\lambda 1.3\,\mathrm{cm}$ H_2O maser emission at radii of 100 to 1000 AU around Source I . The data reported here densely sample the outflow. A self-consistent picture of a high-mass YSO emerges that reinforces the disk-outflow model for 10-100 AU, introduces the prospect of a magnetic field that is dynamically important at radii up to a few $\times 100\,\mathrm{AU}$, and suggests an explanation for excitation of the ground-state SiO and the H_2O masers.

2. Observations and Data Reduction

We observed two transitions of SiO and one of H_2O toward Source I with the NRAO VLA at multiple epochs over 9 years (Table 1).

SiO– At each epoch, we correlated two simultaneous, single-polarization basebands, one tuned to the $v = 0$ transition and the other to the much stronger $v = 1$ transition. We used a 6.25 MHz bandwidth ($-13.7 < \mathrm{V_{lsr}} < 29.4\,\mathrm{km\,s^{-1}}$) and 97.656 kHz ($0.65\,\mathrm{km\,s^{-1}}$) channel spacing. We selected a strong $v = 1$ Doppler component as a reference to self-calibrate antenna gain and tropospheric fluctuations on 10-second time scales. Scans of

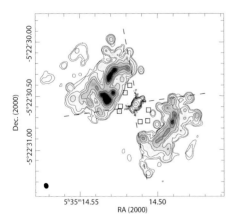

Figure 1. Nested tracers around Source I at $\lambda 7\,\mathrm{mm}$. Velocity-integrated flux from the $v = 0\ J = 1 \rightarrow 0$ transition of SiO (grayscale with contours) brackets continuum emission from Source I - contours at center (Goddi *et al.* 2011a). Extending away from the continuum source in four arms is $v = 1$ and $v = 2$ SiO maser emission, mapped with VLBI (Matthews *et al.* 2010), chiefly at radii $< 50\,\mathrm{AU}$; isolated clumps of emission are outlined for clarity (boxes). The arms delimit the edges of a wide angle outflow directed NE-SW (dashed lines).

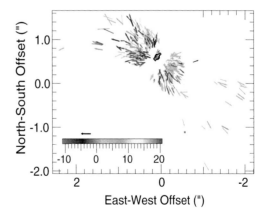

Figure 2. Outflow at radii of 10-1000 AU from Source I, including outward proper motions for SiO $v = 0$ emission clumps from 1999 to 2009 (bars). Contours show $\lambda 7\,\mathrm{mm}$ continuum emission that marks the relative location of Source I as in Figure 1. Colors indicate LSR radial velocity in $\mathrm{km\,s^{-1}}$ (color bar); the systemic velocity is $5\,\mathrm{km\,s^{-1}}$. The horizontal black arrow indicates a proper motion of $30\,\mathrm{km\,s^{-1}}$.

J0541–056 enabled calibration of slowly varying phase offsets between the signal paths for the two observing bands, which enabled us to transfer the antenna and tropospheric calibration to the band containing the (weaker) $v = 0$ line. Absolute and relative astrometry of the maser emission, using close by BN and J0541–0541 as references, was thermal noise limited (Table 1). Registration with respect to Source I is accurate to ~10 mas.

We tracked proper motions for 447 $v = 0$ spots for between 2 and 4 epochs ($> 5\sigma$). Lifetimes were in excess of several years, much stronger than for $v = 1$ and 2 features. To estimate proper motions for each channel map, we fit each spot with a two-dimensional elliptical Gaussian. Images of $v = 0$ emission were noise-limited, and the relative position errors were given by beamwidth divided by 2× SNR per channel, or a few mas for moderately bright emission. Cross-referencing of maser spots among different epochs could be done by eye because the structure of the emission in each velocity channel persisted. Proper motions were calculated using an error-weighted linear least-squares fit to fitted positions. To correct for the motion of the reference $v = 1$ Doppler component, we computed proper motions relative to the strong $v = 0$ feature at +2.7 km s^{-1} (imparting to it an apparent zero motion) and then subtracted the mean motion of all those measured ($\dot{\alpha} = 6.11 \pm 0.02$ km s^{-1} and $\dot{\delta} = 23.26 \pm 0.04$ km s^{-1}).

H$_2$O– We observed H$_2$O maser emission between $V_{lsr} = -138$ and 137 km s^{-1} and report here on emission detected in the so-called H$_2$O Shell (Genzel *et al.* 1981) that is associated with Source I (Table 1). We divided the source spectrum into overlapping 1.56 MHz bands and observed these in pairs with 0.16 km s^{-1} channel spacing after Hanning smoothing. One band within each pair was tuned to include the line emission peak near −4.5 km s^{-1}. We used the emission at −3.86 km s^{-1} (1700 Jy) to obtain self-calibration solutions every 10 seconds. Scans of J0541–056 every 45 minutes enabled calibration of instrumental phase offsets between bands. Registration of emission in the different bands is accurate to < 2 mas. Absolute astrometry was derived from calibration against J0541–0541. The same techniques were used to predict the position of J0605–085, achieving an absolute position uncertainty of 2 mas. From detection of BN in continuum and pseudo-continuum images toward Source I, we

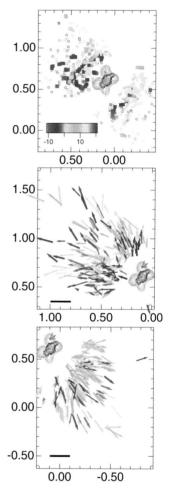

Figure 3. (*top*) SiO $v = 0$ and H$_2$O masers (open circles and squares, respectively), velocity-integrated SiO $v = 1$ masers as mapped with the VLA (orange contours), and the Source I λ7 mm continuum (black contours; see Figure 1). (*middle*) Proper motions for SiO emission clumps, 1999 to 2009, expanded to show proper motions in the lobe northeast of Source I . (*bottom*) Expanded view to the southwest. The horizontal black bar corresponds to a proper motion of 40 km s^{-1} in one year.

obtained for BN a J2000 position of $05^h 35^m 14\overset{s}{.}1131$, $-05°22'22\overset{''}{.}794$ with 3 mas uncertainty (epoch 2001.06). This lies $\sim 1\sigma$ from the trend line at 22 GHz of Rodriguez *et al.* (2005) for epoch 2001.06 (offset 13 and 2 mas in right ascension and declination, respectively). We also detected continuum emission of Source I at 6σ, lying $+5\overset{''}{.}950$ and $-7\overset{''}{.}737$, $\pm 0\overset{''}{.}007$ with respect to BN. The difference from the measurement of Rodriguez *et al.* (2005) is (+2, -6) mas or $< 1\sigma$ (cf. Goddi *et al.* 2011a).

3. Discussion

Morphology & Dynamics– The most intense $v = 0$ SiO maser emission occupies two arcs bracketing Source I, at a radius of ~ 100 AU. This is just outside the maximum radius at which isolated $v = 1$ masers are observed (Figure 1). The arcs subtend about the same opening angle as the nearly radial arms at smaller radii along which $v = 1$ and $v = 2$ emission features move outward. This is believed to trace the edges of a bipolar outflow orthogonal to the edge-on disk whose rotation the vibrationally excited emission also traces (Matthews *et al.* 2010).

The angular and velocity structure of the $v = 0$ emission is suggestive of outflow in the plane of the sky; there is no offset in the mean velocities of the two lobes, at the $O(1)$ km s^{-1} level. The median proper motion for maser spots tracked for >2 epochs is 18 km s^{-1}, and the range of 3D velocities in the local frame ($V_{LSR} = 5$ km s^{-1}) is 4 to 36 km s^{-1}. The H$_2$O emission displays a similar range of line-of-sight velocity. A 20 km s^{-1} expansion in the angular extent of the distribution over ~ 8 years (Greenhill *et al.* 1998) is consistent with the median SiO maser proper motion. The mean position angle measured from the emission locus and separately from the proper motions is 56°.

The linearity of the outflow is notable because (i) Source I lies at the edge of the dense gas associated with the Orion Hot Core (Goddi *et al.* 2011b), (ii) the 18 km s^{-1} outflow is comparable to the 12 km s^{-1} proper motion of the YSO toward the densest portion of the Hot Core, (iii) SiO maser dynamics inside $O(1000)$ AU are indicative of a ~ 500 yr dynamical time for the outflow, and (iv) the crossing time for Source I from the point at which the YSO underwent dynamical interaction with BN (separation 50 AU on the sky) is also ~ 500 yr (i.e., the flow is contemporaneous with the interaction with BN).

For hydrodynamic flow, there being no sweeping back suggests that the momentum flux exceeds that of the ambient medium into which Source I is moving. Ground-state SiO maser emission requires densities of 10^6-10^7 cm^{-3} for collisional pumping (e.g., Goddi *et al.* 2009). Given the comparable velocities (12 vs 18 km s^{-1}), the density of ambient material must be $\ll 10^6$ cm^{-3} (still less if the maser medium reflects enhancement over the mean flow density, as by shocks). However, ambient gas densities in the vicinity of the Hot Core are large, $O(10^6)$ cm^{-3} (e.g., Goddi *et al.* 2011b), and as a result some deflection may be anticipated. Alternatively, there could be a fast flow due to stellar winds ($O(100)$ km s^{-1} for a $10_\odot$ B-star) that fragments and infiltrates the ambient material, generating a trail of maser emission, but limb brightening along leading edges and asymmetry in response to density gradients (which is not seen) would be anticipated.

Magnetocentrifugal Wind– The alternate possibility of a magnetohydrodynamic disk wind is raised by a discernible rotation signature about the major axis of the flow in each lobe. Toward the SE-facing edge, there is a greater preponderance of blueshifted emission; redshifted emission lies preferentially toward the NW. There is considerable scatter about this trend (e.g., there is still relatively blue shifted emission toward the NW face), but it is parallel and only somewhat smaller (15-20 km s^{-1} vs 20-30 km s^{-1}) than the offset seen plainly among vibrationally excited SiO masers at radii of tens of AU and from which rotation is inferred (Matthews *et al.* 2010).

The data presented here are suggestive of these dynamics being communicated from scales of O(10) AU to at least O(100) AU. The most natural mechanism would be action by magnetic field lines threading the flow. This addition also raises the energy density in the flow and more readily enables linear flow in a dense medium. We note that re-assessment of data for the vibrationally excited SiO masers provides some confirming evidence. Matthews *et al.* (2010) conservatively interpreted the maser data in the context of Keplerian motion and the dominant action of gravity (cf. Kim *et al.* 2008), inferring a dynamical mass of $\sim 8\,M_\odot$. However, the value is inconsistent with the more direct estimate of $20\,M_\odot$ obtained by Goddi *et al.* (2011a) under the assumption that BN and Source I are in recoil. Hence, reproducing the SiO dynamics observed with VLBI requires invocation of non-gravitational forces as well.

Super Alfvénic Flow– Why does intense ground-state SiO and H_2O maser emission arise suddenly at 100AU? Why are SiO and H_2O masers apparently intermixed when the densities required for emission differ by (conservatively) an order of magnitude? For a YSO luminosity on the order of $2\times10^4\,L_\odot$ (i.e., $2\times10\,M_\odot$), the sublimation radius of dust will be $\ll 100$ AU, and since maser emission requires a high gas phase abundance, its appearance so far out and over a significant interval of the flow are significant.

We propose that the arcs of maser emission at ~ 100 AU radius indicate the onset of strong shocks in dusty outflowing material. Hydromagnetic C-type shocks as slow as 10-$20\,\mathrm{km\,s^{-1}}$ (comparable to the flow speed) are capable of sputtering grains (Schilke *et al.* 1997), a process that would raise gas phase abundance of SiO and H_2O. Formation of two continuous shock structures subtending broad ranges of polar angle and narrow ranges in radius indicates a systematic change in physical conditions. Transition to a super Alfvénic flow and consequent shock formation may trigger the observed (re)appearance of maser emission in the outflow at ~ 100 AU. Conditions conducive to shocks will arise if the Alfvén velocity declines with radius and crosses the outflow velocity, which does not vary considerably with radius. Where density falls quadratically, this may occur if the field declines at least linearly (not unlikely). One observational prediction of this proposal is that the inner edge of maser emission will not expand on the sky with time.

Maintenance of gas phase abundance well downstream, as indicated by preponderance of maser emission there, suggests that high gas phase abundance and pump excitation persist. In view of flow speed in excess of sound and Alfvén speeds, continued shocking may be expected, although long cooling timescales may also be responsible. In this region, fading of the velocity gradient along the flow minor axis and concomitant redirection of proper motions to be more downstream suggests increasing decoupling of the neutral gas from the field, as anticipated in MHD disk wind models.

References

Bally, J., Cunningham, N. J., Moeckel, N., *et al.* 2011, *ApJ*, 727, 113
Genzel, R., Reid, M. J., Moran, J. M., & Downes, D. 1981, *ApJ*, 244, 884
Goddi, C., Greenhill, L. J., Chandler, C. J., *et al.* 2009, *ApJ*, 698, 1165
Goddi, C., Greenhill, L. J., Humphreys, E. M. L., *et al.* 2011b, *ApJ*, 739, L13
Goddi, C., Humphreys, E. M. L., Greenhill, L. J., *et al.* 2011a, *ApJ*, 728, 15
Gómez, L., Rodríguez, L. F., Loinard, L., *et al.* 2008, *ApJ*, 685, 333
Greenhill, L. J., Gwinn, C. R., Schwartz, C., *et al.* 1998, *Nature*, 396, 650
Greenhill, L. J., Gezari, D. Y., Danchi, W. C., *et al.* 2004, *ApJ*, 605, L57
Kim, M. K., *et al.* 2008, *PASJ*, 60, 991
Matthews, L. D., Greenhill, L. J., Goddi, C., *et al.* 2010, *ApJ*, 708, 80
Rodríguez, L. F., Poveda, A., Lizano, S., & Allen, C. 2005, *ApJL*, 627, L65
Schilke, P., Walmsley, C. M., Pineau des Forets, G., & Flower, D. R. 1997, *A&A*, 321, 293

Cosmic Masers - from OH to H$_0$
Proceedings IAU Symposium No. 287, 2012
R.S. Booth, E.M.L. Humphreys & W.H.T. Vlemmings, eds.

© International Astronomical Union 2012
doi:10.1017/S1743921312006862

The W51 Main/South SFR complex seen through 6-GHz OH and methanol masers

Sandra Etoka, Malcolm D. Gray and Gary A. Fuller

Jodrell Bank Centre for Astrophysics
School of Physics and Astronomy
The University of Manchester, Manchester M13 9PL, UK
email: Sandra.Etoka@googlemail.com

Abstract. W51 Main/South is one of the brightest and richest high-mass star-forming regions (SFR) in the complex W51. It is known to host many ultra-compact HII (UCHII) regions thought to be the site of massive young stellar objects. Maser emission from various species is also found in the region. We have performed MERLIN astrometric observations of excited-OH maser emission at 6.035 GHz and Class II methanol maser emission at 6.668 GHz towards W51 to investigate the relationship between the maser emission and the compact continuum sources in this SFR complex. Here we present the astrometric distributions of both 6.668-GHz methanol and 6.035-GHz excited-OH maser emission in the W51 Main/South region. The location of maser emission in the two lines is compared with that of previously published OH groundstate emission. The interesting coherent velocity and spatial structure observed in the methanol maser distribution as well as the relationship of the masers to infall or outflow in the region are discussed. It appears that the masers are excited by multiple objects potentially at different stages of evolution.

Keywords. astrometry, circumstellar matter, masers, magnetic fields, polarization, stars: formation, ISM: individual objects: W51, radio lines: ISM

1. Introduction

W51 is one of the most massive Giant Molecular Clouds (GMCs) in the Galaxy. This very rich high-mass star-forming region (SFR) environment is located in the Sagittarius spiral arm at 5.4 ± 0.3 kpc (Sato *et al.* 2010). W51 Main/South, one of the brightest regions of the W51 complex, is known to host many ultra-compact (UC) HII regions thought to contain massive young stellar objects at various stages of evolution. The UCHII regions are labelled W51e1, e2, e3 and e4 (Gaume, Johnston & Wilson 1993) and W51e8 (Zhang & Ho 1997). Of these, W51e2 was found to be composed of four more compact continuum sources labelled W51e2-W, e2-E, e2-NW and e2-N (Shi, Zhao & Han 2010a). Zhang, Ho, & Ohashi (1998) found evidence of gravitational infall towards W51e2 and W51e8 and, a CO outflow, oriented NW-SE, has been found associated to W51e2 (Keto & Klaassen 2008, Shi, Zhao & Han 2010b). W51 Main/South has been studied at a wide range of wavelengths both in continuum and molecular (thermal and maser) transitions. In particular, H$_2$O, OH, NH$_3$, CH$_3$OH maser emission has been found in the region (Genzel *et al.* 1981; Menten *et al.* 1990; Imai *et al.* 2002; Fish & Reid 2007; Phillips & van Langevelde 2005).

We have performed MERLIN astrometric observations of excited-OH maser emission at 6.035, 6.030 and 6.049 GHz and Class II methanol maser emission at 6.668 GHz towards W51 to investigate the relationship between the maser emission and the compact continuum sources in this SFR complex (Etoka, Gray & Fuller, 2012). Here we present the results regarding W51 Main/South. Association with the UCHII regions in the W51

Main/South complex and relationship of the masers to infall or outflow in the region are discussed.

2. Results

Overall the bulk of the methanol maser emission, both in terms of flux and number of components, is associated with W51 Main. Only two weak, isolated areas separated by ~ 2.5" are found to be producing methanol maser emission in W51 South, most likely associated with the e3 and e8 sources. Although the work of Fish & Reid (2007) shows that there is significant ground state OH emission from both Main and South, with indeed the brightest OH component being in South, excited OH emission is only detected towards Main (Fig. 1).

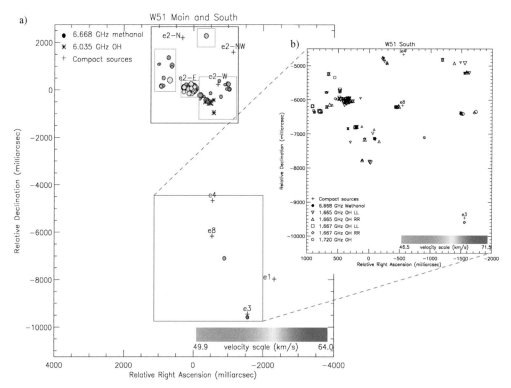

Figure 1. a) Maser components in the methanol 6.668-GHz and the excited OH 6.035-GHz transitions for the overall W51 Main/South region. The velocity colour-code covers the W51 Main and South methanol maser emission velocity range. The size of the symbols is proportional to the log of the intensity. The crosses indicate the positions of the compact continuum sources in the region. **b)** 6.668-GHz methanol Stokes I components from this work with all the ground-state OH maser components at 1.665, 1.667 and 1.720 GHz detected by Fish & Reid (2007) in W51 South. Note the change in the velocity range and hence colour-code.

In W51 Main, there seem to be 4 distinct regions of emission on the plane of the sky (Fig. 1a). Fig 2a presents an expanded view of the 3 most extended structures observed and Fig. 2b presents the velocity distribution of these methanol maser components versus radial distance (excluding the northern component likely associated to e-N/e-NW). The latter figure shows that the red- and blue-shifted maser components are diverging from a common point with an expansion velocity of $\simeq 5$ km s^{-1}. This, added to the general

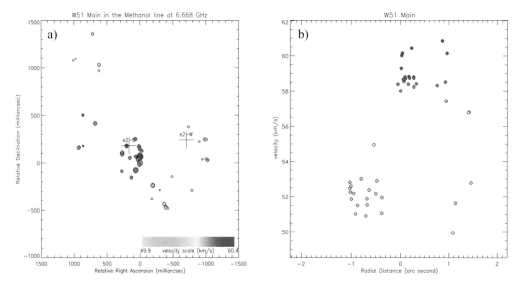

Figure 2. a) Maser components in the methanol 6.668-GHz line towards W51 Main, excluding the northern component. The velocity colour-code is restricted to the W51 Main methanol maser emission velocity range. **b)** The velocity distribution of the maser components versus radial distance. The central position is taken to be [0,+80 mas], inferred from the convergent point of the SW blue-shifted maser components.

distribution of the methanol maser components in the plane of the sky, indicates the presence of a structure with a wide opening angle associated with W51e2.

3. Discussion

As noted by Etoka *et al.* (2005) in the case of the SFR W3(OH), the present astrometric study towards W51 Main/South confirms that associations of individual OH and methanol maser components are rare. Despite all the components identified in the rich maser environment of W51 Main, no overlapping methanol and OH maser components are found, suggesting local variations in the abundance of the species and that both species are found in closely associated, but distinct, pockets. Similarly, even though 6.035-GHz excited-OH and ground-state OH maser components are in similar areas in W51 Main with similar magnetic field strength, they too do not show any overlap when observed at high spatial resolution. The total absence of 6.035-GHz emission and the scarcity of 1.720 GHz emission in W51 South, is suggestive of a lower density in W51 South than in W51 Main (Gray *et al.* 1992, Cragg *et al.* 2002).

Clearly, there are several distinct spatial-kinematic components in the region. These have been identified in Fig 3. The most extreme velocity maser components, which are all ground state OH masers, lie close to a line through W51e2-E at a P.A. $\sim 150°$ with the red-shifted components north of the source and blue-shifted components offset by a similar angular distance (~ 400 mas) south of the source. The location of these components as well as their velocities indicate that they are associated with the outflow from the source which has been imaged by Shi *et al.* (2010b).

There are two possible interpretations for the wide-opening angle structure observed in Fig. 2 which are discussed here below:

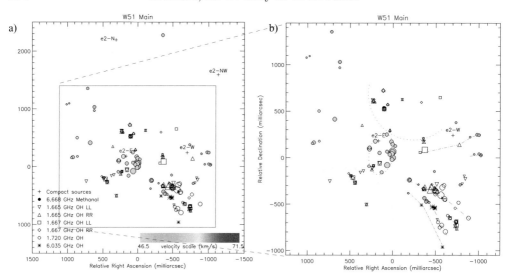

Figure 3. a) 6.668-GHz methanol and 6.035-GHz maser components from this work with all the ground-state OH maser components at 1.665, 1.667 and 1.720 GHz detected by Fish & Reid (2007) in W51 Main. **b)**Magnification on the masers around e2-E and e2-W presenting the various spatial-kinematic components identified.

3.1. Scenario 1: Outflow

According to Churchwell (2002), typically an outflow from a high-mass young stellar object (HMYSO; i.e., $L_{bol} > 10^3$ $L_{\odot}$) lives $\sim 10^4$ yr with a mass outflow rate of $\sim 10^{-3}$ $M_{\odot}$ yr^{-1}. With an expansion velocity of 5 km s^{-1}, only $\sim 5 \times 10^3$ year would be needed to reach 1", corresponding to the extent of the red- and blue-shifted lobe observed in Fig. 2b.

In this scenario, the 2 gaps observed in the distribution of the masers could be explained by 2 episodic events. The ring-like structure closest to e2-E, well defined by an ellipse of long-axis ~ 220 mas, would trace an outflow event $\sim 1.1 \times 10^3$ yr old, assuming no acceleration or deceleration.

3.2. Scenario 2: Accreting flow

The northeast-southwest distribution of the masers is close to perpendicular to the axis of the well collimated outflow of material imaged in CO by Shi *et al.* (2010b). This suggests that alternatively the masers could be tracing material which is part of the static or infalling envelope around the forming star.

In this scenario, the methanol masers closest to e2-E, forming a compact ring-like structure is likely to trace a distinct physical component. Similar structures are seen by Bartkiewicz *et al.* (2009) towards some 29% of the 6.7-GHz methanol maser sources they studied. They interpreted these as a result of the masers being associated with a dense disk or torus around the central source, an interpretation which could also be applied to the masers seen here.

4. Conclusion

We have presented MERLIN astrometric observations towards W51 Main and South of the Class II methanol maser emission at 6.668 GHz and the excited OH maser emission at

6.035 GHz. The 6-GHz maser distributions have been aligned with those of the ground-state OH maser transitions at 1.665, 1.667 and 1.720 GHz from Fish & Reid (2007). Although Main and South have similar number of OH ground-state maser components with the strongest component in South, the bulk of the methanol 6.668 GHz maser emission, and all of the excited OH 6.035 GHz maser emission are found to be associated with e2 in W51 Main.

The methanol masers revealed a wide-opening angle structure centred on e2-E, roughly aligned on a P.A. $\sim 150°$, that is roughly perpendicular to the CO outflow, and showing a clear velocity coherence. The two possible interpretations of this structure are the signature of (**1**) an outflow showing episodic events of $\sim 5 \times 10^3$ yr for the older event and $\sim 1 \times 10^3$ yr for the younger one, assuming an outflow velocity of ~ 5 km s^{-1} or; (**2**) an accretion flow in which two physical components are present: an infalling envelope with the central ring-like structure probing a compact and dense disk or torus around the central object.

Although e2-W is the only continuum source in W51e2 clearly associated with a UCHII region, currently e2-E seems to be the most active source in the region. The presence of methanol masers and the lack of a UCHII region point at a massive central object at an early stage of the star forming process.

References

Bartkiewicz, A., Szymczak, M., van Langevelde, H. J., Richards, A. M. S., & Pihlström, Y. M. 2009, *A&A*, 502, 155

Cragg, D. M., Sobolev, A. M., & Godfrey, P. D. 2002, *MNRAS*, 331, 521

Churchwell E. 2002, *ARA&A*, 40, 27

Etoka, S., Gray, M. D., & Fuller G. A. 2012, *accepted in MNRAS*

Etoka, S., Cohen, R. J., & Gray, M. D. 2005, *MNRAS*, 360, 1162

Fish, V. L. & Reid, M. J. 2007, *ApJ*, 670, 1172

Gaume, R. A., Johnston, K. L., & Wilson, T. L. 1993, *ApJ*, 417, 645

Genzel, R. *et al.* 1981, *ApJ*, 247, 1039

Gray, M. D., Field, D., & Doel, R. C. 1992, *A&A*, 262, 555

Imai, H. *et al.* 2002, *PASJ*, 54, 741

Menten, K. M., Melnick, G. J., Phillips, T. G., & Neufeld, D. A. 1990, *ApJ*, 363, L27

Phillips, C. & van Langevelde, H. 2005, *ASPC*, 340, 342

Keto, E. & Klaasen, P. 2008, *ApJ*, 678, L109

Sato, M., Reid, M. J., Brunthaler, A., & Menten K. M. 2010, *ApJ*, 720, 1055

Shi, H., Zhao, J.-H., & Han, J. L. 2010a, *ApJ*, 710, 843

Shi, H., Zhao, J.-H., & Han, J. L. 2010b, *ApJ*, 718L, 181

Zhang, Q. & Ho, P. T. P. 1997, *ApJ*, 488, 241

Zhang, Q., Ho, P. T. P., & Ohashi, N. 1998, *ApJ*, 494, 636

Cosmic Masers - from OH to H$_0$
Proceedings IAU Symposium No. 287, 2012
R.S. Booth, E.M.L. Humphreys & W.H.T. Vlemmings, eds.
© International Astronomical Union 2012
doi:10.1017/S1743921312006874

What is happening in G357.96-0.16?

Tui R. Britton[1,2], Maxim A. Voronkov[2], and Vanessa A. Moss[2,3]

[1]Dept. of Physics & Astronomy, Macquarie University, Sydney, NSW 2109, Australia
email: tui.britton@csiro.au

[2]CSIRO Astronomy & Space Science, PO Box 76, Epping, NSW 1710, Australia

[3]SIfA, School of Physics, University of Sydney, NSW 2006, Australia

Abstract. In order to answer this question, we examine the relationship between the two sites of maser activity in G357.96-0.16. We also propose future observations for examining the dust properties of this interesting region of massive star formation.

Keywords. masers, stars: formation

G357.96-0.16 in an intriguing region of star formation with various molecular masers and a possible HII region (Figure 1; Voronkov *et al.* 2011). The 6.7 GHz methanol maser data suggest the presence of two massive proto-stars, however, we are unsure as to the mechanism driving the radiation that is exciting the 6.7 GHz maser in the southern site.

1. Is the northern site the source of an outflow?

The 6.7 GHz methanol maser (square) in the northern site indicates the presence of a proto-star. Unusually, the 22 GHz water maser (cross) in the northern site has a large velocity spread over 180 km s^{-1} wide (Breen *et al.* 2010) suggesting an outflow. The weak 25 GHz continuum emission (single contour at the northern site) suggests a possible HII region. Contours filling the 8.0 μm arc represent 16 cm continuum emission (obtained by A. Dicker, C-E. Green, and D. Compton as part of their summer vacation program). This site is considered to be more evolved than the southern site as it contains an OH maser (not shown) within 2" of the other masers (Forster *et al.* 1989).

2. Does a second proto-star exist in the southern site?

The 25 GHz methanol masers (solid circle) indicate a highly energetic shock in the southern site. The 6.7 GHz methanol maser (square) is unexpected here and would suggest the presence of a second proto-star. However, dust heated by the shock may also excite a 6.7 GHz methanol maser. We have recently observed 44 GHz methanol masers in the region and will use these to characterize the shock morphology. A 22 GHz water maser (cross) is also found in the southern site.

3. What is the relationship between the two sites?

Current wide-spread opinion is that a 6.7 GHz maser traces the exact location of a massive young stellar object (Minier *et al.* 2003, Xu *et al.* 2008). Therefore the presence of a 6.7 GHz maser at both sites in G357.96-0.167 would suggest the presence of two massive young stellar objects. Moreover, the velocities of both masers are similar suggesting the two sites might be related.

Methanol masers are also known to be associated with sub-millimetre and millimetre continuum emission, as observed by Walsh *et al.* (2003) and Hill *et al.* (2005). Therefore, we propose observations to estimate the mass at each site by examining the cold (T= 20 - 50 K) molecular dust in the 230 GHz continuum band. We have also searched for multiple 23 GHz ammonia lines using the Australia Telescope Compact Array, which we will use to estimate the temperature of the northern site.

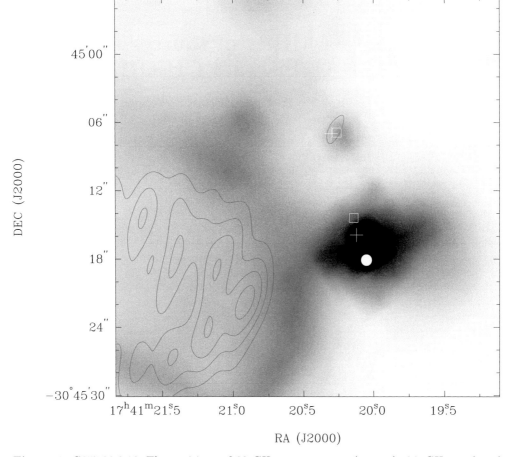

Figure 1. G357.96-0.16. The positions of 22 GHz water masers (crosses), 6.7 GHz methanol masers (squares), 25 GHz methanol masers (solid circle), 16 cm continuum emission (contours) and 25 GHz continuum emission (single contour at northern maser site) over laid on an 8.0 μm Spitzer image (grey scale background; Voronkov *et al.* 2011).

Acknowledgements

TB's travel to this conference was supported by the MQ PGRF and the IAU. The Australia Telescope is funded by the Commonwealth of Australia for operation as a National Facility managed by CSIRO. This research has made use of NASAs Astrophysics Data System Abstract Service.

References

Breen, S. L., Caswell, J. L., Ellingsen, S. P., & Phillips, C. J. 2010, *MNRAS*, 406, 1487

Forster, J. R. & Caswell, J. L., 1989, *A&A*, 213, 339

Hill,T., Burton, M. G., Minier, V., Thompson, M. A., Walsh, A. J., Hunt-Cunningham, M., & Garay, G. 2005 *MNRAS*, 363, 405

Minier V., Ellingsen S. P., Norris R. P., & Booth R. S., 2003, *A&A*, 403, 1095

Voronkov, M. A., Walsh, A. J., Caswell, J. L., Ellingsen, S. P., Breen, S. L., Longmore, S. N., Purcell, C. R., & Urquhart, J. S. 2011, *MNRAS*, 413, 2339

Walsh, A. J., Macdonald, G. H., Alvey, N. D. S., Burton, M. G., & Lee, J., 2003, *A&A*, 410, 597

Xu Y., Li J. J., Hachisuka K., Pandian J. D., Menten K. M., & Henkel C., 2008, *A&A*, 485, 729

Cosmic Masers - from OH to H_0
Proceedings IAU Symposium No. 287, 2012
R.S. Booth, E.M.L. Humphreys & W.H.T. Vlemmings, eds.

© International Astronomical Union 2012
doi:10.1017/S1743921312006886

Infrared characteristics of sources associated with OH, H$_2$O, SiO and CH$_3$OH masers

Jarken Esimbek[1,3], Jian Jun. Zhou[1,3], Gang. Wu[1,3] and Xin Di. Tang[1,2]

[1]Xinjiang Astronomical Observatory, Chinese Academy of Sciences, Urumqi 830011, PR China
email: `jarken@xao.ac.cn`

[2]Graduate University of the Chinese Academy of Sciences, Beijing 100080, PR China
[3]Key Laboratory of Radio Astronomy, Chinese Academy of Sciences, Urumqi 830011,
PR China

Abstract. We collect all published OH, H$_2$O, SiO and CH$_3$OH masers in the literature. The associated infrared sources of these four masers were identified with MSX PSC catalogues. We look for common infrared properties among the sources associated with four masers and make a statistical study. The MSX sources associated with stellar OH, stellar H$_2$O and SiO masers concentrated in a small regions and the MSX sources associated with interstellar OH, interstellar H$_2$O and CH$_3$OH masers also concentrated in a small regions in an [A]-[D].vs.[A]-[E] diagram. These results give us new criterion to search for coexisting stellar maser samples for OH, H$_2$O and SiO masers and interstellar maser samples for OH, H$_2$O and CH$_3$OH masers.

Keywords. Masers, Interstellar, Stellar, Infrared radiation

1. The sample

We collect all 1602 published 43 GHz SiO maser sources (Deguchi *et al.* 2004a, 2004b, 2007, 2010; Fujii *et al.* 2006; Jiang *et al.* 2002; Nakashima *et al.* 2003a, 2003b), 1712 6.7 GHz CH$_3$OH maser sources (Pestalozzi *et al.* 2005; Green *et al.* 2010; Xu *et al.* 2003, 2010), 1417 22.235 GHz H$_2$O maser sources (Esimbek *et al.* 2005) and 3249 1612 MHz OH maser sources (Mu *et al.* 2010). The MSX mission surveyed the entire Galactic belt within $|b| \leqslant 4.5°$ in five infrared bands B,A,C,D and E at 4,8,12,15 and 21 μm (Price *et al.* 2001). 1155 of the 1602 SiO masers, and 885 of the 1712 CH$_3$OH masers are associated with MSX PSC sources within 1'. Of the 743 H$_2$O masers associated with MSX PSC sources within 1', 300 are interstellar, and 155 are stellar masers. Of the 1869 OH masers associated with MSX PSC sources within 1', 63 are interstellar, and 1657 are stellar masers, others are unknown type.

2. Statistical results

OH, H$_2$O, SiO and CH$_3$OH are the strongest and most widespread astrophysical masers. OH and H$_2$O masers could occur in star forming regions and the envelopes of evolved stars. Most of the SiO masers are circumstellar and CH$_3$OH masers are interstellar (Elitzur 1992). Fig. 1 presents color indices [A]-[D] vs. [A]-[E] associated with OH, H$_2$O, SiO and CH$_3$OH masers, respectively. Here, [A]-[D] and [A]-[E] denote $\log(F_D/F_A)$ and $\log (F_E/F_A)$. These mid-IR sources associated with these four masers are concentrated in small regions, while mid-IR sources associated with stellar OH masers are distributed in relatively larger region. Fig. 2a. Presents color indexes [A]-[D] vs. [A]-[E] associated with stellar OH, stellar H$_2$O and SiO masers, there 66%, 77% and 95% mid-IR sources associated with three masers are located in a area $(-0.2,-0.2)$,$(0.1,0.4)$,$(0.4,0.4)$

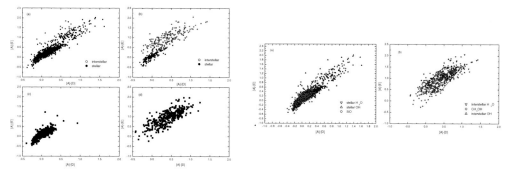

Figure 1. (Fig. 1; left) MSX color-color diagram of [A]-[D] vs. [A]-[E] associated with (a) OH (b) H_2O (c) SiO and (d) CH_3OH masers. (Fig. 2; right) (a) MSX color-color diagram of [A]-[D] vs. [A]-[E] associated with stellar OH, stellar H_2O and SiO masers. (b) MSX color-color diagram of [A]-[D] vs. [A]-[E] associated with interstellar OH, interstellar H_2O and CH_3OH masers.

and $(-0.2, -0.4)$. Fig. 2b presents color indexes [A]-[D] vs. [A]-[E] associated with interstellar OH, interstellar H_2O and CH_3OH masers, there 81%, 75% and 92% mid-IR sources associated with three masers are located in a area $(-0.2, 0.4)$, $(0.5, 1.5)$, $(1, 1.5)$ and $(0.4, 0.4)$.

3. Summary

We collect all published OH maser, H_2O maser, SiO maser and CH_3OH maser in the literature and selected MSX PSC sources within $1'$ as the exciting sources. The results provide us new criterion to search for coexisting stellar OH, stellar H_2O and SiO masers and coexisting interstellar OH, interstellar H_2O and CH_3OH masers.

Acknowledgements

This work was funded by the National Natural Science foundation of China under grant 10778703 and 10873025, and partly supported by China Ministry of Science and Technology under State Key Development Program for Basic Research (2012CB821800).

References

Deguchi, S., Imai, H., Fujii, T. *et al.* 2004a, *PASJ*, 56, 261
Deguchi, S., Fujii, T., Glass, Ian S. *et al.* 2004b, *PASJ*, 56, 765
Deguchi, S., Fujii, T., Ita, Y. *et al.* 2007, *PASJ*, 59, 559
Deguchi, S., Shimoikura, T., & Koike, K. 2010, *PASJ*, 62, 525
Elitzur M., 1992, *Astronomical Masers*, Kluwer Academic Publishers
Esimbek, J., Wu, Y., & Wang, J. 2005 *ChJAA*, 5, 587E
Fujii, T., Deguchi, S., Ita, Y. *et al.* 2006, *PASJ*, 58, 529
Green, J. A., Caswell, J. L., Fuller, G. A. *et al.* 2010, *MNRAS*, 409, 913
Jiang, B. W. 2002, *ApJ*, 566L, 37J
Mu, J., Esimbek, J., Zhou, J. *et al.* 2010, *ChJAA*, 10, 166M
Nakashima, J. & Deguchi, S. 2003a, *PASJ*, 55, 203
Nakashima, J. & Deguchi, S. 2003b, *PASJ*, 55, 229
Pestalozzi, M. R., Minier, V., & Booth, R. S. 2005, *A&A*, 432, 737
Price S., Egan E., Carey S. *et al.*, 2001, *AJ*, 1210, 2819
Xu, Y., Voronkov, M. A., Pandian, J. D. *et al.* 2009, *A&A*, 507, 1117
Xu, Y., Zheng, X. W., & Jiang, D. R. 2003, *ChJAA*, 3, 49

Cosmic Masers - from OH to H$_0$
Proceedings IAU Symposium No. 287, 2012
R.S. Booth, E.M.L. Humphreys & W.H.T. Vlemmings, eds.

© International Astronomical Union 2012
doi:10.1017/S1743921312006898

Massive star-formation toward G28.87+0.07

J. J. Li[1,2], L. Moscadelli[2], R. Cesaroni[2], R. S. Furuya[3], Y. Xu[1], T. Usuda[3], K. M. Menten[4], M. Pestalozzi[5], D. Eliav[5], and E. Schisano[5]

[1] Purple Mountain Observatory, Chinese Academy of Sciences, Nanjing 210008,
email: jjli@pmo.ac.cn

[2] INAF-Osservatorio Astrofisico di Arcetri, Largo E. Fermi 5, I-50125 Firenze, Italy
[3] Subaru Telescope, National Astronomical Observatory of Japan, 650 North A'ohoku Place,
Hilo, HI 96720, USA
[4] Max-Planck-Institut für Radioastronomie, Auf dem Hügel 69, 53121 Bonn, Germany
[5] INAF-Istituto Fisica Spazio Interplanetario, Via Fosso del Cavaliere 100, I-00133 Roma, Italy

Abstract. We investigated the high-mass star-forming region G28.87+0.07 by means of maser kinematics, including H$_2$O, CH$_3$OH, and OH, and radio to infrared, continuum observations. All observational evidence suggests that these masers are associated with the same young star of 20-30 M$_\odot$, still in the main accretion phase and surrounded by a rich stellar cluster.

Keywords. ISM: individual objects (G28.87+0.07) – ISM: kinematics and dynamics – masers – techniques: interferometric

---

Understanding the process of high-mass star formation represents a challenge from both theoretical and observational points of view. A few years ago, we started an observational project to study the high-mass star-forming process by comparing interferometric images of thermal lines of molecular tracers with multi-epoch VLBI studies in three well known maser species (OH, CH$_3$OH and H$_2$O). At present, three sources (IRAS 20126+4104, G16.59−0.05 and G23.01−0.41) have been already analyzed (Moscadelli *et al.* 2011, Sanna *et al.* 2010a, Sanna *et al.* 2010b). The results demonstrate that **the synergy between VLBI, multi-species maser observations and (sub)mm interferometric observations of thermal molecular lines is crucial to achieve a multi-scale picture of the environment of newly formed massive (proto)stars**. Here we represent our study of the high-mass star-forming region (HMSFR) G28.87+0.07.

We observed the HMSFR G28.87+0.07 with the VLBI phase-referencing technique in three powerful maser transitions: 22.2 GHz H$_2$O (VLBA, four epochs), 6.7 GHz CH$_3$OH (EVN, four epochs), and 1.665 GHz OH (VLBA, one epoch). In addition, we also performed VLA observations of the radio continuum emission at 1.3 and 3.6 cm with both the A- and C-arrays, and Subaru observations of the mid-infrared continuum emission. We also made use of data from the Hi-GAL/Herschel, ATLASGAL/APEX, as well as other surveys of the Galactic plane. More details on the observations are described by Li *et al.* 2012. From all these observations, we draw the following conclusions:

(*a*) The bipolar distribution of line-of-sight velocities and the general pattern of observed relative proper motions indicate that the water masers are tracing a (proto)stellar jet along the northeast-southwest direction (Fig. 1), which is also powering the large-scale outflow (Furuya *et al.* 2008).

(*b*) The positions of the CH$_3$OH maser features roughly along the jet axis (Fig. 1), and the similarities in the spectral shape of the water and methanol maser emissions, suggest that also the CH$_3$OH masers could originate in the same jet, albeit associated with more quiescent gas than that traced by the water masers.

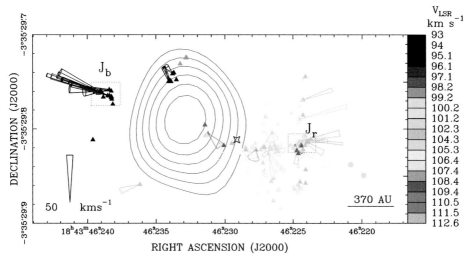

Figure 1. Proper motions of 22 GHz H_2O masers (triangles) relative to their "center of motion" (marked with a cross), and spatial distribution of 6.7 GHz CH_3OH masers (dots) in G28.87+0.07. Different colors are used to indicate the maser LSR velocities, according to the color scale on the righthand side of the plot. The cones indicate the 3-D velocities of the water maser features relative to the "center of motion". Points without an associated cone have been detected only over one or two epochs and the associated proper motion cannot be computed or is considered unreliable. The cone opening angle gives the 1σ uncertainty on the proper motion direction. The length of the cone is proportional to the velocity, with the amplitude scale indicated in the lower left corner of the figure. Open dotted rectangles mark two clusters of water masers located at the northeast and southwest edge of the distribution (labeled "J_b" and "J_r", respectively), moving fast and close to the northeast-southwest direction. The VLA 1.3 cm continuum emission is plotted with solid contours. Contour levels range from 40% to 90% of the peak emission (0.62 mJy beam^{-1}) at multiples of 10%.

(c) The continuum sources "HMC" is spatially associated with the observed H_2O, CH_3OH and OH masers. Whether the free-free emission arising from this source is due to an HII region or a thermal ionized jet cannot be unambiguously established with the present data. However, we believe that the jet hypothesis is more consistent with the direction of the water maser motions, which outline expansion in a bipolar flow.

(d) The combined information obtained from IR observations, permits to estimate with good accuracy the luminosity ($2 \times 10^5\ L_\odot$) and gas mass ($3 \times 10^3\ M_\odot$) of the star forming region of G28.87+0.07, and establish that $\sim$90% of the luminosity is coming from the radio source "HMC". We conclude that this source must contain multiple stars, with the most massive being at least as luminous as $3 \times 10^4\ L_\odot$. The lack of an associated HII region indicates that the 20–30 $M_\odot$ star must be still undergoing heavy accretion from the surrounding envelope.

This work was supported by the Chinese NSF through grants NSF 11133008, NSF 11073054, NSF 10621303, NSF 10733030 and the Key Laboratory for Radio Astronomy, CAS.

References

Furuya, R. S., Cesaroni, R., Takahashi, S., *et al.* 2008, *ApJ*, 673, 363
Li, J. J., Moscadelli, L., Cesaroni, R., *et al.* 2012, *ApJ*, accepted
Moscadelli, L., Cesaroni, R., Rioja, M. J., *et al.* 2011, *A&A*, 526, 66
Sanna, A., Moscadelli, L., Cesaroni, R., *et al.* 2010a, *A&A*, 517, 71
Sanna, A., Moscadelli, L., Cesaroni, R., *et al.* 2010b, *A&A*, 517, 78

Cosmic Masers - from OH to H$_0$
Proceedings IAU Symposium No. 287, 2012
R.S. Booth, E.M.L. Humphreys & W.H.T. Vlemmings, eds.

© International Astronomical Union 2012
doi:10.1017/S1743921312006904

High-mass Star Formation in the Regions IRAS 19217+1651 and 23151+5912

V. Migenes[1], I. T. Rodríguez[2] and M. A. Trinidad[2]

[1]Brigham Young University, Department of Physics and Astronomy
ESC-N145, Provo, UT. 84602 USA
email: vmigenes@byu.edu

[2]University of Guanajuato, Department of Astronomy,
Aptd. Postal 144
Guanajuato, GTO. 36000 Mexico
email: tatiana@astro.ugto.mx, trinidad@astro.ugto.mx

Abstract. We present and discuss VLA-EVLA high-sensitivity and spatial resolution observations of Water Vapor MASERs and continuum emission towards two sources that have been proposed in the literature to be high-mass star forming regions: IRAS 19217+1651 and 23151+5912. Our results indicate the presence of disks which can confirm that these regions are high-mass star forming regions.

Keywords. stars: formation, stars: high mass, ISM: HII regions, masers

1. Introduction

The study of high-mass star formation is complicated because the sources are embedded in regions of dense gas and dust limiting their study to radio and infrared frequencies. In addition, the evolution is much faster than for low-mass proto-stars and are distributed much farther away. Hence, they can only be studied indirectly from the molecular, IR, mm and maser emission associated to the ionized regions in which they form.

These regions were chosen from a large sample of candidates of high-mass protostellar objects (Sridharan *et al.* 2002) embedded in sites with dense gas traced by the CS molecule (J = 2-1) (Bronfman *et al.* 1996). The region IRAS 19217+1651 is located near the galactic plane, at a distance of 10.5 kpc and appears to be contained within a bubble-like structure labeled N109 by Churchwell *et al.* (2006). It has been classified as a region of massive star formation in an early state of evolution (Beuther *et al.* 2004) and it is characterized by its high infrared luminosity $10^{4.9}$ $L_\odot$. In addition, it contains several molecular species, such as SiO, CH$_3$OH and CH$_3$CN, which could trace outflows and hot cores. CS spectral lines have been also reported.

The region IRAS 23151-5912 is located in the molecular cloud of Cepheus. At a distance of 5.7 kpc and characterized by its high infrared luminosity 10^5 $L_\odot$ (Sridharan *et al.* 2002). The H$_2$O maser emission is highly variable (Felli *et al.* 2007). No CH$_3$OH, OH or SiO maser emission has been detected (Shridharan *et al.* 2002; Edris *et al.* 2007; Zapata *et al.* 2009). Strong bipolar outflows have been detected with the CO (2-1) and SiO (2-1) lines. The region is associated with a center of dense dust observed at 1.2mm, 850 and 450 μm (Beuther *et al.* 2002; Williams *et al.* 2004).

2. Observations

The observations were carried out in 2007 June 27 with the VLA-EVLA in transition mode with the configuration A. Continuum at 1.3 cm and H$_2$O maser emission were

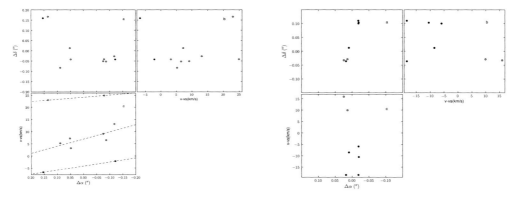

Figure 1. The spatial distribution of water vapor maser emission toward group M1 in IRAS 19217+1651 (left) and IRAS 23151+5912 (right). The axes indicate the positions relative to the average position of the masers for Right Ascension and Declination in 19217+1651 and with respect to the continuum source in 23151+5912. On the bottom-left, the horizontal axis represents the velocity of the masers minus the systematic velocity of the molecular cloud vs relative positions in RA. The dashed-line shows a linear fit to the distribution in each subgroup, which shows a velocity gradient. On the bottom-right, the distribution seems to indicate an elongated-elliptical distribution in RA.

observed simultaneously using two different IFs. The first with a bandwidth of 25 MHz and seven channels for the continuum and the second with a bandwidth of 3.125 MHz and 63 channels for the line emission, both centered at the rest frequency of 22235.080 MHz with V_{LSR}=10.5 & -54.4 km s^{-1}, respectively. RCP and LCP were sampled for both IFs. Data were reduced in AIPS.

3. Results

The source IRAS 19217+1651 A is the most intense and its morphology is consistent with a cometary UC HII region. Associated to it are 3 regions of maser emission which have a velocity gradient in the east-west direction along the elongation of the cometary H II region (see Fig. 1). Source B is extended and appears to be composed of more than one source or stellar component. The second maser cluster is not spatially coincident with any continuum source, but seems to be associated with the molecular outflow in the northeast-southwest direction observed in the region. The 1.3 cm continuum emission of the source IRAS 23151+5912 is consistent with a UC HII region and probably contains an embedded massive protostar of 15 M$_\odot$ of type B1 ZAMS. The data suggest the presence of two circumstellar disks: an expanding one with a radius of $\sim$415 AU and a smaller one with a radius of $\sim$84 AU and a central protostar of $\sim$10 M$_\odot$.

References

Beuther, H., Schilke, P., & Gueth, F. 2004, *ApJ*, 608, 330
Beuther, H., Schilke, P., Menten, K. M., Motte, F., Shridharan, T. K., & Wyrowski, F. 2002, *ApJ*, 566, 945
Bronfman, L., Nyman, L.-A., & May, J. 1996, *A&AS*, 115, 81
Churchwell, E., *et al.* 2006, *ApJ*, 649, 759
Sridharan, T. K., Beuther, H., Schilke, P., Menten, K. M., & Wyrowski, F. 2002, *ApJ*, 566, 931
Williams, S. J., Fuller, G. A., & Shridharan, T. K. 2004, *A&A*, 417, 115
Zapata, L. A., Menten, K. M., Reid, M., & Beuther, H. 2009, *ApJ*, 691, 332

Cosmic Masers - from OH to H_0
Proceedings IAU Symposium No. 287, 2012
R.S. Booth, E.M.L. Humphreys & W.H.T. Vlemmings, eds.

© International Astronomical Union 2012
doi:10.1017/S1743921312006916

325 GHz Water Masers in Orion Source I

Florian Niederhofer[1], Elizabeth Humphreys[1], Ciriaco Goddi[1] and Lincoln J. Greenhill[2]

[1]European Southern Observatory,
Karl-Schwarzschild-Str. 2, 85748 Garching, Germany

[2]Harvard-Smithonian Center for Astrophysics,
60 Garden Street, Cambridge, MA 02138, USA

Abstract. Radio Source I in the Orion BN/KL region provides the closest example of high mass star formation. It powers a rich ensemble of SiO and H_2O masers, and is one of only three star-forming regions known to display SiO maser emission. Previous monitoring of different SiO masers with the VLBA and VLA has enabled the resolution of a compact disk and a protostellar wind at radii <100 AU from Source I, which collimates into a bipolar outflow at radii of 100-1000 AU (see contribution by Greenhill *et al.*, this volume). Source I may provide the best case of disk-mediated accretion and outflow recollimation in massive star formation. Here, we report preliminary results of sub-arcsecond resolution 325 GHz H_2O maser observations made with the SMA. We find that 325 GHz H_2O masers trace a more collimated portion of the Source I outflow than masers at 22 GHz, but occur at similar radii suggesting similar excitation conditions. A velocity gradient perpendicular to the outflow axis, indicating rotation, supports magneto-centrifugal driving of the flow.

Keywords. ISM: individual (Orion KL) - ISM: jets and outflows - stars: formation - masers

1. Introduction

Massive stars affect the composition of the ISM and regulate star formation in molecular cloud complexes, thereby playing an important role in the evolution of galaxies. However, the detailed structures and processes involved in massive star formation remain largely uncharacterised. We have therefore embarked on a high-angular resolution study of massive, nearby YSO radio Source I in Orion BN/KL (d=414±7 pc; Menten *et al.* 2007). Using VLBI, it has been possible to map vibrationally-excited SiO maser emission of Source I at radii between 10-100 AU at 19 epochs over 2 years (around 20% of the crossing timescale) with a resolution of 0.1 AU (Matthews *et al.* 2010). The resulting movie has revealed a rotating and expanding flow that brackets an elongated 7 mm continuum source we interpret as arising from an ionized disk (Reid *et al.* 2007; Goddi *et al.* 2011). At radii >100 AU, proper motions of maser spots from the SiO vibrational ground state indicate a collimated, slow (18 km s^{-1}) outflow along the direction of the Source I rotation axis.

2. Observations and Results

Our 325 GHz water maser data towards Orion Source I were taken on November 30th 2006 with the Submillimeter Array (SMA) in the extended configuration using 6 antennas. This configuration provided maximum baselines of 220 meters corresponding to a resolution of 0.6″ at 325 GHz. The resolution in velocity was 0.37 km s^{-1}. We used CASA to calibrate the data. 3c279 and 0528+134 served as the bandpass and the gain

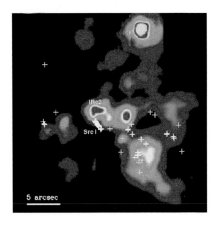 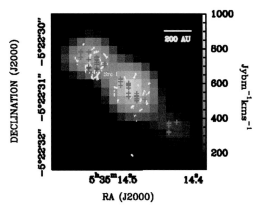

Figure 1. *Left*(Fig. 1a): Orion BN/KL 12.4 μm Gemini mosaic overlaid with fitted positions of 325 GHz H_2O maser emission (white crosses) observed using the SMA. Source I is marked by the clustered maser features south of IRc2; *Right*(Fig. 1b): Comparison of 22 GHz H_2O positions (VLA 1983, 0.1" resolution; Greenhill *et al.* 1998, white dots) with 325 GHz maser spot positions (SMA 2006, crosses). Background color scale is the 325 GHz moment 0 image.

callibrator respectively and flux calibration was performed using Titan. We used the strongest maser feature at 15.3 km s^{-1} to apply self-calibration to the data.

We imaged the 325 GHz H_2O maser emission and identified over 120 components in an area of 15 " around Source I (Figure 1a).The maser features associated with Source I trace a bipolar, collimated outflow for which we estimate a collimation angle of ∼40° and a position angle of ∼60°. The two outflow lobes display almost the same velocity distribution, spanning from -6 to 25 km s^{-1}, indicating the outflow axis lies close to the sky plane. There is a velocity gradient across the lobes perpendicular to the outflow axis which we interpret as rotation. Therefore magnetic fields may play a role in launching and shaping the outflow. We compared the fitted 325 GHz maser positions with the 22 GHz H_2O positions observed with the VLA in 1983 (Greenhill *et al.* 1998; Figure 1b). 325 GHz emission traces a more collimated region than that probed by 22 GHz masers. 22 GHz H_2O masers are excited in shocks with densities of ∼10^9 cm^{-3}, while models disagree on the conditions required to pump 325 GHz masers (10^6 cm^{-3}, Cernicharo *et al.* 1999; 10^9 cm^{-3}, Yates *et al.* 1997). We find that both transitions are excited at similar radii from Source I, suggesting similar excitation conditions for 22 GHz and 325 GHz H_2O masers in the Source I outflow.

References

Cernicharo, J., Pardo, J. R., González-Alfonso, E., *et al.* 1999, *ApJ*, 520, 131

Goddi, C., Humphreys, E. M. L., Greenhill, L. J., Chandler, C. J., & Matthews, L. D. 2011, *ApJ*, 728, 15

Greenhill, L. J., Gwinn, C. R., Schwartz, C., Moran J. M., & Diamond, P. J. 1998, *Nature*, 396, 650

Matthews, L. D., Greenhill, L. J., Goddi, C., *et al.* 2010, *ApJ*, 708, 80

Menten, K. M., Reid, M. J., Forbrich, J., & Brunthaler, A. 2007, *A&A*, 474, 515

Reid, M. J., Menten, K. M., Greenhill, L. J., & Chandler, C. J. 2007, *ApJ*, 664, 950

Yates, J. A., Field, D., & Gray, M. D. 1997, *MNRAS*, 285, 303

Cosmic Masers - from OH to H_0
Proceedings IAU Symposium No. 287, 2012
R.S. Booth, E.M.L. Humphreys & W.H.T. Vlemmings, eds.

© International Astronomical Union 2012
doi:10.1017/S1743921312006928

10 years of 12.2 GHz methanol maser VLBI observations towards NGC 7538 IRS1 N: proper motions and maser saturation

Michele Pestalozzi[1], Anders Jerkstrand[2] and John Conway[3]

[1]IAPS - INAF, via del Fosso del Cavaliere 100, 00133 Roma, Italy
email: michele.pestalozzi@gmail.com

[2]Stockholm Observatory, S - 106 91 Stockholm, Sweden
email: andersj@astro.su.se

[3]Onsala Space Observatory, S - 439 92 Onsala, Sweden
email: jconway@chalmers.se

Abstract. We present the outcomes of the consistent analysis of 6 epochs of VLBA 12.2 GHz data obtained between 1995 and 2005 towards the known high-mass star formation reigon NGC7538 IRS1 N. Our analysis concentrates on the study of the main spectral/spatial feature, which is 20 VLBA synthesized beams in size with a distinct velocity gradient. We looked for proper motion signals relative to the central peak which, in an edge-on disc framework, is expected to be stationary. We also study the peak flux and the spatial brightness profile of the main maser feature searching for maser variability. Our results are twofold: we detect a clear proper motion signal of three spatial features (0.21, 0.1, 0.65 mas yr^{-1}) and conclude that these can be made consistent with previous modelling of a Keplerian disc seen edge-on around a high-mass protostar. We further detect a consistent decrease of the peak flux over the time-span 1995-2005 (~ 5.4 Jy yr^{-1}), confirmed when taking into account earlier data (1986, 1987) as well as by the 6.7 GHz maser emission. Also, the width of the spatial brightness profile of the main feature seems to decrease between 1995 and 2005 by some 50%. We consider these observables as clear signs of partial maser saturation.

Keywords. Star formation, high-mass stars, masers, methanol, saturation, proper motions

1. Introduction

The high-mass star formation region NGC 7538 hosts a number of IR sources, one of them (IRS1 N) showing a prominent methanol maser emitting at 6.7 and 12.2 GHz. The brightest spectral feature of that maser at both frequences has been modelled as a rotating dics seen edge-on (Minier *et al.* 1998, Pestalozzi *et al.* 2009), rotating around a 30 M$_\odot$ protostar. The structure shows an impressive morphological and dynamical coherence across 20 beams, which makes the hypothesis fairly strong. This idea has been challenged recently by e.g. De Buizer & Minier (2005) and Surcis *et al.* (2011).

The present contribution assumes the edge-on rotating disc scenario. Six epochs of VLBA data at 12.2 GHz, taken between 1995 and 2005 build the presented data set.

2. Results

Proper motion. This part of the study had the goal of obtaining an independent measurement of the mass of the central object. In fact, assuming the brightest point to be fix in space, all other distinguishable features followed across the observational epochs would move either away or toward the central peak. Outward motions were registered (see Fig. 2), for the three features away from the central peak, indicating that they must be on the far side of the disc. Their velocity can be made compatible with a central object between 5 and 120 M$_\odot$.

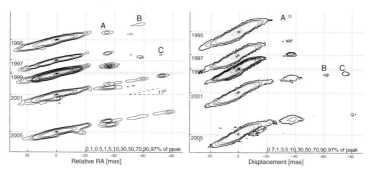

Figure 1. Velocity integrated data (left) and position-velocity diagrams (right) of the six VLBA epochs considered for the present work. Notice the three spatial features A, B, C for which proper motions relative to the central peak have been measured (see Fig. 2).

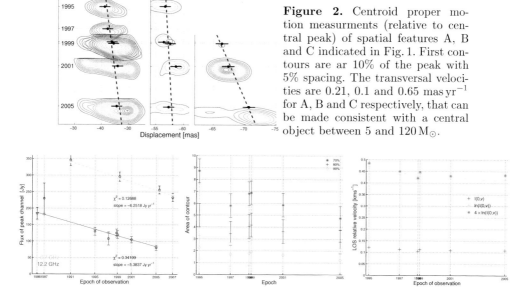

Figure 2. Centroid proper motion measurments (relative to central peak) of spatial features A, B and C indicated in Fig. 1. First contours are ar 10% of the peak with 5% spacing. The transversal velocities are 0.21, 0.1 and 0.65 mas yr^{-1} for A, B and C respectively, that can be made consistent with a central object between 5 and 120 M$_\odot$.

Figure 2. Three observables that have to be modelled with partial saturation along the amplification path. From left to right: peak flux decrease, spatial profile narrowing, spectral profile narrowing.

Variability. We register a constant decrease of the peak flux density of the brightest maser feature, that can be followed over 20 years, both at 12.2 and at 6.7 GHz. Also, we register a narrowing of the spatial and spectral profile (the latter is at the edge of detection, see Fig. 2). These three facts are not compatible with models of *un*saturated maser emission, and the only way to reconcile them with modelling is to introduce partial saturation: shortening of the unsaturated portion of the of amplification path makes it possible to obtain the three effects at once.

References

De Buizer, J. M. & Minier, V. 2005, *ApJ*, 628, L151

Minier, V., Booth, R. S., & Conway, J. E. 1998, *A&A*, 336, L5

Pestalozzi, M., Elitzur, M., & Conway, J. 2009, *A&A*, in press

Surcis, G., Vlemmings, W. H. T., Torres, R. M., van Langevelde, H. J., & Hutawarakorn Kramer, B. 2011, *A&A*, 533, A47

Cosmic Masers - from OH to H$_0$
Proceedings IAU Symposium No. 287, 2012
R.S. Booth, E.M.L. Humphreys & W.H.T. Vlemmings, eds.

© International Astronomical Union 2012
doi:10.1017/S174392131200693X

Internal Proper Motion of 6.7 GHz Methanol Masers in Ultra Compact HII Region S269

S. Sawada-Satoh[1], K. Fujisawa[2], K. Sugiyama[2], K. Wajima[2] and M. Honma[3]

[1]Mizusawa VLBI Observatory, NAOJ, Mizusawa, Oshu, Iwate, 023-0861 Japan
email: satoko.ss@nao.ac.jp

[2]Dept. of Physics, Yamaguchi University, Yoshida, Yamaguchi, 753-8512 Japan
[3]VERA Project, NAOJ, Osawa, Mitaka, Tokyo, 181-8588 Japan

Abstract. We present the internal proper motion of 6.7-GHz methanol masers in S269, an Ultra Compact HII region. The maser distribution in S269 consists of several maser groups, and the spatial structure of the main groups A and B are consistent with the past VLBI image. The remarkable result of comparing the two VLBI maps is that 6.7-GHz methanol maser distribution and velocity range within each group have been kept for eight years. Angular separation between the two groups A and B increases by 3.6 mas, which corresponds to a velocity of 11.5 km s^{-1}.

Keywords. masers – ISM: HII regions – stars: formation

1. Introduction

S269 is an Ultra Compact HII region in the outer Galaxy (Sharpless 1959), and harbors two bright near-IR sources, IRS1 and IRS2. In IRS2, several star-forming activities such as OH, H$_2$O masers and CO wings have been detected. Recent near-IR images imply that several H$_2$ knots are distributed across IRS2, which suggest a bipolar outflow, powered by sources in IRS2 (Jiang *et al.* 2003). The past VLBI observations of the 6.7-GHz methanol maser in S269 detected two groups (A & B), separated by ∼60 mas in 1998 November (Minier *et al.* 2000). The observations have suggested that the 6.7-GHz methanol masers are probably associated with IRS2. We observed the 6.7-GHz methanol masers in S269 in 2006 September using the Yamaguchi 32-m telescope and the Japanese VLBI Network (JVN).

2. Maser distributions and internal proper motion

The spectrum detected with the Yamaguchi 32-m telescope reveals the brightest peak at velocity of 15.2 km s^{-1}, and blue-shifted and red-shifted spectral components at peak velocities of 14.7 and 15.9 km s^{-1}, which are consistent with the past single-dish observations (Szymczak *et al.* 2000, Goedhart *et al.* 2004). Flux densities at the peaks of 14.7, 15.2 and 15.9 km s^{-1} are 9.8, 19.0 and 6.1 Jy, respectively. The flux density values agree well with an extrapolation of the sinusoidal time variation of the methanol maser emission (Goedhart *et al.* 2004).

The maser distribution consists of several maser groups. The most luminous maser spot at 15.2 km s^{-1} belongs to group B. The velocity range of group B is 15.02 to 15.38 km s^{-1}. Group A is the second brightest group, and consists of two maser spots at velocities of 14.67 and 14.85 km s^{-1}, and is located ∼60 mas east from group B. The red-shifted maser component is marginally detected and divided into two spectral channels at velocities of 15.73 and 15.90 km s^{-1} with the JVN. The spots form another

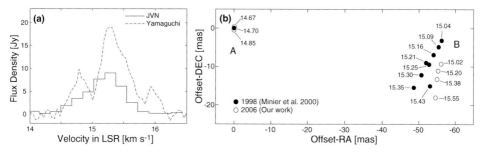

Figure 1. (a) The cross-power spectrum obtained from our JVN observations on 2006 September 10 (Solid line), and the spectral profile measured with the Yamaguchi 32-m telescope averaged spectra observed from 2006 September 4 to 7 (Dashed line). Velocity resolution is 0.178 km s^{-1} and 0.044 km s^{-1}, respectively. (b) Superposed maps of 6.7-GHz methanol maser distribution in 1998 (Minier *et al.* 2000) and 2006 (our work). For the distribution in 1998, Coordinate origin (0,0) is set to (i) the maser spot at velocity of 14.70 km s^{-1} in 1998, and (ii) the middle point of the two spots at velocity of 14.67 and 14.85 km s^{-1} in 2006.

group, -95 mas west from group A. On the other hand, the blue-shifted maser spots at velocities of 14.32, 14.50, 14.67 and 14.87 km s^{-1} have a wider distribution, spanning 250 mas along the southeast-northwest direction. Those maser spots except for groups A and B are not visible in the EVN observations of 1998 November (Minier *et al.* 2000). The maser peaks at 14.7 and 15.9 km s^{-1} could be at the minimum phase in intensity at that time, as Goedhart *et al.* (2004) has shown.

The structures of groups A and B are consistent with the past VLBI image of Minier *et al.* (2000). The remarkable result of comparing the two VLBI maps is that the 6.7-GHz methanol maser distribution and velocity range within each group have been maintained for eight years. Superposition of the two VLBI maps (Figure 1b) reveals that the distance between the two groups A and B increased by 3.6 mas over 7.8 years, which corresponds to a velocity of 11.5 km s^{-1}. The motion is along the direction of position angle $\sim80°$, almost parallel to the east-west direction.

The velocity gradient in the overall distribution could roughly be seen along the direction of position angle of $\sim80°$. However, a more rigorous inspection for the individual maser groups indicates that the velocity gradient is not simple. Velocities of the individual groups spread across the direction at a position angle of $\sim80°$.

The increasing angular separation between groups A and B could trace the outflow powered by IRS2. In order to confirm the movement, we have performed a follow up observation of the 6.7-GHz methanol masers in October 2011.

Acknowledgment

We are grateful to the JVN team for their support during the observations.

References

Goedhart, S., Gaylard, M. J., & van der Walt, D. J. 2004, *MNRAS*, 355, 553

Jiang, Z., Yao, Y., Yang, J., Baba, D., Kato, D., Kurita, M., Nagashima, C., Nagata, T., Nagayama, T., Nakajima, Y., Ishii, M., Tamura, M., & Sugitani, K. 2003, *ApJ*, 596, 1064

Minier, V., Booth, R. S., & Conway, J. E. 2000, *A&A*, 362, 1093

Sharpless, S. 1959, *ApJS*, 4, 257

Szymczak, M., Hrynek, G., & Kus, A. J. 2000, *A&A*, 143, 269

Cosmic Masers - from OH to H₀
Proceedings IAU Symposium No. 287, 2012
R.S. Booth, E.M.L. Humphreys & W.H.T. Vlemmings, eds.
© International Astronomical Union 2012
doi:10.1017/S1743921312006941

The radial velocity acceleration of the 6.7 GHz methanol maser in Mon R2 IRS3

K. Sugiyama[1], K. Fujisawa[2], N. Shino[1], and A. Doi[3]

[1] Dept. of Physics, Yamaguchi Univ., 1677-1 Yoshida, Yamaguchi, Yamaguchi 753-8512, Japan
email: koichiro@yamaguchi-u.ac.jp

[2] The Research Institute of Time Studies, Yamaguchi Univ., 1677-1 Yoshida, Yamaguchi,
Yamaguchi 753-8511, Japan

[3] Institute of Space and Astronautical Science, JAXA, Yoshinodai 3-1-1, Chuo-ku, Sagamihara,
Kanagawa 252-5210, Japan

Abstract. We present the radial velocity acceleration of the 6.7 GHz methanol maser in a high-mass star-forming region Monoceros R2 (Mon R2). The methanol maser is associated with an infrared source IRS3. The methanol maser of Mon R2 shows at least three spectral features having radial velocities (V_{lsr}) of 10.8, 12.7, and 13.2 km s^{-1}. The radial velocity of a feature at V_{lsr} = 12.7 km s^{-1} has changed during ten years from Aug. 1999 to Oct. 2009, corresponding to an acceleration of 0.08 km s^{-1} yr^{-1}. We observed the 6.7 GHz methanol masers of Mon R2 in Oct. 2008 using the Japanese VLBI Network (JVN). Compared with the previous VLBI image obtained in Nov. 1998 using the European VLBI Network (EVN), the maser feature at V_{lsr} = 12.7 km s^{-1} showed relative proper motions of ∼2.5 mas yr^{-1} (about 10 km s^{-1} at 0.83 kpc) toward the intensity peak of IRS3. The radial velocity acceleration could be caused by an inflow from a disk or envelope around a high-mass young stellar object (YSO) at IRS3.

Keywords. Stars: formation, Masers: methanol, Instrumentation: high angular resolution

1. Introduction

Mon R2 is a well-known region consisting of a chain of reflection nebulae, associated with a giant molecular cloud that is one of the closest high-mass star-forming regions to the Sun at a distance of 0.83 kpc (Racine 1968; see the review by Carpenter & Hodapp 2008). There are several near infrared sources identified on the K-band images (Carpenter *et al.* 1997), and the brightest source was labeled as IRS3. The IRS3 source showed a flattened spatial structure in polarimetry observations at position angle ∼140°, which was interpreted as a circumstellar disk around a high-mass YSO (Yao *et al.* 1997). Later high spatial-resolution observations at near infrared bands showed three separated YSOs in the IRS3 field, and each YSO shows micro-jets in the northeast-southwest directions.

The methanol maser at 6.7 GHz in Mon R2 is associated with IRS3, and each maser spot was found to be linearly distributed in the northeast-southwest direction in previous VLBI observations conducted in Nov. 1998 (Minier *et al.* 2000). We observed the 6.7 GHz methanol masers in Mon R2 every year from 2004 to 2009 using the Yamaguchi 32-m radio telescope and in Oct. 2008 using the JVN.

2. Radial Velocity Acceleration and Relative Proper Motions

We conducted the 6.7 GHz methanol maser observations every year from 2004 to 2009 using the Yamaguchi 32-m, and compared with the data obtained in 1999 using the Torun 32-m radio telescope (Szymczak *et al.* 2000). The methanol maser of Mon R2 shows at least three spectral features having radial velocities (V_{lsr}) of 10.8, 12.7, and 13.2 km s^{-1}

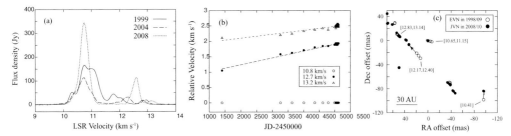

Figure 1. The 6.7 GHz methanol maser in Mon R2. (a) Spectra in 1999 with the Torun 32-m (solid line) and in 2004, 2008 with the Yamaguchi 32-m (dashed and dotted lines). (b) Radial velocity acceleration relative to the feature at $V_{lsr} = 10.8$ km s^{-1}. Each line shows the accelerations obtained from least-squares fits. (c) Overlaid VLBI images in Nov. 1998 (Minier *et al.* 2000, open circle) and Oct. 2008 (our work, filled circle). Attached numbers indicate a range of radial velocities for each spot, and arrows show proper motions relative to the reference spot at $V_{lsr} = 10.8$ km s^{-1}.

in 2008. Uncertainties of absolute radial velocities in both observations, unfortunately, were large of ~ 0.3-0.4 km s^{-1}. We compared radial velocities, therefore, relative to the strong maser feature at $V_{lsr} = 10.8$ km s^{-1}. As shown in Fig. 1(a), the spectral feature at $V_{lsr} = 11.8$ and 12.1 km s^{-1} in 1999 merged to one spectral feature, and changed to the red-shifted direction indicating a radial velocity of 12.3, 12.7 km s^{-1} in 2004 and 2008, respectively. The spectral feature at $V_{lsr} = 12.9$ km s^{-1} in 1999 also changed to the red-shifted direction, but its variation was smaller than that of the former. Each acceleration was estimated as 0.08 and 0.05 km s^{-1} yr^{-1} by a least-squares fit (Fig. 1b).

We overlaid our VLBI image obtained using the JVN in Oct. 2008 on the previous VLBI image obtained using the EVN in Nov. 1998 (Minier *et al.* 2000) using as an origin the reference maser spot at $V_{lsr} = 10.8$ km s^{-1} (Fig. 1c). On the basis of the overlaid VLBI images, the acceleration of maser spots at $V_{lsr} = 12.7$ km s^{-1} moved toward the northeast direction, which was parallel to the direction of the micro-jets. The absolute positional accuracy of the high spatial-resolution map at near infrared bands, unfortunately, was ~ 1 arcsec, which was not enough to understand which YSOs were suitable as an exciting source of the methanol masers. We compared our data with the flattened spatial structure obtained from polarimetry observations and therefore interpreted it as due to a circumstellar disk. The intensity peak of IRS3 in the disk, which could correspond to a position of the exciting YSO, is located at the northeast direction from a cluster of the methanol maser spots. Given the relative proper motions of the methanol maser toward the central exciting YSO obtained in our work, the radial velocity acceleration could be caused by an inflow from a disk or envelope around a high-mass YSO at IRS3.

In order to correctly understand the causes of the radial velocity acceleration, we will measure the absolute proper motions of the methanol masers in Mon R2, and obtain high spatial-resolution maps at near infrared bands with high positional accuracies.

References

Carpenter, J. M. & Hodapp, K. W. 2008, *Handbook of Star Forming Regions*, Volume I, 899
Carpenter, J. M., Meyer, M. R., Dougados, C., *et al.* 1997, *AJ*, 114, 198
Minier, V., Booth, R. S., & Conway, J. E. 2000, *A&A*, 362, 1093
Racine, R. 1968, *AJ*, 73, 233
Szymczak, M., Hrynek, G., & Kus, A. J. 2000, *A&AS*, 143, 2
Yao, Y., Hirata, N., Ishii, M., *et al.* 1997, *ApJ*, 490, 281

Cosmic Masers - from OH to H_0
Proceedings IAU Symposium No. 287, 2012
R.S. Booth, E.M.L. Humphreys & W.H.T. Vlemmings, eds.

© International Astronomical Union 2012
doi:10.1017/S1743921312006953

A Circumstellar Disk toward the High-mass Star-forming Region IRAS 23033+5951

M. A. Trinidad[1], T. Rodríguez[1] and V. Migenes[2]

[1] Depto. Astronomía, Universidad de Guanajuato,
Apdo. Postal 144, Guanajuato, Gto. 36240, Mexico
email: trinidad@astro.ugto.mx

[2] Brigham Young University, Department of Physics and Astronomy,
ESC-N145, Provo, Utah 84602

Abstract. We present water maser observations toward IRAS 23033+5951 carried out with the VLA-EVLA in the A configuration. In order to study the spatio-kinematical distribution of the water masers detected in the region, we made a simple geometrical and kinematical model based on the conical equation. We find that the water masers are tracing a rotating and contracting circumstellar disk of about 110 AU around a very young source of 18 $M_\odot$, which has not enough ionizing photons to be detected at centimeter wavelengths.

Keywords. High-mass stars formation: general — ISM:individual(H II regions, IRAS 23033+ 5951, water masers)

1. Introduction

Molecular outflows and circumstellar disks mark an important phase in the early evolution of low-mass star formation. In order to search for these structures in high-mass star formation regions, we have observed IRAS 23033+5951, which has a bolometric luminosity of 10^4 $L_\odot$ and is located at a distance of 3.5 kpc toward the Cepheus molecular cloud (Sridharan *et al.* 2002). IRAS 23033+5951 has been detected at millimeter and centimeter wavelengths, and a CO(2-1) bipolar outflow has been observed in the region in the northwest-southeast direction. (Beuther *et al.* 2004).

2. VLA-EVLA Observations

This high-mass star formation region was observed with the VLA-EVLA of the NRAO†
(in the transition phase) in the A configuration during 2007 June 27 for five hours. Water maser and 1.3 cm continuum emission were simultaneously observed. The 3.6 cm continuum emission was also observed.

3. Results

Nine water maser spots are detected in the region, of which seven masers are clustered (M1) and two others that appear isolated (see Fig. 1). The maser emission is not spatially associated with any continuum source detected in the region. The water maser spectrum of M1 shows a structure of three peaks, similar to that observed toward several maser sources (e.g. S255, Cesaroni 1990; S140, Lekht *et al.* 1993) and interpreted as tracers of circumstellar disks. We also find that the masers in M1 are distributed, mainly, in two groups, where masers with redshifted velocity are located to the northeast, while that with blueshifted velocity are to the southwest. Moreover,

---

† National Radio Astronomy Observatory is a facility of the National Science Foundation operate under cooperative agreement by Associated Universities

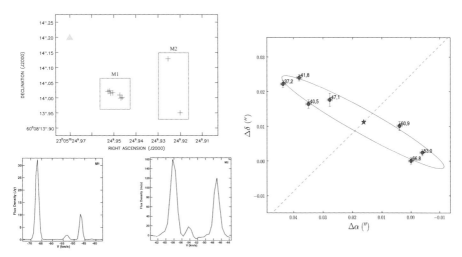

Figure 1. *Left)* Water maser emission toward IRAS 23033+5951. The two minor groupings in M1 are essentially the red-shifted emission to the northeast and the blue-shifted emission towards the southwest. A millimeter source (triangle), shifted about 0.25″ to the northeast, is observed in the region (Schnee & Carpenter 2009). *Right)* Fit of the water maser distribution. The star shows the position of the central source and the dotted line indicates the direction of the CO molecular outflow observed in the region.

the spatial distribution of the masers is almost perpendicular to the CO massive molecular outflow (northwest-southeast direction; e.g. Beuther *et al.* 2004). Based on these results, we suggest that the water maser emission in M1 is tracing a circumstellar disk.

In order to confirm the nature of the water masers in M1, we have built a simple geometrical and kinematic model. The spatial position of the water masers is fitted to the conical equation using the least-squares technique. The physical parameters of the fit indicate that the water masers are tracing an ellipse, which can be interpreted as part of a circumstellar disk of 0.05″ (projected on the plane of the sky; see Fig. 1), corresponding to a linear radius of about 110 AU (assuming a distance of 3.5 kpc), with a position angle of 65 deg. In addition, based on a least-squares fit to the radial velocities of the maser spots, we find that the water masers are rotating and contracting with velocities of 17.2 and 0.7 km s^{-1}, respectively. Using these velocities, a mass of 17.8 M$_\odot$ is calculated for the central object.

Finally, we suggest that the central massive object, the circumstellar disk and the CO outflow observed in the region are forming a disk-YSO-outflow, similar to that found in low-mass stars. Finally, we speculate that the central object could be associated with a HC H II region, which has not enough ionizing photons to be detected at centimeter wavelengths.

References

Beuther, H., Schilke, P., & Gueth, F. 2004, *ApJ*, 608, 330
Cesaroni, R. 1990, *A&A*, 233, 513
Lekht, E. E., *et al.* 1993, *Astronomy Reports*, 37, 367
Schnee, S. & Carpenter, J. M. 2009, *ApJ*, 698, 1456
Sridharan, T. K., Beuther, H., Schilke, P., Menten, K. M., & Wyrowski, F. 2002, *ApJ*, 566, 931

Cosmic Masers - from OH to H_0
Proceedings IAU Symposium No. 287, 2012   © International Astronomical Union 2012
R.S. Booth, E.M.L. Humphreys & W.H.T. Vlemmings, eds.   doi:10.1017/S1743921312006965

Molecular outflows toward methanol masers: detection techniques and their properties.

Helena M. de Villiers[1], **M. A. Thompson**[1], **A. Chrysostomou**[1,2] **and D. J. van der Walt**[3]

[1] Centre for Astrophysics Research, University of Hertfordshire,
College Lane, Hatfield, AL10 9AB, United Kingdom
email: `lientjiedv@gmail.com`

[2] Joint Astronomy Centre, Hilo, Hawaii, U.S.A., [3] USSR, North West University, South Africa

Abstract. Class II methanol masers are thought to trace the brief phase in the evolution of a massive YSO, where outflows are expected to occur. Molecular line maps of the CO isotopes of a subset of 6.7 GHz sources from the MMB catalogue were observed with the JCMT telescope. Utilising optically thick ^{12}CO, a search was done to detect broadened line wings (initially only on the source G20.08-0.13). The physical parameters of these detected lobes were then calculated.

1. Introduction

Class II methanol masers show strong emission at 6.7 GHz and are found in the vicinity of massive young stellar objects (YSO's), uniquely associated with high mass star formation regions (e.g., Menten 1991). These masers are thought to trace the stage immediately before the UC HII region in the development of a massive YSO (e.g., Minier *et al.* 2005). Methanol masers are thus likely to be associated with a high mass star's evolutionary phase when outflows occur (Codella *et al.* 2004). In order to test this hypothesis, we observed a subset of 67 6.7 GHz methanol masers from the MMB survey (Green *et al.* 2009) in ^{13}CO$(3-2)$ and C^{18}O$(3-2)$ using the JCMT, HARP. Matching ^{12}CO$(3-2)$ maps for every methanol maser were obtained from the JCMT HARP ^{12}CO Galactic Plane Survey (Dempsey *et al.* 2012, in prep.).

2. Outflow detections

Two detection methods for one dimensional broadened line wings were applied: (1) the technique followed by Hatchell *et al.* (2007) (called HATCHELL) using fixed velocity limits and (2) a combined method from the techniques used by Codella *et al.* (2004) and van der Walt *et al.* (2007) (called VDWC). In VDWC, the C^{18}O profile was scaled to the same peak temperature of ^{12}CO. A Gaussian profile was then fitted to the scaled C^{18}O peak and overlaid on the ^{12}CO spectrum. All temperature emission $> 3\sigma$ (on the noise level) and outside of the fitted peak were defined as respectively blue or red wings. The results are shown in the top row of Fig. 1. This initial study was only applied to G20.08-0.13, with a fairly good correspondence between the two methods. Repetition on more sources will follow. The integrated emission (intensity) were then calculated over the blue and red wing ranges respectively and shown as contour plots in Fig. 1.

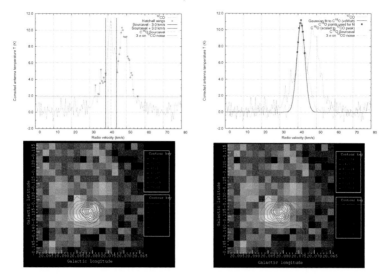

Figure 1. Top: Wing range classification according to (a) HATCHELL and (b) VDWC method for the ^{12}CO emission map associated with maser G20.08-0.13. Bottom: Red and blue wing contours as determined by respectively HATCHELL and VDWC methods.

3. Physical parameters

The contour plots and outflow velocity ranges as determined by method (1), were used to calculate the physical parameters of the outflows associated with G20.08-0.13, following Beuther *et al.* (2002). The H_2 column density were calculated using the approach by Hatchell *et al.* (2007). The resultant parameters were: column density $N_b = 4.9 \times 10^{20}$ and $N_r = 18 \times 10^{20} \text{cm}^{-2}$; total mass: $M_{out} = 114 M_\odot$, momentum: $p = 1120 \ M_\odot \text{kms}^{-1}$; energy $E = 1.35 \times 10^{46}$ erg; characteristical time scale $t = 3.29 \times 10^4$ yr; mass entrainment rate of each molecular outflow: $M_{out}/t = 3.46 \times 10^{-3} M_\odot/\text{yr}$; mechanical force: $F_m = 3.71 \times 10^{-2} M_\odot \text{kms}^{-1}/\text{yr}$; mechanical luminosity: $L_m = 3.37 L_\odot$. Although these listed physical parameters are only preliminary results, all were of the same order as those calculated by Beuther *et al.* (2002) for their 26 high-mass star-forming regions at early stages of their evolution. That is consistent with the hypothesis that methanol masers trace high mass star forming regions in an early evolutionary state. Should the outflow detection methods presented here be refined and the parameter results be expanded for the rest of the available data associated with methanol masers (at least 100 more methanol masers), a more complete picture of the properties of massive young stars in their very early stages of evolution, associated with methanol masers will be obtained.

References

Beuther, H., Schilke, P., Sridharan, T. K., Menten, K. M., Walmsley, C. M., & Wyrowski, F. 2002, *A&AS*, 383, 892

Codella, C., Lorenzani, A., Gallego, A. T., Cesaroni, R., & Moscadelli, L. 2004, *A&A S*, 417, 615

Green, J. A., Caswell, J. L., Fuller, G. A., Avison, A., Breen, S. L., Brooks, K., *et al.* 2009, *MNRAS*, 392, 783

Hatchell, J., Fuller, G. A., & Richer, J. S. 2007, *A&AS*, 472, 187

Menten, K. M. 1991, *ApJ*, 380, L75

Minier, V., Burton, M. G., Hill, M. G., Pestalozzi, M. R., Purcell, C. R., Garay, *et al.* 2005, *A&A S*, 429, 945

van der Walt, D. J., Sobolev, A. M., & Butner, H. 2007, *A&AS* 464, 1015

Sharmila Goedhart (bottom left) made sure the conference ran smoothly

T4:

Stellar Masers *Chair: Elizabeth Humphreys*

Cosmic Masers - from OH to H$_0$
Proceedings IAU Symposium No. 287, 2012
R.S. Booth, E.M.L. Humphreys & W.H.T. Vlemmings, eds.

© International Astronomical Union 2012
doi:10.1017/S1743921312006977

Masers in evolved star winds

Anita M. S. Richards

UK ARC Node, JBCA, School of Physics and Astronomy, University of Manchester, UK
email: amsr@jb.man.ac.uk

Abstract. This review summarises current observations of masers around evolved stars and models for their location and behaviour, followed by some of the many highlights from the past 5 years. Some of these have been the fruition of long-term monitoring, a vital aspect of study of stars which are both periodically variable and prone to rapid outbursts or transition to a new evolutionary stage. Interferometric imaging of masers provide the highest-resolution probes of the stellar wind, but their exponential amplification and variability means that multiple observations are needed to investigate questions such as what drives the wind from the stellar surface; why does it accelerate slowly over many tens of stellar radii; what causes maser variability. VLBI parallaxes have improved our understanding of individual objects and of Galactic populations. Masers from wide range of binary and post-AGB objects are accessible to sensitive modern instruments, including energetic symbiotic systems. Masers have been detected up to THz frequencies with *Herschel* and ALMA's ability to resolve a wide range of maser and thermal lines will provide accurate constraints on physical conditions including during dust formation.

Keywords. masers, surveys, stars: distances, late-type, evolution, AGB and post-AGB, supergiants, binaries: symbiotic, Galaxy: kinematics and dynamics

1. Introduction

Circumstellar masers are most abundant around solitary Asymptotic Giant Branch (AGB) and Red Supergiant (RSG) stars. The former had main sequence precursors < 8 M$_\odot$, the latter are more massive. Both have depleted hydrogen in their outer layers, swollen to between $\sim 1 - 10$ AU or more in diameter, RSG being roughly ten times larger than AGB stars. Their surface temperatures are $\sim 2000 - 3500$ K, RSG being at the hotter end of the range. They exhibit deep pulsations with periods P ranging from $\leqslant 0.5$ yr for low-mass semi-regular (SR) and Mira variables, to several years for the thickest-shelled OH/IR stars and supergiants. Molecules form in the stellar atmosphere and mass loss is copious, from $\sim 10^{-7}$ to $> 10^{-5}$ M$_\odot$ yr^{-1} depending on mass, age and metallicity. In this context 'solitary' means the absence of any companion detectable directly or by a kinematic signature; the distended stellar surfaces do not show any rotation and the gently expanding winds allow masers to propagate very efficiently in all directions. It has been traditional to present a cartoon of stellar mass loss traced by masers, but it is now possible to illustrate the process directly from observations, Fig. 1.

There are many open questions even in solitary star evolution, discussed in Section 2. Some of the advances in precise distance measurements are summarised in Section 3. Masers have also been detected from binary or possibly binary objects and I will describe some dramatic recent discoveries in Section 4. The masers accessible to cm-wave radio telescopes are those found in O-rich circumstellar envelopes (CSEs), i.e. SiO, H$_2$O and OH. (Sub-)mm wave masers include higher transitions of SiO and H$_2$O and also species found in C-rich CSEs, summarised in Section 5.

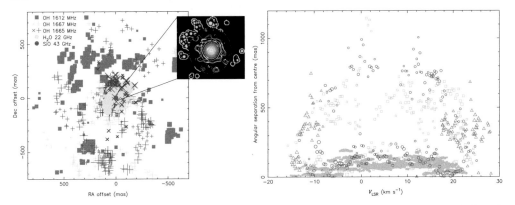

Figure 1. The inset shows the stellar disc of VX Sgr, resolved at 2μm by Chiavassa *et al.* (2010) overlaid with SiO masers mapped by Chen *et al.* (2006). The larger-scale structure traced by H_2O (Murakawa *et al.* 2003) and OH masers is shown on the left. The angular separation of the H_2O and OH masers from the assumed stellar position, as a function of V_{LSR}, is shown on the right.

2. The origin and acceleration of stellar winds

2.1. *Acceleration*

Bowen (1988) modelled the AGB mass loss process initiated by the levitation of material by stellar pulsations, to the point where dust forms and radiation pressure on grains accelerates the wind away from the star, imparting momentum to the gas by collisions. However, Woitke (2006) cast doubt on whether silicate-based grains could form close enough to the star for this process to be effective in O-rich stars. IR interferometry (e.g. Danchi *et al.* 1994) showed the inner rim of the silicate dust shell is typically at $\geqslant 5$ stellar radii ($R_\star$). The dust must form promptly, since the collision rate is too slow for grains to grow outside this radius, but Chapman & Cohen (1986) found that the wind traced by all maser species accelerates gradually out to $\sim 100R_\star$ in VX Sgr (Fig. 1). This was also seen for e.g. S Per (Richards *et al.* 1999a), IK Tau, RT Vir, U Her and U Ori (Bains *et al.* 2003). The wind is accelerated through escape velocity as it traverses the H_2O maser shell (Yates & Cohen 1994) and larger shells have higher velocities but follow a similar gradient, Fig. 2. *Herschel* results also provide evidence for gradual acceleration. Lines from higher excitational states, emanating from closer to the star, have narrower velocity widths than those produced under cooler conditions (Decin *et al.* 2010).

2.2. *Maser beaming*

The position and size of individual maser spots is measured by fitting 2-D Gaussian components, uncertainty proportional to (beamsize)/(signal-to-noise ratio), and hence can achieve sub-mas (often sub-AU) accuracy for bright masers. The measured size is often much smaller than the true size of the emitting region. The spots often form series, with a Gaussian spectral profile and systematic position increments. If the spot positions are spatially resolved (as is the case for MERLIN observations of 22-GHz H_2O masers) the total extent of such features represent the physical size of maser clumps. In such cases, (component size)/(feature size) gives a direct estimate of the maser beaming angle (Strelniskii these proceedings; (Richards *et al.* 2011)). The arrangements of components within clouds can sometimes be traced from epoch to epoch in proper motion studies (Richards *et al.* 1999b; 2012), twisting and distorting at approximately the local sound speed.

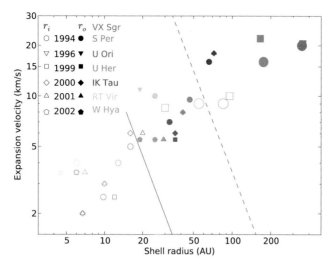

Figure 2. Inner and outer H_2O maser shell limits for 7 stars, showing consistent increase of velocity with radial separation from the star. The solid and dashed lines show the escape velocities for 1 and 20 $M_\odot$ stars, respectively.

Elitzur *et al.* (1992) predicted that unsaturated, 'amplification bounded' maser beaming from spherical clouds produces maser components of measured size s which appears smaller near the line centre, according to the relationship $s_\nu \propto 1/\sqrt{\ln(I_\nu)}$ where I is the flux density at frequency ν. This relationship was investigated for several epochs of observation of 22-GHz masers around five evolved stars (Richards *et al.* 2011). s is inversely proportional to I for most observations of S Per, RT Vir and IK Tau but a more random scatter or an increase in size of the brightest masers was seen in a minority of cases, particularly for U Her and U Ori. The latter behaviour is consistent with 'matter bounded' beaming from shocked slabs. There is additional evidence that these objects might be more prone to shocks. Anomalous OH 1612-MHz flares have been detected (Pataki & Kolena 1974; Chapman & Cohen 1985; Etoka & Le Squeren 1997). Moreover, U Her and U Ori have regular optical pulsations, amplitudes 5–6 magnitudes, in contrast to e.g. RT Vir, with amplitude < 2 mag or S Per, amplitude < 4 mag although it is an RSG. These results are summarized in Fig. 3.

2.3. *Cloud size and mass loss from the stellar surface*

The size of 22-GHz H_2O maser clouds is proportional to the stellar radius (as measured by NIR interferometry), Fig. 4 (see Richards *et al.* 2012 for references). Although there is a tenfold range in $R_\star$, from the smallest AGB stars to the largest RSG, in each case, if the maser cloud size was scaled back to the star assuming radial expansion, the birth size would be 5–10% $R_\star$. This is not what would be seen if cloud size was determined by local micro-physics, such as cooling related to dust formation, since this would operate on the same scale for all stars. Some stellar property must determine the cloud size. Models by Chiavassa *et al.* (2010) and Freytag & Höfner (2008) produce convection cells on suitable scales and optical/IR interferometry (Haubois 2009) shows stellar surface features which are consistent with these models.

VLTI observations reveal an AL_2O_3 layer at the stellar surface, the depth and temperature of which depend on stellar phase (Wittkowski *et al.* 2007; Wittkowski, Cotton these proceedings). These could provide the nuclei for silicate dust condensation at larger radii. Coordinated VLBA imaging of SiO masers shows that these are generally

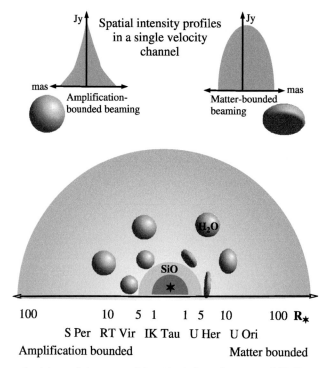

Figure 3. Cartoon (not to scale) summarising the inferred nature of H_2O maser clumps. Their appearance is mostly consistent with random, on average spherical clouds, but some clouds, round some CSEs, show properties consistent with shock-flattening.

brightest during the first part of the stellar phase, as predicted by Gray *et al.* (2009) and this technique could be used to show any link between dust production and enhanced acceleration.

2.4. *SiO masers tracing conditions near the star*

Comparisons between models and observations of SiO maser variability, the relative locations of the various transitions and other properties are also presented by Assaf, Cotton, Gonidakis, Ramsted and Richer (these proceedings), and so I will not repeat the wealth of SiO research undertaken in the last 5 years. There has been some success e.g. Soria-Ruiz *et al.* (2007) but also discrepancies such as the coincidence of $v = 2$ and $v = 3$ emission (Desmurs, these proceedings). Testing the prediction that many, bright maser spots at early phase become fewer, larger ones is hampered by the resolution of the VLBA; up to 90% of the total intensity flux is missing from images of R Cas at 176 pc (Assaf *et al.* 2011; Vlemmings *et al.* 2003) and this effect is likely to be felt out to 500 pc or further. Interestingly, the polarized emission seems to have structure on smaller scales, since some R Cas features appear to be $\gg 100\%$ linearly polarized (Assaf these proceedings).

A few other observational highlights include evidence for disruption of the SiO shell by stellar flares from R Cas. The $v = 1, J = 2 - 1$ emission formed a one-sided arc at one of two epochs observed by Phillips *et al.* (2001); several stellar periods later, Assaf *et al.* (2011) found the $J = 1 - 0$ emission was dominated by an eastern arc during the first stellar cycle monitored, and by a western arc during the second. This behaviour is seen occasionally in some stars at some epochs, but not at all epochs nor in all stars. RX Boo shows an analogously one-sided H_2O shell (Winnberg *et al.* 2008).

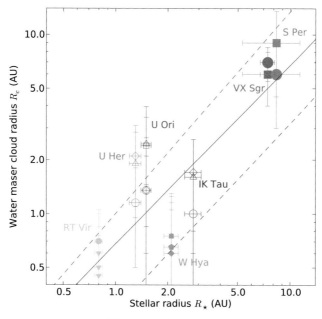

Figure 4. Cloud size

The development of improved calibration techniques (Cotton *et al.* 2011; Kemball &
Richter 2011) has led to striking confirmation of the Goldreich *et al.* (1973) model for
the behaviour of polarization when the angle between the magnetic field and the line
of sight goes through $\arcsin(\sqrt{2/3}$. The fractional linear polarization passes through a
minimum and the polarization angle abruptly flips 90°, Fig. 5 (Assaf, Gonidakis these
proceedings, Kemball *et al.* 2011). This behaviour is also seen in H_2O masers, covered in
the polarization review by Vlemmings (these proceedings).

There is mounting evidence for the Zeeman interpretation of SiO maser polarization,
e.g. Richer, these proceedings, but it is still debatable how significant a role magnetic
fields play in the mass loss process. Kemball *et al.* (2011) showed that the proper motions
of SiO maser clumps round TX Cam are non-linear and not consistently radial, yet
the polarization vectors remain approximately orthogonal to the direction of motion.
These authors infer that the masers are tracing material accelerated along magnetic field
lines, rather than dragging a frozen field in maser clumps (Hartquist & Dyson 1996). In
contrast, Matsumoto *et al.* (2008) fitted ballistic trajectories to SiO masers around IK
Tau, consistent with pulsation-driven ejection followed by deceleration under the star's
gravity. Infall was observed in R Cas (Assaf *et al.* 2011) since faint red-shifted components
were seen at several epochs projected against the centre of the SiO ring. These can only
be on the near side, since R Cas (radius 13 mas, Weigelt *et al.* 2000) is indubitably
optically thick.

An original interpretation of SiO polarization variability was suggested by Wiesemeyer
et al. (2009), who studied 77 AGB stars with the IRAM 30-m telescope. Of these, only V
Cam and R Leo showed quasi-periodic fluctuations of circular polarization, each in just
one, individual spectral feature. The variability period was 5–6 hr, and total intensity and
linear polarization were not affected. They suggest that this is due to local interaction
with the precessing magnetosphere of a Jupiter-like planet. The required magnetic field
strength, degree of misalignment of the planetary rotational and magnetic axes, timescale
and orbital period/radius, are all consistent with a planet of greater than twice the mass

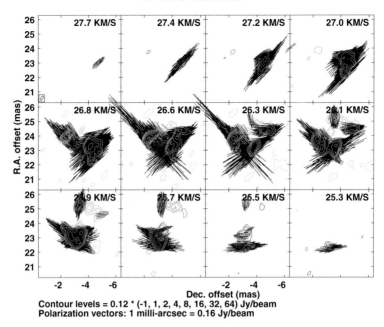

Contour levels = 0.12 * (-1, 1, 2, 4, 8, 16, 32, 64) Jy/beam
Polarization vectors: 1 milli-arcsec = 0.16 Jy/beam

Figure 5. Linear polarization angle of an R Cas SiO feature, showing an abrupt $90°$ change.

of Jupiter at greater than 3 AU radius. Alternative explanations relying solely on a stellar field either require much too high a field strength at the stellar surface, or cannot explain the locality of the effect, or would be tied to the imperceptibly slow stellar rotation. It would be very interesting to look for similar effects in other monitoring data.

2.5. Variability in the H_2O and OH mainline maser zone

The SiO masers at $2 - 4R_\star$ are very probably strongly affected by stellar pulsations, both shocks and changes in radiation. Conversely, the radiatively-pumped OH 1612-MHz masers at, typically, $\geqslant 100R_\star$, can show variations following the stellar light curve but are otherwise usually stable. 22-GHz H_2O masers, at intermediate distances, are collisionally pumped and highly variable. In some cases, this can be attributed to a discrete spectral and spatial region, likely to be the result of cloud overlap (Kartje et al. 1999) e.g. W Hya in 2002 (Richards et al. 2012).

Decades of single-dish monitoring has shown long-term variations affecting the whole 22-GHz maser profiles of many stars. These can appear to be correlated with the stellar pulsations, with a lag. Rudnitskii & Chuprikov (1990) suggested that this was due to shocks transmitted through a quasi-stationary layer at $\sim 6 - 7$ AU. This model was applied to Pushchino data for a number of stars, e.g. R Cas (Pashchenko & Rudnitskii 2004). Shintani et al. (2008) monitored 242 evolved stars at 22 GHz using the Iriki telescope and found a correlation between the stellar luminosity and the time lag between stellar and maser maxima, optimised if the time lag was constrained to be greater that half a stellar period by adding one (or potentially more) stellar periods. Such a relationship would be due to the longer periods of higher luminosity stars, which also exert more radiation pressure and possess larger, faster-expanding CSEs. However, the detailed models for the impact of shocks on the 22-GHz masers involve a thin shell expanding at constant velocity, whereas in reality the shells are generally thick and accelerating (Fig. 1). Reid & Menten (1997) and Gray et al. (2009) showed that pulsation shock speeds are unlikely

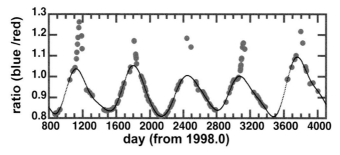

Figure 6. The ratio of the flux densities from the blue- and red-shifted 1612-MHz OH maser peaks of WX Psc, with a curve fitted to the data excluding the cusps overplotted (Lewis 2011).

to exceed $5 - 7$ km s^{-1} which is smaller than the maximum H_2O maser velocity in most CSEs (Fig. 2).

Rudnitskij (2008) proposed an alternative hypothesis, the effects of a low-mass companion. Intriguingly, one of the stars contributing to this model is R Leo, see Section 2.4 (Wiesemeyer *et al.* 2009). However, proper motion studies show no general rotation in most CSEs. It is possible that both these mechanisms play a limited rôle, with shocks impacting on the inner H_2O masers in some objects e.g. U Her and U Ori, Fig. 3.

Additional potential contributions to variability include changes in the mass loss rate and in dust heating. Evidence for sporadic dust formation comes from Aricebo and Nançay monitoring of WX Psc at 1612 MHz by Lewis (2011). Section 3.1 summarises the expected variability, with the blue-shifted flux density peak leading the red-shifted, so that the ratio is a smooth sinusoid. However, WX Psc shows a sharp cusp at the time of the blue-shifted peak, Fig. 6, due to additional pump photons produced by a thin, enhanced dust shell near the star.

OH masers also exhibit flares which seem generally to be confined to a small region, e.g. Szymczak *et al.* (2010), Wolak *et al.* (2012), Chapman (these proceedings). One of the best-known, *o* Cet, has SiO and fairly weak H_2O masers, but OH appears rarely, only seen around the time of optical maxima on a few occasions. The best-observed previous event was reported by Gerard & Bourgois (1993). The most recent flare (noticed in Nançay monitoring) started in 2010 and has been imaged using the EVN (Etoka & Diamond 2010). Combined with e-MERLIN astrometry of Mira A and B, this will reveal whether the compact companion has any influence on the occurrence of the flares and what the mechanism is.

Two issues are clear; firstly, that long-term monitoring and wide surveys are invaluable (see e.g. Cho, Kim these proceedings) and secondly, multi-epoch imaging is also required to avoid being misled by sporadic asymmetries and brightness fluctuations (periodic or localised).

3. Distance measurements

3.1. *Phase-lag*

OH 1612-MHz shells around OH/IR stars can be several light-weeks in diameter. The maser pump requires 53 μm emission from dust which waxes and wanes depending on the stellar period. Thus, the maser intensity follows the stellar light curve, but we detect the brightening of the blue-shifted peak before that of the red-shifted peak, leading to a sinusoidal variation in the red-blue flux density ratio. Comparing this phase-lag with the angular diameter of a thin, spherical shell gives the distance. However, there were large discrepancies between early measurements (Herman & Habing 1985; van Langevelde *et al.* 1990) probably because the maser shells were neither thin nor spherical. For example, MERLIN imaging shows that OH 26.5+0.6 is a prolate spheroid, elongated along he magnetic field axis (Etoka & Diamond 2010). Its phase-lag measurements are discussed by Engels (these proceedings), who is leading a campaign (using Nançay, the EVN, e-MERLIN and the EVLA) to obtain adequately complex models to determine the distances to a large sample of OH/IR stars.

3.2. *VLBI parallax*

VERA, the VLBA and the EVN have been used to measure the parallax of many evolved stars as well as star-forming regions (e.g. review by Reid, these proceedings). Individual maser features were tracked by e.g. Vlemmings & van Langevelde (2007), providing distances accurate to 10% or better, in some cases differing by $\sim 30\%$ from *Hipparcos* positions for objects within a few hundred pc and even more for remoter stars.

Statistical parallax – measuring the motion of the centroid of a group of masers – has been used at the distance of the Galactic centre and further. Implications for Galactic dynamics and understanding populations are discussed by Choi, Honma and Oyama (these proceedings) and Whitelock *et al.* (2008) has recalibrated the period-luminosity relationship for Miras using parallax results. The consequences for the classification of individual objects can be very significant, see Imai, Zhang (these proceedings). For example, IRAS 22480+6002 had a dynamical distance of 5 kpc but is actually at 1 kpc, reclassified as an unusual Population II supergiant (Imai *et al.* 2008). IRC-10414 'moved' from 0.7 to > 2 kpc (Maeda *et al.* 2008), thus trebling the proper motion velocities of its masers and turning a gentle outflow into a water fountain.

4. Asymmetry, binarity and post-AGB evolution

These topics are covered by Vlemmings, Amiri, Desmurs, Gomez and Suarez (these proceedings); I briefly note some of the surprises emerging at the meeting. Amiri *et al.* (2011) describe very clearly the distribution of OH and H_2O masers in some water fountain sources. OH12.8-0.9 fits nicely to a biconical model for the OH masers surrounding a relatively young H_2O jet; the higher-velocity jet of W43A has suppressed OH masing which is now confined to an equatorial ring. However, the seemingly well-established scenario in which the H_2O masers trace jets at speeds which can exceed 100 km s^{-1} was dealt a blow by Claussen (these proceedings) who reported that one of the twin outflows from IRAS16432-3814 has apparently come to an abrupt stop. Most models assume that binarity is needed to produce the highly collimated, magnetised jets, but since companions have not been detected in many objects it is still possible that a solitary star's core, revealed in the last stages of envelope loss, possesses sufficient angular momentum.

Symbiotic stars are interacting binaries showing the signature of accretion from an extended star, e.g. a Mira, onto a compact main sequence or white dwarf. Optical and higher frequency lines indicate an accretion disc and some undergo periodic nova outbursts or even possess a jet, e.g. R Aqr, one of only 3 such objects in which masers had previously been detected (Ivison *et al.* 1994). These are D-type symbiotics, showing evidence for dust but S-types have bright stellar spectra and in all cases there is evidence for an ionised nebular e.g. radio continuum. Amazingly, a survey by Cho & Kim (2010) of 47 symbiotics revealed that 19 (including one S-type) have SiO and/or H_2O masers. This suggests that dense, shielded clumps persist despite the hostile environment, which has implications for the survival of molecular material in AGB winds generally.

5. Meermasers

High-resolution imaging of the well-known 43, 22, and 1.6 GHz circumstellar masers, e.g. Fig. 1, has shown that these are generally found at increasing distances from the star, consistent with their decreasing excitation temperatures from species to species. However, the location of different transitions of the same species, e.g. SiO, does not always agree with predictions. Either the maser theory is incomplete or the CSE structure is more clumpy and inhomogeneous than has so far been modelled. Occasional detections of excited OH transitions (e.g. 6 and 1.7 GHz, Desmurs *et al.* 2010) around post-AGB or later stars, suggest pockets of conditions similar to star-forming regions, perhaps associated with collimated outflows, but these detections are very rare and seem to be transient. Millimetre and sub-mm masers, covered in this meeting by Menten, Perez Sanchez, Vlemmings and Wooten, are predicted to sample a wider range of physical conditions, and hence to provide much better constraints on the detailed temperature and density structure of the wind. ALMA will provided unprecedented imaging capability, producing great advances in at least three areas:

(*a*) C-rich star CSEs will be imaged in as good detail as O-rich, using lines such as HCN and SiS (first reported masing by Lucas & Cernicharo (1989) and Fonfría Expósito *et al.* (2006), respectively).

(*b*) Multiple H_2O transitions will constrain on the conditions from $\sim 5 - 100R_*$, in particular straddling the dust formation zone. An equally rich range of SiO lines will allow even further refinement of detailed models for conditions close to the star.

(*c*) Simultaneous detection of the star, dust and thermal lines will settle many old arguments about expansion mechanisms, maser pumping and chemistry.

Herschel is already providing valuable data, such as overall H_2O abundances and the ortho:para ratio (e.g. IK Tau, Decin *et al.* 2010), and tests of predictions for maser lines. 13 predicted masing H_2O lines have been detected above 1 THz in VY CMa (Royer *et al.* 2010), and new OH masers should also appear in this range.

Acknowledgements

I thank the many people who contributed ideas, information and figures to this review, including Marcello Agundez, Nikta Amiri, Khudhair Assaf Al-Muntafki, Valentin Bujarrabal, Dieter Engels, Sandra Etoka, Malcolm Gray, Liz Humphreys, Murray Lewis, Mikako Matsuura, Huib van Langevelde, Georgij Rudnitskii, Wouter Vlemmings, Markus Wittkowski and Jeremy Yates. I apologise for any ommisions or inaccuracies which are entirely mine.

References

Amiri, N., Vlemmings, W., & van Langevelde, H. J. 2011. *A&A*, **532**, A149.
Assaf, K. A., Diamond, P. J., Richards, A. M. S., & Gray, M. D. 2011. *MNRAS*, **415**, 1083.
Bains, I., Cohen, R. J., Louridas, A., Richards, A. M. S., Rosa-Gonzaléz, D., & Yates, J. 2003. *MNRAS*, **342**, 8.
Bowen, G. H. 1988. *ApJ*, **329**, 299.
Chapman, J. M. & Cohen, R. J. 1985. *MNRAS*, **212**, 375.
Chapman, J. M. & Cohen, R. J. 1986. *MNRAS*, **220**, 513.
Chen, X., Shen, Z.-Q., Imai, H., & Kamohara, R. 2006. *ApJ*, **640**, 982.
Chiavassa, A. *et al.* 2010. *A&A*, **511**, A51.
Cho, S.-H. & Kim, J. 2010. *ApJ*, **719**, 126.
Cotton, W. D., Ragland, S., & Danchi, W. C. 2011. *ApJ*, **736**, 96.
Danchi, W. C., Bester, M., Degiacomi, C. G., Greenhill, L. J., & Townes, C. H. 1994. *AJ*, **107**, 1469.
Decin, L., *et al.* 2010. *A&A*, **521**, L4.

Desmurs, J.-F., Baudry, A., Sivagnanam, P., Henkel, C., Richards, A. M. S., & Bains, I. 2010. *A&A*, **520**, A45.

Elitzur, M., Hollenbach, D. J., & McKee, C. F. 1992. *ApJ*, **394**, 221.

Etoka, S., & Diamond, P. J. 2010. *MNRAS*, **406**, 2218.

Etoka, S., & Le Squeren, A. M. 1997. *A&A*, **321**, 877.

Fonfría Expósito, J. P., Agúndez, M., Tercero, B., Pardo, J. R., & Cernicharo, J. 2006. *ApJLett.*, **646**, L127.

Freytag, B., & Höfner, S. 2008. *A&A*, **483**, 571.

Gerard, E., & Bourgois, G. 1993. Page 365 of: A. W. Clegg & G. E. Nedoluha (ed), *Astrophysical Masers*. Lecture Notes in Physics, Berlin Springer Verlag, vol. 412.

Goldreich, P., Keeley, D. A., & Kwan, J. Y. 1973. *ApJ*, **179**, 111.

Gray, M. D., Wittkowski, M., Scholz, M., Humphreys, E. M. L., Ohnaka, K., & Boboltz, D. 2009. *MNRAS*, **394**, 51.

Hartquist, T. W. & Dyson, J. E. 1996. *AP&SS*, **245**, 263.

Haubois, X. *et al.* 2009. *A&A*, **508**, 923.

Herman, J., & Habing, H. J. 1985. *A&AS*, **59**, 523.

Imai, H., Fujii, T., Omodaka, T., & Deguchi, S. 2008. *PASJ*, **60**, 55.

Ivison, R. J., Seaquist, E. R., & Hall, P. J. 1994. *MNRAS*.

Kartje, J. F., Königl, A., & Elitzur, M. 1999. *ApJ*, **513**, 180.

Kemball, A. J., & Richter, L. 2011. *A&A*, **533**, A26.

Kemball, A. J., Diamond, P. J., Richter, L., Gonidakis, I., & Xue, R. 2011. *ApJ*, **743**, 69.

Lewis, B. M. 2011. Page 629 of: F. Kerschbaum, T. Lebzelter, & R. F. Wing (ed), *Why Galaxies Care about AGB Stars II*. ASP Conference Series, vol. 445.

Lucas, R. & Cernicharo, J. 1989. *A&A*, **218**, L20.

Maeda, T. *et al.* 2008. *PASJ*, **60**, 1057.

Matsumoto, N. *et al.* 2008. *PASJ*, **60**, 1039–.

Murakawa, K., Yates, J. A., Richards, A. M. S., & Cohen, R. J. 2003. *MNRAS*, **344**, 1. M+03.

Pashchenko, M. I., & Rudnitskii, G. M. 2004. *Astronomy Reports*, **48**, 380.

Pataki, L., & Kolena, J. 1974. *Bull. Am. astr. Soc.*, **6**, 340.

Phillips, R. B., Sivakoff, G. R., Lonsdale, C. J., & Doeleman, S. S. 2001. *AJ*, **122**, 2679.

Reid, M. J., & Menten, K. M. 1997. *ApJ*, **476**, 327.

Richards, A. M. S., Yates, J. A., & Cohen, R. J. 1999a. *MNRAS*, **306**, 954.

Richards, A. M. S., Yates, J. A., Cohen, R. J., & Bains, I. 1999b. Page 315 of: Le Bertre, T, Lèbre, A, & Waelkens, C (eds), *IAU Symp. 191: Asymptotic Giant Branch Stars*. ASP.

Richards, A. M. S., Elitzur, M., & Yates, J. A. 2011. *A&A*, **525**, A56.

Richards, A. M. S. *et al.* 2012. *A&A*. in prep.

Royer, P. *et al.* 2010. *A&A*, **518**, L145.

Rudnitskii, G. M., & Chuprikov, A. A. 1990. *Soviet Astronomy*, **34**, 147.

Rudnitskij, G. M. 2008. *Journal of Physical Studies*, **12**, 1301.

Shintani, M. *et al.* 2008. *PASJ*, **60**, 1077.

Soria-Ruiz, R., Alcolea, J., Colomer, F., Bujarrabal, V., & Desmurs, J.-F. 2007. *A&A*, **468**, L1.

Szymczak, M., Wolak, P., Gérard, E., & Richards, A. M. S. 2010. *A&A*, **524**, A99.

van Langevelde, H. J., van der Heiden, R., & van Schooneveld, C. 1990. *A&A*, **239**, 193.

Vlemmings, W. H. T., & van Langevelde, H. J. 2007. *A&A*, **472**, 547–533.

Vlemmings, W. H. T., van Langevelde, H. J., Diamond, P. J., Habing, H. J., & Schilizzi, R. T. 2003. *A&A*, **407**, 213–224.

Weigelt, G. *et al.* 2000. Page 617 of: P. Léna & A. Quirrenbach (ed), *SPIE Conference Series*, vol. 4006.

Whitelock, P. A., Feast, M. W., & van Leeuwen, F. 2008. *MNRAS*, **386**, 313.

Wiesemeyer, H., Thum, C., Baudry, A., & Herpin, F. 2009. *A&A*, **498**, 801.

Winnberg, A., Engels, D., Brand, J., Baldacci, L., & Walmsley, C. M. 2008. *A&A*, **482**, 831.

Wittkowski, M., Boboltz, D. A., Ohnaka, K., Driebe, T., & Scholz, M. 2007. *A&A*, **470**, 191.

Woitke, P. 2006. *A&A*, **460**, L9.

Wolak, P., Szymczak, M., & Gérard, E. 2012. *A&A*, **537**, A5.

Yates, J. A., & Cohen, R. J. 1994. *MNRAS*, **270**, 958.

Cosmic Masers - from OH to H$_0$
Proceedings IAU Symposium No. 287, 2012
R.S. Booth, E.M.L. Humphreys & W.H.T. Vlemmings, eds.

Radio and IR interferometry
of SiO maser stars

Markus Wittkowski[1], **David A. Boboltz**[2], **Malcolm D. Gray**[3],
Elizabeth M. L. Humphreys[1] **Iva Karovicova**[4], **and Michael Scholz**[5,6]

[1] ESO, Karl-Schwarzschild-Str. 2, 85748 Garching bei München, Germany

[2] US Naval Observatory, 3450 Massachusetts Avenue, NW, Washington, DC 20392-5420, USA

[3] Jodrell Bank Centre for Astrophysics, Alan Turing Building, University of Manchester, Manchester M13 9PL, UK

[4] Max-Planck-Institut für Astronomie, Königstuhl 17, 69117 Heidelberg, Germany

[5] Zentrum für Astronomie der Universität Heidelberg (ZAH), Institut für Theoretische Astrophysik, Albert-Ueberle-Str. 2, 69120 Heidelberg, Germany

[6] Sydney Institute for Astronomy, School of Physics, University of Sydney, Sydney NSW 2006, Australia

Abstract. Radio and infrared interferometry of SiO maser stars provide complementary information on the atmosphere and circumstellar environment at comparable spatial resolution. Here, we present the latest results on the atmospheric structure and the dust condensation region of AGB stars based on our recent infrared spectro-interferometric observations, which represent the environment of SiO masers. We discuss, as an example, new results from simultaneous VLTI and VLBA observations of the Mira variable AGB star R Cnc, including VLTI near- and mid-infrared interferometry, as well as VLBA observations of the SiO maser emission toward this source. We present preliminary results from a monitoring campaign of high-frequency SiO maser emission toward evolved stars obtained with the APEX telescope, which also serves as a precursor of ALMA images of the SiO emitting region. We speculate that large-scale long-period chaotic motion in the extended molecular atmosphere may be the physical reason for observed deviations from point symmetry of atmospheric molecular layers, and for the observed erratic variability of high-frequency SiO maser emission.

Keywords. masers, radiative transfer, turbulence, techniques: interferometric, stars: AGB and post-AGB, stars: atmospheres, stars: circumstellar matter, stars: fundamental parameters, stars: mass loss, supergiants

---

1. Introduction

Low to intermediate mass stars, including our Sun, evolve to red giant and subsequently to asymptotic giant branch (AGB) stars. An AGB star is in the final stage of stellar evolution that is driven by nuclear fusion. Mass loss becomes increasingly important during the AGB evolution, both for the stellar evolution, and for the return of material to the interstellar medium. Depending on whether or not carbon has been dredged up from the core into the atmosphere, AGB stars appear to have an oxygen-rich or a carbon-rich chemistry. A canonical model of the mass-loss process has been developed for the case of the carbon-rich chemistry, where atmospheric carbon dust has a sufficiently large opacity to be radiatively accelerated and driven out of the gravitational potential of the star and where it drags along the gas (e.g., Wachter *et al.* 2002, Mattsson *et al.* 2010). For the case of an oxygen-rich chemistry, the details of this process are not understood, and are currently a matter of debate (e.g., Woitke 2006, Höfner 2008). Questions remain also for the carbon-rich case, such as regarding the recent observational evidence that the oxygen-bearing molecule H_2O is ubiquitous in C-stars (Neufeld *et al.* 2011). Another unsolved problem in stellar physics is the mechanism by which (almost) spherically symmetric stars on the AGB evolve to form axisymmetric planetary nebulae (PNe). Currently, a consensus seems to form that single stars cannot trivially evolve towards non-spherical PNe, by mechanisms such as magnetic fields or rotation, but that a binary companion is required in the majority of the observed PN morphologies (e.g. de Marco 2009).

In order to solve these open questions, it is important to observationally establish the detailed stratification and geometry of the extended atmosphere and the dust formation region, and to compare it to and constrain the different modeling attempts. This includes the following questions: How is the mass-loss process connected to the stellar pulsation? Which is the detailed radial structure of the atmosphere and circumstellar envelope? At which layer do inhomogeneities form? Which are the shaping mechanisms? Which is the effect of inhomogeneities on the further stellar evolution?

2. Synergy of radio and infrared interferometry

Radio and infrared interferometry are both well suited to probe the structure, morphology, and kinematics of the extended atmospheres and circumstellar environment of evolved stars, because of their ability to spatially resolve these regions. In fact, evolved stars have been prime targets for radio and infrared interferometry for decades, because they match well the sensitivity and angular scale of available facilities. Interferometric observations continue to provide new observational results in the field of evolved stars thanks to newly available spectro-interferometry at infrared wavelengths, new radio interferometric facilities, larger samples of well studied targets, and comparisons to newly available theoretical models. Here, radio and infrared interferometry provide complementary information. From the perspective of infrared studies, radio interferometry of SiO, OH, and H_2O maser emission adds information on the morphology and kinematics at different scales from a few stellar radii (SiO maser) to a few hundred stellar radii (OH maser). From the perspective of radio interferometry, infrared interferometry provides information on the environment of the astrophysical masers, including the radiation field, the radii at wavelengths of maser pumping, and constraints on the stratification of the temperature, density, and number densities.

3. Project outline

We have established a project of coordinated interferometric observations of evolved stars at infrared and radio wavelengths. Our goal is to constrain the radial structure and kinematics of the stellar atmosphere and the circumstellar environment to understand better the mass-loss process and its connection to stellar pulsation. We also aim at tracing asymmetric structures from small to large distances in order to constrain shaping processes during the AGB evolution. We use two of the highest resolution interferometers in the world, the Very Large Telescope Interferometer (VLTI) and the Very Long Baseline Array (VLBA) to study AGB stars and their circumstellar envelopes from near-infrared to radio wavelengths. For some sources, we have coordinated near-infrared broad-band photometry obtained at the South African Astronomical Observatory (SAAO) in order to derive effective temperatures. We have started to use the Atacama Pathfinder Experiment (APEX) to investigate the line strengths and variability of high frequency SiO maser emission in preparation of interferometric observation of SiO emitting regions using the ALMA facility.

4. Observations

Our pilot study included coordinated observations of the Mira variable S Orionis including VINCI K-band measurements at the VLTI and SiO maser measurements at the VLBA (Boboltz & Wittkowski 2005). For the Mira variables S Ori, GX Mon, RR Aql and the supergiant AH Sco, we obtained long-term mid-infrared interferometry covering several pulsation cycles using the MIDI instrument at the VLTI coordinated with VLBA SiO (42.9 GHz and 43.1 GHz transitions) observations (partly available in Wittkowski *et al.* 2007, Karovicova *et al.* 2011). For the Mira variables R Cnc and X Hya, we coordinated near-infrared interferometry (VLTI/AMBER), mid-infrared interferometry (VLTI/MIDI), VLBA/SiO maser observations, VLBA/H_2O maser, and near-infrared photometry at the SAAO (first results in Wittkowski *et al.* 2008). Most recently, we obtained measurements of the $v = 1$ and $v = 2$ $J = 7 - 6$ SiO maser transitions toward a sample of evolved stars at several epochs using the APEX telescope.

5. Modeling

We used the P & M model series by Ireland *et al.* (2004a,b) and most recently the CODEX models by Ireland *et al.* (2008, 2011) to describe the dynamic model atmospheres of Mira variable AGB stars. Wittkowski *et al.* (2007) and Karovicova *et al.* (2011) added ad-hoc radiative transfer models to these dynamic model atmosphere series to describe the dust shell. They employed the radiative transfer code `mcsim_mpi` by Ohnaka *et al.* (2007). Gray *et al.* (2009) combined these hydrodynamic atmosphere plus dust shell models with a maser propagation code in order to model the SiO maser emission.

Wittkowski *et al.* (2007) conducted coordinated mid-infrared interferometry using the VLTI/MIDI instrument and observations of SiO maser emission using the VLBA of the Mira variable S Ori. Based on the modeling of the mid-infrared interferometry as outlined above, they showed that the maser emission is located just outside the layer where the molecular layer becomes optically thick at mid-infrared wavelengths ($\sim 10\mu m$), roughly at two photospheric radii, and is close to, and possibly co-located with Al_2O_3 dust. One of the questions that led to the modeling effort by Gray *et al.* (2009) was whether the combined dynamic model atmosphere and dust shell model that was successful to describe the mid-infrared interferometric observations of S Orionis by Wittkowski *et al.* (2007) would also lead to model-predicted locations of SiO maser emission that is consistent with the simultaneous VLBA observations. The input to the maser propagation model included the temperature and density stratification of the successful model of the mid-infrared interferometric data, the radii of the $1.04\,\mu m$ continuum layer, and of optically thick layers at IR pumping bands of SiO ($8.13\,\mu$, $4.96\,\mu$, $2.71\,\mu$, and $2.03\,\mu m$), the IR radiation field of the dust model, and the number densities of SiO and its main collision partners (assuming LTE chemistry). Modeled masers indeed formed in rings with radii of 1.8–2.4 photospheric radii, which is consistent with the S Ori VLBA observations, with other observations in the literature, and with earlier such maser propagation models (Gray *et al.* 1995, Humphreys *et al.* 1996). The new models confirm the $v = 1$ ring at larger radii than the $v = 2$ ring. Maser rings, a shock front, and the $8.13\,\mu m$ layer appear to be closely related, suggesting that collisional and radiative pumping are closely related spatially and therefore temporally.

6. Infrared and radio interferometry of the Mira variable AGB star R Cnc

R Cancri (R Cnc) is a Mira variable AGB star with a V magnitude between 6.1 and 11.8 and a period of 362 days (Samus *et al.* 2009) at a distance of 280 pc based on the period-luminosity relation by Whitelock *et al.* (2008). We obtained two epochs of observations (23 Dec 2008 to 10 Jan 2009 and 25 Feb 2009 to 3 Mar 2009) that were coordinated between near-infrared spectro-interferometry obtained with VLTI/AMBER, mid-infrared spectro-interferometry obtained with VLTI/MIDI, near-IR $JHKL$ photometry obtained at the SAAO, and VLBA observations of the SiO maser emission and H_2O maser emission. The details of these observations will be available in Wittkowski *et al.* (in preparation). The first epoch of near-infrared spectro-interferometry of R Cnc is available in Wittkowski *et al.* (2011).

6.1. *Near-infrared spectro-interferometry of R Cnc*

Fig. 1 shows the near-infrared squared visibility amplitudes and closure phases of R Cnc obtained with the VLTI/AMBER instrument (from Wittkowski *et al.* 2011). The visibility amplitudes show a characteristic *bumpy* shape that has also been observed in VLTI/AMBER observations of S Ori in Wittkowski *et al.* (2008) and that is interpreted

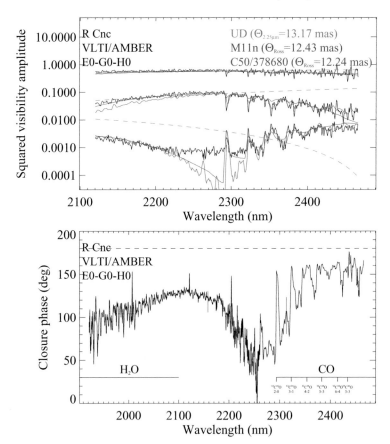

Figure 1. Near-IR squared visibility amplitude (top) and closure phase (bottom) of R Cnc obtained with the VLTI/AMBER instrument from Wittkowski *et al.* (2011). The upper panel also shows predictions by dynamic model atmospheres (M series by Ireland *et al.* 2004a,b, and CODEX series by Ireland *et al.* 2008, 2011).

as being indicative of the presence of molecular layers lying on top of the continuum-forming photosphere and that extend to a few photospheric radii. The visibilities are well consistent with predictions by the latest dynamic model atmosphere series by Ireland *et al.* (2008, 2011) that include such molecular layers. The wavelength-dependent closure phases indicate deviations from point symmetry at all wavelengths and thus a complex non-spherical stratification of the atmosphere. In particular, Wittkowski *et al.* (2008) discuss that the strong closure phase signal in the water vapor and CO bandpasses can be a signature of large-scale inhomogeneities/clumps of the molecular layers. These might be caused by pulsation- and shock-induced chaotic motion in the extended atmosphere as theoretically predicted by Icke *et al.* (1992) and Ireland *et al.* (2008, 2011). We note that these extended atmospheric layers correspond roughly to the radii where SiO maser emission is observed.

6.2. *Mid-infrared spectro-interferometry of R Cnc*

Fig. 2 shows the mid-IR visibility amplitude of R Cnc obtained with VLTI/MIDI from Karovicova (2011). The MIDI data can well be re-produced with a dynamic model atmosphere, which naturally includes molecular layers, and an ad-hoc radiative transfer model of an Al_2O_3 dust shell with an inner radius of ~ 2.2 photospheric radii and an optical depth $\tau_V \sim 1.4$. As in the case of MIDI observations of S Ori (Wittkowski *et al.* 2007), R Cnc does not show an indication of an additional silicate dust shell. The inner radius of the Al_2O_3 dust shell is located at a radius that is close to the radius where SiO maser emission is observed. Oher sources that have been studied by Karovicova (2011) show either Al_2O_3 dust, silicate dust, or both dust species. Karovicova (2011) discussed an indication that the dust content of stars with low mass-loss rates is dominated by Al_2O_3, while the dust content of stars with higher mass-loss rates predominantly exhibit significant amounts of silicates, as suggested by Little-Marenin & Little (1990) and Blommaert *et al.* (2006).

6.3. *VLBA observation of the SiO maser emission toward R Cnc*

Fig. 3 shows our two epochs of VLBA images of the $J = 1 - 0$, $v = 2$ (42.8 GHz) and $J = 1 - 0$, $v = 1$ (43.1 GHz) SiO maser emission toward R Cnc. The two epochs are separated by about 7 weeks. The two transitions were registered to each other by transferring the calibration from one transition to the other. Consistently with earlier

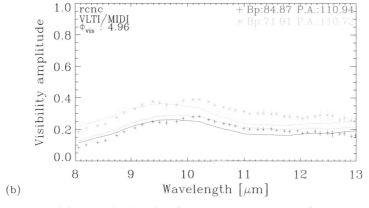

Figure 2. Mid-IR visibility amplitude of R Cnc obtained with VLTI/MIDI from Karovicova (2011). Also shown is a model of a dynamic model atmosphere combined with a radiative transfer model of the dust shell (see text).

observations, the $v = 1$ transition is located at larger radii than the $v = 2$ transition, but with more overlap at epoch 1 and a clearer separation at epoch 2. The morphology is more ring-like at epoch 2 compared to epoch 1.

7. Monitoring of high-frequency SiO maser emission toward evolved stars

Observations of high-frequency SiO maser emission has been shown to be more variable than centimeter SiO maser emission, indicating that high-frequency maser emission depends more strongly on the environmental conditions of the masers, and that

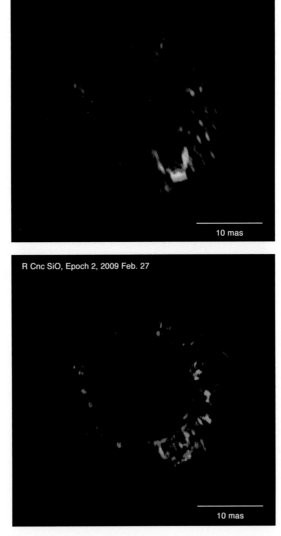

Figure 3. VLBA observations of the SiO emission toward R Cnc at two epochs (top: 4 Jan 2009, bottom: 27 Feb 2009). The red maps denotes the $J = 1 - 0$, $v = 2$ transition and the green maps the $J = 1 - 0$, $v = 1$ transition.

thus their observations may provide stronger observational constraints (Gray *et al.* 1995, Humphreys *et al.* 1997, Gray *et al.* 1999).

Following these studies, we have established a monitoring of high-frequency $J = 7 - 6$, $v = 1, 2, 3$ SiO maser emission toward a sample of AGB stars in order to compare their variability to the predictions by Gray *et al.* (2009). This represents also an important precursor of ALMA imaging studies of the SiO emitting regions of evolved stars.

Our observational results suggest that the variability is erratic. It does not appear to be correlated with the stellar phase and is also not consistent among the different sources of our sample. For, example, o Cet showed $v = 1$ emission at all phases between stellar minimum and post-maximum with increasing intensity, and $v = 2$ emission only at post-maximum (phase 1.1) with lower intensity compared to $v = 1$ emission at this phase (Fig. 4). To the contrary, R Hya, for instance, shows $v = 1$ emission only at post-maximum (1.1) but not between minimum and maximum, but $v = 2$ emission at all phases between 0.5 and 1.1 and at phase 1.1 with higher intensity compared to $v = 1$. R Leo showed both $v = 1$ and $v = 2$ emission at phases between 0.5 and 0.8 with alternating ratio between the strength of the $v = 1$ and $v = 2$ emission.

We hypothesize that large-scale (a few cells across the stellar surface) long-period (times scales corresponding to a few pulsation cycles) chaotic motion in the extended atmosphere, induced by the interaction of pulsation and shock fronts with the extended atmosphere, and a possibly related erratic variability of the SiO abundance, may be the reason for our observed erratic maser variability.

8. Summary

Near-infrared interferometry of oxygen-rich evolved stars indicates a complex atmosphere including extended atmospheric molecular layers (in the IR most importantly H_2O, CO, SiO), which is consistent with predictions by the latest dynamic model atmospheres. Near-IR closure phases indicate a complex non-spherical stratification of the atmosphere, indicating asymmetric/clumpy molecular layers. These are possibly caused by chaotic motion in the extended atmosphere, which may be triggered by the pulsation in the stellar interior. Mid-infrared interferometry constrains dust shell parameters including Al_2O_3 dust with inner radii of typically two photospheric radii and silicate dust with inner radii of typically four photospheric radii. SiO masers lie in the extended atmosphere as seen by infrared interferometry. They are located just outside the radius where the molecular layer becomes optically thick at mid-IR wavelengths. There are located

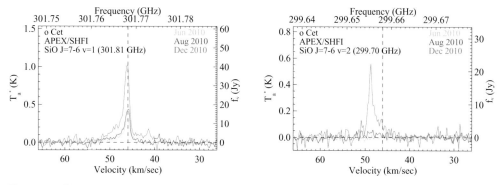

Figure 4. APEX observations of the $J = 7-6$, $v = 1$ (left) and $v = 2$ (right) SiO maser emission of o Cet at three epochs in June 2010 (phase 0.5), August 2010 (phase 0.7), and December 2010 (phase 1.1).

close-to, possibly co-located, with Al_2O_3 dust. Their location is consistent with dynamic model atmospheres combined with a maser propagation code. They are also located at the distances where the near-IR interferometry indicates a clumpy morphology. APEX detects a strong and erratic variability of high-frequency maser emission. We speculate that the erratic variability may be connected to chaotic motion in the extended atmosphere, i.e. to the same mechanism that may lead to the observed clumpyness of extended atmospheric molecular layers.

Acknowledgements

This article is based on a project of coordinated VLTI and VLBA observations of evolved stars to which several people have contributed during the last years in addition to the authors of this article, including Carlos de Breuck, Thomas Driebe, Eric Fossat, Michael Ireland, Keiichi Ohnaka, Anita Richards, Francois van Wyk, Patricia Whitelock, Peter Wood, and Albert Zijlstra.

References

Blommaert, J. A. D. L., Groenewegen, M. A. T., Okumura, K., *et al.* 2006, *A&A*, 460, 555
Boboltz, D. A. & Wittkowski, M. 2005, *ApJ*, 618, 953
De Marco, O. 2009, *PASP*, 121, 316
Gray, M. D., Ivison R., Yates J., Humphreys E., Hall P., & Field D., 1995, *MNRAS*, 277, L67
Gray, M. D., Humphreys, E. M. L., & Yates, J. A. 1999, *MNRAS*, 304, 906
Gray, M. D., Wittkowski, M., Scholz, M., *et al.* 2009, *MNRAS*, 394, 51
Höfner, S. 2008, *A&A*, 491, L1
Humphreys E. M. L., Gray M., Yates J., Field D., Bowen G., & Diamond P., 1996, *MNRAS*, 282, 1359
Humphreys, E. M. L., Gray, M. D., Yates, J. A., & Field, D. 1997, *MNRAS*, 287, 663
Icke, V., Frank, A., & Heske, A. 1992, *A&A*, 258, 341
Ireland, M. J., Scholz, M., & Wood, P. R. 2004a, *MNRAS*, 352, 318
Ireland, M. J. Scholz, M. Tuthill, P. G., & Wood, P. R. 2004b, *MNRAS*, 355, 444
Ireland, M. J., Scholz, M., & Wood, P. R. 2008, *MNRAS*, 391, 1994
Ireland, M. J., Scholz, M., & Wood, P. R. 2011, *MNRAS*, 418, 114
Karovicova, I., Wittkowski, M., Boboltz, D. A., *et al.* 2011, *A&A*, 532, A134
Karovicova, I., 2011, *PhD thesis*, University of Nice
Little-Marenin, I. R. & Little, S. J. 1990, *AJ*, 99, 1173
Mattsson, L., Wahlin, L., & Höfner, S. 2010, *A&A*, 509, A14
Neufeld, D. A., González-Alfonso, E., & Melnick, G. J., *et al.* 2011, *ApJL*, 727, L28
Ohnaka, K., Driebe, T., Weigelt, G., & Wittkowski, M. 2007, *A&A*, 466, 1099
Samus, N. N., Durlevich, O. V., *et al.* 2009, *VizieR Online Data Catalog*, 2025
Wachter, A., Schröder, K.-P., Winters, J., Arndt, T., & Sedlmayr, E. 2002, *A&A*, 384, 452
Whitelock, P., Feast, M. W., & van Leeuwen, F. 2009, *MNRAS*, 386, 313
Wittkowski, M., Boboltz, D. A., Ohnaka, K., Driebe, T., & Scholz, M. 2007, *A&A*, 470, 191
Wittkowski, M., Boboltz, D. A., Driebe, T., *et al.* 2008, *A&A*, 479, L21
Wittkowski, M., Boboltz, D. A., Ireland, M., *et al.* 2011, *A&A*, 532, L7
Woitke, P. 2006, *A&A*, 452, 537

Cosmic Masers - from OH to H_0
Proceedings IAU Symposium No. 287, 2012
R.S. Booth, E.M.L. Humphreys & W.H.T. Vlemmings, eds.

© International Astronomical Union 2012
doi:10.1017/S1743921312006990

Maser emission during post-AGB evolution

J.-F. Desmurs

Observatorio Astronómico Nacional, Madrid, Spain
email: desmurs@oan.es

Abstract. This contribution reviews recent observational results concerning astronomical masers toward post-AGB objects with a special attention to water fountain sources and the prototypical source OH 231.8+4.2. These sources represent a short transition phase in the evolution between circumstellar envelopes around asymptotic giant branch stars and planetary nebulae. The main masing species are considered and key results are summarized.

Keywords. Maser, stars: AGB and Post AGB

1. Introduction

After leaving the main sequence, stars of low and intermediate mass travel across the Hertzsprung-Russell diagram and reach the asymptotic giant branch (AGB). During this phase, the star ejects matter at a very high rate (up to 10^{-4} solar masses per year) in form of a slow (5 to 30 km s^{-1}), dense, isotropic wind. The resulting circumstellar envelope often exhibits maser emission from several molecules, the most common being SiO, H_2O and OH. These masers arise at different distances from the star from different layers in the envelope, tracing different physical and chemical conditions. SiO masers are found close to the star (at few stellar radii), the water maser a little farther out (up to a few hundreds of stellar radii) and the OH masers even farther out, the (at up to a few thousands of stellar radii) see Habing (1996). As the star follows its evolution to the planetary nebulae (PN) phase, mass-loss stops and the envelope begins to become ionized, such that the masers emission disappear progressively. The SiO masers are supposed to disappear first, the H_2O masers may survive a few hundreds of years and OH masers remain for a period of ~1000 years and can even be found in the PN phase.

The evolution of the envelopes around AGB stars toward PNe, through the phase of Proto Planetary Nebulae (pPNe) is yet poorly know, in particular the shaping of PNe. While during the AGB phase the star exhibits roughly spherical symmetry, about 10000 years later, PNe are often asymmetrical (about 75% see Manchado *et al.* 2000), showing axial symmetry, including multi-polar or elliptical symmetry, and very collimated and fast jets. Bipolarity appears very early in the post-AGB or pPN stage evolution.

Sahai & Trauger (1998) surveyed young PNe with the Hubble Space Telescope and found that most of them were characterized by multi polar morphology with collimated radial structures, and bright equatorial structures indicating the presence of jets and disks/tori in some objects. They propose that during the late AGB or early PPN stages the high-speed collimated outflows carve out an imprint in the spherical AGB wind, which provides the morphological signature for the development of asymmetric PNe.

The mechanism explaining how axial symmetry appears during this evolutionary phase is still an open question. Several models have been proposed, that in general involve the interaction of very fast and collimated flows, ejected by means of magneto-centrifugal launching, with the AGB fossil shell. Interferometric observations of maser emission in such sources allow us to get access to the formation and evolution of these jets and with a very high spatial resolution.

This review will report mainly on recent results published since the previous IAU 242 maser symposium, held in Alice Springs in 2007.

2. Surveys

During the last years, several surveys have been conducted to discover new maser emission toward post-AGB objects. Deacon *et al.* (2007), searched for water masers (at 22 GHz) and SiO masers (at 86 GHz) using respectively the telescopes of Tibdinbilla-70m and Mopra-22m. They observed a list of 85 sources in total (11 of them in SiO), selected on the basis of their OH 1612 MHz spectra, and get 21 detections (3 in SiO), out of which 5 sources present high velocity profiles and one source show a wide double peak profile of the SiO maser. Suárez *et al.* (2007, 2009), searched for water maser in the northern hemisphere (with the Robledo-70m antenna) and in the southern hemisphere (Parkes-64m). They surveyed 179 sources (mostly pPNe & PNe) and detected 9 sources (4 pPNe and 5 PNe, one of these, IRAS 15103-5754, has water fountain characteristics, see Figure 1)

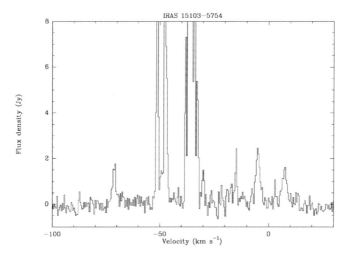

Figure 1. IRAS 15103-5754, first water fountain detected in a planetary nebula (Suárez *et al.* 2009).

Using the 100 m telescope of Effelsberg, Desmurs *et al.* (2010) undertook a high sensitivity discrete source survey for the first excited state of OH maser emission (J = 5/2, $^2\Pi_{3/2}$ at 6 GHz) in the direction of 65 PNe and pPNe exhibiting 18 cm OH emission (main and/or satellite lines). They detected two sources at 6035 MHz (5 cm), both of them young (or very young) PNe. And very recently, Amiri *et al.* (2012), conducted a sensitive survey with the Effelsberg antenna of water maser emission toward 74 post-AGBs, and found 6 new water fountain candidates showing a double peak profile.

3. Water Fountains

In the class of the proto-planetary nebulae, there is a very interesting sub-class of young pPNe called the water fountains. They present both hydroxyl and water maser emission,

Table 1. Confirmed water fountains[a]

PNe	Other name	OH Velocity range (in km s^{-1})	H$_2$O Velocity range (in km s^{-1})	References
IRAS 15445-5449	OH 326.5-0.4		90	Deacon *et al.* (2007)
IRAS 15544-5332	OH 325.8-0.3		74	Deacon *et al.* (2007)
IRAS 16342-3814	OH 344.1+5.8		260	Claussen *et al.* (2009)
IRAS 16552-3050	GLMP 498			Suárez *et al.* (2007)
IRAS 18043-2116	OH 0.9-0.4		400	Walsh *et al.* (2009)
IRAS 18113-2503	PM 1-221		500	Gómez *et al.* (2011)
IRAS 18139-1816	OH 12.8-0.9	23	42	Boboltz & Marvel (2007)
IRAS 18286-0959	OH 21.79-0.1		200	Yung *et al.* (2011)
IRAS 18450-0148	W 43A/OH 31.0+0.0		180	Imai *et al.* (2002)
IRAS 18460-0151	OH 31.0-0.2	20	300	Imai *et al.* (2008)
IRAS 18596+0315	OH 37.1-0.8	30	60	Amiri *et al.* (2011)
IRAS 19067+0811	OH 42.3-0.1	20	70	Gómez *et al.* (1994)
IRAS 19134+2131	G054.8+4.6		105	Imai *et al.* (2007)
IRAS 19190+1102	PM 1-298		100	Day *et al.* (2010)
IRAS 15103-5754	G320.9-0.2		80	Suárez *et al.* (2009)

[a] +6 new water fountains candidates (see Amiri *et al.* 2012).

however their characteristics differ from those typical of AGB stars†. First of all, H$_2$O and OH maser spectra exhibit a double peaked profile with the peaks symmetrically distributed about the star velocity. But, unlike in AGB stars, H$_2$O maser are spread over a larger velocity range than the OH masers (see Imai *et al.* 2008 for example) and display higher velocity (up to 400 km s^{-1} see Figure 2) than the OH masers or AGB radial expansion wind (10-20 km s^{-1} te Lintel *et al.* 1989). High spatial resolution observations of the water masers emission in these objects reveals bipolar distribution and highly collimated outflows (hence the name of water fountain). The first member of this subclass of pPN, W 43A, was first observed by Imai *et al.* (2002) with the VLBA, and showed a well collimated and precessing jet with an outflow velocity of $\sim$145 km s^{-1}. When optical images are available, the masers appear coincident with the optical bipolar structures (see for example IRAS 16342-3814, Claussen *et al.* 2009 and this proceeding).

Table 1 shows a list of the confirmed‡ water fountains. Up to now, 14 sources have been identified as water fountains and 6 new sources have been recently found by Amiri *et al.* (2012) and are good candidates to be classified as such (they all present a double peak spectra and high velocity profiles). The most recent source confirmed to belong to this sub-class is IRAS 18113-2503 (Gómez *et al.* 2011, see Figure 2). The source shows the typical double peak spectra, with a very large velocity dispersion and the peaks of water emission are separated by about 500 km s^{-1} (from -150 to +350 km s^{-1} LSR). It is likely to be the fastest outflows observed up to now with a velocity of at least 250 km s^{-1}. The e-VLA map clearly shows a bipolar spatial distribution with a blueshifted part to the north and a redshited lobe to the south.

IRAS 19067+0811 is also an interesting source, observations in 1988 clearly detected a double peak H$_2$O spectrum covering a velocity range of twice of the velocity range of the OH spectra, but Gómez *et al.* (1994) only detected OH maser emission. And finally,

---

† OH masers in AGB star generally exhibit double-peaked profiles covering up to 25 km/s (e.g. te Lintel *et al.* 1989), and H$_2$O maser spectra present a velocity range within the OH maser one and their profiles are more irregular.

‡ The candidates sources IRAS 07331+0021 and IRAS 13500-6106 from Suárez *et al.* (2009) turned to be "classical" proto-planetary nebulae and not water fountain.

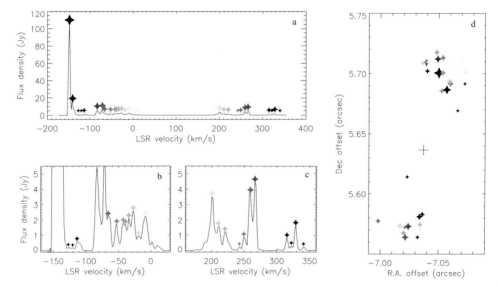

Figure 2. e-VLA spectrum and map of the water fountain IRAS 18113-2503, see Gómez *et al.* (2011) for details.

I would like to mention IRAS 15103-5754 a very peculiar source as it is the first confirmed water fountain that is not a pPN but a PN (see Suárez *et al.* 2009 and these proceedings).

3.1. *Outflows Proper motion*

A careful analysis of the 3D structure of jets (including proper motions studies) is a fundamental element for the development of theoretical models of the PN shaping mechanisms and of the nature of the outflows that give rise to the complex structure found in many pPNe and PNe.

In W 43A, VLBA observations of H_2O maser emission show very well collimated jets with a velocity of the order of $\sim$145 km s^{-1} (Imai *et al.* 2002, 2005). The proper motion analysis also demonstrates that the jets exhibit a spiral pattern and is precessing with a period of 55 years. In the source IRAS 16342-3814 (e.g. Claussen *et al.* 2009 and these proceedings) it is found that water masers lie on opposite sides of the optical nebula and their distribution is generally tangential to the inferred jet axis. Proper-motion measurements give a velocity of the jet at the position of the extreme velocity maser components of at least 155 to 180 km s^{-1} but no direct evidence for precession was found, maser features appear to follow a purely radial motion (no curve trajectories are observed). A detailed morpho-kinematical structure analysis of the H_2O masers from the water fountain IRAS 18286-0959 has been carried on by Yung *et al.* (2011) (see Figure 3). Observations are best interpreted by a model with two precessing jets (or "double helix" outflow pattern) with velocities of up to 138 km s^{-1} and a precessing period of less than 60 years.

This proper motion studies allow also to estimate other parameters like the dynamical age of these outflows and they are found to be surprisingly young, of the order of few tens of years, up to 150 years, which suggests that the evolutionary stage that these water-fountain sources represent is likely to be very short: 50 years for W 43A (see Imai *et al.* 2005), $\sim$30 years for IRAS 18286-0959 (Yung *et al.* 2011), $\sim$59 years for IRAS 19190+1102 (Day *et al.* 2010) and about 120 years for IRAS 16342-3814 (Claussen *et al.* 2009). VLBI astrometry of H_2O masers also allow to derive more accurate distance of these sources.

3.2. *Polarization and magnetic collimation*

The origin of the jet collimation is still an open question but several models (Chevalier & Luo 1994, García-Segura *et al.* 1999) have shown that the magnetic field could be a dominant factor in jet collimation (Blackman *et al.* 2001, García-Segura *et al.* 2005) in post-AGB stars. The Zeeman effect produces a shift in frequency between the two circular polarizations (LCP and RCP) that is directly proportional to the strength of the magnetic field (projected on the line of sight). Hence, by measuring this shift, we can deduce the value of the magnetic field. Several molecules giving rise to maser emission are sensible to this effect (like OH and H_2O), and then provide a unique tool for studying the role of the magnetic field in the jet collimation (of water fountains for example).

Full polarization observations measuring both linear and circular polarization toward the archetype water fountain W 43A have been conducted (see Vlemmings *et al.* 2006, Amiri *et al.* 2010). A strong toroidal magnetic field has been measured, with an estimated strength on the surface of the star as high as 1.6 G (see Vlemmings *et al.* 2006). Such a magnetic field is strong enough to actively participate in the collimation of the jets.

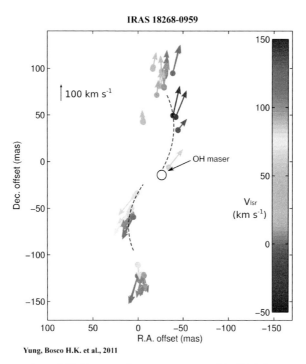

Figure 3. Proper motion of H_2O maser features measured in IRAS 18286-0959 (from Yung *et al.* 2011)

Recently, Wolak *et al.* (2011) published a single dish survey conducted with the Nancay radio telescope toward 152 late type stars, out of which 24 were post AGB sources. In more than 75% of the sample, they detected polarization features and a magnetic field strength of 0.3 to 2.3 mG. In summary, strong magnetic field are observed, strong enough to play a major role in shaping and driving the outflows in water fountains.

4. SiO PPN

SiO maser are very rare in pPNe, very few sources have been detected, OH 15.7+0.8 (tentative detection), OH 19.2-1.0, W 43A, OH 42.3-0.1, IRAS 15452-5459, IRAS 19312+ 1950 and OH 231.8+4.2. The last one discovered is IRAS 15452-5459 (Deacon *et al.* 2007). The SiO maser emission of two of these sources has been mapped, W 43A and OH 231.8+4.2. In the first case, the spatial distribution was found to be compatible with a bi-conical decelerating flow and in the case of OH 231.8+4.2, the distribution can be described by a torus with rotation and infalling velocities. of the same order and within a range between ~ 7 and ~ 10 km s^{-1} (see Sánchez Contreras *et al.* 2002).

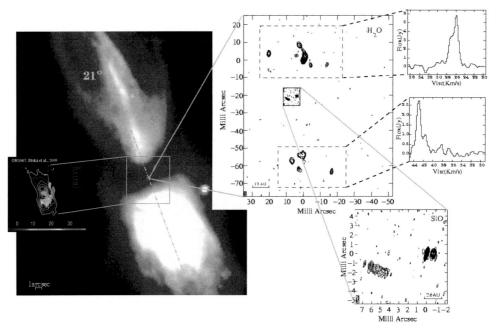

Figure 4. Composition image summarizing interferometric observations OH, H$_2$O and SiO toward OH 231.8+4.2

5. The prototype of bipolar proto Planetary Nebulae: OH 231.8+4.2

OH 231.8+4.2 (also known as the Calabash nebula or QX Pup) is well studied and prototype of bipolar pPNe. It is located in the open cluster M46 at a distance of 1.5 kpc ±0.05 (Choi *et al.* 2012) and the inclination of the bipolar axis with respect to the plane of the sky, $\sim 36^{o}$.. The central source is a binary system formed by an M9-10 III Mira variable (i.e. an AGB star) and an A0 main sequence companion, as revealed from optical spectroscopy by Sánchez Contreras *et al.* (2004). This remarkable bipolar nebula shows all the signs of post-AGB evolution: fast bipolar outflows with velocities $\sim$200–400 km s^{-1}, shock-excited gas and shock-induced chemistry. Mid-infrared MIDI observations (Matsuura *et al.* 2006) show the presence of a compact circumstellar region with an inner radius of 40-50 AU. An equatorial torus is observed at distances greater than 1 arcsec, however, no trace of rotation is found at this scale and the gas is in expansion, as shown by CO and OH emission data (Alcolea *et al.* 2001, Zijlstra *et al.* 2001). Hubble space telescope observations clearly show two extended lobes and PdBI

CO observations measured a traveling speed of the molecular outflow of the order of 400 km s^{-1}.

OH 231.8+4.2 still shows intense SiO masers, contrarily to what happens in the majority of pPNe. The SiO maser emission arises from several compact, bright spots forming a structure elongated in the direction perpendicular to the symmetry axis of the nebula.

Figure 4 is a composition image summarizing OH, H$_2$O and SiO maser emission observations compared with the HST image of the nebula (taken with the WFPC2 Bujarrabal *et al.* 2002). The left panel presents the velocity map of the OH maser emission at 1667-MHz aligned over the L-band image obtained at the VLT by Matsuura *et al.* (2006). Top right panels show the total intensity map of the H$_2$O maser for the two main regions. The small square map at bottom right indicates the position of the map of SiO maser obtained by Sánchez Contreras *et al.* (2002). OH 231.8+4.2 is a strong emitter in the OH ground state line at 1667 MHz. This strong maser emission, radiated by the circumstellar material around OH 231.8+4.2, was mapped with MERLIN by Etoka *et al.* (2009) The OH maser distribution (4 arcsecond) traces a ring-like structure presenting a velocity gradient that is explained by the authors as the blueshifted rim of the bi-conical outflow. The distribution of the polarization vectors associated with the maser spots attests a well-organized magnetic field which seems to be flaring out in the same direction as the outflow. H$_2$O maser emission is distributed in two distinct regions of $\sim$20 mas in size, spatially displaced by 60 milli-arcs (less than 100 AU, comparable to the size of the AGB envelopes) along an axis oriented nearly north-south, similarly to the axis of the optical nebula. The expansion velocity of the H$_2$O masers spots is very low compared to water fountain jets and lower than that of the OH maser spots. Proper motion observations (Leal-Ferreira *et al.* 2012, Desmurs *et al.* 2012) derived velocities on the sky of the order of 2–3 mas/year. Taking into account the inclination angle of the source, this corresponds to an average separation velocity of 15 km s^{-1}. Moreover, the H$_2$O emission is not as well collimated as in water-fountains. Linear polarization of H$_2$O maser yields a value of the magnetic field, assuming a toroidal structure, of 1.5–2.0 G on the stellar surface (see Leal-Ferreira *et al.* 2012). SiO masers are tentatively placed between the two H$_2$O maser emitting regions and rise from several compact features tracing an elongated structure in the direction perpendicular to the symmetry axis of the nebula. Probably this is a disk rotating around the M-type star. The distribution is consistent with an equatorial torus with a radius of $\sim$6 AU around the central star. A complex velocity gradient was found along the torus, which suggests rotation and infall of material towards the star with velocities of the same order and within a range between $\sim$7 and $\sim$10 km s^{-1} (see Sánchez Contreras *et al.* 2002).

Acknowledgments

I would like to acknowledge financial support from the Visiting Scientist grant from the National Research Foundation of South Africa. I would like also to acknowledge V.Bujarrabal for his very useful comments and careful reading of the manuscript. Thank's POD!

References

Alcolea, J., Bujarrabal, V., Sánchez Contreras, C., Neri, R. & Zweigle, J. 2001, *A&A* 373, 932.
Amiri N., Vlemmings W., & van Langevelde H. J. 2010, *A&A*, 509, 26.
Amiri N., Vlemmings W., & van Langevelde H. J. 2011, *A&A*, 532, 149.
Amiri, N., Vlemmings, W. H. T. & van Langevelde, H. J. 2012, *A&A in preparation*

Boboltz, D. A. & Marvel, K. B. 2007, *ApJ*, 665, 680.

Blackman, E. G., Frank, A., Markiel, J. A., Thomas, J. H., & Van Horn, H. M. 2001, *Nature* 409, 485.

Bujarrabal, V., Alcolea, J., Sánchez Contreras, C., & Sahai, R. 2002, *A&A*, 389, 271.

Chevalier, R. A. & Luo, D. 1994,*ApJ* 421, 225.

Choi, Y. K. *et al.* 2012, *this proceeding*

Claussen, M. J., Sahai, R., & Morris, M. R. 2009, *ApJ*, 691, 219.

Day, F. M., Pihlstrm, Y. M., Claussen, M. J., & Sahai, R. 2010, *ApJ*, 713, 986.

Deacon, R. M., Chapman, J. M., Green, A. J., & Sevenster, M. N. 2007, *ApJ*, 658, 1096.

Desmurs, J.-F., Baudry, A., Sivagnanam, P., Henkel, C., Richards, A. M. S., & Bains, I. 2010, *A&A*, 520, 45.

Desmurs, J.-F. *et al.* 2012 *A&A in preparation*

Etoka, S., Zijlstra, A., Richards, A. M., Matsuura, M. & Lagadec, E. 2009 *ASPC* 404, 311.

García-Segura, G., Langer, N., Rózuczka, M., & Franco, J. 1999, *ApJ*, 517, 767.

García-Segura, G., López, J. A., & Franco, J. 2005, *ApJ*, 618, 919.

Gómez, Y., Rodríguez, L. F., Contreras, M. E., & Moran, J. M., 1994, *RMxAA*,28, 97.

Gómez, J. F., Rizzo, R. J., Suarez, O., & iranda, L. F. 2011, *ApJL*, 739, L14.

Habing, H. J. 1996, *A&A Rev.*, 7, 97.

Imai, H., Obara, K., Diamond, P. J., Omodaka, T., & Sasao, T. 2002, *Nature* 417, 829.

Imai, H., Nakashima, J. I., Diamond, P. J., Miyazaki, A., & Deguchi, S. 2005, *ApJ* 622, L125.

Imai, H., Sahai, R., & Morris, M 2007, *ApJ*, 669, 424.

Imai, H., Diamond, P., Nakashima, J. I., Kwok, S., & Deguchi, S. 2008, *Proceedings of the 9th European VLBI Network Symposium on The role of VLBI in the Golden Age for Radio Astronomy and EVN Users Meeting. September 23-26, 2008. Bologna, Italy*, 60

Imai 2009,

Leal-Ferreira, M. L., Vlemmings, W. H. T., Diamond, P. J., Kemball, A., Amiri, N., & Desmurs, J.-F. 2012, *A&A accepted (arXiv:1201.3839v1)*

Manchado, A., Villaver, E., Stanghellini, L., & Guerrero, M. A. 2000, *in ASP Conf. Ser. 199, Asymmetrical Planetary Nebulae II: From Origins to Microstructures*, ed. J. H. Kastner *et al.* (*San Francisco: ASP*, 17).

Matsuura, M., Chesneau, O., Zijlstra, A. A., Jaffe, W., Waters, L. B. F. M., Yates, J. A., Lagadec, E.; Gledhill, T., Etoka, S., & Richards, A. M. S. 2006, *ApJ* 646, 123.

Sahai, R. & Trauger, J. T. 1998, *AJ*, 116, 1357

Sánchez Contreras, C., Desmurs, J.-F., Bujarrabal, V., & Alcolea, J., Colomer, F. 2002, *A&A*, 385, L1.

Sánchez Contreras, C., Gil de Paz, A., & Sahai, R. 2004, *ApJ*, 616, 519.

Suárez, O., Gómez, J. F., & Morata, O. 2007, *A&A* 467, 1085.

Suárez, O., Gómez, J. F., & Miranda, L. F. 2007, *ApJ* 689, 430.

Suárez, O., Gómez, J. F., Miranda, L. F., Torrelles, J. M.; Gómez, Y., Anglada, G., & Morata, O. 2009, *A&A* 505, 217.

te Lintel Hekkert, P., Versteege-Hensel, H. A., Habing, H. J., & Wiertz, M. 1989, *A&AS*, 78, 399.

Vlemmings, W. H. T., Diamond, P. J. & Imai, H.*Nature* 440, 58.

Walsh, A. J., Breen, S. L., Bains, I. & Vlemmings, W. H. T., 2009, *MNRAS*, 394, 70.

Wolak, P., Szymczak, M., & Gerard, E. 2011, *A&A*

Yung, Bosco H. K., Nakashima, J., Imai, H., Deguchi, S., Diamond, P. J., & Kwok, S. 2011, *ApJ*, 741, 94

Zijlstra, A., Chapman, J. M., te Lintel Hekkert, P., Likkel, L., Comeron, F., Norris, R. P., Molster, F. J., & Cohen, R. J. 2001, *MNRAS* 322, 280.

Cosmic Masers - from OH to H_0
Proceedings IAU Symposium No. 287, 2012
R.S. Booth, E.M.L. Humphreys & W.H.T. Vlemmings, eds.

© International Astronomical Union 2012
doi:10.1017/S1743921312007004

Water Fountains in Pre-Planetary Nebulae: The Case of IRAS16342−3814

Mark Claussen[1], Raghvendra Sahai[2], Mark Morris[3], and Hannah Rogers[1,4]

[1]National Radio Astronomy Observatory, P.O. Box O, 1003 Lopezville Rd., Socorro, NM 87801
[2]Jet Propulsion Laboratory, 4800 Oak Grove Drive, Pasadena, CA 91109
[3]University of California, Los Angeles, CA 90095
[4]Augustana College, 2001 S. Summit Ave, Sioux Falls, SD 57197

Abstract. We present a brief review of Very Long Baseline Array (VLBA) observations of water masers in the so-called water fountain pre-planetary nebulae, and report on new VLBA and Very Large Array (VLA) data for the water masers in the prototypical water fountain source IRAS16342−3814, taken approximately monthly in 2008−2009. A new and very strong water maser is found at an LSR velocity of -3 km s^{-1}, which is offset from the central (likely stellar) velocity. The new VLBA observations still show a similar general structure as was observed in 2002, but there are details which are difficult to explain.

Keywords. Stars: AGB and post-AGB — Masers

1. Introduction

Water fountain pre-planetary nebulae (PPN) are a particularly interesting subclass of PPN whose original distinguishing characteristic in the presence of very high-velocity red and blueshifted H_2O and OH maser features. Originally the velocity separations of discovered water fountains were in the range of $50-150$ km s^{-1}, but recently water fountains have been discovered that extend to as much as 500 km s^{-1} (Gómez *et al.* 2011). When examined with sufficiently high angular resolution (using Very Long Baseline Interferometry - VLBI), or with the VLA, the blue- and red-shifted water masers are typically displaced from each other and show large, opposed, proper motions (e.g. Claussen, Sahai, & Morris 2009). Lifetimes of the water fountain stage have been estimated to be 50 - 100 years, based on the size and outflow velocity of the maser distribution.

IRAS16342−3814 is a prototypical water fountain source, one of the first three discovered (Likkel & Morris 1988; Likkel, Morris, & Maddalena 1992). The discovery water maser spectra are shown in Figure 1. The central LSR velocity is thought to be +43 km s^{-1}, from analysis of water maser "pairs" by Likkel, Morris & Maddalena 1992 and OH analysis by Sahai *et al.* 1999. Observations using the VLBA in 2002 show that the red- and blue-shifted water masers are on opposite sides of the bipolar nebula as seen in scattered light (see Sahai *et al.* 1999, and are comprised of "bow-shock" features (Figure 2). The groups of extreme velocity masers (both red- and blue-shifted) are situated just outside the optical lobes.

The separation of these extreme velocity features expands with time (see Figure 2). In particular the line joining the most extreme red- and blue-shifted masers keeps its orientation and the length of the line increased over the 5 months of observation (Figure 3). A linear least squares fit to the data shown in Figure 3 gives a two-sided expansion proper motion of 63±2 μas day^{-1} (which translates to 23 mas yr−1). Claussen, Sahai & Morris 2009 suggested that these detailed structures were actually bow shocks on either side of a

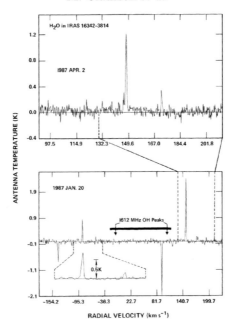

Figure 1. Discovery spectra of water masers from water fountain IRAS16342−3814 (Likkel & Morris 1988).

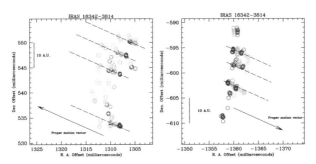

Figure 2. The motion of water masers for 5 monthly epochs in 2002 (the colors denote the different epochs). Left: the motion of water masers for the $+178.7$ km s^{-1} radial velocity features. The proper motion vector here represents 14.4 mas yr^{-1}. Right: the motion of water masers for the -66.0 km s^{-1} radial velocity features. The proper motion vector represents 10.8 mas yr^{-1}.

highly collimated jet. The three-dimensional velocities were found to be $\sim\pm180$ km s^{-1}, which leads to a very short dynamical timescale for this source of ~100 years.

2. VLBA and VLA Data from 2008−2009

We observed the water masers from IRAS16342−3814 for twelve epochs beginning in March 2008, spaced approximately monthly. Observations were made with the VLBA and the VLA. For the VLA observations we used seven 6.25 MHz wide frequency settings in the VLA correlator to cover the emission, from LSR velocities -140.0 to $+80.0$ km s^{-1} and velocity resolution 0.66 km s^{-1}. For the VLBA, we used four 8 MHz baseband channels covering LSR velocities from $+230.0$ to -140.0 km s^{-1} and velocity resolution 0.21 km s^{-1}. A combined spectrum from the VLA taken on March 26, 2008 is shown in Figure 4.

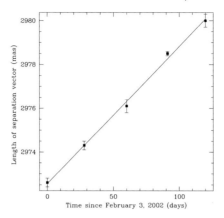

Figure 3. Expansion of line connecting extreme velocity water masers of IRAS16342−3814 in 2002 (Claussen *et al.* 2009).

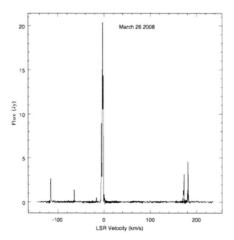

Figure 4. VLA spectrum of water masers of IRAS16342−3814 on March 26, 2008.

3. Discussion and Future Directions

In comparing Figure 4 with Figure 1 we see that, in general, the spectra are similar, with red-shifted masers around LSR velocities of $+155$ $\mathrm{km\,s^{-1}}$and $+180$ $\mathrm{km\,s^{-1}}$, and blue-shifted masers near an LSR velocity of -66 $\mathrm{km\,s^{-1}}$. But there are obvious new features as well: the strong masers around -3 $\mathrm{km\,s^{-1}}$and a feature at -118 $\mathrm{km\,s^{-1}}$. The strong feature at -3 $\mathrm{km\,s^{-1}}$does not lie at the putative stellar velocity ($+43$ $\mathrm{km\,s^{-1}}$with respect to the LSR). Given the new features in this spectrum, however, one might argue that the stellar velocity is closer to $\sim+30$ $\mathrm{km\,s^{-1}}$(centered between the most blue-shifted and the most red-shifted water masers). The new masers at -3 $\mathrm{km\,s^{-1}}$are located in approximately the same region as a group of blue-shifted masers (-21 to -33 $\mathrm{km\,s^{-1}}$) were found in 2002 (Claussen, Sahai, & Morris 2009), approximately halfway out from the center to the edge of the southwest optical (scattered light) lobe (Sahai *et al.* 1999 and Claussen, Sahai, & Morris 2009).

Although space limits the number of figures allowed for this contribution, the maser structures and distributions for the 2008−2009 observations are similar to those found in

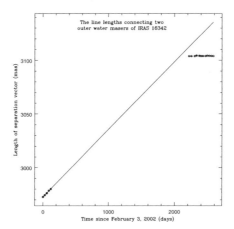

Figure 5. Expansion of line connecting extreme velocity water masers of IRAS16342−3814 in 2002 and 2008−2009.

2002 (Claussen, Sahai & Morris 2009). The bow-shocked shaped structures seen in the 2002 data, however, are not as clearly defined in the 2008-9 data.

We once again measured the length and position angle of the line joining the +154.2 km s^{-1} and −65.2 km s^{-1} maser features for each epoch in 2008−2009. (We did this because neither project - in 2002 or 2008−2009 used absolute astrometry.) As in 2002, the position angle of this line was unchanging over the 12 months of observation and had the same value as in 2002, at 66.1°. The results of the line length measurements both in 2002 and 2008−2009 are shown in Figure 5. A very surprising and completely unexpected result was obtained. The length of the line essentially did not change for the 12 epochs in 2008−2009 (the data points in the upper right hand area of Figure 5). The straight line shown in Figure 5 is the extension of the least-squares fit line (of the 2002 data) seen in Figure 3. (In the lower left corner of the figure, the 2002 data can be seen.)

Claussen, Sahai, & Morris 2009 argue that the proper motion of the water masers in IRAS16342-3814 results from either actual physical motion of the emitting material, or to a "theatre marquee lights" effect, so that dense material at progressively larger distances "light up" sequentially in maser emission as, for example, a passing shock wave produces an outward-moving compression front. These authors argue that for either case, the proper motions of the water masers reflects the physical motion of the dense gas in the jet head, either directly or indirectly. The unexpected result from the 2008−2009 data, that the length of the line joining the (apparently) same two masers is no longer increasing as it was in 2002, may call into question the argument that the masers trace physical motion.

We turn to possible explanations for this unexpected result. We believe that there is no mistake in the reduction of the data; indeed we cannot imagine a problem in data processing that would cause an apparent stoppage of the line length increase. Systematic errors in data processing could cause either a random change in the length of the line (added to the *real* change in length) or perhaps a large increase in the error itself.

Some other possibilities for the explanation of the change in the (apparent) motion of the water masers are

• The masers really do not reflect physical motion of dense gas, either directly or indirectly. Although this is certainly possible, other observations of masers in different environments have all seemed to be tracing physical motions, and other evidence seems to support this idea (e.g. Bloemhof, Moran, & Reid 1996).

• Sometime in the interval between 2002 and 2008, the shock driving the masers reached the edge of the very dense gas in the bipolar nebula. This is a definite possibility, as the positions of the extreme velocity masers, overlaid on the optical nebula, already appeared nearly outside the optical emission. This might explain the effect of an apparent halt to the expansion since the masers seen now are just remnants of earlier shock excitation. In this case one might expect to see some deceleration of the remaining masers. It's not clear that we have seen such deceleration.

Clearly we don't really understand the phenomenon that we have reported in this contribution. Further high angular resolution observations of the water masers in IRAS16342−3814 are likely needed to completely understand this phenomenon that we have seen in 2008−2009. If the second bullet above is a possible explanation, we might expect to see a complete fade-out of the extreme velocity masers.

The National Radio Astronomy Observatory (NRAO) is a facility of the National Science Foundation operated under cooperative agreement by Associated Universities, Inc. Part of this research was carried out under the auspices of the National Science Foundation's Research Experience for Undergraduates (REU) program at the NRAO, and we gratefully acknowledge the funding for this program.

References

Claussen, M. J., Sahai, R. S., & Morris, M. R. 2009, *ApJ*, 691, 219
Gómez, J. F., Rizzo, J. R., Suárez, O., Miranda, L. F., Guerrero, M. A., & Ramos-Larios, G. 2011, *ApJ*, 739, 14
Likkel, L. & Morris, M. 1988, *ApJ*, 329, 914
Likkel, L., Morris, M. & Maddalena, R. 1992, *A&A*, 256, 581
Sahai, R., te Lintel Hekkert, P., Morris, M., Zijlstra, A., & Likkel, L. 1999, *ApJ*, 514, L115

Cosmic Masers - from OH to H_0
Proceedings IAU Symposium No. 287, 2012
R.S. Booth, E.M.L. Humphreys & W.H.T. Vlemmings, eds.

© International Astronomical Union 2012
doi:10.1017/S1743921312007016

The first water fountain in a planetary nebula

Olga Suárez[1], José Francisco Gómez[2], Philippe Bendjoya[1], Luis. F. Miranda[3,4], Martín. A. Guerrero[2], Gerardo Ramos-Larios[5], J. Ricardo Rizzo[6], and Lucero Uscanga[2,6,7]

[1] Lab. Lagrange, UMR7293, UNSA, CNRS, Observatoire de la Côte d'Azur, F-06300 Nice, France
[2] Instituto de Astrofísica de Andalucía (IAA-CSIC), Glorieta de la Astronomía, s/n, E-18008 Granada, Spain
[3] Departamento de Física Aplicada, Facultade de Ciencias, Campus Lagos-Marcosende s/n, Universidade de Vigo, E-36310 Vigo, Spain (present address)
[4] Consejo Superior de Investigaciones Científicas (CSIC), c/ Serrano, 117, E-28006 Madrid, Spain
[5] Instituto de Astronomía y Meteorología, Av. Vallarta No. 2602, Col. Arcos Vallarta, 44130 Guadalajara, Jalisco, Mexico
[6] Centro de Astrobiología (INTA-CSIC), Ctra. M-108, km. 4, E-28850 Torrejón de Ardoz, Spain
[7] Observatorio Astronómico Nacional (IGN), E-28014 Madrid, Spain

Abstract. Water fountains are evolved stars showing water masers with velocity spanning more than ∼100 km/s. They usually appear at the end of the Asymptotic Giant Branch (AGB) phase or at the beginning of the post-AGB phase, and their masers trace the first manifestation of axisymmetric collimated mass-loss. For the first time, masers with water fountain characteristics have been detected towards a PN (IRAS 15103−5754), which might require a revision of the current theories about jet formation and survival times. IRAS 15103-5754 was observed using the ATCA interferometer at 22 GHz (both continuum and water maser). The main results of these observations are summarized here. The evolutionary classification of this object is also discussed.

Keywords. masers, stars: AGB and post-AGB, stars: evolution, ISM: planetary nebulae: individual (IRAS 15103-5754)

1. Introduction

Water fountains are evolved stars showing water masers with high velocity components ($\geqslant 100$ km s^{-1}). The name "water fountain" was first used by Likkel (1989) to refer to IRAS 16342−3814. In the last 10 years, the number of known water fountains has increased to more than 10 (see the review by J.F. Desmurs in these proceedings). The importance of these objects lies on the jets that are traced by their water maser emission. These jets could be the first manifestation of axisymmetric, collimated emission in evolved stars.

According to Sahai & Trauger (1998), the shapping of bipolar/multipolar planetary nebulae (PNe) is due to the presence of jets produced during the post-Asymptotic Giant Branch (post-AGB) phase. The mechanism that drives these jets is still unknown. Water fountains are probably the best objects to study the onset of these jets, since their dynamical ages are shorter than 100 yr (see, for example Imai *et al.* 2002, 2007).

All the water fountains discovered up to now have been found in either post-AGB or late AGB stars. The mechanism driving the jet is believed to act during the post-AGB phase (Imai *et al.* 2007) and, water masers are expected to survive up to a maximum of

$\simeq$ 100 yr after the end of the AGB mass-loss (Gómez *et al.* 1990; Lewis *et al.* 1990). Therefore, if this scenario is correct, the presence of water fountain characteristics in a PN is extremely unlikely.

During a single-dish survey for water masers in evolved stars (Suárez *et al.* 2009), we detected water masers with high velocity components ($\sim$80 km s^{-1}) towards a PN candidate: IRAS 15103−5754 (I15103, hereafter). The velocity spread of the water maser components makes I15013 the first PN - water fountain candidate.

2. Observations and results

To confirm the physical association between water maser emission and I15103 and to confirm the PN nature of this source, we simultaneously observed water maser emission and radio continuum (both at $\simeq$ 22 GHz), using the Australia Telescope Compact Array on August 2011. We self-calibrated the maser emission, and then applied the same phase and amplitude correction to the continuum. Such procedure allowed us to obtain a very high relative positional accuracy between maser and continuum emission, with uncertainties < 50 mas. We found the maser and radio continuum emission to be spatially coincident, thus confirming their association.

The water maser spectrum and the spatial distribution of the different components are shown in Fig. 1. The phase center of these observations is R.A.(J2000) = $15^h 14^m 18.4^s$, Dec(J2000) = $-58°05'21.0''$. By "maser components" we refer to individual intensity peaks in the spectrum. The positional information has been obtained by fitting elliptical gaussians to the maser components, only in the channels in which a spectral peak is present. The total velocity span of the maser emission is $\simeq$80 km s^{-1}, from $V_{\rm LSR} \simeq -70$ to +10 km s^{-1}.

There has been an important increase in the maximum flux density of the water maser components with respect to the spectrum obtained in 2007 with the Parkes telescope (Suárez *et al.* 2007). In 2007, the maximum flux density was observed at $V_{\rm LSR} \sim -37$ km s^{-1} and it reached $\sim$ 60 Jy. In 2011 the maximum flux density was also observed at $V_{\rm LSR} \sim -37$ km s^{-1} but reaching $\sim$ 1700 Jy. Thus, this source displays the most variable water maser emission among all known water fountains. However, we note that the global velocity span of the maser components has not changed significantly between 2007 and 2011.

3. Discussion

3.1. *The nature of the source*

This source was classified as a PN mainly based on the presence of radio continuum emission (van de Steene & Pottasch 1993). In principle, a young stellar object (YSO) could show both radio continuum and water maser emission, but there are reasons to rule out this interpretation for I15103:

• High density tracers detected toward this source are weak (T_a < 0.2 K), as seen in the CS and NH$_3$ spectra from the RMS (Urquhart *et al.* 2008) and HOPS (Walsh *et al.* 2011) surveys, respectively. Strong emission of these tracers are normally associated with active YSOs (Anglada *et al.* 1989, 1996)

• Moreover, molecular lines detected towards I15103 are narrow ($\simeq$ 0.6 km s^{-1} for CS, Urquhart *et al.* 2008), which is not consistent with arising from the environment around a YSO, specially if the object is so active that it pumps water maser emission. This would induce significant heating and turbulence in the gas, thus widening the molecular lines ($\Delta V \geqslant 2$ km s^{-1}, Anglada *et al.* 1996). Therefore, it seems that the CS and NH$_3$ emission traces dense molecular gas that is not associated with this source.

• If the source were a YSO out of the parental cloud, it would be detected at optical wavelengths, but this is not the case.

Assuming that I15103 is an evolved object, the reasons that support its classification as a PN are:

• Presence of radio continuum emission, which indicates ionisation.

• IR images show a bipolar morphology, consistent with that observed in developed PN (Lagadec *et al.* 2011; Ramos-Larios *et al.* 2012)

• The detection of significant [NeII] emission at 12.8 μm (M.Blanco, private communication), typically associated to photoionisation. This implies that I15103 has reached the temperature necessary to become a PN.

3.2. *The morphology of the source*

The water maser components in Fig. 1 show a different spatial distribution from that typically found in the rest of water fountains, where they follow a bipolar pattern. In this case, there seems to be a trend, with more blueshifted water maser components towards the NW, while the more redshifted ones are located to the SE. However, it is difficult to define a clear bipolar structure in the water maser distribution, and its preferential orientation.

The mid-IR images of this PN obtained by Lagadec *et al.* (2011) with VISIR at the VLT show a bipolar structure oriented NE - SW, in nearly the same direction as the near-IR images shown in Ramos-Larios *et al.* (2012), taken with the NTT telescope at La Silla

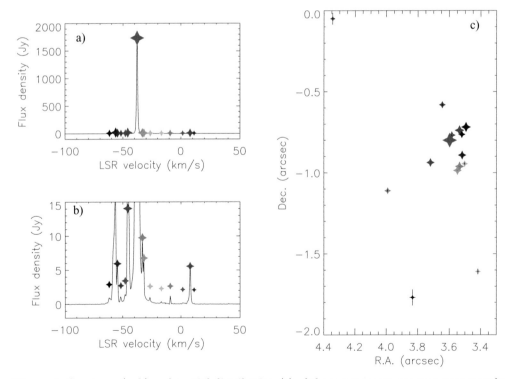

Figure 1. Spectrum (a, b) and spatial distribution (c) of the water maser components towards I15103. A zoom of the spectrum is shown in b) to display the weaker components. Symbols are proportional to the logarithm of the flux density of the individual maser components

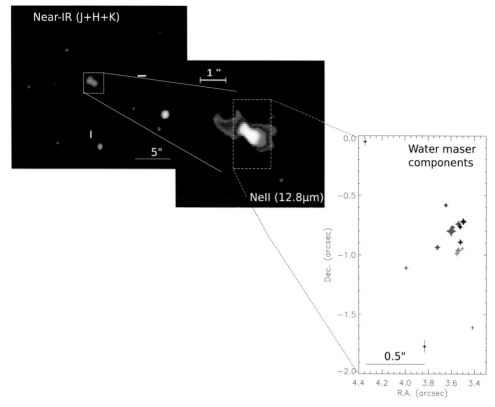

Figure 2. Left: near-IR image from Ramos-Larios *et al.* (2012); center: mid-IR image from Lagadec *et al.* (2011); right: water maser distribution

(Chile). These images are shown in Fig. 2 in comparison with the spatial distribution of the masers. The maser emission does not seem to follow the orientation of the IR images. In order to understand the real orientation of the maser emission with respect to the structure of the nebula, the information about the proper motion of the masers would be helpful.

3.3. *The evolutionary scenario for water fountain-PNe*

The discovery of the first PN with water fountain characteristics leads us to revise the scenario for the evolution of evolved sources harbouring water masers. This source is the fourth PN found showing water maser emission, but the other three: K3−35 (Miranda *et al.* 2001), IRAS 17347−3139 (de Gregorio-Monsalvo *et al.* 2004) and IRAS 18061−2505 (Gómez *et al.* 2008), do not show high velocity maser components.

The presence of water maser jets (which typically show very short dynamical ages) in an already bipolar-shaped PN supports the idea that jets in water fountains could be a ballistic and repetitive phenomenon. Episodic jets, ejected from the central star, would be traced by the water masers. These jets could be first launched at the end of the AGB or the beginning of the post-AGB stages but, as we are witnessing, they could, in the most extreme cases, be also launched, or at least maintained, when the object has already become a PN. Moreover, the outflows traced by water masers in IRAS 15103 are completely misaligned with respect to the bipolar axis of symmetry, what could indicate a different outflow episode at a different orientation.

4. Conclusion

We have shown the water maser distribution of the first known PN with water fountain characteristics. The existence of such a source supports the idea that the jets of water fountains are ejected as successive episodes.

References

Anglada, G., Estalella, R., Pastor, J., Rodriguez, L. F., & Haschick, A. D. 1996, *ApJ*, 463, 205

Anglada, G., Rodriguez, L. F., Torrelles, J. M., *et al.* 1989, *ApJ*, 341, 208

de Gregorio-Monsalvo, I., Gómez, Y., Anglada, G., *et al.* 2004, *ApJ*, 601, 921

Gómez, J. F., Suárez, O., Gómez, Y., *et al.* 2008, *AJ*, 135, 2074

Gómez, Y., Moran, J. M., & Rodríguez, L. F. 1990, *Rev. Mexicana AyA*, 20, 55

Imai, H., Obara, K., Diamond, P. J., Omodaka, T., & Sasao, T. 2002, *Nature*, 417, 829

Imai, H., Sahai, R., & Morris, M. 2007, *ApJ*, 669, 424

Lagadec, E., Verhoelst, T., Mekarnia, D., *et al.* 2011, *A&A* in press

Lewis, B. M., Eder, J., & Terzian, Y. 1990, *ApJ*, 362, 634

Likkel, L. 1989, *ApJ*, 344, 350

Miranda, L. F., Gómez, Y., Anglada, G., & Torrelles, J. M. 2001, *Nature*, 414, 284

Ramos-Larios, G., Guerrero, M., Suárez, O., Miranda, L. F., & Gómez, J. F. 2012, Submitted to *A&A*

Sahai, R. & Trauger, J. T. 1998, *AJ*, 116, 1357

Suárez, O., Gómez, J. F., Miranda, L. F., *et al.* 2009, *A&A* 505, 217

Suárez, O., Gómez, J. F., & Morata, O. 2007, *A&A*, 467, 1085

Urquhart, J. S., Hoare, M. G., Lumsden, S. L., Oudmaijer, R. D., & Moore, T. J. T. 2008, in Astronomical Society of the Pacific Conference Series, Vol. 387. Ed. H. Beuther, H. Linz, & T. Henning, 381

van de Steene, G. C. M. & Pottasch, S. R. 1993, *A&A*, 274, 895

Walsh, A. J., Breen, S. L., Britton, T., *et al.* 2011, *MNRAS*, 416, 1764

Cosmic Masers - from OH to H$_0$
Proceedings IAU Symposium No. 287, 2012
R.S. Booth, E.M.L. Humphreys & W.H.T. Vlemmings, eds.

© International Astronomical Union 2012
doi:10.1017/S1743921312007028

Polarization properties of R Cas SiO masers

K. A. Assaf[1], P. J. Diamond[2], A. M. S. Richards[1] and M. D. Gray[1]

[1]University of Manchester, Oxford Road, Manchester, M13 9PL, UK
email: kam@jb.man.ac.uk

[2]CSIRO Astronomy and Space Sciences, PO Box 76, Epping, NSW 1710, Australia
email: philip.diamond@csiro.au

Abstract. Silicon monoxide maser emission has been detected in the circumstellar envelopes of many evolved stars. It is a good tracer of the wind dynamics within a few stellar radii of the central star. We investigated the polarization morphology in the circumstellar envelope of an AGB star, R Cas, by using the VLBA to map the linear and circular polarization of the v=1, J=1-0 SiO maser transition during 23 epochs over two stellar cycles. The average fractional circular polarization is a few percent. The average fractional linear polarization per epoch is 11–58%, but some isolated features exceed 100%, probably because the total intensity emission is smoother and more resolved-out. The maser electric polarization vector angle has a preferrential tendency to be either parallel or perpendicular to the radial direction to the star.

Keywords. masers – polarization – star: AGB – star: late-type star: individual: R Cas.

1. Introduction

The circumstellar envelope (CSE) of an asymptotic giant branch (AGB) star is a very active region. Mass loss from AGB stars is an important means of enriching the stellar medium with processed material. The nature of material returned to the ISM (molecular, dusty and/or ionised) is affected by the inhomogeneity and asymmetry of the stellar wind, and the magnetic field may have a significant influence over these properties. If maser polarization is magnetic in origin, the polarization morphology and linear polarization position angle provides information about the structure of the magnetic field in the CSE.

43-GHz SiO maser images typically show a ring with a radius of a few stellar radii (within the dust formation zone). The maser emission is significantly linearly polarized but circular polarization is weaker, as expected for a non-paramagnetic molecule. The linear and circular polarization percentages are given by:

$$m_\ell = \frac{P}{I} = \frac{\sqrt{(Q^2 + U^2)}}{I} \qquad (1.1)$$

$$m_c = \frac{V}{I} \qquad (1.2)$$

Herpin *et al.* (2006) found that Mira variables have average values of $m_\ell \sim 30\%$ and $m_c \sim 0.9\%$, respectively. The typical values of m_c suggest a magnetic field strength of few Gauss, if the standard Zeeman interpetation is adopted (Elitzur 1996, as applied to TX Cam by Kemball & Diamond 1997). This model predicts that a radial, stellar-centred magnetic field would produce tangential polarization vectors from emission originating in the plane of the sky containing the star.

Alternatively, even in the absence of a magnetic field, maser pumping by an anisotropic, stellar radiation field could produce strongly polarized maser emission (Western *et al.* 1983). This mechanism might cause the tangential polarization seen in VLBA SiO maser images (Desmurs *et al.* 2000). High-resolution full polarization VLBA imaging of SiO

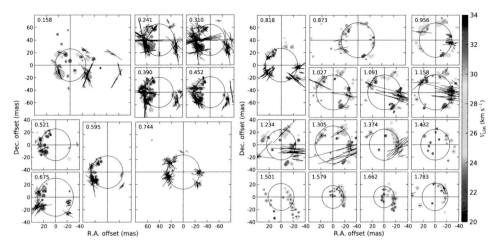

Figure 1. Polarization morphology of R Cas SiO masers. Each pane is labelled with the stellar phase. Symbol size is proportional to total intensity. The vectors show the orientation of the EVPA, length proportional to P.

masers resolves the winds of nearby stars on scales less than 0.1 AU, at high spectral resolution, testing the predictions of these models. We monitored the Mira R Cas, at $\sim$ 176 AU (Vlemmings *et al.* 2003), over 2 stellar cycles, in the 43-GHz SiO $v = 1$ $J = 1 - 0$ line. The data analysis, variability and morphology of the total intensity emission were published by Assaf *et al.* (2011). These results showed broad agreement with the models of Gray *et al.* (2009). The maser peak lags the optical by stellar phase $\phi \sim 0.2$, the brightest masers occuring during ϕ 0.1 − 0.4. We summarise here the preliminary polarization results.

2. Results

R Cas was observed in full polarization, to provide Stokes I, Q, U and V data cubes at spatial and spectral resolutions of approximately $(40 \times 20 \ \mu\text{as}^2, 0.2 \ \text{km s}^{-1})$. Polarization calibration and measurements were made following the methods described by Kemball *et al.* (2009). The polarization detection threshold for individual components is $5\sigma_{\text{rms}}$ and $m_\ell > 5\%$ or $m_c > 15\%$; lower thresholds are possible when averaging over larger spectral or spatial regions (including correction for Ricean bias).

2.1. *Linear Polarization*

Fig. 1 shows the orientation of the electric vector position angles, EVPA = 0.5 $\arctan(U/Q)$; the length of the vectors is proportional to the linearly polarized intensity $P = (Q^2 + U^2)^{1/2}$. The SiO maser emission is significantly linearly polarized. The percentage linear polarization (averaged over all features per epoch) is $m_{ell} \sim 11 \rightarrow 58\%$. m_{ell} exceeds 100% in some isolated features (Section 3.2).

We investigated the relationship between the EVPA and the radial direction with respect to the star, defined by the position angle in the plane of the sky, θ. Fig. 2 shows the proportion of the polarized emission within bins of (EVPA$-\theta$) in the whole SiO shell and in the inner shell (within the radius enclosing 25% of the total maser flux at each epoch). The thickness of the line is proportional to the logarithm of the total linearly polarized flux. For the first cycle (ϕ = 0.158 to ϕ = 1.158, 15 epochs), the polarized flux in the whole shell is dominated by emission with EVPA either parallel

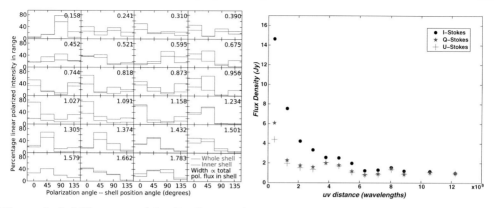

Figure 2. *Left* Histogram of the distribution of polarized emission with respect to the deviation of the polarization angle from the radial direction (EVPA$-\theta$). *Right* Stokes I, Q and U flux density as a function of baseline length for channels averaged from $24.8 - 29.4$ km s^{-1}, $\phi = 0.744$.

(10 epochs) or perpendicular (5 epochs) to θ. However, in the inner shell, only half the epochs have a high proportion of emission with parallel (6) or perpendicular (2) EVPA; it is at intermediate angles for the remaining 7 epochs. Seven out of the remaining 8 epochs (from $\phi = 1.234$ to $\phi = 1.783$), are dominated by emission with intermediate EVPA in the whole shell but this comes mainly from a single feature. At $\phi = 1.662$ the EVPA is predominantly parallel. The inner shell has more emission with a parallel EVPA at epochs $\phi = 1.501, 1.662$, otherwise behaving similarly to the whole shell.

The polarization structure during first stellar cycle can be summarised as a bimodal distribution. Most of the linear polarization vectors are either radial (parallel to the position angle of the location of the emission in the projected shell) or tangential (perpendicular). However, the polarization in the inner part of the shell is somewhat less ordered. The later parts of the second stellar cycle do not show any clear pattern, but the emission generally was noisier with fewer significantly polarized features.

2.2. *Circular Polarization*

Previous observations have shown that the mean degree of circular polarization in SiO masers is small but not zero, e.g. m$_c$ =1%-3% (Barvainis *et al.* 2009; Kemball& Diamond 1997). The average fractional circular polarization per epoch in R Cas is $m_c \sim 0.4 \rightarrow 6\%$. A few features showed 'S' shaped Stokes V profiles and we attempted to fit these with the function $V(\nu) = aI(\nu) + b(dI/d\nu)(\nu)$ ((Elitzur 1998)), but this did not succeed as V is very faint and the channel width is large compared with the Zeeman splitting.

3. Discussion

3.1. *Linear polarization angle*

In Section 2.1, we found that the linear polarization angle was predominantly parallel or perpendicular to the radial direction with respect to the star, in the plane of the sky. Fig. 3 shows a single maser feature (spanning ~ 2 km s^{-1}) with an abrupt transition in EVPA of approximately $\frac{\pi}{2}$. At this point, the linearly polarized intensity is near its minimum. The polarization vectors are tangential to the projected ring in the inner part of the feature and radial in its outer region. Goldreich *et al.* (1973) predicted this discontinuity in EVPA when the magnetic field direction with respect to the line of sight passes through θ_F, where $\sin^2 \theta_F = 2/3$. See Kemball *et al.* (2011) for a fuller description of this behaviour, observed in TX Cam.

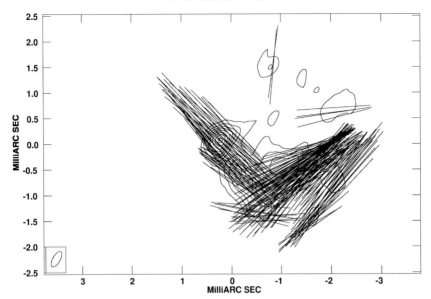

Figure 3. The blue-shifted maser feature seen in the NE, at around (20, 20) mas, averaged over V_{LSR} 25.6 – 27.43 km.s^{-1}, at ϕ = 0.818 (Fig. 1). The vectors show the EVPA, length proportional to the polarized intensity.

3.2. *Fractional polarization*

We found that some isolated features are more than 100% linearly polarized. For instance, the clump at (25, 12) mas at $\phi = 0.744$ has 286% polarization. Fig. 2 (*right*) shows the visibility amplitudes for the channels including this clump (also containing others with smaller fraction polarization). The Stokes I flux density rises far more steeply on shorter baselines than does the polarized intensity. This suggests that a higher proportion of total intensity emission is on scales too large to be imaged on the shortest interferometer spacings, whilst a smaller proportion of polarized intensity is resolved-out. Polarization exceeding 100% has also been measured in the Galactic ISM (Haverkorn 2003a, b), explicable if the polarized emission has structure on smaller scales than the total intensity.

Kemball *et al.* (2009, 2011) explain why a magnetic origin is most likely for SiO polarization; the bulk kinetic energy density and the magnetic energy density are likely to be comparable, giving $B \sim 0.725$ G. Previously, it has been presumed that Faraday rotation ψ_F is negligible at the SiO wavelength $\lambda = 7$mm. We used the relation below from Garcia-Barreto *et al.* (1988):

$$\psi_F = 0.5 \left(\frac{n_e}{10^6 \text{cm}^{-3}} \right) \left(\frac{B_{||}}{\text{mG}} \right) \left(\frac{L}{10^{13}\text{m}} \right) \left(\frac{\lambda}{0.18\text{m}} \right)^2 \quad (3.1)$$

where $L \approx 2 \times 10^{11}$ m is the path length in the maser region (maser shell thickness). We estimated the electron density using the fractional ionisation model of Reid & Menten (1997) at number densities suitable for SiO masing, giving $n_e \sim 1.5 \times 10^9$ m^{-3}. The Faraday rotation is thus about 16°, small enough not to affect our inferences about the EPVA but sufficient to provide structure in the polarized emission.

4. Conclusions

The orientation of linear polarization vectors from the SiO masers around R Cas is consistent with a radial magnetic field. Where the masers come from material close to the plane of the sky, the magnetic field is approximately perpendicular to the line of sight and the masers. Goldreich *et al.* (1973) predicts that the polarization angle would be perpendicular to the magnetic field direction. However, if the masing region has a significant depth, such that in some places the magnetic field is at $< 55°$ to the line of sight, the linear polarization angle would become parallel to the magnetic field. There is a systematic tendency for the maser polarization angle to be either parallel or perpendicular to the radial direction, and in some cases a local 90° change of direction in polarization angle seems to reflect the transition. The magnetic field is not more ordered in the inner shell, compared with the shell as a whole. This provides more evidence against anisotropic pumping as the main agent of maser polarization, since that would be most effective nearest the star.

The total intensity maser emission contains components which are smooth on scales larger than the maximum (< 1 AU) to which the VLBA is sensitive; the single dish flux density is several times greater than the correlated flux (Assaf *et al.* 2011). A few features with apparent linear polarization $\gg 100\%$ suggests that some factor affecting the propagation of polarized emission, e.g. the magnetic field, has structure on smaller scales. The magnetic field in the SiO maser region would produce Faraday rotation $\sim 15°$ (if the magnetic and bulk kinetic energy densities are similar) and local compression could enhance this.

References

Assaf, K. A., Diamond, P. J., Richards, A. M. S., & Gray, M. D. 2011, *MNRAS*, 415, 1083
Barvainis, Richard, McIntosh, Gordon, & Predmore, C. Read 1987, *Nature*, 329, 613.
Desmurs, J. F., Bujarrabal, V., Colomer, F., & Alcolea, J. 2000, *A&A*, 360, 189.
Elitzur, M. *ApJ*, 457, 415.
Elitzur, M. *ApJ*, 504, 390.
Garcia-Barreto, J. A., Burke, B. F., Reid, M. J., Moran, J. M., Haschick, A. D. & Schilizzi, R. T. 1988,*ApJ*, 326, 954.
Goldreich, P., Keeley, D. A., & Kwan, J. Y. 1973, *ApJ*, 179, 111.
Gonidakis, I., Diamond, P. J., & Kemball, A. J. 2010, *MNRAS*, 406, 395.
Gray M. D., Wittkowski, M., Scholz, M., Humphreys,E. M. L., Ohnaka K. & Boboltz, D. 2009,*MNRAS*, 394, 51.
Habing H. J. 1996,*A&AR*, 7, 97.
Haverkorn M. Katgert, P. & de Bruyn, A. G. 2003, *A&A*, 403, 1031.
Haverkorn M. Katgert, P. & de Bruyn, A. G. 2003, *A&A*, 404, 233.
Herpin, F., Baudry, A., Thum, C., Morris, D. & Wiesemeyer, H. 2006, *A&A*, 450, 667.
Kemball A. J. & Diamond P. J. 1997, *ApJ*, 481, L111.
Kemball, A. J., Diamond, P. J., Gonidakis, I., *et al.* 2009, *ApJ*, 698, 1721.
Kemball, A. J., Diamond, P. J., Richter, L., Gonidakis, I., & Xue, R. 2011, *ApJ*, 743, 69.
Reid, Mark J. & Menten, Karl M. 1997*ApJ*, 476, 327.
Troland, T. H., Heiles, C., Johnson, D. R., & Clark, F. O. 1979, *ApJ*, 232, 143.
Western, L. R. & Watson, W. D1983, *ApJ*, 275, 195.
Vlemmings, W. H. T., van Langevelde, H. J., Diamond, P. J., Habing, H. J., & Schilizzi, R. T. 2003, *A&A*, 407, 213.

Cosmic Masers - from OH to H₀
Proceedings IAU Symposium No. 287, 2012
R.S. Booth, E.M.L. Humphreys & W.H.T. Vlemmings, eds.

© International Astronomical Union 2012
doi:10.1017/S174392131200703X

The final 112-frame movie of the 43 GHz SiO masers around the Mira Variable TX Cam

Ioannis Gonidakis[1] Philip J. Diamond[1] and Athol J. Kemball[2]

[1] CSIRO Astronomy and Space Science
Vimiera and Pembroke Roads, Marsfield NSW 2122, Australia
email: ioannis.gonidakis@csiro.au

[2] Dept. of Astronomy, University of Illinois at Urbana-Champaign
1002 W. Green Street, Urbana, IL 61801, USA

Abstract. The proximity of the SiO masers to the star, makes them a powerful tool for studying the properties of the extended stellar atmosphere. This project is a long monitoring campaign of the 43 GHz SiO masers (v=1 J=1→0) around the Mira Variable TX Cam. The target source was observed with the Very Long Baseline Array (VLBA) from the 24th of May 1997 to the 25th of January 2002 in bi-weekly or monthly intervals, covering 3.06 stellar cycles. The time-span and frequency of observations helped us examine the long and short-term properties of the emission and study the dynamics of the gas, the existence of shock waves and their contribution to the morphology and kinematics. Maps from each epoch were concatenated into a 112-frame movie showing the evolution of the emission around TX Cam.

Keywords. masers, shock waves, techniques:high angular resolution, techniques:interferometric, stars:AGB and post-AGB, stars:imaging, stars:winds and outflows, stars:circumstellar matter, radio lines:stars

1. Introduction

The first high resolution images of the 43 GHz SiO masers around late-type stars were produced by Diamond *et al.* (1994). Their VLBA observations of the Miras TX Cam and U Her revealed that the masers are confined in well defined rings, overruling the until then prevailing belief that SiO masers form chaotic structures in variable stars. They showed that they form at 2-4 $R_\star$, placing the masers in the extended atmosphere of the stars, within the dust formation zone. The main kinematic behaviour seemed to be that of outflow, with the gas confined in an ellipsoidal region; the maser effect is dominant on the thicker parts of the ellipsoid along our line of side, thus the emission appeared confined in a projected ring.

The first movie of the TX Cam monitoring campaign was published by Diamond & Kemball in 2003. This 44-frame version was covering a complete stellar pulsation cycle at an angular resolution of ∼0.1 mas and revealed the gross kinematic properties of the SiO maser emission. The morphology of the shell appeared to vary with time and some of its properties appeared dependant on the stellar phase. Individual maser features persisted over many epochs and the predominant kinematic behaviour of the ring was that of expansion. Contrary to the models that assume spherical symmetry, the structure and evolution of the ring revealed a high degree of asymmetry. There was also evidence of ballistic deceleration and proper motion analysis revealed motions between ∼5-10%, distributed randomly around the ring.

The 73-frame movie by Gonidakis *et al.* (2010) uncovered more properties of the morphology and the kinematics of the masering shell. Covering two pulsation cycles, the movie revealed another kinematic motion of the ring; contraction was following expansion during the second cycle. The time-span of the movie allowed the study of short- and long-term variability properties, i.e. changes not only within a cycle but also from one cycle to another. The 43 GHz flux variability follows that of the optical with a ~10% lag but the fluxes are uncorrelated. The width of the ring was also correlated with the stellar pulsation and there was no correlation between the velocities and position angle of the ring. The lifetime of individual components followed a Gaussian distribution with a peak between 150 and 200 days.The spectra were dominated by blue and red-shifted peaks that formed at different times in the stellar cycle and had different lifetimes.

2. Observations and Data Reduction

We observed the $v=1$, $J=1{\rightarrow}0$ masers on TX Cam with the VLBA and one antenna from the VLA. Observations started on the 24th of May 1997 and until the 9th of September 1999 were conducted in biweekly intervals; from that date until the end of the project on the 25th of January 2002 observations occurred every month. In total 80 individual data-sets were collected that corresponded to a coverage of 3.06 pulsation periods, given a period of 557 days for TX Cam.

The rest frequency of observations for all epochs was 43.122027 GHz, centred at an LSR velocity of 9.0 $\mathrm{km\,s^{-1}}$. Within each 6-8 hours scan several sources were observed; 2.5-3.5 hours of each scan were devoted to TX Cam and the rest to calibrators and the other target sources. In order to get all 4 Stokes polarisation data after correlation with the VLBA correlator, we observed in dual polarisation over a 4 MHz bandwidth. The output was 128 spectral channels, thus a spectral resolution of 31.25 kHz corresponding to a velocity resolution of 0.217 $\mathrm{km\,s^{-1}}$.

There were a number of problems and limitations that made the compilation of the movie a very challenging task. Firstly, we had to ensure that data were recorded and analysed in a consistent and uniform manner. The former was dealt by keeping the same configuration during each experiment as described in the previous paragraph. The latter demanded the creation of an automated procedure, that was developed as a POPS script within the AIPS package and was based on the technique described by Kemball, Diamond and Cotton (1995) and Kemball & Diamond (1997). The pipeline was divided into logical steps and demanded minimum interaction by the user, which was limited to

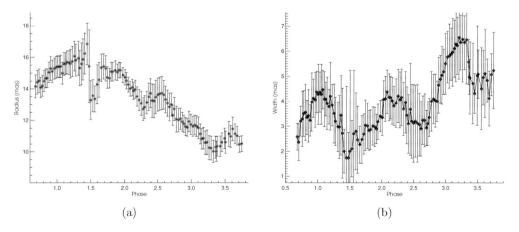

(a)   (b)

Figure 1: (a) Plot showing the inner shell radius with the stellar phase. (b) Plot showing the change in the ring's width with the stellar phase.

the examination of the results at the end of each step and the editing. The result was a 128-channel cube of 1024×1024 pixels in R.A and Dec. with a pixel separation of 0.1 mas. However, due to self-calibration any information on the absolute position was lost and the frames were aligned using a two-step approach. During the first step the frames were superimposed and shifted in order to be aligned on a by-eye estimate. Then we used task XYCOR in AIPS that provided a sub-pixel approximation of the needed shift. Once all the images were aligned we had to introduce frames to insure the time-coherence of the movie. For missing epoch or bad data, frames were interpolated with task COMB in AIPS.

3. The Movie

The movie consists of 112 frames and a 10% sample is plotted in Fig. 2. The plots correspond to every tenth frame of the movie that covers three complete stellar cycles spread over four pulsation periods.

3.1. *General Characteristics*

As previously observed, radiation is confined in a ring-like structure around the star that changes in luminosity and shape. The change in luminosity follows the stellar pulsation but the changes in shape are not correlated with the pulsation of the star. The masering zone starts as a ring in the first cycle, resembles an ellipsoid in cycles two and three end ends up as a ring again in cycle four. During the first half of the movie maser emission is mainly located in bright spots, in the second half though bright filaments appear to be the main hosts of emission. Diffuse emission is occasionally apparent, especially during the first cycle and is located at the outer parts of the structure. The flow of the material appears to follow ordered motions, favouring a specific kinematic behaviour along the ring perimeter. Despite this uniform behaviour, there are features that deviate and their effect in the overall appearance of the ring is evident. Characteristic examples are the gash in the eastern part of the ring (ϕ=1.47) and the split in the northeast (ϕ=2.02).

3.2. *Variability*

The characteristics of the masering region appear to change with the stellar phase, some of them in correlation while some randomly. Fig. 1 is a plot of the inner shell radius with respect to the stellar phase. This is a justification of what was visually observed in the movie; the first cycle is dominated by expansion but in the remaining cycles both expansion and contraction are apparent. The radius at which the ring forms is not correlated to the phase so in each cycle the inner boundary can reach significantly different maximum and minimum distances from the center of the star. For example, in the first cycle the ring appears further from the center of the star than in the last. The radius is correlated with the intensity and during more intense cycles the rings form closer to the star.

Fig. 1 shows the change of the ring width with the stellar phase. Despite a lag of ∼10%, the width follows the pulsation of the star so, it appears wider at maxima and narrower at minima. The width of the ring is correlated to the intensity of the masers so stronger cycles have wider rings. This explains the much wider ring at the final maximum compared to the others. The width of the ring is also dependant on the radius at which it is formed. Thus, rings closer to the star (which are brighter according to the previous paragraph) are wider too.

It is apparent from the above that the radius of the ring, its width and its intensity are correlated. Results show that the variability of the intensity is correlated with the

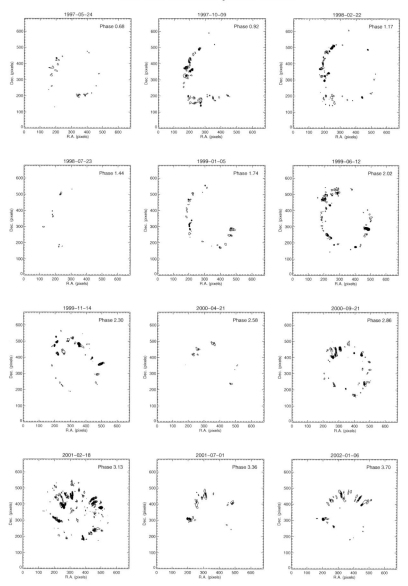

Figure 2: A collage of the total intensity frames. The phase of the frame is written at the top write of each contour plot and the date that the data were taken at the top of each frame. The maps correspond to frames 1,11,21,31,41, 51,61,71,81,91,101 and 111 from the movie. Lower contour value is 1 Jy and the step between each contour is 2.22 Jy until a maximum value of 45.5484 Jy.

pulsation of the star, exhibiting a similar lag to the width. On the other hand the maser and optical fluxes are not correlated. Although the maser intensity follows the pulsation of the star, it does not follow the changes in the optical light; a more intense cycle in the radio is not accompanied by stronger optical fluxes. As a matter of fact, all cycles appear quite uniform with regard to their flux variations in the optical while the maser emission becomes stronger from cycle to cycle.

3.3. *Velocities*

Each cycle is quite different kinematically, and velocity maps reveal the complicated kinematics of the extended atmosphere in the region where the masers are located. During the first and second cycles the maser features appear to move along the line of sight with velocities close to the systemic. This behaviour changes dramatically during the last two cycles where blue- and red-shifted masers seem to dominate the perimeter of the ring and systemic components are absent; there is no gradient, so rotation is not evident. Velocities tend to be blended and in the same portion of the ring both blue and red-shifted masers can be dominant. A characteristic property of the filaments is a velocity gradient along their axes always toward the systemic velocity.

3.4. *Shock Waves*

It is generally believed that shock waves are created in each cycle, when the star is at its maximum. In order to examine the existence of shock waves in the extended atmosphere of TX Cam and their contribution to the overall structure of the ring, we examined the relative position of a shock wave to some key features. As mentioned before there are features that deviate from the ordered flow and they appear at several occasions in the movie as bouncing components or splits in the ring structure. We used for the shock velocity the value graphically calculated by Gonidakis *et al.* (2010), a distance of 390 pc (Olivier, Whitelock & Marang 2001) and a value for the radius of TX Cam of $R_\star = 7.3 \times 10^{13}$ cm at $\phi = 0.939$ (Pegourie 1987). Our results show that the kinematics of all these features can be attributed to their encounter with a shock wave.

References

Diamond, P. J., Kembal, A. J., Junor, W., Zensus, A., Benson, J., & Dhawan, V. 1994, *ApJ*, 430, 61
Diamond, P. J. & Kembal, A. J. 2003, *ApJ*, 599, 1372
Gonidakis, I., Diamond, P. J., & Kembal, A. J. 2010, *MNRAS*, 406, 395
Kemball, A. J. & Diamond, P. J. 1997, *ApJ*, 481, L111
Kemball, A. J., Diamond, P. J., & Cotton, W. D. 1995, *A&AS*, 110, 383K
Olivier, E. A., Whitelock, P., & Marang F. 2001, *MNRAS*, 326, 490
Pegourie, B. 1987, *Ap&SS*, 136, 133

Cosmic Masers - from OH to H$_0$
Proceedings IAU Symposium No. 287, 2012
R.S. Booth, E.M.L. Humphreys & W.H.T. Vlemmings, eds.

© International Astronomical Union 2012
doi:10.1017/S1743921312007041

High Resolution Radio and IR Observations of AGB Stars

W. Cotton[1], G. Perrin[2] R. Millan-Gabet[3], O. Delaa[4], and B. Mennesson[5]

[1] National Radio Astronomy Observatory, 520 Edgemont Rd., Charlottesville, VA 202903, USA
email: bcotton@nrao.edu

[2] LESIA, Observatoire de Paris, CNRA, UPMC, Université Paris-Diderot, Paris Sciences et Letters, 5 place Jules Janssen, 92195, Meudon, France
email: guy.perrin@obspm.fr

[3] California Institute of Technology, NASA Exoplanet Science Institute, Pasadena, CA 91125, USA
email: R.Millan-Gabet@caltech.edu

[4] Laboratoire Lagrange, UMR 7293 UNS-CNRS-OCA, Boulevard de l'Observatoire, B.P. 4229 F, 06304 NICE Cedex 4, France
email: omar.delaa@oca.eu

[5] Jet Propulsion Laboratory, California Institute of Technology, 4800 Oak Grove Drive, Pasadena, CA 91190, USA
email: bertrand.mennesson@jpl.nasa.gov

Abstract. Asymptotic Giant Branch Stars (AGB) are evolved, mass losing red giants with tenuous molecular envelopes which have been the subject of much recent study using infrared and radio interferometers. In oxygen rich stars, radio SiO masers form in the outer regions of the molecular envelopes and are powerful diagnostics of the extent of these envelopes. Spectroscopically resolved infrared interferometry helps constrain the extent of various species in the molecular layer. We made VLBA 7 mm SiO maser, Keck Interferometer near IR and VLTI/MIDI mid IR high resolution observations of the stars U Ari, W Cnc, RX Tau, RT Aql, S Ser and V Mon. This paper presents evidence that the SiO is depleted from the gas phase and speculate that it is frozen onto Al_2O_3 grains and that radiation pressure on these grains help drive the outflow.

Keywords. masers, stars: AGB and post-AGB, stars: imaging, stars: winds, outflows

1. Introduction

Asymptotic Giant Branch Stars (AGB) are low to intermediate mass stars that have exhausted their nuclear fuel, have become pulsating red giants and are losing most of their mass to become planetary nebulae. The extended, cool envelopes of these stars contain a variety of molecules some of which eventually condense into dust grains. In oxygen rich AGB stars, SiO masers form in the outer parts of the molecular envelope interior to where the silicate dust forms; see Reid & Menton (1997), Danchi *et al.* (1994). Observations by Perrin *et al.* (2004) and Wittkowski *et al.* (2008) have shown that molecules in the envelope occur in shells.

Open questions about AGB stars are how do they sometimes form very asymmetric planetary nebulae and how is the mass loss driven. The observed silicate dust forms too far out in the envelope for radiation pressure on this dust to help drive the outflow. Recent speculation has centered on the role of Al_2O_3 dust which can form at a relatively high temperature ($\sim$1700 K). Wittkowski *et al.* (2007) presented SiO maser and mid-IR

Table 1. Primary IR Opacity Sources

wavelength	Opacity sources
near-IR	photosphere, inner molecular region
7.80 – 9.30 μm	SiO gas
9.37 – 11.50 μm	Silicate dust
11.55 – 13.26 μm	H2O gas

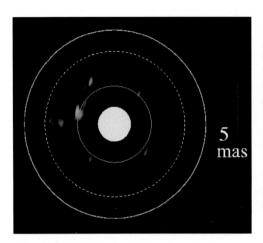 

Figure 1. S Ser. On the left is the image in the SiO ν=2, J=1-0 transition. The green circle gives the fitted diameter on the maser ring, the light blue line the fitted size of the IR molecular shell and the dashed white line, the radius at which Al_2O_3 could condense. On the right is the image in the SiO ν=1, J=1-0 transition.

interferometric observations of the AGB star S Ori which they interpreted as showing Al_2O_3 dust forming just interior to the SiO masers.

2. Observations

A sample of AGB stars was observed using the VLBA, VLTI/MIDI and the Keck Interferometer. Multiple snapshot observations of the 7 mm SiO masing ν=1,J=1-0 and ν=2,J=1-0 transitions near 43 GHz were made on 1 July 2007 and 24 February 2008. Observations were analyzed as described in Cotton *et al.* (2006). The diameters of the masing regions were characterized by fitting circular rings.

Observations with the VLTI/MIDI (Leinert *et al.* 2003) measured between 7 and 14 μm; The Keck Interferometer observed S Ser at 2.2 μm. Models were fitted to the IR data to derive sizes of the photosphere and one or more layers needed to characterize the envelope seen in the mid IR. The mid-IR spectrum was modeled by wavelength ranges; the expected principle opacity sources of these are given in Table 1.

3. Results

The observational results are given for S Ser in Fig. 1, W Cnc in Fig. 2, RX Tau in Fig. 3, U Ari in Fig. 4, V Mon in Fig. 5 and RT Aql in Fig. 6. For S Ser, only a single mid IR visibility was measured and a single layer model was fitted to the data. Too few maser spots were detected in U Ari and V Mon to reliably fit ring sizes. However, the minimum ring diameter for V Mon is the separation of the two spot groups which is very nearly the same as derived in the mid IR. No IR interferometric data were obtained

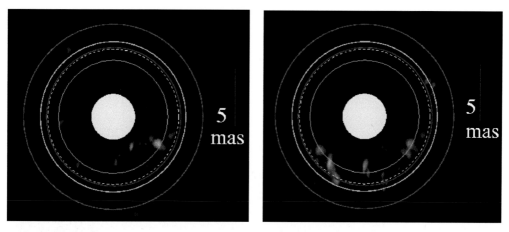

Figure 2. W Cnc. Like figure 1 except that the light blue circle shows the fitted size of the 7.8–9.3 μm data; magenta, the size of the 9.4–11.5 μm and red, 11.6–13.3 μm.

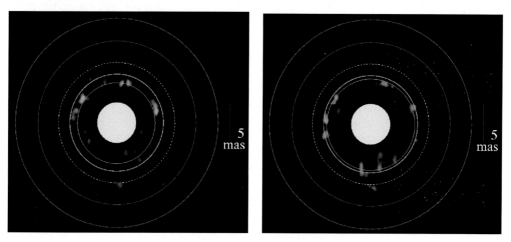

Figure 3. RX Tau. Like figure 2.

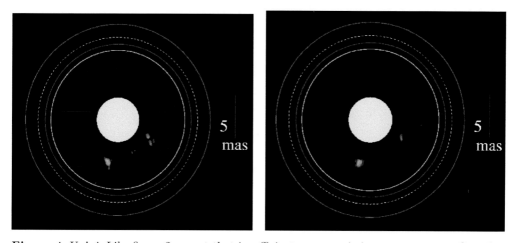

Figure 4. U Ari. Like figure 2 except that insufficient maser emission was present to fit a ring and align with the IR data.

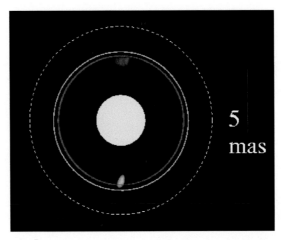

Figure 5. V Mon. Like figure 3 except that maser emission was only detected in the $\nu=1$, J=1-0 transition.

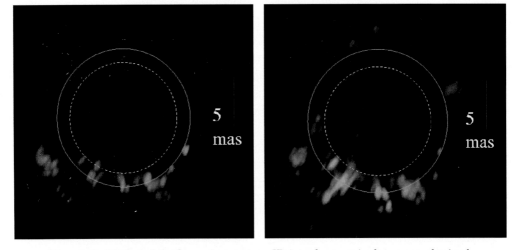

Figure 6. RT Aql. Like figure 2 except no IR interferometric data were obtained.

for RT Aql and the size of the photosphere, hence the anticipated Al_2O_3 condensation distance was based on photometric measurements from the literature.

4. Discussion

In none of the visibility spectra do we detect the strong 9.8 μm silicate feature. Conversion of SiO to silicate must take place exterior to the region probed by this data.

There are two indicators of SiO gas in our data, the SiO masers and the SiO lines in the 7.80 – 9.30 μm spectral region. With the possible exception of RT Aql, none of the stars observed show evidence for either of these diagnostics significantly exterior to the radius at which Al_2O_3 is expected to form (dashed circles in Figures 1–6). There was no IR interferometric data obtained for RT Aql and the photospheric size and the radius of Al_2O_3 condensation were estimated from photometric measurements in the literature. The ratio of the SiO maser ring size to the photospheric size is quite different for RT Aql and the values near 2.0 found in well constrained cases.

Following the suggestion of Verhoelst *et al.* (2006), we propose that the gas phase SiO is condensing onto Al_2O_3 grains soon after these grains form. If this is the case, then the chemical conversion from SiO to silicates must take place on the dust grains. The radiation pressure on these dust grains with SiO mantles relatively close to the photosphere could then help drive the outflow.

References

Cotton, W. D., Vlemmings, W., Mennesson, B., *et al.* 2006, *A&A*, 456, 339

Perrin, G., Coude Du Foresto, V., Ridgway, S. T., *et al.* 2004, *A&A*, 426, 279

Reid, M. J. & Menton, K. M. 1997, *ApJ*, 476, 327

Danchi, W. C., Bester, M., Degaicomi, C. G., Greenhill, L. J., & Townes, C. H. 1997, *AJ*, 107, 1469

Leinert, C., Graser, U, Richichi, A., *et al.* 2003, *The Messenger*, 112, 13

Verhoelst, T., Decin, L., van Malderen, R., *et al.* 2006, *A&A*, 447, 311

Wittkowski, M., Boboltz, D. A., Ohnaka, K., Driebe, T., & Scholz, M. 2007, *A&A*, 470, 191

Wittkowski, M., Boboltz, D. A., Driebe, T., *et al.* 2008, *A&A*, 479, L21

Cosmic Masers - from OH to H$_0$
Proceedings IAU Symposium No. 287, 2012
R.S. Booth, E.M.L. Humphreys & W.H.T. Vlemmings, eds.

© International Astronomical Union 2012
doi:10.1017/S1743921312007053

OH mainline maser polarisation properties of post-AGB stars

Jessica M. Chapman[1] Ioannis Gonidakis[1], Rachel M. Deacon[2] and Anne Green[2]

[1] CSIRO Astronomy and Space Science
PO Box 76, Epping, NSW 2122, Australia
email: jessica.chapman@csiro.au

[2] University of Sydney, School of Physics
NSW 2006, Sydney, Australia

Abstract. The Parkes 64-m telescope was used to study the OH mainline polarisation properties at 1665 and 1667 MHz for a sample of 36 evolved stars, identified by their far-infrared and OH 1612 MHz maser properties as likely post-AGB stars.

Keywords. masers, polarization, stars:AGB and post-AGB, radio lines:stars

1. Observations and results

We observed thirty-six sources that exhibit 1612 MHz OH emission and according to their infrared properties can be categorized as likely post-AGB stars. The sources were bright and evenly spread over four infrared selection groups (Deacon *et al.* 2004, Deacon 2006).

Observations were taken in 2004 using the Parkes telescope in two 12-hour observing sessions, using two orthogonal linear feeds. A bandwidth of 4 MHz was centered at 1666.4 MHz to cover both the OH 1665 and 1667 MHz mainline transitions, with 4096 spectral channels, giving a velocity resolution of 0.175 km s^{-1}. Each source was observed for ten two-minutes scans to cover a range of paralactic angles.

OH mainline polarization was detected in 21 out of the 36 sources observed, with 14 detections at 1665 MHz and 13 at 1667 MHz. For the former emission, 13 sources were found to be circularly and 11 linearly polarized, while for the latter the number of detections were 15 and 18 respectively (Table 1). In total, circular polarization was observed in 42% of the sources and linear in 50% of them.

2. Linear polarisation from post-AGB stars

As seen in Fig. 2 linear polarisation from post-AGB stars is typically detected as a small number of very narrow features. These are not always associated with corresponding circularly polarised features; in some cases only linear polarisation is detected. For three

Table 1: OH mainline polarisation detection statistics

OH line	Total Observed	Polarised	Circular	Linear
1665	24	14	13	11
1667	31	13	7	10
Total	36	21	15	18

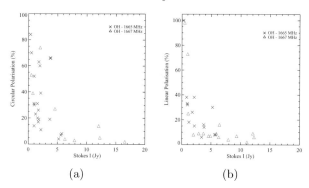

(a) (b)

Figure 1: Percentage polarisation for maser features at 1665 MHz (crosses) and 1667 MHz (triangles). (a) Circular polarisation. (b) Linear polarisation.

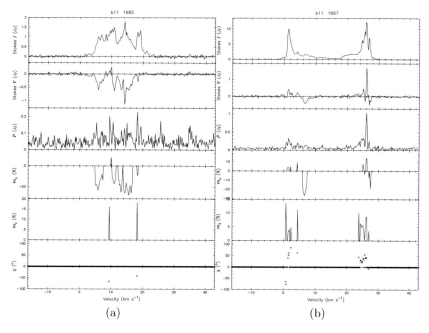

(a) (b)

Figure 2: An example of the polarisation spectra obtained for source b11. From top to bottom the panels are: Total Intensity (I), Circular Polarisation (V), Total Linear Polarisation (P), Percentage Circular Polarisation (m_c), Percentage Linear Polarisation (m_l) and Polarisation Position Angle (χ).

sources in this sample, b292 (1665 MHz), v87 (1665 MHz) and v189 (1667 MHz) we have detected spectral features that are ~100% linearly polarised (Fig. 1). These features may correspond to Zeeman 'π' components.

References

Deacon, R. M., Chapman, J. M., & Green, A., 2004, *ApJS*, 155, 595

Deacon, R. M., 2006, PhD Thesis, The University of Sydney

Cosmic Masers - from OH to H_0
Proceedings IAU Symposium No. 287, 2012
R.S. Booth, E.M.L. Humphreys & W.H.T. Vlemmings, eds.

© International Astronomical Union 2012
doi:10.1017/S1743921312007065

Preliminary results on SiO $v=3$ $J=1$–0 maser emission from AGB stars

J.-F. Desmurs[1], V. Bujarrabal[1], M. Lindqvist[2], J. Alcolea[1], R. Soria-Ruiz[1], and P. Bergman[2]

[1] Observatorio Astronómico Nacional, Madrid, Spain.
email: [desmurs, bujarrabal, alcolea, r.soria] @oan.es

[2] Onsala Space Observatory, Chalmers Univ. of Technology, Sweden
email: [michael.lindqvist, pbergman] @chalmers.se

Abstract. We present the results of SiO maser observations at 43 GHz toward two AGB stars using the VLBA. Our preliminary results on the relative positions of the different $J=1$–0 SiO masers ($v=1,2$ and 3) indicate that the current ideas on SiO maser pumping could be wrong at some fundamental level. A deep revision of the SiO pumping models could be necessary.

Keywords. Maser, AGB stars

1. Introduction

Many stars have been mapped in SiO emission $J=1$–0 $v=1$ and 2, particularly using the VLBA (Diamond *et al.* 1994, Desmurs *et al.* 2000, Cotton *et al.* 2006, etc). The maser emission is found to form a ring of spots at a few stellar radii from the center of the star. In general, both distributions are similar, although the spots are very rarely coincident and the $v=2$ ring is slightly closer to the star (see e.g. Desmurs *et al.* 2000).

The similar distributions of the $v=1$, 2 $J=1$–0 transitions were first interpreted as favoring collisional pumping, because the radiative mechanisms tend to discriminate somewhat more strongly both states. But the lack of coincidence was used as an argument in favor of radiative pumping, leading to the well-known, long-lasting discrepancy in the interpretation of the $v=1$, 2 $J=1$–0 maps in terms of pumping mechanisms (see discussion in e.g. Desmurs *et al.* 2000).

The discussion on this topic has dramatically changed when the first comparisons between the $v=1$ $J=1$–0 and $J=2$–1 maser distributions were performed (see Soria-Ruiz *et al.* 2004, 2005, 2007). In contradiction with predictions, from both radiative and collisional models, the $v=1$ $J=2$–1 maser spots systematically occupy a ring with a significantly larger radius ($\approx30\%$) than that of $v=1$ $J=1$–0, both spot distributions being completely unrelated. Soria-Ruiz *et al.* (2004) interpreted these unexpected results invoking line overlap between the ro-vibrational transitions $v=1\,J=0$ – $v=2\,J=1$ of SiO and $\nu_2=0$ $J_{k_a k_c}=12\,_{7,5}$ – $\nu_2=1$ $J_{k_a k_c}=11_{6,6}$ of H_2O. This phenomenon would also introduce a strong coupling of the $v=1$ and $v=2$ $J=1$–0 line, explaining their similar distribution.

If our present theoretical ideas are correct (e.g. Bujarrabal & Nguyen-Q-Rieu 1981, Bujarrabal 1994, Locket & Elitzur 1992, Humphreys *et al.* 2002), the $v=3$ $J=1$–0 emission should require completely different excitation conditions than the other less excited lines. No pair of overlapping lines is expected to couple the $v=3$ $J=1$–0 inversion with any of the other SiO lines. The $v=3$ $J=1$–0 spatial distribution should be different compared to the $J=1$–0 $v=1$, 2 ones and, of course, of the $J=2$–1 $v=1$ maser, and placed in a still smaller ring than $v=2$.

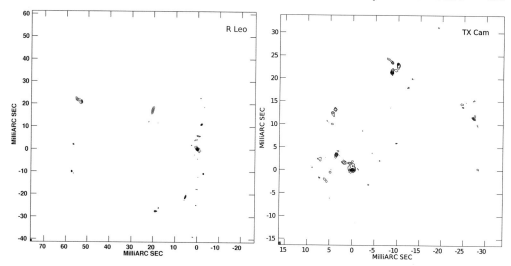

Figure 1. VLBA map of SiO $J=1$–0 $v=2$ (in black) and $v=3$ (in red) maser emission from R Leo (right) and TX Cam (left)

2. Preliminary results toward R Leo and TX Cam

The $v=3$ $J=1$–0 line is sometimes quite intense (Alcolea *et al.* 1989), and bright enough to be mapped with the VLBA, but it is strongly variable, both in time (with characteristic times scales of a few months) and from object to object. With the 20-m antenna of Onsala, we monitored a number of AGB stars to select the best candidates to be mapped with the VLBA, observing simultaneously the $v=1,2$ and 3 $J=1$–0 SiO masers at 42-43 GHz.

In the figure above, we show a preliminary map of the brightness distribution of ^{28}SiO $v=2$ and 3, $J=1$–0 obtained toward R Leo (on the left) and TX Cam (on the right). These are the first VLBA maps ever of the $v=3$ $J=1$–0 maser (in red). Although the alignment between the maps of the two lines is just indicative (the observations were not done in phase referencing mode, and the proposed alignment is based on the similarity in velocity and the spatial distribution of some spots).

These preliminary results show a surprising similar distribution in the $v=3$ $J=1$–0 and $v=1, 2$ $J=1$–0 masers. Would this result be confirmed, our ideas on SiO maser pumping scheme must be wrong at some fundamental level and a deep revision of the SiO pumping models will be necessary.

References

Alcolea, J., Bujarrabal, V., & Gallego, J. D., 1989, *A&A* 211, 187
Bujarrabal, V. & Nguyen-Q-Rieu, 1981, *A&A* 102, 65
Bujarrabal, V., 1994, *A&A* 285, 953
Cotton, W. D., Vlemmings, W., Mennesson, B. *et al.*, 2006, *A&A* 456, 339
Desmurs J.-F., Bujarrabal V., Colomer F., & Alcolea J., 2000, *A&A* 360, 189
Diamond P. J., Kemball A. J., Junor W. *et al.*, 1994, *ApJ* 430, L61
Humphreys, E. M. L., Gray, M. D., Yates, J. A. *et al.*, 2002, *A&A* 386, 256
Lockett P. & Elitzur M., 1992, *ApJ* 399, 704
Soria-Ruiz, R., Alcolea, J., Colomer, F. *et al.*, 2004, *A&A* 426, 131
Soria-Ruiz, R., Colomer, F., Alcolea, J. *et al.*, 2005, *A&A* 432, L39
Soria-Ruiz, R., Colomer, F., Alcolea, J. *et al.*, 2007, *A&A* 468, L1

Cosmic Masers - from OH to H₀
Proceedings IAU Symposium No. 287, 2012
R.S. Booth, E.M.L. Humphreys & W.H.T. Vlemmings, eds.

© International Astronomical Union 2012
doi:10.1017/S1743921312007077

1612 MHz OH maser monitoring with the Nançay Radio Telescope

D. Engels[1], E. Gérard[2], and N. Hallet[2]

[1]Hamburger Sternwarte, Gojenbergsweg 112, D–21029 Hamburg, Germany
email: dengels@hs.uni-hamburg.de

[2]Observatoire de Paris, 5 Place J Janssen, F–92195 Meudon Cedex, France
email: eric.gerard@obspm.fr, nicole.hallet@obspm.fr

Abstract. 20 OH/IR stars are monitored in the 1612 MHz OH maser line with the Nançay Radio Telescope. The program started in 2008 with monthly observations of the full sample and will last at least until end of 2012. The aim is the determination of the linear diameter of the circumstellar shell using the phase lag between the light curves of the varying OH maser lines. To use them for distance determinations, angular diameters are obtained by interferometric measurements while the stars pass the maximum of their OH maser flux density variations. The periods of the OH/IR stars monitored are between 425 and >2000 days.

Keywords. masers, stars: AGB and post-AGB, stars: late-type

1. Phase-lag distances

OH/IR stars have optically thick circumstellar dust and gas envelopes and were discovered first by their intense OH maser emission. They pulsate similarly to Miras, but with much longer periods (>2 years). The general understanding is that OH/IR stars have more massive main-sequence progenitors than Mira variables. However, because of the uncertain distances of OH/IR stars, the evidence for mass segregation between Mira and OH/IR stars is concluded from indirect arguments (Habing 1996, Chen *et al.* 2001).

A promising technique to determine distances to OH/IR stars is the use of phase-lags between the two varying OH maser peaks originating from the front and back sides of the shell. These lags yield linear diameters and combined with angular diameters, obtained from interferometric observations, distances can be derived. This technique was explored in the 1980ies by Herman & Habing (1985) finding that the majority of their sample of OH/IR stars has distances between 2 and 10 kpc. The accuracy of these distances

Table 1. Periods and amplitudes determined after four years of observations (≈ 1500 days) of the stars observed. P is the period in days and Amp. is the relative amplitude defined as amplitude/mean flux density.

Object	P	Amp.	Object	P	Amp.	Object	P	Amp.
IRAS 01037+1219	650	0.48	OH 20.7+0.1	1720	0.49	OH 55.0+0.7	1270	0.44
OH 127.8+0.0	1590	0.51	OH 26.5+0.6	1589	0.44	IRAS 20234−1357	425	0.37
OH 138.0+7.2	1580	0.20	OH 30.1−0.7	2000	0.38	OH 75.3−1.8	1652	0.45
OH 141.7+3.5	3500	0.42	OH 32.0−0.5	1410	0.40	OH 83.4−0.9	1497	0.47
IRC +50137	635	0.61	OH 32.8−0.3	1690	0.60	IRAS 21554+6204	1400	0.43
IRAS 05131+4530	1050	0.49	OH 39.7+1.5	1260	0.40	OH 104.9+2.4	1750	0.50
OH 16.1−0.2	2000	0.24	OH 44.8−2.3	534	0.40			

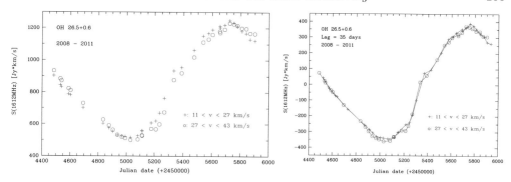

Figure 1. *Left:* Lightcurves of the integrated maser flux of the blue and the red maser peak of OH 26.5+0.6 between 2008 and 2011. The fluxes of the red peak were scaled to match the blue peak in mean flux and amplitude. The red lightcurve (o) lags behind the blue lightcurve (+). The period of the star is 4.35 years. *Right:* Result of the correlation of the normalized lightcurves. The difference between the lightcurves is minimized for $\tau_0 = 35^{+5}_{-8}$ days, where the error interval is defined by a 10% degradation of the fit.

is however not well determined. Van Langevelde *et al.* (1990) re-determined the phase-lags and hence linear diameters and revised their values to a much greater extent, than expected from the errors quoted.

2. Observations and first results

We therefore decided to remeasure with the Nançay Radio Telescope the phase-lags for a number of sources and to apply this technique to new sources with bright OH masers. The list of sources is given in Table 1. The observations are made every month with the Low-Frequency receiver. The digital autocorrelator is split into two banks with a bandwidth of 0.78125 MHz centered on 1612.231 MHz to observe both linear polarizations simultaneously. The velocity resolution is 0.035 km/s. Typical integration times are 8 minutes on source, which yield a noise level of ≈ 0.1 Jy and an S/N > 100.

OH 26.5+0.6 has the strongest maser in the sample and is used to experiment with different strategies to determine the phase-lag. Fig. 1 shows the phase-lag determination using integral fluxes for the two peaks. This gives the best signal-to-noise ratio, but averages the differing phase-lags arising from different velocity intervals. The lightcurves of the two peaks were normalized, after that the data was edited and smoothed to get rid of scatter. After scaling the red peak in amplitude and shifting it along the time-axis, the lightcurves were correlated. A phase-lag of $\tau_0 = 35^{+5}_{-8}$ days was determined by minimizing the differences of the fluxes (Fig. 1).

This phase-lag is in excellent agreement with the result of van Langevelde *et al.* (1990), who obtained $\tau_0 = 37 \pm 7$ days. It confirms the consistency of the lightcurve analysis methods and shows that no major changes in the structure of the OH masing shell of this star occurred during the last 30 years. For most of our sample at least one period will be covered end of 2012 and phase-lag distances of the sample will be available as soon as the interferometry observations are completed.

References

Chen, P. S., Szczerba, R., Kwok, S., & Volk, K. 2001, *A&A* 368, 1006

Habing, H. J. 1996, *A&A Reviews* 7, 97

Herman, J. & Habing, H. J. 1985, *A&AS* 59, 523

van Langevelde, H. J., van der Heiden, R., & van Schooneveld, C. 1990, *A&A* 239, 193

Cosmic Masers - from OH to H$_0$
Proceedings IAU Symposium No. 287, 2012
R.S. Booth, E.M.L. Humphreys & W.H.T. Vlemmings, eds.

© International Astronomical Union 2012
doi:10.1017/S1743921312007089

The Hamburg Database of Circumstellar OH Masers

Dieter Engels

Hamburger Sternwarte, Gojenbergsweg 112, D–21029 Hamburg, Germany
email: dengels@hs.uni-hamburg.de

Abstract. A new version of the Hamburg Database of Circumstellar OH Masers at 1612, 1665, and 1667 MHz was released in 2012 January. The database now lists 13170 OH maser observations of stars in the Milky Way. They belong to 6318 different objects and 2324 of them were detected in at least one of the transitions. The database contains flux densities and velocities of the two strongest maser peaks, the expansion velocity of the shell and the radial velocity of the star. Compared to the first version presented in 2007 at the IAU Symposium 242 in Alice Springs new observations published 2008–2011 are included. Interferometric observations and monitoring programs of the maser emission were also added. Access to the database is possible over the Web (www.hs.uni-hamburg.de/maserdb), allowing cone searches for individual objects and lists of objects. A general object search is possible in selected regions of the sky and by defining ranges of flux densities and/or velocities.

Keywords. masers, catalogs, stars: AGB and post-AGB, stars: late-type

1. Introduction

Since the discovery of masers 40 years ago several thousand observations to detect masers in circumstellar shells have been made. Comfortable tools to access these observations were lacking, until we released the first version of the Database of Circumstellar OH Masers (Engels & Bunzel 2008). Prior to this release, catalogues listing 1612 MHz OH masers in AGB stars (OH/IR stars) were published by te Lintel Hekkert *et al.* (1989) and Benson *et al.* (1990). The te Lintel Hekkert *et al.* catalog contains detected OH masers with their flux densities and velocities and covers the literature until 1984. The Benson *et al.* catalog lists references to OH maser observations and covers the years until 1989. The number of detected 1612 OH Masers listed are 439 and 713 respectively. Since then the number of detected masers has been almost tripled.

In the last decade however, no larger surveys for OH masers have been made anymore, and research has shifted towards interferometric and variability studies of known stellar masers. The interferometric studies focus on astrometry and polarization properties of the masers. To take this development into account, we started to add such publications to the database.

2. The database

The literature searched in refereed journals for OH maser observations cover the years 1984 - 2011. Earlier discoveries are included via the catalogue of te Lintel Hekkert *et al.* (1989). Therefore the database is considered to be (almost) complete for 1612 MHz detections, but contains no non-detections published prior to 1984. For the main lines only measurements published after 1984 are contained. The primary table of the database contains the coordinates of the masers, as well as flux densities and velocities of the two strongest peaks. For the 1612 MHz masers these are usually the outermost peaks, from which the radial velocity of the stars and the expansion velocity of the circumstellar shells are calculated. For more complex spectra, as usually seen for main-line masers, radial and expansion velocities are calculated from the outermost peaks listed in the reference paper. These informations are drawn mostly from observations of single-dish telescopes.

Compared to the first release of the database the number of observations increased from 10774 to 13170 (+22%), while only 50 (+2%) new detections were reported. The marginal increase of newly discovered masers is due to the absence of sensitive surveys in this area.

The interferometric and monitoring observations are kept in separate tables, because they usually provide different parameters than the single-dish observations. For the interferometric observations the database contains entries for the instrument, the spatial resolution, the sensitivity, a flag to mark polarization observations, and the observing date. For the monitoring observations the start and end dates of the observations are given and the range in peak flux densities reported. Currently these tables contain 155 interferometric observations from 18 papers, and 50 monitoring observations from 6 papers.

Updates of the database will be made at least once a year to incorporate observations from upcoming publications. To access the database a web tool is available at www.hs.uni-hamburg.de/maserdb.

Acknowledgements

This work is supported by the Deutsche Forschungsgemeinschaft under grant number En 176/33.

References

Benson, P. J., Little-Marenin, I. R., Woods, T. C., Attridge, J. M., Blais, K. A., Rudolph, D. B., Rubiera, M. E., & Keefe, H. L. 1990, *ApJS* 74, 911

Engels, D. & Bunzel, F. 2008, in J. M. Chapman, W. A. Baan (eds.), *Astrophysical masers and their environments*, Proc. IAU Symposium 242 (San Francisco: ASP), p. 316

te Lintel Hekkert, P., Versteege-Hensel, H. A., Habing, H. J., & Wiertz, M. 1989, *A&AS* 78, 399

Cosmic Masers - from OH to H$_0$
Proceedings IAU Symposium No. 287, 2012
R.S. Booth, E.M.L. Humphreys & W.H.T. Vlemmings, eds.

© International Astronomical Union 2012
doi:10.1017/S1743921312007090

Imaging the water masers toward the H$_2$O-PN IRAS 18061−2505

Yolanda Gómez[1]†, **Daniel Tafoya**[2]‡, **Olga Suárez**[3], **Jose F. Gómez**[4],
Luis F. Miranda[5], **Guillem Anglada**[4], **Jóse M. Torrelles**[6],
and Roberto Vázquez[7]

[1] Centro de Radioastronomía y Astrofísica, UNAM
Antigua Carretera a Pátzcuaro # 8701 Ex-Hda. San José de la Huerta
Morelia, Michoacán. México. C.P.58089 Apartado Postal 3-72 (Xangari).

[2] Department of Physics and Astronomy, Kagoshima University, Japan.
email: dtafoya@milkyway.sci.kagoshima-u.ac.jp

[3] Université de Nice Sophia Antipolis, France.

[4] Instituto de Astrofísica de Andalucía, Spain.

[5] Departmento de Física Aplicada, Universidade de Vigo, Spain.

[6] Facultat de Fisica, Instituto de Ciencias del Espacio, Spain.

[7] Observatorio Astronómico Nacional, UNAM, México.

Abstract. It has been suggested that the presence of disks or tori around the central stars of pre Planetary Nebulae and Planetary Nebulae is related to the collimation of the jet that are frequently observed in these sources. These disks or tori can be traced by the maser emission of some molecules such as water. In this work we present Very Large Array (VLA) observations of the water maser emission at 22 GHz toward the PN IRAS 18061−2505, for which the masers appear located on one side of the central star. For comparison with the observations, we present a simple kinematical model of a disk rotating and expanding around the central star. The model matches qualitatively the observations. However, since the masers appear only on one side of the disk, these results are not conclusive.

Keywords. (ISM:) planetary nebulae: individual (IRAS 18061−2505), ISM: jets and outflows, masers

1. Introduction

The standard scenario for explaining the last stages of the evolution of low- and intermediate-mass stars includes the presence of a massive stellar wind that could reach mass-loss rates as high as 10^{-4} M$_\odot$ yr^{-1} at the end of the Asymptotic Giant Branch phase. Subsequently, a tenuous fast stellar wind takes over. The latter interacts hydrodynamically with the former, piling up the gas and forming a circumstellar shell. When the central star becomes hot enough, the UV radiation ionizes the circumstellar shell. At this point, it is said that the star enters the planetary nebula (PN) phase. This model explains the formation of spherical PNe, however it fails to explain the formation of narrow-waist bipolar PNe. Many of this type of PNe exhibit a dark lane in the equatorial region that usually is traced by molecular gas and dust emission. It has been proposed that this high-density equatorial structure works as a nozzle that would squeeze the fast stellar wind toward the polar directions, forming the bipolar morphology. On the other hand, it has also been proposed that it is possible that, as a result of a binary system, an accretion disks forms in the central region of the PN. Similarly to the star forming

† Passed away 16 February 2012
‡ author

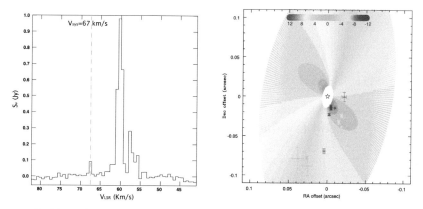

Figure 1. (Left) VLA water maser spectrum of IRAS 18061-2505. (Right) Modeling of the maser emission in terms of a rotating and expanding disk.

regions, the accretion disks leads to the formation of fast collimated jets that interact with the circumstellar envelope, giving origin to bipolar morphologies (e.g. Reyes-Ruiz & Lopez 1999).

The maser emission has proven to be a very useful tool to trace and study structures around evolved stars. So far, water maser emission has been detected in three young PNe: K 3–35 (Miranda *et al.* 2001), IRAS 17347–3139 (de Gregorio-Monsalvo *et al.* 2004) and IRAS 18061–2505 (Gómez *et al.* 2008). All three PNe exhibit a bipolar morphology and the first two objects seem to have disks traced by the water masers (Uscanga *et al.* 2008; Tafoya *et al.* 2009). IRAS 18061–2505 shows well defined bipolar lobes seen in Hα, separated by a clear narrow waist (Miranda *et al.* 2012). In this work, we present the results of interferometric observations of the water masers toward IRAS 18061–2505.

2. Results and Conclusions

The observations of the water masers (22 GHz) were carried out in 2008 with the VLA in its A configuration. The maser emission appears over a velocity range of $\sim$15 km s^{-1} with the strongest maser feature at V(LSR)= 61.6 km s^{-1} and weak maser components at 55.7 and 57.7 km s^{-1} (Figure 1). From optical observations it has been estimated that the systemic velocity of IRAS18061–2505 is v_{sys}= +67 km s^{-1}. This indicates that almost all the masers features are blue-shifted with respect of the systemic velocity. The spectrum and maser positions are in agreement with the previous results presented by Gómez *et al.* (2008). Spatially, the masers appear located at one side of the central star, which is assumed to coincide with the peak of the radio continuum. In the right panel of Figure 1 we present a kinematical model of an expanding and rotating disk that fits the data qualitatively.

References

de Gregorio-Monsalvo, I., Gómez, Y., Anglada, G., Cesaroni, R., Miranda, L. F., Gómez, J. F., & Torrelles, J. M. 2003, *ApJ*, 601, 921
Gómez, J. F., Suárez, O., Gómez, Y., Miranda, L. F., Torrelles, J. M., Anglada, G., & Morata, O. 2008, *AJ*, 135, 2074
Miranda, L. F., Gómez, Y., Anglada, G., & Torrelles, J. M. 2001, *Nature* 414, 284
Reyes-Ruiz, M. & López, J. A. 1999, *ApJ*, 524, 952
Tafoya, D., Gómez, Y., Patel, N. A., Torrelles, J. M., Gómez, J. F., Anglada, G., Miranda, L. F., & de Gregorio-Monsalvo, I. 2009, *ApJ* 691, 611
Uscanga, L., Gómez, Y., Raga, A. C., Cantó, J., Anglada, G., Gómez, J. F., Torrelles, J. M., & Miranda, L. F. 2008, *MNRAS*, 390, 1127

Cosmic Masers - from OH to H₀
Proceedings IAU Symposium No. 287, 2012
R.S. Booth, E.M.L. Humphreys & W.H.T. Vlemmings, eds.

© International Astronomical Union 2012
doi:10.1017/S1743921312007107

TWINKLING STARS
The disappearing SiO masers of W Aql

Sofia Ramstedt[1], Wouter Vlemmings[2], Shazrene Mohamed[3], Yoon Kyung Choi[4] and Hans Olofsson[2]

[1] Argelander Institut für Astronomie, Bonn, Germany
email: sofia@astro.uni-bonn.de

[2] Onsala Space Observatory, Onsala, Sweden
[3] South African Astronomical Observatory, P.O. Box 9, 7935 Observatory, South Africa
[4] Max Planck Institut für Radioastronomie, Bonn, Germany

Abstract. W Aql is a binary S-type AGB star showing SiO maser emission. The dust distribution around the star is asymmetric, possibly indicating gravitational interaction between the binary pair. There are indications that the gas distribution also exhibits asymmetries. To investigate the source of the circumstellar structures, we applied for and were rewarded VLBI time to map the distribution of the SiO masers around this source and to constrain the presence and distribution of a possible magnetic field. Using VERA observations we also aim at measuring an accurate parallax to determine the binary separation, however, from showing peak emission of 21 Jy in June 2010, the SiO(1-0) v=1 line at 43 GHz has now disappeared. We find no correlation with the stellar pulsational period.

Keywords. masers, stars: AGB and post-AGB, stars: mass loss

1. Binary interaction and the shaping of planetary nebulae

The late evolution of low- to intermediate-mass stars is still subject to several open questions. The physical conditions required for AGB stars to become planetary nebulae (PNe), and which processes govern the shaping of the circumstellar environment, are currently being studied. A binary companion could explain circumstellar asymmetries

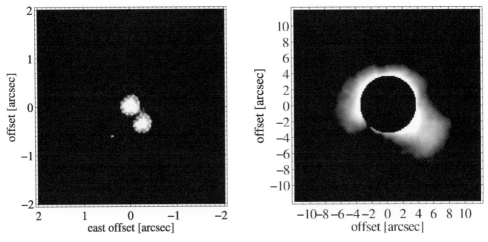

Figure 1. *Left:* The high-resolution B-band HST image of W Aql. The binary pair is clearly resolved and separated by 0.46″. *Right:* The circumstellar envelope around W Aql imaged in dust-scattered polarized light (Ramstedt *et al.* 2011). Note the different scales.

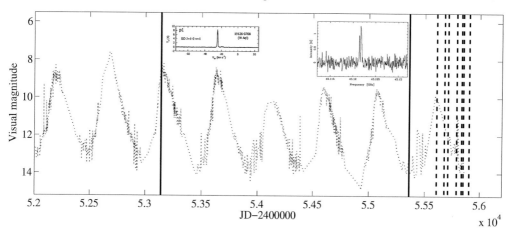

Figure 2. The AAVSO light curve of W Aql showing the visual magnitude versus the Julian date in JD-2400000. The dates of the SiO maser observations are marked by horizontal lines (solid for detections and dashed for non-detections). The two ∼ 21 Jy detections from the Nobeyama 45 m telescope (Nakashima & Deguchi 2007) and the Onsala 20 m telescope are shown as subplots.

both directly through its gravitational influence, as well as indirectly by being the necessary component required to sustain a large scale magnetic field around the AGB star. The different shaping scenarios will be strongly dependent on the separation between the binary pair and the resulting morphologies can range from slight distortions or bubbles for the very wide binaries (100 AU), to bi-polar or even multi-polar outflows, jets and disks, for the closer binaries. W Aql is an ideal source to study in order to reach a better understanding of these very important issues for stellar evolution. It is relatively nearby (D=200 pc), binary, and the fact that it is known to have an asymmetric dust distribution (Fig. 1) makes the source particularly intriguing.

2. SiO maser observations and the visual light curve

The SiO(J=1-0, v=1) maser emission was observed in May 2004 with the Nobeyama 45 m telescope (Nakashima & Deguchi 2007) and with the Onsala 20 m telescope in June 2010. On both occasions the peak intensity was around 21 Jy. VERA interferometric observations of the source started in February 2011 and around the same time the source was observed with the VLBA interferometer. VERA observations have been performed on with a three-month interval throughout 2011. Unfortunately, we found that the source was not detected during several epochs. We observed the source again with the Onsala 20m in October 2011. This observation confirmed that the maser emission from the source has disappeared. We find no correlation with the visual light curve of the source (tracing the stellar pulsations, Fig. 2) and the reason for the sudden decline is not understood. The emission from the source is continuously being monitored.

References

Ramstedt, S., Maercker, M., Olofsson, G., Olofsson, H., & Schöier, F. L. 2011, *A&A*, 531, 148
Nakashima, J. & Deguchi, S. 2007, *ApJ*, 669, 446

T5:

Maser Surveys *Chair: Jessica Chapman*

Cosmic Masers - from OH to H$_0$
Proceedings IAU Symposium No. 287, 2012
R.S. Booth, E.M.L. Humphreys & W.H.T. Vlemmings, eds.

© International Astronomical Union 2012
doi:10.1017/S1743921312007119

SiO Maser Surveys of Nearby Miras and their Kinematics in the Galaxy

Shuji Deguchi[1,2]

[1]Nobeyama Radio Observatory, National Astronomical Observatory,
Minamimaki, Minamisaku, Nagano 384-1305, Japan
email: deguchi@nro.nao.ac.jp

[2]Graduate University for Advanced Studies, National Astronomical Observatory,
Minamimaki, Minamisaku, Nagano 384-1305, Japan

Abstract. I review recent developments of SiO maser observations of miras, and discuss their kinematics in the Galaxy. They are deeply related to problems in other fields, such as the noncircular motion of Galactic spiral arms, streaming motions of stars in the solar neighborhood, and the interface between the interstellar and circumstellar medium revealed by Far-IR imaging of some miras, as well as the bar structure of the Galaxy. Recent automated surveys of variable stars in optical bands have increased the number of red variables enormously. Most of these stars were not observed in the radio before because of their blue colors. We have obtained radial velocities of these long-period variables with SiO masers. The longitude-velocity diagrams have revealed interesting deviations of stellar motions from the circular motion of the Galaxy, some of which show spatial motion similar to that of the Hercules group of stars. Important developments of SiO maser observations have been made for symbiotic stars, one of which showed a nova eruption with concurrent γ ray emission in 2010 (V407 Cyg). SiO masers have been monitored for this star after the nova eruption. I summarize the SiO maser observations of individual stars also.

Keywords. masers, stars: kinematics, stars: late-type, surveys

1. Introduction

Almost 40 years have passed since the discovery of SiO masers in miras (Kaifu *et al.* 1975). So far, we have performed many maser surveys of silicon monoxide at ~43 GHz (see a summary in Deguchi 2007). We have surveyed approximately 4000 stars and detected half of them. Why do we need to add more? The previous surveys were made mostly for infrared bright, optically faint stars located at relatively distant places in the Galactic disk or bulge. At this time, we are focused more on the nearby blue objects, i.e., optically well observed mira variables, because a lot of new red variables were catalogued recently by automated optical photometric surveys. Therefore, we have made a new survey of nearby mira variables. This subject is related to the Galactic kinematics of stars, streaming motions of stars in the solar neighborhood, and the bar structure of the Milky Way Galaxy, or possibly related to streaming motions of stars in merging dwarf galaxies (Deguchi *et al.* 2007a; Deguchi *et al.* 2010).

The stars we can observe using SiO masers are mostly aged stars, i.e., stars in the Asymptotic-Giant-Branch phase of stellar evolution. Their ages are typically a few Giga years old. Because they rotate in our Galaxy a few tens of times, their motions are perturbed by the gravitational potential of the bulge bar periodically. As a result, their spatial motions deviate from the circular motions of the Galaxy considerably.

SiO masers occur in the rotational transitions of vibrationally excited states of silicon monoxide. The masers require hot gas of more than 1000 K. They are emitted in the very inner circumstellar shell of late-type stars, i.e, at 2–3 stellar radii. They have two strong lines at 42.820 and 43.122 GHz, the $J = 1$–0 $v = 1$ and 2 transitions. The two transitions are observable simultaneously by some radio telescopes, and for many stars, the intensities of these two transitions are approximately equal (see Nakashima & Deguchi 2007). It is nice to observe both transitions and secure detections. In our previous surveys, we took our sample from infrared catalogs, for example, IRAS, MSX, and 2MASS, Spitzer Glimpse etc. (Deguchi *et al.* 2004). However, recently the Automated Optical Sky Surveys have been made and several catalogs were published. One is the NSVS: Northern Sky Variability Survey (Williams *et al.* 2004), and another is the ASAS: All Sky Automated Survey, which was mainly focused on the Southern sky (Pojmanski *et al.* 2012). These two catalogs listed up ~ 50 thousands of variable stars, most of which are new. We have chosen our targets from the long-period variable stars in these catalogs, and have made a survey of them (Deguchi *et al.* 2012). We have also made a small complementary survey based on the AAVSO variable star catalog, which will be published in a future paper.

2. Overview

We surveyed about 580 stars and detected 330 of them in total. Our sample consists of optical miras and semiregulars. The detection rate of semiregulars is very low, i.e., about 10%, so that it is negligible in a discussion. A histogram of the period of the sample (see a similar one in figure 3 of Deguchi *et al.* 2012) indicates that the SiO maser detections peak at the period bin between 300 and 400 days, and that SiO maser emitting miras have slightly longer periods on average than those with no detections. The color-magnitude diagrams show that most of the stars have bluer colors than the usual sample (see a similar one in figure 2 of Deguchi *et al.* 2012). The median value of the color, $H - K$ or logarithmic ratio of IRAS 25 to 12 micron flux density falls at a much bluer part than that of the previous samples.

The bluer color of the star indicates that the circumstellar envelope is optically thin in the infrared bands and they are stars with very low mass-loss rates. However, we detected SiO masers at a detection rate similar to those in the previous surveys. Does this mean that the SiO maser intensity of this sample is stronger than that of the previous red

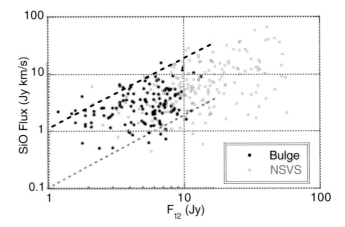

Figure 1. IRAS 12 μm intensity versus SiO maser integrated intensity ($J = 1$–0 $v = 2$). The red filled circles indicates the current sample of nearby miras, while black dots indicate SiO maser sources in the Galactic bulge taken from Jiang *et al.* (1995). See a similar plot in Jewell *et al.* (1991).

samples? No, it is not. In Figure 1, we plotted maser intensity against IRAS 12 micron flux density, and compared the present sample with the bulge stars. It indicates that the upper boundary of the SiO maser intensity is proportional to the IRAS 12 micron flux density (with an inclination of approximately unity). This means that the intrinsic SiO maser intensity does not vary between the present optical mira sample and the bulge IRAS sample, if normalized by IRAS 12 micron flux density.

3. Kinematics of SiO maser sources

Figure 2 shows the longitude-velocity diagram of the current sample of optical miras. Filled circles are disk miras at low Galactic latitudes and unfilled are miras at high-latitudes. Note that these stars are bright optical miras, which are relatively close, i.e., at a distance within about 3 kpc from the Sun. Distances were well estimated using the Period-Luminosity relation for this sample of miras with a period shorter than 400 d (Whitelock *et al.* 2008) and for periods longer than 400 d (Ita & Matsunaga 2011). Several groups of stars in this diagram show characteristic deviations from circular rotation, which are indicated by green ellipses. One group at a Galactic longitude of about 30° shows a radial velocity of -50 km s^{-1}, which deviates by 100 km s^{-1} from Galactic circular rotation. The other group is located at a longitude range between 90° to 140° at a radial velocity of -60 km s^{-1}, and the third group is located at $V_{lsr} = 60$ km s^{-1} at a longitude of about 30°. The two broken curves in Figure 2 are the expected velocities for known moving groups: the Hercules moving group and the Arcturus moving group. The Arcturus moving group is considered to be a remnant of a hypothetical dwarf galaxy merged into our galaxy in the past.

The same diagram, if overlaid on the CO $J = 1$–0 map (Dame *et al.* 2001), contrasts the difference between the distribution and the spiral arms (see Figure 6 of Deguchi *et al.*

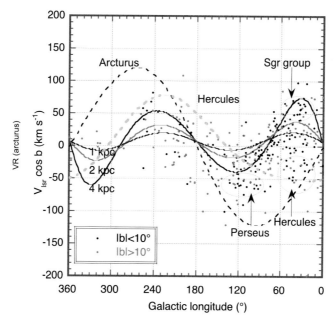

Figure 2. Longitude-velocity diagram of nearby miras with SiO masers. Thick curves indicate the expected radial velocities of stars at distances of 1, 2, and 4 kpc from the Sun for a circular orbit with a flat rotation curve of 220 km s^{-1} and a Sun-Galactic center distance of 8 kpc. The green ellipses indicate the groups of stars deviant from circular rotation.

Table 1. Distance statistics for Perseus and Sgr deviant groups of stars

Group (number)	Quantity	D_k (kpc)	D_{Lc}^{*} (kpc)
Perseus	average	5.27	2.19
(18)	standard dev.	1.04	1.26
	probability[†]	—	$< 10^{-4}$
Sgr	average	3.46	2.65
(17)	standard dev.	0.62	1.22
	probability[†]	—	0.02

Notes:

[*]: luminosity distances calculated from the period–luminosity relations given by Whitelock et al. (2008) and Ita & Matsunaga (2011) for the periods below and above 400d, respectively.

[†]: a probability of the Student's t-test for the averages of two sets of D_k and D_{Lc} being generated by the same distribution function.

2012). The first group, that deviates about 100 km s^{-1} from circular rotation is called here the Hercules group, because this velocity is explained by the known moving group called the Hercules moving group originally discovered by Olin Eggen, or later a more extended one for blue short-period miras by Feast & Whitelock (2000). The second group is called the Perseus group, because the distribution of these stars in the l–v diagram fall on the Perseus spiral arm feature of CO emission. It has been well known that the Perseus spiral arm has a 30 km s^{-1} inward motion from the circular rotation of the Galaxy (Xu et al. 2006; Reid et al. 2009; Asaki et al. 2010). We calculated the distances of these stars using the period-luminosity relation and obtained an average distance of about 1.9 kpc. So that it is consistent with the Perseus spiral arm. Another deviant group is called the Sgr group, which deviates from the circular rotation by about 30 km s^{-1}. The average luminosity distance is about 2.7 kpc, while the kinematic distance is 3.5 kpc. The difference is statistically significant with a 98 percent confidence level as shown in Table 1. Therefore, the Sgr group is also explained by the noncircular motion.

The largest velocity deviation from circular motion is the Hercules group of stars. It is explained by the resonance effect at the outer Lindblad resonance. The bar gravitational potential moves at the bar-pattern speed, which is considered to be approximately 55 km s^{-1} kpc^{-1}, while the speed of the Sun is about a half of the bar pattern speed. Therefore, the bar potential minimum passes the Sun every 0.2 Gyr or so (twice if the bar is symmetric). If we take a rotational coordinate which is moving with the average circular motion of the Galaxy, the star slightly deviated from the circular rotation exhibits epicyclic motion in an ellipsoidal orbit in the rotational coordinate frame. The outer Lindblad resonance occurs at the radius of 7 to 8 kpc, which makes the major axis of the ellipse toward the Sun (see Appendix 3 of Deguchi et al. 2010). Therefore, we observe a large negative radial velocity at a Galactic longitude of about 30 degrees. This occurs for all of the stars as far as the stars with ages more than the rotational period of the bar, say about 0.2 Gyr. It has been known that a few percent of stars in the solar neighborhood belong to this moving group (Famaey et al. 2005). This is like the winter Olympic game, called "half pipe" Here a player is a star, and the half pipe is the bar potential. In the case of the Galaxy, the gravitational potential is rotating with a pattern speed, so that the pipe must be rotating. At the radius of the Sun, the bar gravitational potential is represented by a very shallow dip. The player can have a large outward motion if it is resonant with the rotation of the bar. This is a simple explanation for the Hercules group. However, for the other two groups, the stars are in a spiral arm, and they have longer periods, and they are much younger compared with the time scale of the bar-pattern rotational period. So that gas friction and magnetic field make the motion of spiral arms slightly different. We need a more sophisticated explanation for the other two groups of stars.

A 3D N-body numerical simulation of the stars in a barred spiral galaxy has been made by a number of people (e.g., see Baba *et al.* 2009; Monari, *et al.* 2011). Large deviations from circular motion can be found on the UV plane in this simulation, as well as where further minor groups of stars are.

4. Implications

The kinematics of nearby miras are directly related with the proper motion measurements by VLBI astrometry. Systematic observations have been made by the Japanese VLBI system, called VERA (Kobayashi *et al.* 2008). This telescope measures angular separations between a maser source and a continuum reference source with two-beam receivers and obtains the parallaxes and proper motions of maser sources. One example is the case of VY Canis Majoris observed by Choi *et al.* (2008). With this telescope, the accurate parallax and proper motions have been obtained for about 20 miras, and more will come in the near future (Honma 2012). Though current progress is limited to the very bright objects only, they will reveal some spatial motions of SiO maser sources, and will give precise complete 3D spatial motions of these stars in the Galaxy, as well as the motion of the maser clouds in the spiral arms.

Another related issue comes from the field of infrared astronomy. Recently infrared satellites, Spitzer and Akari, took far-infrared images of some miras, which have a large-angular-size circumstellar envelope. I mention here only an example of R Cas, which is one of the strong SiO maser sources (Cotton *et al.* 2004). In the far infrared images between 60 and 160 microns, this star exhibts an extended far-IR emission of the size of about a few arcminutes (Ueta *et al.* 2010). In their FIR images, the stellar position does not fall at the center of the emission, but slightly shifted to the East. In fact, some stars show a clear bow-shock. This occurs due to the motion of the star relative to interstellar medium. From a number of these observations, we can obtain the stellar motions in the interstellar medium (Ueta *et al.* 2011). We can also obtain the proper motions of the central star from VLBI measurements. Therefore, we get the proper motion of the interstellar medium in the Galaxy.

I have to mention a recent development made by the Korea Yonsei 21m telescope. Cho *et al.* (2010) made an interesting survey of symbiotic stars using the 43 GHz SiO maser lines. They can observe 43 GHz SiO and 22 GHz H_2O masers simultaneously. The symbiotic star is a spectroscopic binary. If the counterpart is a mira, it has an extended envelope and the secondary is often a white dwarf showing high temperature characteristics. They found SiO masers in 27 stars in 47 surveyed stars. The SiO masers usually show a central peak in the spectrum, and double peaks are relatively rare. However, in the case of the symbiotic stars, SiO maser spectra often seem to show double peaks. This is likely due to gas flow in a binary system.

5. Individual objects

5.1. *V407 Cyg*

I would like to talk about individual objects, especially about the V407 Cygni, which is a SiO maser source, but made a nova eruption in 2010 March. This is the first SiO maser source which is observed to be a classic nova, and this is also the first nova ever detected in γ rays. The optical light curve indicates that it is a typical classical nova; a rapid rise and gradual decline in the light curve. A very interesting aspect of this nova is that the H_α line showed a very wide width for the first one week (Munari *et al.*2011). The half width was initially up to 3000 km s^{-1}, and gradually decreased down to several 100 km s^{-1} by three weeks after the eruption. The Fermi-LAT (Large Area Telescope) detected γ ray toward this object concurrently (Abdo *et al.* 2010).

Classical novae are a thermonuclear run-away, which occurs on the surface of a white dwarf in a binary system. The donor star in this case is a mira-type variable. The separation between the two stars is enough large to compose a detached binary. The shock which has started from the surface of the white dwarf expands first to the radius of about 1 AU and makes an optically thick hot envelope around white dwarf, when the optical light curve becomes maximum. It gradually interacts with the cool envelope of the AGB star, when X-rays are emitted. It took about 20 days in the case of V407 Cyg. The γ-ray emission appeared in the first 1 week or so, and died out quickly. The X-ray emission comes later. It has been known that the donor is a mira of a period of 745 days with SiO masers. It is interesting to know how the SiO maser emitting regions of the donor star are influenced by the nova outburst. If we assume the shock speed is 1000 km s^{-1}, it crosses a distance of 10 AU in 2 weeks. It is a good time scale for measuring the variation of SiO maser intensities in this star.

In fact, three months before the nova eruption, this object was observed with the Yonsei 21m telescope (Cho *et al.* 2010). The spectra exhibited clear emission in the SiO $J = 1$–0 $v = 1$ and 2 lines and the H_2O 6_{16}–5_{23} line. Later, we recognized that this is the source in which we detected SiO masers 6 years ago (Deguchi *et al.* 2005a). A week after the eruption, we luckily had observation time at the Nobeyama 45m telescope. We detected SiO maser emission in the $J = 1$–0 $v = 2$ line, but not in the $v = 1$ line. Since then, we continued to observe the time variation of SiO masers. The spectral variation of the $v = 2$ emission in the first two months is shown in figure 2 of Deguchi *et al.* (2011). Initially the profile exhibited the emission at the red-shifted side only. It continued for

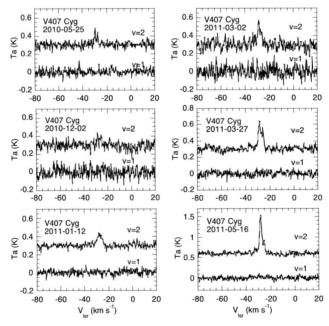

Figure 3. Spectral variations of SiO masers from V407 Cyg two months after the 2010 nova outburst. The SiO $J = 1$–0 $v = 1$ emission has never been detected after the nova outburst.

the first one week and the intensity decreased very much. Then, it disappeared 20 days after the nova eruption (Nelson *et al.* 2012). I thought the emission would never come back for a year or so. Contrary to my expectation, new emission appeared at the lower velocity a month later.

How do we interpret this result? We consider the geometry of the binary as shown in Figure 3 of Deguchi *et al.* (2011). The SiO masers are usually emitted at about 2–3 stellar radii from the red giant. The nova shock is decelerated due to its ploughing of the circumstellar material of the mira and is going down gradually. Initially the material released by the explosion was estimated to be about 10^{-6} $M_\odot$ and the shock speed is about 3000 km s^{-1} (Abt *et al.* 2010; Orlando & Drake 2012). In 12 days, it reaches a distance of about 20 AU from the white dwarf. In such a case, most of the SiO masers near the mira vanish except in the shadow of the mira. Therefore, it emits a receding (red-shifted) part of SiO masers. After 3 weeks, all the SiO maser terminates because soft X-ray emission penetrated into the whole envelope, and quenched SiO masers. We can explain the disappearance of SiO masers very well with this model, except one fact that the lower velocity SiO maser appeared a month later.

We monitored the SiO maser emission since then. In Figure 3, I show the SiO maser spectra one year after. The $v = 2$ emission became stronger now. It probably means that the region of the SiO $v = 2$ emission is replenished by the gas from the inner part, and the maser intensity recovered after one year. However, we have not yet detected any $v = 1$ emission, nor H$_2$O masers at 22.2 GHz so far. It indicates that the $v = 1$ emission region and H$_2$O emission region are located at a much outer side of the envelope than the $v = 2$ emission. Therefore, it takes more time to refresh with unaffected gas from inside to reach the $v = 1$ masing region. We recognized that the radial velocity of $v = 2$ emission is gradually increasing with time. This is likely a gas motion due to pulsation of the mira, or equatorial flow of the gas around the binary.

We also tried to detect CO emission from the star. When we pointed at the star, we found broad CO emission, which looked like an inverse-parabolic line shape from the expanding envelope. However, the radial velocity is shifted by about 5 km s^{-1} from the stellar velocity. Therefore we mapped the CO emission using the $40''$ grid, 5 point

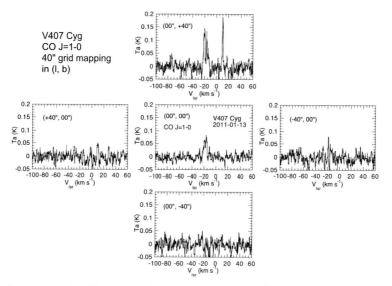

Figure 4. Spectra of the CO $J = 1$–0 line obtained with $40''$ grid mapping toward V407 Cyg.

mapping mode. The results are shown in Figure 4. We found that the strongest emission comes from $40''$ north of the star. It indicates that this CO emission is an interstellar component toward this direction. Another sharp interstellar component was also seen at $40''$ north. So far, we have not yet found any CO emission from this star.

5.2. R Aqr

The most famous, brightest symbiotic star is R Aqr. The circumstellar maser of this star has been well-investigated (see, e.g., Cotton *et al.* 2006). A 7mm radio continuum mapping of this star simultaneously with the SiO maser line was made with the VLA (Hollis *et al.* 1997). The overlaid map showed that the radio continuum comes from a secondary (probable white dwarf) 55 mas apart. Kamohara *et al.*(2010) measured the trigonometric parallax and proper motions of this star with VERA using a phase reference source. The white dwarf is likely receding from us. The proper motion is about 45 mas yr^{-1} to the south east, and the distance is 210 pc, which indicates a spatial motion of about 50 km s^{-1} in the direction away from the Galactic plane. These values obtained with VLBI astrometry coincide well with the past measurements by Hipparcos. It is likely that geometry of the white dwarf and SiO maser star is revealed in these observations. This should provide some hint on the geometry of V407 Cyg previously noted, or symbiotic stars in general.

5.3. V838 Mon

We have also to pay attention to the totally different idea of stellar evolution of red giants, which has been proposed recently for some peculiar stars. The R CrB-type stars are supergiants with hydrogen-deficient carbon-rich atmospheres. Recently they have been considered to be a product of two merged white dwarfs (Tisserand 2011). We have previously known another living example of a merged star, V838 Mon, a famous star with a light echo. One model merging two main-sequence stars with 5–10 $M_\odot$ and 0.1–0.5 $M_\odot$ can create this supergiant (Soker & Tylenda 2003; Tylenda & Soker 2006). We detected the SiO masers of this star in 2005, five years after its nova eruption (Deguchi *et al.* 2005b). Nearby CO emission, which is somewhat related to but is not from the circumstellar envelope of the star, has been mapped (Kamiński *et al.* 2011).

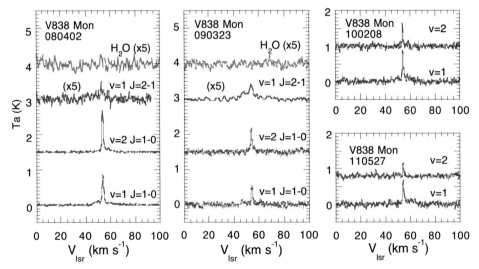

Figure 5. Spectral variations of SiO maser emission from V838 Mon after 2008. The date of observations is shown on the upper left under the star name in *yymmdd* format.

Since then we have monitored this star every year (Deguchi *et al.* 2007b). Furthermore, we tried to detect the H_2O maser and SiO maser lines at 22 and 86 GHz. In 2009, we finally detected clear emission of the SiO $J = 2$–1 $v = 1$ line at 86.2 GHz (Fig. 5). Therefore the $J = 2$–1 $v = 1$ emission has formed recently in this star. For this star, a sudden mass loss started after the merging event when the supergiant was created. Therefore, our late detection of the $J = 2$–1 $v = 1$ line indicates that the maser emitting region of the $J = 2$–1 emission is much on the outer side of the $J = 1$–0 emission region. We also found that the intensity of the $J = 1$–0 $v = 2$ is decreasing gradually. How long does it take for the disappearance of these lines? This is an interesting problem, if the stellar merging can create OH/IR or SiO maser stars without an AGB phase of stellar evolution. If this time scale is long enough, say, more than 100 years, about 0.1 percent of the observed SiO maser stars can be made in the merging process (see the discussion given in Deguchi *et al.* 2005b). So that to measure how long it continues is an important issue.

6. Conclusion

In the last 10 years a number of unusual phenomena have been observed in AGB and post-AGB stars using maser lines. Most of which are more or less related to binary phenomena. "Water Fountains" may also be included in this category. A nice example of binary nature is the case of IRAS 18286−0959. The H_2O maser proper motion map of this star shows clear double jets, and the axes of both jets seem to show precession on a time scale of ∼ 10 years (Yung *et al.* 2011). This is certainly a binary phenomenon. We are not very sure whether the jet is created in an accretion disk around the compact star, or in the atmosphere of the AGB star, or in the equatorial disk in the binary. The previous example of R Aqr clearly indicates that the jet is created by an accretion disk. Therefore, it is highly likely that these jets are created in the accretion disk of the compact star associated with the AGB or post-AGB star. We have also to pay attention to the totally different idea of the stellar evolution of red giants, which has been proposed recently for some peculiar stars such as V838 Mon. The binary phenomena and stellar mergers seem to produce rare but interesting deviations in stellar evolution. In contrast, understanding the kinematics of evolved stars in the Galaxy seems to be considerably difficult, but the field has matured sufficiently to explore this issue using masers as a tool. Information on the complete 3D velocities of stellar motions is absolutely necessary and will be obtained by future VLBI measurements.

The author thanks all the collaborators of their SiO maser survey observations for preparing this manuscript.

References

Abt *et al.* 2010, *Science*, 329, 817 (and Supporting Online Material)

Asaki, Y., Deguchi, S., Imai, H., Hachisuka, K., Miyoshi, M., & Honma, M. 2010, *ApJ*, 21, 267

Baba, J., Asaki, Y., Makino, J., Miyoshi, M., Saitoh, T. R., & Wada, K. 2009, *ApJ*, 706, 471

Cho, S. & Kim, J. 2010, *ApJ*, 719, 126

Choi, Y. K., Hirota, T., Honma, M., Kobayashi, H., Bushimata, T. *et al.* 2008, *PASJ*, 60, 1007

Cotton, W. D., Mennesson, B., Diamond, P. J., Perrin, G. Coudé du Foresto, V. *et al.* 2004, *A&A*, 414, 275

Cotton, W. D., Vlemmings, W., Mennesson, B., Perrin, G., Coudé du Foresto, V. *et al.* 2006, *A&A*, 456, 339

Dame, T. M., Hartmann, D., & Thaddeus, P. 2001, *ApJ*, 547, 792

Deguchi, S. 2007, *IAUS*. 242, p200 (Ed. Chapman, J. M., & Baan, W. A.; Cambridge Univ.)

Deguchi, S., Fujii, T., Ita, Y., Imai, H., Izumiura, H., *et al.* 2007a, *PASJ*, 59, 559

Deguchi, S., Koike, K., Kuno, N., Matsunaga, N. Nakashima, J., & Takahashi, S. 2011, *PASJ*, 63, 309

Deguchi, S., Matsunaga, N., & Fukushi, H. 2005, *PASJ*, 57, L25

Deguchi, S., Matsunaga, N., & Fukushi, H. 2007b, *ASPC*, 363, 81

Deguchi, S., Sakamoto, T., & Hasegawa, T. 2012, *PASJ*, 64, 4

Deguchi, S., Shimoikura, T., & Koike, K. 2010, *PASJ*, 62, 525

Famaey, B., Jorissen, A., Luri, X., Mayor, M., Udry, S., Dejonghe, H., & Turon, C. 2005, *A&A*, 430, 165

Feast, M. W. & Whitelock, P. A. 2000, *MNRAS*, 317, 460

Hollis, J. M., Pedelty, J. A., & Lyon, R. G. 1997, *ApJ*, 482, L85

Honma, M. 2012, a talk given in this conference "Maser astrometry with VERA and Galactic structure"

Ita, Y. & Matsunaga, N. 2011, *MNRAS*, 412, 2345

Jewell, P. R., Snyder, L. E., Walmsley, C. M., Wilson, T. L., & Gensheimer, P. D. 1991, *A&A*, 242, 211

Jiang, B. W., Deguchi, S., Izumiura, H., Nakada, Y., & Yamamura, I. 1995,*PASJ*, 47, 815

Kamiński, T., Tylend, R., & Deguchi, S. 2011, *A&A*, 529, A48

Kaifu, N. Buhl, D. & Snyder, L. E. 1975, *ApJ* 195, 359

Kamohara, R., Bujarrabal, V., Honma, M., Nakagawa, A., Matsumoto, N. *et al.* 2010, *A&A*, 510, 69

Kobayashi, H., Kawaguchi, N., Manabe, S., Shibata, K. M., Honma, M. *et al.* 2008, *IAUS*, 248, 148

Nakashima, J., Deguchi, S., Imai, H., Kemball, A., & Lewis, B. M. 2011, *ApJ*, 728, 76

Orlando, S. & Drake, J. J. 2012, *MNRAS*, 419, 2329

Mohamed, S. & Podsiadlowski, P. 2007, Proc. Asymmetrical Planetary Nebulae IV. #78

Monari, G., Antoja, T., & Helmi, A. 2011, *arXiv*1110.4505

Munari, U., Joshi, V. H., Ashok, N. M., Banerjee, D. P. K., Valisa, P., *et al.* 2011, *MNRAS*, 410, L52

Nakashima, J. & Deguchi, S. 2007, *ApJ*, 649, 446

Nelson, T., Donato, D., Mukai, K., Sokoloski, J., & Chomiuk, L. 2012, *arXiv*1201.5643

Pojmanski, G., Maciejewski, G., Pilecki, B., & Szczygiel, D. 2006, *yCat*, 2264

Reid, M. J., Menten, K. M., Zheng, X. W., Brunthaler, A., Moscadelli, L. *et al.* 2009, *ApJ*, 700, 137

Soker, N. & Tylenda, R. 2003, *ApJ*, 582, L105

Tisserand, P. 2011, *arXiv*1110.6579

Tylenda, R. & Soker, N. 2006, *A&A*. 451, 223

Ueta, T., Stencel, R. E., Yamamura, I., Geise, K. M., Karska, A. *et al.* 2010, *A&A*, 514, A16

Ueta, T., 2011, *ASPC*, 445, 295

Williams, P. R., Wozniak, S. J., Vestrand, W. T., & Gupta, V. 2004, *A. J.* 128, 2965

Whitelock, P. A., Feast, M. W., & van Leeuwen, F. 2008, *MNRAS*, 386, 313

Xu, Y., Reid, M. J., Zheng, X. W., & Menten, K. M. 2006, *Science*, 311, 54

Yung, B. H. K., Nakashima, J., Imai, H., Deguchi, S., Diamond, P. J., & Kwok, S. 2011, *ApJ*, 741, 94

Cosmic Masers - from OH to H_0
Proceedings IAU Symposium No. 287, 2012
R.S. Booth, E.M.L. Humphreys & W.H.T. Vlemmings, eds.

© International Astronomical Union 2012
doi:10.1017/S1743921312007120

Water maser follow-up of the Methanol Multi-Beam Survey.

Anita Titmarsh[1,2], Simon Ellingsen[1], Shari Breen[2], James Caswell[2] and Maxim Voronkov[2]

[1] School of Mathematics and Physics, University of Tasmania,
Private Bag 37, Hobart, Tasmania 7001, Australia
email: Anita.Titmarsh@utas.edu.au

[2] CSIRO Astronomy and Space Science,
PO Box 76, Epping, NSW 1710

Abstract. The Australia Telescope Compact Array has been used to observe all the 603 6.7 GHz methanol masers detected in the Methanol Multi-Beam survey between $l = 310° - 20°$. To date we have measured positions with arcsecond accuracy for all the observations in the $l = 6° - 20°$.

Keywords. masers, stars: formation, radio lines: ISM.

1. Introduction

Common masers such as the 22 GHz water and 6.7 GHz methanol masers are important tools for studying the formation of high-mass stars. They are common, intense, and being observable at radio frequencies they allow us to probe deep into the heart of the dusty molecular envelope where high-mass stars are forming. Water masers are the most common maser species known, tracing shocked gas, outflows and dense circumstellar shells around evolved stars. They are found at sites of both low and high-mass star formation. In contrast, some other species (e.g. 6.7 GHz methanol masers) are observed exclusively at sites of high-mass star formation (Minier *et al.*, 2003; Xu *et al.*, 2008). The presence and/or absence of various maser transitions are thought to trace different evolutionary stages in their formation (Breen *et al.*, 2010); however, further work is required to quantify the timescale over which the very important water maser transition occurs.

Previous targeted observations of water masers have been carried out with single dish telescopes or have searched 'special' sources. For example Szymczak *et al.* (2005) observed 79 6.7 GHz methanol masers with the Effelsberg telescope achieving a spatial resolution of ~40 arcseconds. Because water masers are very common around regions of star formation, a positional accuracy of at least a few arcseconds is required to reliably identify if the water and methanol masers are coincident with the same object. Beuther *et al.* (2002) observed a sample of young stellar objects with the VLA (achieving the necessary spatial resolution) and found a detection rate of 62%. However, their sample of 6.7 GHz methanol masers were chosen using *IRAS*-based selection criteria, which are known to miss a substantial fraction of 6.7 GHz methanol masers (Ellingsen *et al.*, 1996).

The Methanol Multi-Beam (MMB) survey is an unbiased survey of the Galactic Plane $0° < l < 360°$ and $b \pm 2°$ for the 6.7 GHz methanol 6.035 GHz excited OH. The southern portion of the MMB survey was completed in March 2009 (Green *et al.*, 2009) and $l = 186° - 20°$ have been published (Caswell *et al.*, 2010, 2011; Green *et al.*, 2010, 2012). The Parkes 64 m dish performed the initial search of the southern Galactic Plane and accurate positions for the sites of maser emission detected here were then followed up with the ATCA.

Here we report preliminary results of follow-up 22 GHz water maser observations towards the MMB detections. Although these observations are targeted (rather than an unbiased search), we will be able to combine the results of these sensitive, high resolution observations, with the less sensitive, but statistically complete HOPS survey (Walsh *et al.*, 2011), to properly quantify the water maser transition with the maser-based evolutionary scheme.

2. Water maser observations

The observations were made between November 2010 and August 2011 with the ATCA in various antenna configurations. The observations on the 2nd and 3rd of November 2010 were carried out in the H214 array configuration, and the observations on the 9th and 10th of August 2011 were in the H168 configuration. These hybrid array configurations have both East - West and North - South baselines and are better for observations of equatorial sources, although at the cost of a larger synthesised beam. The primary beam of the compact array at 22 GHz is 2.1 arcminutes and the synthesised beams of the H214 and H168 configurations are ~ 9.6 and ~ 12.4 arcminutes respectively. It is important to realise that the astrometric accuracy for an connected-element interferometer such as the ATCA depends upon both the size of the synthesised beam, and the quality of the phase calibration. For observations at 22 GHz observations with longer baselines (and hence smaller synthesised beams), will not necessarily lead to better astrometric accuracy if the quality of the phase calibration for those longer baselines is poor. In good observing conditions the absolute astrometric accuracy for the ATCA is around 0.4 arcseconds (set by the astrometric accuracy of the phase calibrators and the degree to which they are point sources for the array configuration and the frequency of observation). For our observations the astrometric accuracy is in the range 0.4 - 2.0 arcseconds, as some of the observations were made in relatively poor weather. This accuracy is comparable to the maximum angular extent observed for water maser clusters by Breen *et al.* (2011).

Targeted observations of each of the 6.7 GHz methanol masers were performed with at least four observations of 1.5 minutes duration spread over an hour angle range of 6 hours to ensure sufficient u-v coverage for imaging. The sensitivity in an individual spectral channel for these observations ranged from $\sim$40 mJy in good weather conditions to $\sim$80 mJy in poor weather. The Compact Array Broadband Backend (CABB) with two zoom bands of 64 MHz and 32 kHz resolution were used to observe the 22 GHz water maser transition in the first zoom band and the the ammonia (1,1) and (2,2) transitions in the second zoom band. The velocity coverage in the zoom bands was > 800 kms^{-1} with velocity resolution of 0.42 kms^{-1}. 2 × 2 GHz continuum bands were also available for the August 2011 observations.

3. Results for $l = 6° - 20°$

Here we report the water maser observations for the Galactic longitude range $l = 6° - 20°$. Approximately 40% of the 6.7 GHz methanol masers in this region have an associated water maser. The criteria we used to determine association between the water and methanol masers was if they have an angular separation of less than two arcseconds. This criterion was used for consistency with previous high resolution, large surveys of water masers (eg. Breen *et al.* 2010). Our detection rate is lower than other targeted surveys of water masers such as Szymczak *et al.* (2005) who used the Effelsberg telescope

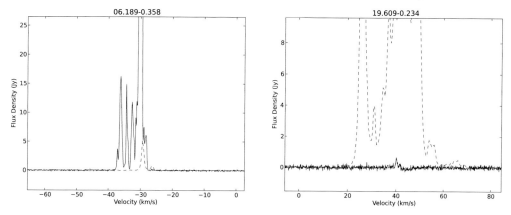

Figure 1. Spectra of associated 22 GHz water masers taken with the ATCA (shown with dashed lines) and 6.7 GHz methanol masers from the MMB survey (shown with solid lines).

(half power beam width of 40 arcseconds at 22 GHz) and found 52% of the 6.7 GHz methanol masers had an associated water maser. Since our survey is also more sensitive we would expect to have a higher detection rate. However, if we relax the association criteria to include masers within 10 arcseconds then our detection rate increases to 79%.

Comparing the 6.7 GHz methanol masers with and without associated water masers we found that the average peak flux densities of the methanol masers were higher in those without water masers (average of 92 Jy for sources without water and 36 Jy for those with water). However, this result is strongly affected by the famous and unusual source G09.621+0.196 which had a peak flux density of 5196 Jy at the time of the MMB observations, it has also been observed to have periodic flares (eg. Goedhart *et al.* 2003). Disregarding this source changes the result so methanol masers with associated water masers have greater average peak flux densities (24 Jy for sources without water and 49 Jy with water) which would be expected if water masers occurred at a later stage during star formation.

Breen *et al.* (2010) showed that the luminosity of the 6.7 GHz methanol masers increases as they evolve. Comparing the 6.7 GHz methanol maser peak flux densities with their associated water maser peak flux densities (Fig. 2) suggests a correlation between the flux densities and is consistent with the scenario that water masers also increase in intensity as they evolve. This is a loose correlation, and the Pearson correlation coefficient is 0.37. However, this is strongly affected by one very strong source and without it the correlation coefficient reduces to 0.26. The sources observed in this survey with the most extreme differences between water and methanol peak flux densities are shown in Figure 1. Source G06.189-0.358 has a strong methanol maser (221.6 Jy) and weak water maser (5.2 Jy) while G19.609-0.234 is the reverse with strong water (33.4 Jy) and weak methanol (1 Jy).

Water masers are well known for sometimes exhibiting high velocity emission, offset from the systemic emission of the region by 100 kms^{-1} or more. Figure 3 shows a comparison of the velocities of the peak emission of the two maser species. In most masers the water and methanol masers have peak velocities within a few kms^{-1} of each other. This suggests that for water masers associated with 6.7 GHz methanol masers the peak emission is in most cases close to the systemic velocity of the region and any high-velocity emission is generally weaker. One notable exception is the methanol maser source G18.999-0.239. The 6.7 GHz methanol emission peaks at 69.4 kms^{-1} whereas the

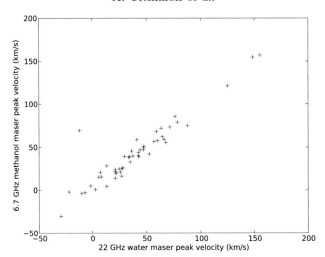

Figure 2. 22 GHz water maser peak velocity (kms^{-1}) vs. 6.7 GHz methanol maser peak velocity (kms^{-1}).

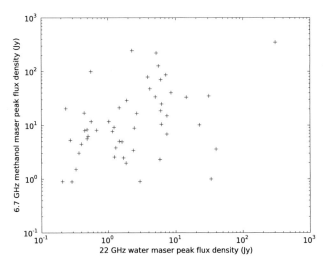

Figure 3. 22 GHz water maser peak flux density (Jy) vs. 6.7 GHz methanol maser peak flux density (Jy).

water maser emission peaks at -11.8 kms^{-1}. They are only separated by 1.4 arcseconds and neither maser species shows any emission around the others peak velocity.

References

Breen, S. L., Caswell, J. L., Ellingsen, S. P., & Phillips, C. J. 2010, *MNRAS*, 406, 1487
Breen, S. L. & Ellingsen, S. P. 2011, *MNRAS*, 416, 178
Beuther *et al.* 2002, *A&A*, 390, 289
Caswell *et al.* 2010, *MNRAS*, 404, 1029
Caswell *et al.* 2011, *MNRAS*, 417, 1964
Ellingsen *et al.* 1996, *MNRAS*, 280, 378
Goedhart S., Gaylard M. J., & van der Walt D. J. 2003, *MNRAS*, 339, L33
Green *et al.* 2009, *MNRAS*, 392, 783

Green *et al.* 2010, *MNRAS*, 409, 913

Green *et al.* 2012, *MNRAS*, 420, 3108

Minier, V., Ellingsen, S. P., Norris, R. P., & Booth, R. S. 2003, *A&A*, 403, 1095

Szymczak, M., Pillai, T., & Menten, K. M. 2005, *A&A*, 434, 613

Walsh *et al.* 2011, *MNRAS*, 416,1764

Xu *et al.* 2008, *A&A*, 485, 729

Cosmic Masers - from OH to H$_0$
Proceedings IAU Symposium No. 287, 2012
R.S. Booth, E.M.L. Humphreys & W.H.T. Vlemmings, eds.
© International Astronomical Union 2012
doi:10.1017/S1743921312007132

Identification of Class I Methanol Masers with Objects of Near and Mid-Infrared Bands and the Third Version of the Class I Methanol Maser (MMI) Catalog

Olga Bayandina, Irina Val'tts and Grigorii Larionov

Astro Space Center of the Lebedev Physical Institute, Moscow, Russia
email: bayandix@yandex.ru

Abstract. An identification has been conducted of class I methanol masers with 1) short-wave infrared objects EGO (extended green objects) - tracer bipolar outflow of matter in young stellar objects, and 2) isolated pre-protostellar gas-dust cores of the interstellar medium which are observed in absorption in the mid-infrared in the Galactic plane. It is shown that more than 50% of class I methanol masers are identified with bipolar outflows, considering the EGO as bipolar outflows (as compared with the result of 22% in the first version of the MMI catalog that contains no information about EGO). 99 from 139 class I methanol masers (71%) are identified with SDC. Thus, it seems possible that the MMI can be formed in isolated self-gravitating condensations, which are the silhouette of dark clouds - IRDC and SDC.

Keywords. Masers, catalogues, etc.

1. MMI CATALOG DESCRIPTION

The catalog is a table in electronic form at http://www.asc.rssi.ru/MMI with a file 'readme' and a file of references (101 ref items). In our analysis, the identification was carried out by comparing the equatorial coordinates of sources. Since class I methanol masers usually were observed with a single dish with a beam of about 2′, identified sources were considered if the distance between the coordinates did not exceed 2′.

2. RESULTS OF STATISTICAL ANALYSIS

Identification was made with a help of TOPCAT (http://www.star.bris.ac.uk/~mbt/topcat/) for the catalogues presented in:

206 MMI – http://www.asc.rssi.ru/MMI

> 300 objects EGO – http://vizier.cfa.harvard.edu/viz-bin/VizieR?-source=J/AJ/136/2391 (Cyganowski *et al.* 2008),

10931 objects IRDC – http://vizier.cfa.harvard.edu/viz-bin/VizieR?-source=J/ApJ/639/227 (Simon *et al.* 2006),

11303 objects SDC – http://vizier.cfa.harvard.edu/viz-bin/VizieR?-source=J/A+A/505/405 (Peretto & Fuller 2009).

Stastistical estimates were made for 206 class I methanol masers (MMI), except for identification with EGO and SDC, because only 139 MMI fall in the longitude interval of Spitzer. Results: in star-forming regions class I methanol masers have been identified

• in 42% of cases (59 sources from 139) with EGO;

- in 16% of cases (33 sources from 206) with infrared dark clouds IRDC from MSX;
- in 71% of cases (99 sources from 139) with infrared dark clouds SDC from Spitzer.

3. CONCLUSIONS

- The third version of the class I methanol maser MMI/SFR catalog was compiled and presented in the Internet: http://www.asc.rssi.ru/MMI.
- 206 class I methanol maser were catalogued detected mostly at 44 GHz in the direction of well-known star-forming regions (MMI/SFR) until the end of 2011 (results from Chen *et al.* 2011 - EGO survey at 95 GHz in 2011 - were not included).
- Many methanol maser sources are objects of mixed type, combining classification features of both classes.
- In the last version of the catalog more than 50% of class I methanol masers are associated with bypolar outflows - if outflows traced by EGO.
- It is shown that only 33 MMI are identified with IRDC from MSX survey, but for SDC (from Spitzer), this number increases to 99, that is 71% of MMI.
- Thus, it seems possible that MMI can be formed in the isolated self-gravitating condensations, which at certain stages of evolution could be IRDC/SDC. Sample of SDC may be a new list for to study in order to detect new MMI. It is noteworthy that this work has been done positive in the direction of IRDC in the center of the Galaxy (see Deguchi *et al.* 2011).

Support for this work was provided by Russian Foundation of Basic Research (project number 10-02-00147-) and Federal National Scientific and Educational Program (project number 16.740.11.0155).

References

Chen, X., Ellingsen, S. P., Shen, Z-Q., Titmarsh, A. *et al.* 2011, *ApJSS*, 196, Issue 1, article id.9
Cyganowski, C. J., Whitney, B. A., Holden E., Braden, E. *et al.* 2008, *Astron. J.*, 136, 2391
Deguchi, S., Tafoya, D., Shino N. *et al.* 2011, *arXiv1109.0677D*
Peretto, N. & Fuller, G. A. 2009, *A&A* 505, 405
Simon, R., Jackson, J. M., Rathborne, J. M., Chambers, E. T. *et al.* 2006, *ApJ*, 639, 227

Cosmic Masers - from OH to H₀
Proceedings IAU Symposium No. 287, 2012
R.S. Booth, E.M.L. Humphreys & W.H.T. Vlemmings, eds.

© International Astronomical Union 2012
doi:10.1017/S1743921312007144

25 GHz methanol masers in regions of massive star formation

Tui R. Britton[1,2] and Maxim A. Voronkov[2]

[1]Dept. of Physics & Astronomy, Macquarie University,
Sydney, NSW 2109, Australia
email: tui.britton@csiro.au

[2]CSIRO Astronomy & Space Science,
PO Box 76, Epping, NSW 1710, Australia
email: maxim.voronkov@csiro.au

Abstract. The bright 25 GHz series of methanol masers is formed in highly energetic regions of massive star formation and provides a natural signpost of shocked gas surrounding newly forming stars. A systematic survey for the 25 GHz masers has only recently been carried out. We present the preliminary results from the interferometric follow up of 51 masers at 25 GHz in the southern sky.

Keywords. masers, stars: formation

1. Introduction

Class I methanol masers are widely accepted to be associated with outflows from young stellar objects. However, this association has only been observed in a small number of sources (Plambeck & Menten 1990, Kurtz *et al.* 2004, Voronkov *et al.* 2006) due to the lack of accurate absolute positions of class I masers. A proper test on the association between class I methanol masers and outflows is needed on a larger sample of sources for which accurate positions are known.

The bright 25 GHz masers are rare but models suggest they trace higher density and temperature regions in outflows than other class I methanol masers (Sobolev *et al.* 2005). Voronkov *et al.* (2007) searched for the $J = 5$ line in the 25 GHz E methanol maser series towards the majority of southern regions of massive star formation. We followed up 51 of these detections by observing the $J = 2$ to $J = 9$ lines in the same series. We used the new broadband correlator on the Australia Telescope Compact Array, which allows observation of multiple molecular lines simultaneously.

2. Preliminary Results

We find that all 51 sources exhibit emission in at least one of the eight maser lines allowing us to determine their exact positions. Our preliminary results show that at least eight of the sources exhibit emission in five or more transitions. We also find that five of the emission sources are associated with previously reported HII regions and other molecular masers. We have also obtained polarization data on all our sources.

Preliminary analysis of one source selected from our list as an example, known as G351.24 + 0.67, shows the relative flux across the transitions to be similar to that found in sources such as G343.12 − 0.06 and G357.96 − 0.16 (Britton *et al.* in this proceedings). Figure 1 shows the spectra of the $J = 3$, 5 and 6 lines for G351.24 + 0.67. This is an interesting source as there are multiple components within the individual spectra and it is known to lie within the complex star forming region NGC 6334.

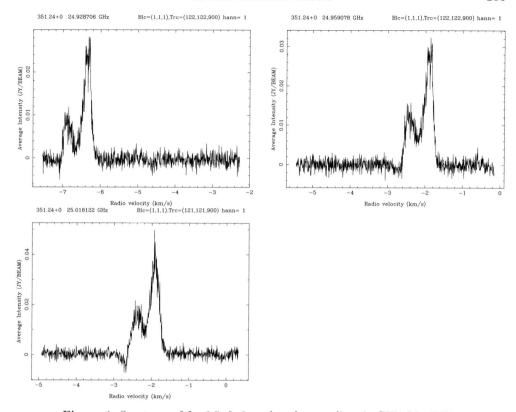

Figure 1. Spectrum of J = 3,5, & 6 methanol maser lines in G351.24 + 0.67.

We intend to release the positions of all 51 sources in a new catalogue providing the first accurate positions of 25 GHz methanol masers in the southern sky.

Acknowledgements

Travel to this conference was supported by the MQ PGRF and the IAU. TB is supported by the MQ Research Excellence Scholarship. The Australia Telescope is funded by the Commonwealth of Australia for operation as a National Facility managed by CSIRO. This research has made use of NASAs Astrophysics Data System Abstract Service.

References

Kurtz, S., Hofner, P., & Álvarez, C. V. 2004, *ApJS*, 155, 149

Plambeck R. L. & Menten K. M. 1990, *ApJ*, 364, 555

Sobolev A. M., Ostrovskii A. B., Kirsanova M. S., Shelemei O. V., Voronkov M. A., & Malyshev A. V. 2005, in: R. Cesaroni, E. B. Churchwell, M. Felli, & C. M. Walmsley (eds.), *Massive Star Birth: A Crossroads of Astrophysics.*, Proc. IAU Symposium No. 227 (Cambridge: CUP), p. 174

Voronkov M. A., Brooks K. J., Sobolev A. M., Ellingsen S. P., Ostrovskii A. B., & Caswell J. L. 2006, *MNRAS*, 373, 411

Voronkov M. A., Brooks K. J., Sobolev A. M., Ellingsen S. P., Ostrovskii A. B., & Caswell J. L. 2007, in: J. Chapman & W. A. Baan (eds.), *Astrophysical Masers and their Environments*, Proc. IAU Symposium No. 242 (Cambridge: CUP), p. 182

Cosmic Masers - from OH to H_0
Proceedings IAU Symposium No. 287, 2012   © International Astronomical Union 2012
R.S. Booth, E.M.L. Humphreys & W.H.T. Vlemmings, eds.   doi:10.1017/S1743921312007156

44-GHz class I methanol maser survey towards 6.7-GHz class II methanol masers

Do-Young Byun, Kee-Tae Kim and Jae-Han Bae

Korea Astronomy and Space Science Institute,
776, Daedeokdae-ro, Yuseong-gu, Daejeon, Republic of Korea
email: bdy@kasi.re.kr

Abstract. The Class II 6.7-GHz methanol maser is a tracer of high mass young stellar objects. We present results of a 44-GHz class I methanol maser and 22-GHz water maser survey using the KVN (Korean VLBI Network) 21-m single dish radio telescopes towards 284 6.7-GHz maser sites. Class I methanol maser and water maser emission is detected towards 116 (41%) and 136 (48%) sources, respectively. About 50 sources have a peak flux density higher than 10 Jy at 44-GHz. They are candidates for VLBI studies using the KVN.

Keywords. masers, surveys, stars: formation, stars: outflows

1. Introduction

Since high mass young stellar objects (HMYSOs) are deeply embedded in natal molecular clouds and their distances are usually large, it is hard to study their environment in detail. Consequently, the physical processes of HMYSOs, such as jet formation, mass infall and disk formation, are not yet fully understood. Class I methanol maser emission is excited in the interface region between the outflow of a central object and the surrounding molecular material (Kurtz *et al.* 2004 etc.). This methanol maser line is therefore a very useful tool to study the outflows of HMYSOs.

We present the results of a 44-GHz class I methanol maser and 22-GHz water maser survey towards 6.7-GHz methanol sites. The 6.7-GHz class II methanol maser is known to occur only near HMYSOs. The primary aim of this study is to build a data base of 44-GHz methanol masers for future VLBI studies of HMYSOs using the KVN. Additionally, a statistical analysis using the survey results is used to study the relation between maser occurence and the physical condition of HMYSOs.

2. Observations and Results

We selected 284 targets with a declination higher than -30 degree among 519 sources in the catalog of 6.7-GHz methanol masers of Pestalozzi *et al.* (2005). 44-GHz methanol maser observations towards the selected targets were made with the KVN 21-m radio telescopes at Yonsei and Ulsan during October and November 2009. The on-source integration time of 20 or 30 minutes on each source resulted in a typical rms noise level of 0.7Jy at 0.2 km s^{-1} velocity resolution. The 22-GHz water maser observations towards the same targets were made during May and June 2010. The rms noise levels of the water maser spectra are $\sim$1.2Jy. The beam size of the 21-m telescope at the 44-GHz and 22-GHz is $65''$ and $125''$, respectively (Lee *et al.* 2011).

The 44-GHz methanol masers and water masers are detected towards 116 and 136 of 284 targets equating to detection rates of 41% and 48%, respectively. Fig. 1 shows that more than 50 sources have a peak flux density at 44-GHz methanol stronger than 10 Jy,

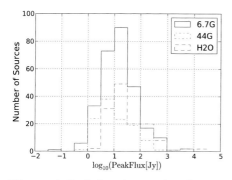

Figure 1. Peak flux density distribution.

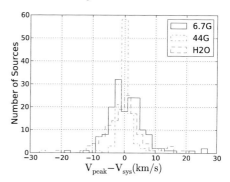

Figure 2. Peak velocity distribution.

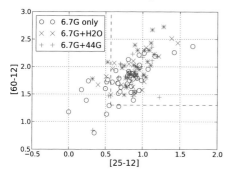

Figure 3. IRAS color-color plot.

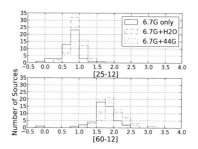

Figure 4. IRAS color distribution.

which is around the 7-σ detection level of the KVN baseline. We find that 3/4 of the class I methanol maser sources are associated with water masers. This implies that class I methanol masers and water masers share a common evolutionary stage. We compared the peak velocities of the maser emission with the systematic velocities of HMYSOs determined from thermal molecular lines tracing dense regions. Fig. 2 shows that the peak velocities of 44-GHz methanol masers are concentrated around the systematic velocities more tightly than the other masers.

We investigated the maser occurence rates with the IRAS colors. For this investigation, 116 IRAS associated sources with reliable colors were selected. Figs. 3 and 4 show that the maser occurence rates of 44-GHz methanol and 22-GHz water masers increase systematically with IRAS colors. This seems to imply that 44-GHz methanol maser and 22-GHz water maser emission increases with luminosity as the central object evolves from protostar to UCHII, considering the UCHII have larger IRAS colors.

It should be noted, however, that the IRAS association could be mistaken due to the limited positional accuracies of the single dish observations of masers and the IRAS data, since massive star forming regions often contain multiple HMYSOs within a few arcsec scale. This can contaminate the maser occurence rates with IRAS colors. Study with higher angular resolution is neccessary to confirm the relation between the maser occurence and the evolutionary stage of HMYSO.

References

Kurtz, S., Hofner, P., & Álvarez, C. V. 2004, *ApJS*, 155, 149

Lee, S. S. *et al.* 2011, *PASP*, 123, 1398

Pestalozzi, M., Minier, V., & Booth, R. S. 2005, *A&A.*, 432, 737

Cosmic Masers - from OH to H$_0$
Proceedings IAU Symposium No. 287, 2012
R.S. Booth, E.M.L. Humphreys & W.H.T. Vlemmings, eds.

© International Astronomical Union 2012
doi:10.1017/S1743921312007168

Water Masers Toward Star-Forming Regions in the Bolocam Galactic Plane Survey

Miranda K. Dunham[1] and The BGPS Team

[1]Department of Astronomy, Yale University,
P. O. Box 208101, New Haven, CT, USA
email: miranda.dunham@yale.edu

Abstract. We present preliminary results of a search for 22 GHz water masers toward 1400 star-forming regions seen in the Bolocam Galactic Plane Survey (BGPS) using the Green Bank Telescope (GBT). The BGPS is a blind survey of the Northern Galactic plane in 1.1 mm thermal dust emission that has cataloged star-forming regions at all evolutionary stages. Further information is required to determine the stage of each BGPS source. Since water masers are produced by outflows from low and high-mass star forming regions, their presence is a key component of determining whether the BGPS sources are forming stars and which evolutionary stage they are in. We present preliminary detection statistics, basic properties of the water masers, and correlations with physical properties determined from the 1.1 mm emission and ammonia observations obtained concurrently with the water masers on the GBT.

Keywords. ISM: molecules, masers, radio lines: ISM, stars: formation, surveys

---

One of the main outstanding problems of high-mass star formation is the lack of an evolutionary sequence similar to that of low-mass star formation (e.g. Shu, Adams & Lizano 1987). Several evolutionary sequences have been proposed recently (e.g. Battersby *et al.* 2011; Chambers *et al.* 2009) that are based on the presence or absence of 8 and 24 μm emission, H$_2$O and CH$_3$OH masers, 4.5 μm outflows, ultra-compact HII regions, as well as a determination of the dust temperature. A detailed study of the physical properties of a large sample of star-forming regions with different combinations of these evolutionary indicators is necessary to solidify an evolutionary sequence. A large scale, unbiased catalog of star-forming regions is of paramount importance in such a study.

The BGPS is a blind survey of the northern Galactic plane in optically thin 1.1 mm thermal dust continuum. The BGPS has detected 8,358 sources (Rosolowsky *et al.* 2010) and follow-up observations in dense gas tracers return high detection rates (72% in NH3, Dunham *et al.* 2010, 2011; 77% in HCO+, Schlingman *et al.* 2011) demonstrating that the high column density features in the BGPS correspond to high volume density regions. The Bolocam Galactic Plane Survey (BGPS) provides an ideal catalog of high-mass star-forming regions to solidify an evolutionary sequence.

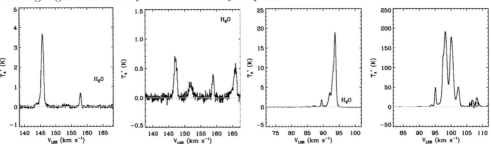

Figure 1. H$_2$O spectra for four BGPS sources, from left to right 1311, 1316, 4760, and 4764.

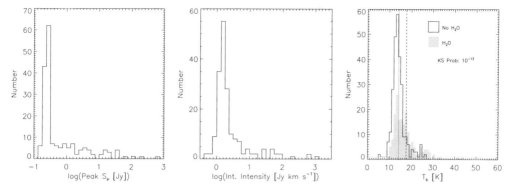

Figure 2. Distrbution of the peak flux densities (left) and integrated intensity (middle) of the H_2O masers observed toward the BGPS sources. Right: Distribution of gas kinetic temperatures as measured from NH_3 of BGPS sources with (gray filled) and without (solid line) H_2O maser emission (gray filled). The dashed and dotted lines denote the mean temperatures of BGPS sources with and without H_2O, respectively.

We have used the Green Bank Telescope (GBT) to observe single pointings of $NH_3(1,1)$, (2,2), (3,3) and the 22 GHz H_2O line toward $\sim$1400 BGPS sources. The NH_3 observations provide a determination of the gas temperature and the H_2O observations address the presence/absence of H_2O masers within the $30''$ GBT beam. Here we present preliminary H_2O results from a subset of 456 BGPS sources.

Of the 456 BGPS sources considered, 182 (40%) exhibit H_2O emission above a 5σ integrated intensity and a 4σ peak flux density threshold. Figure 1 shows example H_2O spectra from four BGPS sources spanning a range of maser strength and number of components. Figure 2 shows distributions of various observed properties of the H_2O masers, as well as the distributions of gas kinetic temperatures for BGPS sources with and without H_2O maser emission. Sources with H_2O masers have a warmer mean kinetic temperature (17.6 K) compared to sources without H_2O masers (14.2 K), suggesting that sources with H_2O masers are more evolved since kinetic temperature increases along the proposed evolutionary sequences.

In the future, we will expand this analysis to include all $\sim$1400 BGPS sources with GBT observations. With the large sample size provided by the BGPS, we will explore trends in physical properties as a function of presence/absence of H_2O masers. Beyond this study of H_2O masers, we will compare the BGPS sources with catalogs of the remaining evolutionary indicators in order to solidify an evolutionary sequence for high-mass star formation.

References

Aguirre, J. E., Ginsburg, A. G., Dunham, M. K., *et al.* 2011, *ApJS*, 192, 4

Battersby, C., Bally, J., Ginsburg, A., *et al.* 2011, *A&A*, 535, A128

Chambers, E. T., Jackson, J. M., Rathborne, J. M., & Simon, R. 2009, *ApJS*, 181, 360

Dunham, M. K., Rosolowsky, E., Evans, N. J., II, *et al.* 2010, *ApJ*, 717, 1157

Dunham, M. K., Rosolowsky, E., Evans, N. J., II, Cyganowski, C., & Urquhart, J. S. 2011, *ApJ*, 741, 110

Rosolowsky, E., Dunham, M. K., Ginsburg, A., *et al.* 2010, *ApJS*, 188, 123

Schlingman, W. M., Shirley, Y. L., Schenk, D. E., Rosolowsky, E., Bally, J., *et al.* 2011, *ApJS*, 195, 14

Shu, F. H., Adams, F. C., & Lizano, S. 1987, *ARAA*, 25, 23

Cosmic Masers - from OH to H_0
Proceedings IAU Symposium No. 287, 2012
R.S. Booth, E.M.L. Humphreys & W.H.T. Vlemmings, eds.
© International Astronomical Union 2012
doi:10.1017/S174392131200717X

The VLBI mapping survey of the 6.7 GHz methanol masers with the JVN/EAVN

K. Fujisawa[1], K. Hachisuka[2], K. Sugiyama[3], A. Doi[4], M. Honma[5], Y. Yonekura[6], T. Hirota[5], S. Sawada-Satoh[5], Y. Murata[4], K. Motogi[7], H. Ogawa[8], X. Chen[2], K.-T. Kim[9] and Z.-Q. Shen[2]

[1] The Research Institute of Time Studies, Yamaguchi Univ., 1677-1 Yoshida, Yamaguchi, Yamaguchi 753-8511, Japan
email: kenta@yamaguchi-u.ac.jp

[2] Shanghai Astronomical Observatory, Chinese Academy of Sciences, Shanghai 200030, China

[3] Dept. of Physics, Yamaguchi Univ., 1677-1 Yoshida, Yamaguchi, Yamaguchi 753-8512, Japan

[4] Institute of Space and Astronautical Science, JAXA, Yoshinodai 3-1-1, Chuo-ku, Sagamihara, Kanagawa 252-5210, Japan

[5] Mizusawa VLBI Observatory, NAOJ, Hoshigaoka-cho 2-12, Oshu, Iwate 023-0861, Japan

[6] Center for Astronomy, Ibaraki Univ., 2-1-1 Bunkyo, Mito, Ibaraki 310-8512, Japan

[7] Dept. of Cosmosciences, Hokkaido Univ., N10 W8, Sapporo 060-0810, Japan

[8] Graduate School of Science, Osaka Pref. Univ., 1-1 Gakuen-cho, Nakaku, Sakai, Osaka 599-8531, Japan

[9] Korea Astronomy and Space Science Institute, Hwaam-Dong, Yuseong-Gu, Daejeon 305-348, Korea

Abstract. We present VLBI maps of the 6.7 GHz methanol maser emission in 32 sources obtained using the Japanese VLBI Network (JVN) and the East-Asian VLBI Network (EAVN). All of the observed sources provide new VLBI maps, and the spatial morphologies have been classified into five categories similar to the results obtained from European VLBI Network observations (Bartkiewicz *et al.* 2009). The 32 methanol sources are being monitored to measure the relative proper motions of the methanol maser spots.

Keywords. Stars: formation, Masers: methanol, Instrumentation: high angular resolution

1. Introduction

Methanol masers at 6.7 GHz are good tracers for investigating the three dimensional dynamics around high-mass young stellar objects (YSOs). This maser could be associated with an accretion disk around a high-mass YSO because of its ring/elliptical spatial morphology (e.g., Bartkiewicz *et al.* 2009) and rotational/infall proper motions in some sources (Sanna *et al.* 2010a, b; Moscadelli *et al.* 2011; Goddi *et al.* 2011).

We started a VLBI monitoring project of the 6.7 GHz methanol maser in 36 target sources in 2010 using the JVN/EAVN for the investigation of the evolution of accretion disks around high-mass YSOs. This project is still ongoing to measure the relative proper motions of each maser spot around high-mass YSOs. Here, we present each VLBI map for the 32 sources obtained in 2010 Aug. and 2011 Oct. All of these target sources in our VLBI project provided new VLBI maps.

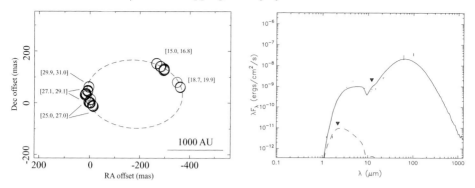

Figure 1. VLBI map of the 6.7 GHz methanol maser spots (left) and SED estimated from the online fitting tool (right; Robitaille *et al.* 2006) in G6.795−0.257. In the left panel, positions of each spot are relative to the reference spot at the origin, and attached numbers indicate a range of radial velocities for each spot. A dashed ellipse shows the model fitted to the maser distribution.

2. Observations and Results

The target 6.7 GHz methanol maser sources were selected from the methanol maser catalog of Pestalozzi *et al.* (2005) and the Methanol Multibeam Survey catalog (Caswell *et al.* 2010; Green *et al.* 2010) using the following criteria: 1) source declination $-40° < \delta < +30°$ (observable with ALMA); 2) total flux densities > 65 Jy in either catalog; 3) no previous VLBI observations. Applying these criteria, 36 sources were selected, and 95% of the target sources (34/36) were located in southern hemisphere (declination $< 0°$).

We observed 22 sources in 2010 Aug. 28-30 and 10 sources in 2011 Oct. 27, 28 using the EAVN, which consists of the Yamaguchi 32-m, Ibaraki (Hitachi) 32-m, Usuda 64-m, VERA 20-m four stations, and Shanghai 25-m radio telescope. As a result, we obtained new VLBI maps for all of the observed sources, and the spatial morphology was classified into five categories on the basis of the classification by Bartkiewicz *et al.* (2009): three sources in elliptical; five in arched; six in linear; nine in isolated pairs; and nine in a complex morphology. In three sources, the methanol maser in G6.795−0.257 showed a clear elliptical morphology (left-panel in figure 1) with a radial velocity gradient on the ellipse. On the basis of the rotation model fitting suggested by Bartkiewicz *et al.* (2009), this maser was well fitted as rotation with infall components of 2.1 km s^{-1}, and may be the best target to detect rotational/infall proper motions.

On the basis of spectral energy distributions (SEDs) estimated from the online fitting tool (Robitaille *et al.* 2006) using infrared data (e.g., right-panel in figure 1), our observed sources are distributed in each evolutionary phase of central YSOs in a range of $10^3 - 10^6$ yr. Our future measurement of the three-dimensional dynamics will yield information about the evolution of the accretion disk around high-mass YSOs.

References

Bartkiewicz, A., Szymczak, M., van Langevelde, *et al.* 2009 *A&A*, 502, 155
Caswell, J. L., Fuller, G. A., Green, J. A., *et al.* 2010, *MNRAS*, 404, 1029
Goddi, C., Moscadelli, L., & Sanna, A. 2011, *A&A*, 535, L8
Green, J. A., Caswell, J. L., Fuller, G. A., *et al.* 2010, *MNRAS*, 409, 913
Moscadelli, L., Cesaroni, R., Rioja, M. J., *et al.* 2011, *A&A*, 526, A66
Pestalozzi, M. R., Minier, V., & Booth, R. S. 2005, *A&A*, 432, 737
Robitaille, T. P., Whitney, B. A., Indebetouw, R., *et al.* 2006, *ApJS*, 167, 256
Sanna, A., Moscadelli, L., Cesaroni, R., *et al.* 2010a, *A&A*, 517, A71
Sanna, A., Moscadelli, L., Cesaroni, R., *et al.* 2010b, *A&A*, 517, A78

Cosmic Masers - from OH to H_0
Proceedings IAU Symposium No. 287, 2012
R.S. Booth, E.M.L. Humphreys & W.H.T. Vlemmings, eds.
© International Astronomical Union 2012
doi:10.1017/S1743921312007181

Simultaneous observations of SiO and H$_2$O masers toward known stellar SiO and/or H$_2$O maser sources

Jaeheon Kim[1,2,3], Se-Hyung Cho[1,2] and Sang Joon Kim[3]

[1] Yonsei University Observatory and Department of Astronomy, Yonsei University, Seongsan-ro 262, Seodaemun, Seoul, 120-749, Republic of Korea
e-mail: jhkim@kasi.re.kr

[2] Korean VLBI Network, Korea Astronomy and Space Science Institute, P.O. Box 88, Yonsei University, Seoul, 120-749, Republic of Korea
e-mail: cho@kasi.re.kr

[3] Department of Astronomy and Space Science, Kyung Hee University, Yongin, Gyeonggi-Do, 446-701, Republic of Korea

Abstract. We present the results of simultaneous observations of SiO $v = 1$, 2, ^{29}SiO $v = 0$, $J = 1-0$ and H$_2$O $6_{16} - 5_{23}$ maser lines toward 318 known stellar SiO and/or H$_2$O maser sources using the Yonsei 21-m radio telescope of the Korean VLBI Network. Toward 166 known SiO and H$_2$O maser sources, both SiO and H$_2$O maser emissions were detected from 112 sources giving a detection rate of 67.5 %. On the other hand, toward 152 known H$_2$O-only maser sources, both SiO and H$_2$O maser emissions were detected from 62 sources, giving a detection rate of 40.8 %. Characteristics of all observed sources in the IRAS two-color diagram is investigated including their evolutionary sequence and mutual relations between SiO and H$_2$O maser properties.

Keywords. circumstellar matter, masers, surveys, stars: AGB and post-AGB

1. Introduction

The SiO and H$_2$O masers, which display very compact structures and high brightness temperatures in oxygen-rich AGB stars, are good probes to study the physical conditions and dynamics in the circumstellar shells. In order to investigate mutual relations between SiO and H$_2$O maser properties, and the dynamical connection from the pulsating atmosphere to the inner circumstellar envelope through dust forming layers in relation with mass-loss processes, we have performed simultaneous observations of SiO and H$_2$O masers using the KVN single dish.

2. Observations

For 166 sources with both SiO and H$_2$O masers, which are selected from Cho et al. (1996) and Takaba et al. (2001), observations were performed in 2009 June. For 152 objects, which were previously detected only in H$_2$O maser lines (43 sources were detected only in the 22 GHz H$_2$O maser line in spite of SiO maser observations and 109 sources were not observed in the SiO masers), the observations were performed from 2009 June to 2011 January. The half power beam widths and aperture efficiencies were measured to be $122''$, 0.65 (at 22 GHz) and $64''$, 0.67 (at 43 GHz), respectively (Lee et al. 2011). The conversion factor from the antenna temperature, T_A^*, to the flux density is about 12.27 Jy K^{-1} at 22 GHz and 11.90 Jy K^{-1} at 43 GHz.

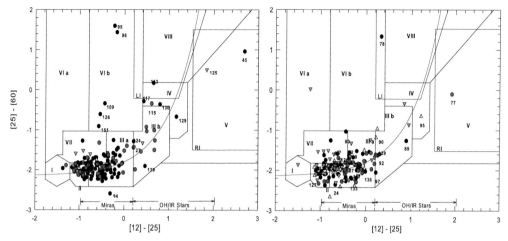

Figure 1. IRAS two-color diagram with the position of the observed sources in our observations toward 166 both SiO and H$_2$O maser sources (left) and 152 H$_2$O maser sources (right). The red line is the evolutionary track for AGB stars. The black circles indicate both SiO and H$_2$O maser detected sources, and the red circles and the yellow triangles indicate SiO-only and H$_2$O-only detected sources, respectively. Undetected sources are marked with the green inverted triangles. The numbers indicate identification numbers in Table 2 of Kim *et al.* (2010)

3. Results

Detailed observational results of the 166 known SiO and H$_2$O maser sources were described in Kim *et al.* (2010). Toward 152 H$_2$O-only maser sources, both SiO and H$_2$O masers were detected from 62 sources. Furthermore, we have identified 19 new detections of SiO maser emission for previously non-detected sources and 51 new detections of SiO maser for previously not observed sources. Most of the SiO maser emission peaks near the stellar velocity, while the peak of H$_2$O maser shows a wide spread compared with that of SiO. We examined the distribution of the sources with single, double, and multiple peaks of H$_2$O maser lines in the IRAS two-color diagram (Fig. 1) because they can be associated with an asymmetric wind and bipolar outflows commonly seen in PPNe and PNe (Engels 2002). These single and double peak sources are distributed in Regions IIIb, IV, V, and VIb with a relatively high percentage compared with those of Regions II, IIIa, and VII stars (Kim *et al.* 2010). The Regions IV and V are thought to be main areas of PPNe. However, candidates for young PPNe can be distributed in Region IIIb and the bipolar structure can already appear in the AGB stage (Zijlstra *et al.* 2001; Engels 2002) as an earliest transition phase from AGB stars to PPNe. The distribution of H$_2$O double peak sources, V1366 Aql and OH83.42−0.89 etc. in Region IIIb may support these facts. Statistical analyses based on these homogeneous data (intensity ratios, peak and mean velocities between SiO and H$_2$O masers etc.) are in progress.

References

Cho, S.-H., Kaifu, N., & Ukita, N. 1996, *A&AS*, 115, 117
Engels, D. 2002, *A&A*, 388, 252
Kim, J., Cho, S.-H., Oh, C. S., & Byun, D.-Y. 2010, *ApJS*, 188, 209
Lee, S.-S., *et al.* 2011, *PASP*, 123, 1398
Takaba, H., Iwata, T., Miyaji, T., & Deguchi, S. 2001, *PASJ*, 53, 517
Zijlstra, A. A., *et al.* 2001, *MNRAS*, 322, 280

Cosmic Masers - from OH to H_0
Proceedings IAU Symposium No. 287, 2012
R.S. Booth, E.M.L. Humphreys & W.H.T. Vlemmings, eds.

© International Astronomical Union 2012
doi:10.1017/S1743921312007193

SHOOTING STARS
Masers from red giants

Sofia Ramstedt[1], Wouter Vlemmings[2], Elizabeth Humphreys[3] and Felipe Alves[1]

[1] Argelander Institut für Astronomie, Bonn, Germany
email: sofia@astro.uni-bonn.de

[2] Onsala Space Observatory, Onsala, Sweden
[3] European Southern Observatory, Garching, Germany

Abstract. SiO maser emission probes the region close to the stellar surface where the wind is formed and is observed to better constrain the physical conditions in this region. We have started a long-term project where high-excitation ^{28}SiO maser lines (i.e., J = 5-4, v = 1 and 2) are observed in a large sample of southern AGB stars. The primary goals are to put constraints on the physical conditions in the extended atmospheres, and to achieve a better understanding of the maser excitation process. Since the maser emission is strong and often highly linearly polarized, the detected sources could also complement the polarization calibrator catalogue for ALMA. Preliminary results show a high detection rate and that in approximately 20% of the sources, the v = 2 transition emits stronger than the v = 1 transition. We interpret this as possibly indicative of a hot dust shell very close to the stars.

Keywords. masers, polarization, catalogs, stars: AGB and post-AGB, stars: mass loss

1. Maser excitation and physical conditions in evolved stars

SiO maser emission will trace the physical conditions just above the stellar surface where the mass loss is initiated. The pumping of the masers can be either collisional or radiative and the presence and temperature distribution of the dust can have a strong effect on the excitation of the molecules through radiative excitation by the dust emission radiation field (Gray *et al.* 2009). Theory predicts that masers from separate vibrational levels will be emitted at different radii from the evolved star (e.g., Elitzur *et al.* 1983, Langer & Watson 1984, Bujarrabal 1994a, Bujarrabal 1994b, Doel *et al.* 1995) and our preliminary analysis of the available data already hints at very interesting results. Besides a high detection rate, just under 20% of the sources exhibit stronger emission in the $v = 2$ transition. Recent modelling work (Gray *et al.* 2009), indicates that for the much stronger

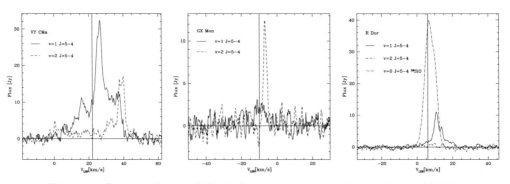

Figure 1. Spectra of some of the high-excitation masers detected with APEX.

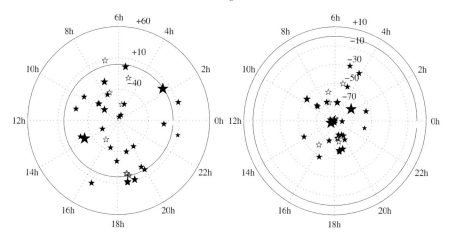

Figure 2. *Left:* The source distribution in RA and Dec. *Right:* The source distribution in RA and maximum elevation at the ALMA site. The strength of the $v = 1$ maser emission is indicated by the size of the star; the smallest correspond to ~ 1 Jy, the medium size correspond to ~ 10 Jy, and the largest correspond to ~ 100 Jy. Open stars mark non-detections.

amplification in the $v = 2$ transition, radiative effects due to a hot dust shell of small inner radius, are required. Further modelling of this effect is ongoing.

2. Polarization calibrators for ALMA

Polarization observations will be part of the capabilities of ALMA with the goal of detecting polarization degrees of 0.1%. Initial testing will be performed during Cycle 1 in 2013. The needed correction will depend on the mount of the antennas (Az-Alt for ALMA) and on whether the feeds are measuring orthogonal linear or circular polarization (linear for ALMA). In all cases, the polarization of the source, and the instrumental polarization, will depend differently on the parallactic angle. To be able to chose calibrators close to different science targets, to be observed across large ranges in parallactic angle, a long list of calibrator sources is desirable, covering a large part of the sky (Fig. 2).

Since SiO maser emission from red giants is often strong and highly linearly polarized (e.g. Vlemmings *et al.* 2011), these sources are ideal for polarization calibration measurements, because a short exposure time will provide good calibration. The emission is expected to vary (see e.g. Ramstedt *et al.*, this volume), however, the correlation with e.g. pulsational period remains to be understood. Observations at different epochs show examples where the overall spectral shape stays essentially the same. The variability of the polarization characteristics over the pulsational cycle also remains to be evaluated.

References

Bujarrabal, V. 1994, *A&A*, 285, 953

Bujarrabal, V. 1994, *A&A*, 285, 971

Doel, R. C., Gray, M. D., Humphreys, E. M. L., Braithwaite, M. F., & Field, D. 1995, *A&A*, 302, 797

Elitzur, M., Watson, W. D., & Western, L. R. 1983, *ApJ*, 274, 210

Gray, M. D., Wittkowski, M., Scholz, M., Humphreys, E. M. L., Ohnaka, K., & Boboltz, D. 2009, *MNRAS*, 394, 51

Langer, S. H. & Watson, W. D. 1984, *ApJ*, 284, 751

Vlemmings, W. H. T., Humphreys, E. M. L., & Franco-Hernández, R. 2011, *ApJ*, 728, 149

Cosmic Masers - from OH to H_0
Proceedings IAU Symposium No. 287, 2012
R.S. Booth, E.M.L. Humphreys & W.H.T. Vlemmings, eds.

© International Astronomical Union 2012
doi:10.1017/S174392131200720X

New OH Observations toward Northern Class I Methanol Masers

I. E. Val'tts[1], I. D. Litovchenko[1], O. S. Bayandina[1], A. V. Alakoz[1], G. M. Larionov[1], D. V. Mukha[2], A. S. Nabatov[2], A. A. Konovalenko[2], V. V. Zakharenko[2], E. V. Alekseev[2], V. S. Nikolaenko[2], V. F. Kulishenko[2] and S. A. Odincov[3]

[1]Astro Space Center of the Lebedev Physical Institute, Moscow, Russia
email: ivaltts@asc.rssi.ru,

[2]Institute of Radio Astronomy of National Academie of Science, Kharkov, Ukraine,

[3]National Center for Management and Testing of Space Resources, Evpatoriya, Ukraine

Abstract. Maser emission of OH(1720) is formed, according to modern concepts, under the influence of collisional pumping. Class I methanol masers (MMI) are also formed by a collisional mechanism of the inversion of the molecular levels. It is not excluded in this case that physical conditions in the condensations of the interstellar medium where masers are formed may be similar for MMI and OH(1720) masers, and they can associate with each other. To establish a possible association between these two kinds of masers, and obtain reliable statistical estimates, a survey of class I methanol masers at a frequency of 1720 MHz has been carried out.

Keywords. interstellar medium, masers, molecules

1. Observations and Results

The observations were made at a frequency of OH 1720.530 MHz with the 70-m radio telescope of the National Academy of Sciences of Ukraine (NASU). The sample included 111 objects from the class I methanol maser catalogue (http://www.asc.rssi.ru/MMI), available for observations in Evpatoria, Crimea (declination $>-35°$). The antenna beam width was $9'$. A digital autocorrelation spectrometer, fabricated in the Institute of Radio Electronics of NASU, Kharkov, Ukraine, was used - with a frequency resolution of 4.028 kHz (4096 channels, 16.5 MHz bandwidth) or a radial velocity of 0.7 km s^{-1} at 1720 MHz.

There are 72 spectra without obvious interference, 27 (38%) have no OH emission or absorption lines, 30 sources have both emission and absorption details. All spectra exhibit notable circular polarization. Gaussian parameters for the line profiles were obtained. Many emission lines are narrow ($\Delta V < 2$ km s^{-1}). Flux densities in narrow features greater than 100 mJy were obtained for a significant number of sources and for some more than 500 mJy.

2. Statistical Analysis and Discussion

OH maser emission at 1720 MHz in the direction of SNR are observed in 10% cases, i.e. in 20 SNR out of 200 (Hewitt & Yusef-Zadeh 2008 and references therein).

OH maser emission at 1720 MHz in the direction of star forming regions (SFR) from observations in the southern hemisphere is present in 11% of cases, i.e. in 28 SFR out of 200 (Caswell 2004).

OH maser emission at 1720 MHz in the direction of class II methanol masers observed in the northern hemisphere is present in 6% of cases, i.e. in 6 SFR from 100 (Szymczak & Gerard 2004).

Narrow emission lines of OH(1720) in the direction of class I methanol masers in the observational data of this survey is present in 38% of cases, i.e. in 27 MMI from 72.

3. Conclusions

• The detection of emission in OH(1720) lines and their correlation with the positions and velocities of class I methanol masers may be considered as an indirect indication of the presence of shock waves in the observed regions: this idea was also discussed in Frail (2008), Pihlström *et al.* (2011) and Litovchenko *et al.* (2011).

• The width of the observed OH(1720) features means that they may be maser lines, and values of the observed fluxes enable us to propose these sources as prospective candidates for VLBI experiments to determine the sizes and brightness temperatures of the emitting condensations.

Acknowledgements

Support for this work was provided by the Russian Foundation of Basic Research (project number 10-02-00147-) and the Federal National Scientific and Educational Program (project number 16.740.11.0155).

References

Caswell, J. L. 2004, *A&A*, 349, 99
Frail, D. A. 2008, *Mem. Soc. Astron. Italiana* 75, 282 2011, arXiv:1108.4137v1 [astro-ph.HE]
Hewitt, J. W. & Yusef-Zadeh, F., 2008, *ApJ*, 683, 189
Litovchenko, I. D., Alakoz, A. V., Valtts, I. E., & Larionov, G. M. 2011, *Astronomicheskii Zhurnal* 88, 1061
Pihlström, Y. M., Sjouwerman, L. O., & Fish, V. L. 2011, arXiv:1105.4377v1 [astro-ph.GA]
Szymczak, M. & Gerard, E. 2004, *A&A*, 414, 235

Cosmic Masers - from OH to H$_0$
Proceedings IAU Symposium No. 287, 2012
R.S. Booth, E.M.L. Humphreys & W.H.T. Vlemmings, eds.

© International Astronomical Union 2012
doi:10.1017/S1743921312007211

22 GHz Water Maser Survey of the Xinjiang Astronomical Observatory

Jian-jun Zhou[1,2] Jarken Esimbek[1,2] and Gang Wu[1,2]

[1] Xinjiang Astronomical Observatory, CAS
150 Science 1 street Urumqi, Xinjiang 830011, China
email: zhoujj@xao.ac.cn

[2] Key Laboratory of Radio Astronomy, CAS,
150 Science 1 street Urumqi, Xinjiang 830011, China

Abstract. Water masers are good tracers of high-mass star-forming regions. Water maser VLBI observations provide a good probe for studying high-mass star formation and galactic structure. We plan to make a blind survey toward the northern Galactic plane in future years using the 25 m radio telescope of the Xinjiang Astronomical Observatory. We will select some water maser sources discovered in the survey and perform high resolution observations to study the gas kinematics close to high-mass protostars.

Keywords. Masers, surveys, stars: formation.

1. Introduction

High-mass star forming regions are usually distant and heavy obscuration makes it difficult to observe them. Water masers are good probes of the physical conditions and dynamics of star forming regions. Maser VLBI observations are the unique means by which one can explore the gas kinematics close (within tens or hundreds of AU) to forming high-mass protostars (Moscadelli *et al.* 2011). Measurements of the trigonometric parallaxes and proper motions of water masers found in high-mass star-forming regions using a VLBI phase reference method can provide very accurate distances to them. Combining positions, distances, proper motions and radial velocities yields complete 3-dimensional kinematic information on the Galaxy (Xu *et al.* 2006; Reid *et al.* 2009). Water masers are very rich in the Galaxy and they are reliable tracers of high-mass star-forming regions (Caswell *et al.* 2011). Therefore, it is valuable to discover more water masers associated with high-mass star-forming regions.

Earlier water maser searches have chiefly been made toward targeted sources, and many masers may yet be undiscovered. There are only a few unbiased water maser surveys (Breen *et al.* 2007; Walsh *et al.* 2008; Caswell & Breen 2010). Recently, one much larger blind survey toward 100 square degrees of the southern Galactic plane has been completed successfully (Walsh *et al.* 2011). However, no large blind water maser survey has been performed toward the northern Galactic plane. We will make a blind survey toward 90 square degrees of the northern galactic plane using our 25 m radio telescope. We hope to discover a large sample of water masers and high-mass star-forming regions at earlier stages, and study high-mass star formation and galactic structure.

2. 25 m radio telescope of Xinjiang Astronomical Observatory

The Nanshan 25 m radio telescope of the Xinjiang Astronomical Observatory was built in 1992 as a station for the Chinese Very Long Baseline Interferometry network. It is located at the Nanshan mountains west of Urumqi city at an altitude of 2080m. Its front-end receiver system includes several receivers working at 18, 13, 6, 3.6 and 1.3 cm. At 1.3 cm, one dual-polarization cryogenic receiver has been installed on the telescope recently, the noise temperature of the receiver is better than 20 K. When the weather is good, the system temperature is better than 50 K. We built a molecular spectrum observing system in 1997. One digital filter bank (DFB) system is employed as the spectrometer. It is is capable of processing up to 1 GHz of bandwidth with 8192 channels. Our telescope now can observe several molecules such as OH, H_2O, NH_3, H_2CO and $H_{110\alpha}$.

3. Our plan

We will make a large scale blind survey toward the Northern Galactic plane in order to have the opportunity to detect most water masers concentrated in the region along the galactic plane ($|b| < 0.5°$). We plan to survey 90 square degrees of the northern galactic plane, covering the region between $l = 30°$ and $l = 120°$, and $|b| < 0.5°$. In order to complete the project in a reasonable time, a scanning observation mode (on the fly) will be used for our observations. The final sensitivity of the survey is about 1.4 Jy.

On the other hand, many surveys at millimeter, submillimeter and infrared wavelengths have discovered a large sample of possible star-forming regions, e.g. Bolocam, Planck, Glimpse and MIPS. These sources provide us with good candidates for our water maser search. We can therefore also select some sources to make a targeted survey.

Acknowledgements

This work was funded by the National Natural Science Foundation of China under Grant 10778703, the China Ministry of Science and Technology under the State Key Development Program for Basic Research (2012CB821800) and The National Natural Science Foundation of China under Grant 10873025.

References

Breen, S. L., Ellingsen, S. P., Johnston-Hollitt, M., *et al.* 2007, *MNRAS*, 377, 491
Caswell, J. L. & Breen, S. L. 2010, *MNRAS*, 407, 2599
Caswell, J. L., Breen, S. L., & Ellingsen, S. P. 2011, *MNRAS*, 410, 1283
Moscadelli, L., Sanna, A., & Goddi, C. 2011, *A&A*, 536, 38
Reid, M. J., Menten, K. M., Zheng, X. W., Brunthaler, A., *et al.* 2009, *ApJ*, 700, 137
Walsh, A. J., Lo, N., Burton, M. G., *et al.* 2008, *PASA*, 25, 105
Walsh, A. J., Breen, S. L., Britton, T., *et al.* 2011, *MNRAS*, 416, 1764
Xu, Y., Reid, M. J., Zheng, X. W., & Menten, K. M. 2006, *Science*, 311, 54

T6:

Cosmology and the Hubble constant
Chair: Willem Baan

Cosmic Masers - from OH to H_0
Proceedings IAU Symposium No. 287, 2012
R.S. Booth, E.M.L. Humphreys & W.H.T. Vlemmings, eds.

© International Astronomical Union 2012
doi:10.1017/S1743921312007223

Cosmology and the Hubble Constant: On the Megamaser Cosmology Project (MCP)

C. Henkel[1], J. A. Braatz[2], M. J. Reid[3], J. J. Condon[4], K. Y. Lo[5], C. M. Violette Impellizzeri[6] and C. Y. Kuo[7]

[1]Max-Planck-Institut für Radioastronomie, Auf dem Hügel 69, 53121 Bonn, Germany
Astron. Dept., King Abdulaziz University, P.O. Box 80203, Jeddah, Saudi Arabia
email: `chenkel@mpifr-bonn.mpg.de`

[2]National Radio Astronomy Observatory, 520 Edgemont Road,
Charlottesville, VA 22903, USA
email: `jbraatz@nrao.edu`

[3]Harvard-Smithsonian Center for Astrophysics, 60 Garden Street,
Cambridge, MA02138, USA
email: `reid@cfa.harvard.edu`

[4]National Radio Astronomy Observatory, 520 Edgemont Road,
Charlottesville, VA 22903, USA
email: `jcondon@nrao.edu`

[5]National Radio Astronomy Observatory, 520 Edgemont Road,
Charlottesville, VA 22903, USA
email: `flo@nrao.edu`

[6]National Radio Astronomy Observatory, 520 Edgemont Road, Charlottesville,
VA 22903, USA
Joint ALMA Observatory, Alonso de Córdova 3107, Vitacura, Santiago, Chile
email: `violette@nrao.edu`

[7]Dept. of Astronomy, University of Virginia, Charlottesville, VA 22904, USA
AASIA, Astron.-Math. Building, Roosevelt Rd, Taipei 10617, Taiwan
email: `ck2v@virginia.edu`

Abstract. The Hubble constant H_0 describes not only the expansion of local space at redshift $z \sim 0$, but is also a fundamental parameter determining the evolution of the universe. Recent measurements of H_0 anchored on Cepheid observations have reached a precision of several percent. However, this problem is so important that confirmation from several methods is needed to better constrain H_0 and, with it, dark energy and the curvature of space. A particularly *direct* method involves the determination of distances to local galaxies far enough to be part of the Hubble flow through water vapor (H_2O) masers orbiting nuclear supermassive black holes. The goal of this article is to describe the relevance of H_0 with respect to fundamental cosmological questions and to summarize recent progress of the "Megamaser Cosmology Project" (MCP) related to the Hubble constant.

Keywords. masers, galaxies: active, galaxies: ISM, galaxies: nuclei, cosmology: cosmological parameters, cosmology: distance scale, radio lines: galaxies

1. Cosmological Background

For 85 years, it has been known that our universe is expanding (Lemaître 1927). This expansion was believed to slow down in time because of gravitational attraction. However, based on observations of luminous standard candles (Type Ia supernoave) Riess *et al.*

(1998) and Perlmutter *et al.* (1999) suggested instead accelerated expansion, turning cosmology upside down and winning the most recent Nobel Prize in physics. More than a decade after this discovery, accelerated expansion is well established. A de-acceleration during the initial few billion years after the "Big Bang", when densities of matter and radiation were much higher than today, is followed by accelerated expansion. The cause of the accelerated expansion is so far unknown and is described by the term "Dark Energy". Following the standard model, it should account for the majority of the energy density of the universe and retards the formation of large scale structure. Understanding dark energy may be the most important problem existing in physics today.

Dark Energy dominates the energy budget, accelerates the expansion of the universe, and affects large scale structure. What is its nature? There are three classes of potential explanations: (1) a cosmological constant, which has been proposed already in the early days of general relativity (Einstein 1917) as a kind of repulsive gravity, (2) a scalar field, somewhat analogous to that proposed to explain inflation at a much earlier time (e.g., Wetterich 1988; Ratra & Peebles 1988), and (3) modified gravity (e.g., Tsujikawa 2010), which will not be considered here.

Assuming that the universe is homogeneous and isotropic (as approximately suggested by the large scale matter distribution and the 3 K microwave background), the space-time metric can be written in the following form

$$ds^2 = dt^2 - a^2(t) \times [dr^2/(1 - kr^2) + r\,d\theta^2 + r^2 \sin^2\theta\,d\phi^2], \tag{1.1}$$

with t and $a \propto (1 + z)^{-1}$ being time and cosmic scale factor, r, θ, and ϕ denoting comoving spatial coordinates, and k representing the curvature of 3-dimensional space. The field equations of general relativity, applied to this Friedmann-Robertson-Walker metric, lead to the so-called Friedmann equations,

$$H = \left(\frac{\dot{a}}{a}\right)^2 = \frac{8\,\pi\,G}{3c^2}\,\rho - k\,\frac{c^2}{a^2} + \frac{\Lambda}{3} \tag{1.2}$$

and

$$\frac{\ddot{a}}{a} = -\frac{4\,\pi\,G}{3c^2}\,(\rho + 3p) + \frac{\Lambda}{3}. \tag{1.3}$$

H is the Hubble parameter for a given time (H_0 stands for redshift $z = 0$), ρ is the density of matter and radiation, p denotes pressure, and Λ represents the traditional cosmological constant, which (like dark energy) can be subsumed into the density and pressure term,

$$\frac{\ddot{a}}{a} = -\frac{4\,\pi\,G}{3c^2}\,\sum_i^n (\rho_i + 3p_i). \tag{1.4}$$

Defining $w = p/\rho$, the different (in part putative) components yield:

Matter: $w = 0$

Radiation: $w = 1/3$

"Quintessential" scalar field: $-1 < w < -1/3$

Cosmological constant: $w = -1$

Phantom energy: $w < -1$

The resulting values of w determine the normalized acceleration $\ddot{a}/a$. For gravity (matter) we obtain the expected negative value. This also holds for radiation, which has dominated, according to the standard model, at redshifts $\gtrsim 10^4$. The $(\rho + 3p)$ term in Eq. 1.4 directly infers that w values smaller than $-1/3$ are required for accelerated expansion,

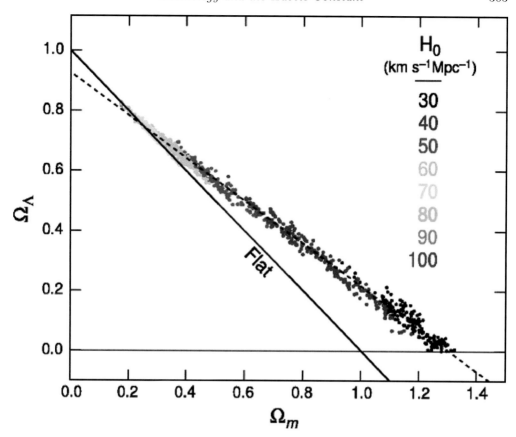

Figure 1. Constraints from the cosmic microwave background (Spergel *et al.* 2007). The different colors correspond to different values of the Hubble constant H_0. Ω_m is the energy density of baryonic and dark matter in units of the critical density, Ω_Λ is the corresponding parameter for dark energy. For a "flat" universe, $\Omega_m + \Omega_\Lambda = 1$. The data are consistent with a wide range of H_0 values.

which is therefore the possible w range for dark energy. w and H, and thus also H_0, are obviously related, emphasizing the cosmological importance of the Hubble constant.

Fig. 1 shows the range of not necessarily flat cosmological cold dark matter (CDM) models consistent with Wilkinson Microwave Anisotropy Probe (WMAP) data (Spergel et al. 2007). Assuming that the universe is flat would provide a rather accurate solution, but is it really flat as suggested by inflationary models? The present value of the curvature radius, R_0, is related to H_0 and $\Omega_0 = \Omega_m + \Omega_\Lambda$ (see Fig. 1 for definitions) by

$$R_0 = (a/2) \times |k|^{-1/2} = (c/2) \times H_0^{-1} \times (\Omega_0 - 1)^{-1/2}. \qquad (1.5)$$

c is the speed of light and $(c/2) \times H_0^{-1} \sim 2.1\,\mathrm{Gpc}$ is known as the Hubble radius. Obviously, with our present precision to determine Ω_0, curvature radii as small $10\,\mathrm{Gpc}$ cannot be excluded and we are still far from being able to state that the universe is truly flat. All this implies that the cosmic microwave background alone does not provide stringent limits. This is not unexpected, since the CMB provides information at a single early epoch, when dark energy did not play a role. Its importance can only be deduced by a combination of CMB observations with data from the much younger universe.

There are several lines of observational activity to constrain the wide range of models permitted by the CMB: These include (1) Type Ia supernovae, the standard candles, where luminosity distances and redshifts can be compared; (2) galaxy clusters, where redshift independent ratios between baryonic and dark matter masses can only be obtained with a small subset of possible cosmologies; (3) gravitational lensing or cosmic shear, where the dark matter distribution can be determined, isolating dark energy; and (4) baryon acoustic oscillations (BAO), which left an imprint on the cosmic microwave background, which is clearly seen in CMB power spectra (e.g., Spergel *et al.* 2003, 2007; Komatsu *et al.* 2011). This imprint is still visible affecting structure at moderate redshifts and providing a local size scale of order 140 Mpc. All these tracers are observed at significant redshifts. However, it is at redshift zero where dark energy is most dominant, and only here its energy density is significantly higher than that of baryonic and dark matter combined. Thus it is the Hubble constant, providing a measure of the *local* universe, which provides the longest lever arm with respect to the CMB to measure the effects of dark energy (Hu 2005).

2. Constraining H_0

Assuming a Λ cold dark matter universe and excluding curvature, exotic neutrino or specific early universe physics, Komatsu *et al.* (2011) derive from WMAP data $H_0 = 71.0 \pm 2.5 \, \mathrm{km \, s^{-1} \, Mpc^{-1}}$. While studies based on gravitational lens time delays (e.g., Treu & Koopmans 2002; Cardone *et al.* 2002) and on X-ray and Sunyaev-Zel'dovich data of galaxy clusters (e.g., Bonamente *et al.* 2006) have been used to constrain H_0, such deductions from redshifted objects depend on the chosen cosmological model and are no substitute for a measurement of H_0 in the local universe.

With the HST key project to measure the Hubble constant, Freedman *et al.* (2001) obtained from Cepheids in nearby galaxies $H_0 = 72 \pm 3_{\mathrm{r}} \pm 7_{\mathrm{s}} \, \mathrm{km \, s^{-1} \, Mpc^{-1}}$, estimating both random and systematic errors. This was based on the extragalactic distance ladder using Cepheid variable stars in the Large Magellanic Cloud to calibrate the measurements. However, the LMC has a low metallicity and significant depth. A potential dependence of Cepheid luminosities on metallicity and uncertainties in the distances to individual stars complicate the calibration. This was highlighted by Sandage *et al.* (2006), who used similar methods but a different metallicity correction to obtain $H_0 = 62 \pm 1.3_{\mathrm{r}} \pm 5.0_{\mathrm{s}} \, \mathrm{km \, s^{-1} \, Mpc^{-1}}$ (Fig. 2). While the H_0 value by Freedman *et al.* (2001) is consistent with a flat universe, the Sandage *et al.* (2006) result challenges it.

To date the most ambitious program to determine H_0 is that led by A.G. Riess (see also Freedman & Madore 2010 for a recent review). Based on three anchors, (1) the parallax determinations of Galactic Cepheids, (2) Cepheides in the LMC with distances deduced from eclipsing binaries, (3) the distance to NGC 4258 (see Sect. 3) and with new *HST* (Hubble Space Telescope) data from galaxies with Cepheids *and* Type Ia supernovae, Riess *et al.* (2011) derive $H_0 = 73.8 \pm 2.4 \, \mathrm{km \, s^{-1} \, Mpc^{-1}}$. This estimate also makes use of Cepheids measured in the near infrared, which helps to reduce both systematic and random errors with respect to optical observations. Note that the distance to NGC 4258 has been slightly revised in the meantime (E.M.L. Humphreys, priv. comm.).

While all these measurements are highly encouraging and pave the way to a more precise knowledge of our universe, a totally independent measure of H_0 is essential to either confirm the above cited results or to hint at problems that may have been overlooked so far.

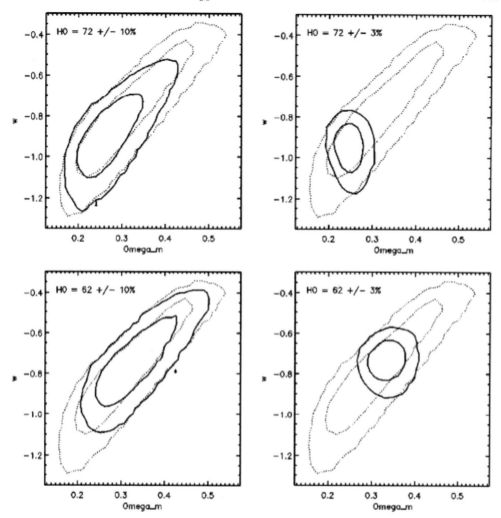

Figure 2. WMAP (Wilkinson Microwave Anisotropy Probe) 1σ and 2σ likelihood surfaces for $\Omega_{\rm m}$ (see Fig. 1) and w (Eq. 1.4), with priors on H_0 (upper panels: $72\,{\rm km\,s^{-1}\,Mpc^{-1}}$; lower panels: $62\,{\rm km\,s^{-1}\,Mpc}$). Right versus left panels (solid lines) demonstrate the improvements gained by reducing the uncertainty in H_0 from 10% to 3%. Dotted lines: "wcdm + no perturbations" model from Spergel *et al.* (2007); solid lines: the same, but with constraints from H_0 incorporated.

3. The Megamaser Cosmology Project (MCP)

The MCP is an NRAO (National Radio Astronomy Observatory)† key project to determine H_0 by measuring geometric distances with an accuracy of $\sim$10% to $\sim$10 galaxies in the local Hubble flow. Following the experience gained by studying the prototypical source, NGC 4258 (Miyoshi *et al.* 1995; Herrnstein *et al.* 1999), it includes (1) a GBT (Green Bank Telescope) survey to identify suitable circumnuclear 22 GHz H_2O maser disks (see Fig. 3 for a spectrum), (2) direct imaging of these sub-pc disks, using the *VLBA* (Very Long Baseline Array), the *GBT*, and, for northern sources, also the

† The National Radio Astronomy Observatory is a facility of the National Science Foundation operated under cooperative agreement by Asscociated Universities, Inc.

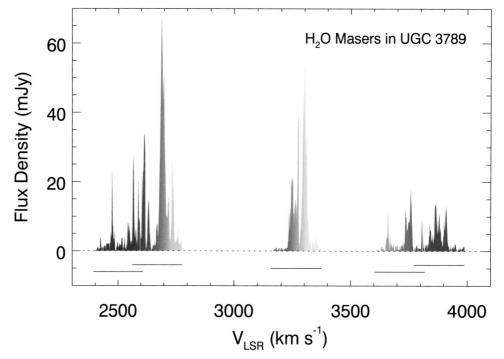

Figure 3. One of the newly found maser disks with systemic (green-yellow) and high velocity (violet-blue and orange-red) components.

Effelsberg telescope‡, (3) *GBT* monitoring to measure accelerations of the spectral components, and (4) model calculations to simulate the maser disk dynamics. So far published articles include Reid *et al.* (2009), Braatz *et al.* (2010), Greene *et al.* (2010), and Kuo *et al.* (2011).

The 22 GHz H_2O "megamasers", luminous masers associated with active galactic nuclei, are mostly found in Seyfert 2 and LINER (Low Ionization Nuclear Emission Line Region) galaxies with high column densities (Braatz et al. 1997; Zhang *et al.* 2006; Madejski *et al.* 2006; Greenhill *et al.* 2008), relatively high optical luminosity, velocity dispersion, and [OIII]λ5007 luminosity (Zhu *et al.* 2011) as well as relatively strong Fe Kα lines in those sources which are Compton thick (Zhang *et al.* 2010). Low X-ray/[OIV]λ25890 ratios (Ramolla *et al.* 2011), and high nuclear radio continuum luminosities (Zhang *et al.* 2012) with respect to H_2O undetected galaxies are also statistically obtained.

The H_2O masers reside in thin, edge-on gaseous annuli. Emission near the systemic velocity of the parent galaxy originates from the near side of the disk and red- and blue-shifted satellite lines come from the two tangent points (see Figs. 3–5). Assuming an ideal circular, warpless thin disk, seen perfectly edge-on, and in Keplerian motion, the mass M_{AGN} enclosed by the disk is

$$M_{\mathrm{AGN}} = 1.12 \times \left[\frac{V_{\mathrm{rot}}}{\mathrm{km\,s^{-1}}}\right]^2 \times \left[\frac{R}{\mathrm{mas}}\right] \times \left[\frac{D}{\mathrm{Mpc}}\right] \ \mathrm{M_\odot}, \qquad (3.1)$$

with V_{rot} denoting the rotation velocity at angular radius R and D representing the

‡ Based on observations with the 100-m telescope of the MPIfR (Max-Planck-Institut für Radioastronomie) at Effelsberg.

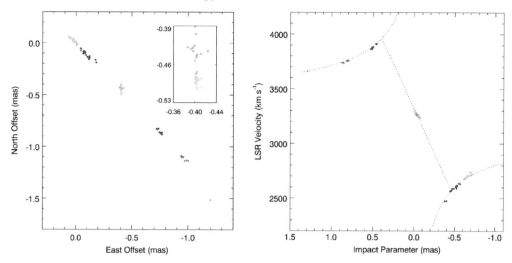

Figure 4. Left panel: H_2O image of the maser disk in UGC 3789 (from Reid et al. 2009). The insert presents a magnification of the systemic features. Right panel: Radial velocity versus impact parameter (also Reid *et al.* 2009). For the blue- and red-shifted high velocity components, Keplerian $r^{-1/2}$ rotation curves are also displayed.

distance. Very long baseline interferometry maps allow us to directly measure V_{rot} for various values of R. From the Keplerian rotation curve we then obtain the constant

$$C_1 = \left[\frac{V_{rot}}{km\,s^{-1}}\right] \times \left[\frac{R}{mas}\right]^{1/2}. \tag{3.2}$$

The velocity gradient of the systemic features as a function of impact parameter provides another constant,

$$C_2 = \left[\frac{V_{rot}}{km\,s^{-1}}\right] \times \left[\frac{R}{mas}\right]^{-1}. \tag{3.3}$$

$C_1/C_2 = R^{3/2}$ then gives the angular radius R_s of the systemic features as viewed from a direction in the plane of the disk, but perpendicular to the line of sight. The total distance to the galaxy is then determined by the centripetal acceleration

$$dV_s/dt = \frac{V_{rot}^2}{r_s}, \tag{3.4}$$

with the index "s" denoting the systemic maser components. With dV_s/dt being measured, the linear scale r_s can be compared with the angular scale R_s to provide the preliminary distance estimate.

While such a procedure leads to a rough first estimate, disks may be warped (see Miyoshi *et al.* 1995 for the first such case, NGC 4258), orbits may be eccentric, and inclinations may not equal $90°$. To model the circumnuclear disks as seen in H_2O as detailed as possible, a Bayesian fitting procedure has been developed (M.J. Reid), using a Markov Chain Monte Carlo approach. A Metropolis Hastings algorithm is applied to choose successive trial parameters covering the parameter space. Fig. 6 displays such a simulation for NGC 6264. These very preliminary simulations indicate low eccentricities ($e < 0.1$).

For UGC 3789, Fig. 5 indicates two groups of systemic maser components, one with a higher acceleration than the other. Unlike in NGC 4258, the systemic features do not arise from a single ring segment with specific galactocentric radius. The distance to the

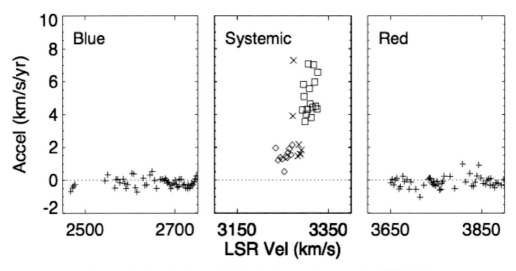

Figure 5. Acceleration of individual maser features in UGC 3789

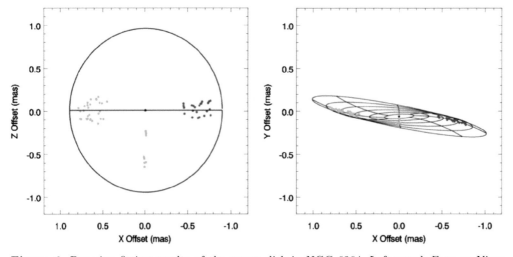

Figure 6. Bayesian fitting results of the maser disk in NGC 6264. Left panel: Face-on View onto the disk with systemic (green), approaching (blue) and receding (red) components. Right panel: Model of the warped disk viewed approximately edge-on as observed from Earth.

galaxy can be derived separately for the two rings. These yield $D_1 = 50.2 \pm 7.7\,\mathrm{Mpc}$ and $D_2 = 48.1 \pm 17.4\,\mathrm{Mpc}$. The weighted mean is $D_{\mathrm{UGC3789}} = 49.9 \pm 7.0\,\mathrm{Mpc}$ (14%). With a peculiar radial velocity relative to the cosmic microwave background of $-151 \pm 163\,\mathrm{km\,s^{-1}}$ and $V_{\mathrm{CMB}} = V_{\mathrm{LSR}} + 60\,\mathrm{km\,s^{-1}}$, the relativistic, recessional flow velocity becomes $3481 \pm 163\,\mathrm{km\,s^{-1}}$. This yields with standard ΛCDM parameters $H_0 = 69 \pm 11\,\mathrm{Mpc^{-1}}$ (Braatz *et al.* 2010).

For NGC 6264, a source at a distance of 150 Mpc, a similar analysis results in $H_0 = 65.8 \pm 7.2\,\mathrm{km\,s^{-1}}$ (Kuo 2011). Combining both sources, our present best estimate for the Hubble constant becomes $H_0 = 67 \pm 6\,\mathrm{km\,s^{-1}}$. This preliminary result is so far consistent with all previously (Sect. 2) mentioned values.

Figure 7. The maser disk of NGC 4258 (Argon *et al.* 2007), extended east-west, and below the maser disk of NGC6323, extended approximately north-south, on the same angular scale. The given beam shows a typical synthesized beam for a high-declination target at 22 GHz using the *VLBA*, the *GBT*, and *Effelsberg*. The sketch emphasizes the great progress achieved in recent years when mapping H_2O maser emission in distant sources.

4. Prospects

So far, two sources yielded publishable results. Increasing this number to 10 and accounting for the fact that the targets are located in different parts of the sky, the 1σ error of $6\,\mathrm{km\,s^{-1}\,Mpc^{-1}}$ obtained so far should decrease by a factor of $(2/10)^{1/2}$ to $\sim2.7\,\mathrm{km\,s^{-1}\,Mpc^{-1}}$ or 4%. Longer monitoring and more interferometric maps can reduce this uncertainty further. To demonstrate the degree of sensitivity required for these measurements, Fig. 7 shows the size of the prototypical nuclear disk in NGC 4258 on the same angular scale as the disk toward NGC 6323. While NGC 4258 is with $V \sim 500\,\mathrm{km\,s^{-1}}$ not yet in the Hubble flow and therefore not useful for a direct H_0 estimate (its maser lines are nevertheless essential to calibrate the distance scale defined by Cepheids), NGC 6323 has the potential to probe the distance scale with its recesseional velocity of almost $7800\,\mathrm{km\,s^{-1}}$. Toward NGC 6323, NGC 1194, NGC 2273, and Mrk 1419 the maser disks have also been mapped, demonstrating that the technique to derive distances, first tried out on NGC 4258, can also be used for much more distant galaxies.

Figs. 3, 5, and 7 directly demonstrate the importance of sensitivity. While in UGC 3789 many features have flux densities of $\sim5\,\mathrm{mJy}$ or higher, which can be readily analyzed, more distant sources reveal a plethora of components below this critical level. Getting these components as well would greatly facilitate any analysis. Thus the inclusion of a phased *Jansky VLA* (Very Large Array) is highly desirable. The completion of the Sardinia telescope may also help in the foreseeable future. Furthermore, as Fig. 7 indicates, angular resolution is another essential point. While an SKA-high would guarantee extreme sensitivity, the small angular extent of the H_2O maser disks requires a world-wide array, with space-VLBI providing another significant improvement.

Aside of the maser sources mentioned above, new targets have been detected, which look promising when analyzing their single dish spectra. While it remains to be seen how useful they will be for detailed mapping, it is worth mentioning that so far no systemic feature has been detected that shows a secular drift to the blue side. Either the nuclei are opaque at 22 GHz or the radiation is so highly beamed that maser photons from the backside of the disks have no chance to reach us.

References

Argon, A. L., Greenhill, L. J., Reid, M. J., *et al.* 2007, *ApJ*, 659, 1040
Bonamente, M., Joy, M. K., LaRoque, S. J., *et al.* 2006, *ApJ*, 647, 25
Braatz, J. A., Wilson, A. S., & Henkel, C. 1996, *ApJS*, 110, 321

Braatz, J. A., Reid, M. J., Humphreys, E. M. L. *et al.* 2010, *ApJ*, 718, 657

Cardone, V. F., Capozziello, S., Re, V., & Piedipalumbo, E. 2002, *A&A*, 382, 792

Einstein, A. 1917, *Sitzungsber. Königl. Preuß. Akad. der Wiss.*, 6, 142

Freedman, W. L. & Madore, B. F. 2010, *ARA&A*, 48, 673

Freedman, W. L., Madore, B. F., Gibson, B. K., *et al.* 2001, *ApJ*, 553, 47

Greenhill, L. J., Tilak, A., & Madejski, G. 2008, *ApJ*, 686, L13

Greene, J. E., Peng, C. Y., Kim, M. *et al.* 2010, *ApJ*, 721, 26

Herrnstein, J. R., Moran, J. M., Greenhill, L. J. *et al.* 1999, *Nature*, 400, 539

Hu, W. 2005, *ASP Conf. Ser.* 339, Observing Dark Energy, eds. S. C. Wolff & T. R. Lauer (San Francisco, ASP), 215

Komatsu, E., Smith, K. M., Dunkley, J., *et al.* 2011, *ApJS*, 192, 1

Kuo, C. Y. 2011, *Ph.D. Thesis*, Univ. of Virginia, Charlottesville

Kuo, C. Y., Braatz, J. A., Condon, J. J. *et al.* 2011, *ApJ*, 727, 20

Lemaître, G. 1927, *Annales de la Société Scientifique de Bruxelles*, 47, 49

Madejski, G., Done, C., & Zycki, P. T. 2006, *ApJ*, 636, 75

Miyoshi, M., Moran, J., Herrnstein, J. *et al.* 1995, *Nature*, 373, 127

Perlmutter, S., Aldering, G., Goldhaber, G., *et al.* 1999, *ApJ*, 517, 565

Ramolla, M., Haas, M., Bennert, V. N., & Chini, R., 2011 *A&A*, 530, 147

Ratra, B. & Peebles, P. J. E. 1988 *Phys. Rev. D*, 37, 3406

Reid, M. J., Braatz, J. A., Condon, J. J., *et al.* 2009, *ApJ*, 695, 287

Riess, A. G., Filippenko, A. V., Challis, P., *et al.* 1998, *AJ*, 116, 1009

Riess, A. G., Macri, L., Casertano, S., *et al.* 2011, *ApJ*, 730, 119

Sandage, A., Tammann, G. A., Saha, A., *et al.* 2006, *ApJ*, 653, 843

Spergel, D. N., Verde, L., Peiris, H. V., *et al.* 2003, *ApJS*, 148. 175

Spergel, D. N., Bean, R., Doré, O., *et al.* 2007, *ApJS*, 170. 377

Treu, T. & Koopmans, L. V. E. 2002, *MNRAS*, 337, L6

Tsujikawa, S. 2010, *Lect. Notes in Phys.*, 800, 99

Wetterich, C. 1988 *Nucl. Phys. B*, 302, 668

Zhang, J. S., Henkel, C., Kadler, M., *et al.* 2006 *A&A*, 450, 933

Zhang, J. S., Henkel, C., Gui, Q., *et al.* 2010 *ApJ*, 708, 1582

Zhang, J. S., Henkel, C., Gui, Q., & Wang, J. 2012 *A&A*, 538, 152

Zhu, G., Zaw, I., Blanton, M. R., & Greenhill, L. J. 2011 *ApJ*, 742, 73

Cosmic Masers - from OH to H_0
Proceedings IAU Symposium No. 287, 2012
R.S. Booth, E.M.L. Humphreys & W.H.T. Vlemmings, eds.

© International Astronomical Union 2012
doi:10.1017/S1743921312007235

Mrk 1419 - a new distance determination

C. M. Violette Impellizzeri[1,2], James A. Braatz[1], Cheng-Yu Kuo[3], Mark J. Reid[4], K. Y. Lo[1], Christian Henkel[5] and James J. Condon[1]

[1]National Radio Astronomy Observatory, 520 Edgemont Road, Charlottesville, USA
email: vimpelli@alma.cl

[2]Joint Alma Observatory, Alónso de Cordova, Vitacura, Santiago, Chile

[3]Institute of Astronomy and Astrophysics, Academia Sinica, Taipei 106, Taiwan

[4]Harvard-Smithsonian Center for Astrophysics, 60 Garden Street, Cambridge, USA

[5]Max-Planck-Institut für Radioastronomie, Auf dem Hügel 69, 53121 Bonn, Germany

Abstract. Water vapor megamasers from the center of active galaxies provide a powerful tool to trace accretion disks at sub-parsec resolution and, through an entirely geometrical method, measure direct distances to galaxies up to 200 Mpc. The Megamaser Cosmology Project (MCP) is formed by a team of astronomers with the aim of identifying new maser systems, and mapping their emission at high angular resolution to determine their distance. Two types of observations are necessary to measure a distance: single-dish monitoring to measure the acceleration of gas in the disk, and sensitive VLBI imaging to measure the angular size of the disk, measure the rotation curve, and model radial displacement of the maser feature. The ultimate goal of the MCP is to make a precise measurement of H_0 by measuring such distances to at least 10 maser galaxies in the Hubble flow. We present here the preliminary results from a new maser system, Mrk 1419. Through a model of the rotation from the systemic masers assuming a narrow ring, and combining these results with the acceleration measurement from the Green Bank Telescope, we determine a distance to Mrk 1419 of 81 ± 10 Mpc. Given that the disk shows a significant warp that may not be entirely traced by our current observations, more sensitive observations and more sophisticated disk modeling will be essential to improve our distance estimation to this galaxy.

1. Introduction

The Megamaser Cosmology Project (MCP) aims to determine the Hubble Constant to within 3%, by accurately measuring the distance to 10 galaxies in the Hubble flow. The Hubble Constant is an important complement to CMB data for constraining the nature of Dark Energy, the geometry (flatness) of the universe, and the fraction of the critical density contributed by matter. The technique used by the MCP, first pioneered by Herrnstein *et al.* (1999) to measure the distance to NGC 4258, uses water megamaser emission at 22 GHz from the center of active galaxies to trace the inner disk geometries at high angular resolution (mas-scale), and thus determine their angular size. The angular size of this inner disk is then compared to the linear size measured through single dish observations, yielding the distance to the galaxy. The ability to image these objects at such high angular resolution comes through the very high brightness provided by the maser process, the maser disks observed by the MCP are usually extremely compact, extending to $\ll 1$ pc from the central black hole. The spectral signature of such a maser disk is a cluster of systemic H_2O features, and two additional H_2O clusters, one red- and one blue-shifted with respect to the cluster of systemic features. Maser disks suitable for the distance technique require a special geometry (the nuclear accretion disk has to be edge-on for significant maser amplification; Lo, 2005), and therefore are extremely rare, so many galaxies must be surveyed to find good candidates, which are then followed up with VLBI.

The precision obtained by the MCP in determining H_0 depends on the quality of the individual measurements, but also on the number of galaxies that can be measured, their distance distribution, and distribution on the sky. An overall 3% precision in H_0 therefore can be achieved by measuring the distances to 10 galaxies if each distance could be measured to 10% precision, assuming the individual distance measurements are uncorrelated. There are currently about 150 galaxies detected in water vapor maser emission, of which about one third show some evidence of disk origin. NGC 4258 is the only galaxy with a 10% or better distance determination, but it not suitable to constrain H_0 directly as it is too close and could be mostly affected by peculiar motion. On the other hand, the MCP is currently studying in detail six H_2O maser disks in galaxies which are well into the Hubble flow (e.g Braatz *et al.* 2010, Kuo *et al.* 2011). However, because a broad distribution of megamaser sources in the sky is essential for reducing measurement uncertainties, surveys to find more such galaxies remain crucial for the success of the project.

We present here recent results on Mrk 1419, which is one of galaxies studied within the MCP with the aim of determining an accurate geometric distance. The megamaser in Mrk 1419 was discovered in 2002 by Henkel *et al.* with the 100 m Effelsberg telescope, and it was the first maser after NGC 4258 to display the characterestic "disk signature", but is ten times farther away. Through single dish monitoring with Effelsberg, the authors could already measure the secular acceleration of the systemic components, and therefore concluded that the maser emission in Mrk 1419 must arise from an almost edge-on circumnuclear disk.

2. Observations and calibration

We observed Mrk 1419 with global VLBI between May 2009 and January 2011, for a total of six epochs of 12 hours each. The global VLBI array comprises the VLBA, the GBT, and Effelsberg. In two of these epochs, we also added the EVLA, tuned to the frequency of the systemic masers, in order to improve the signal-to-noise level in this part of the spectrum.

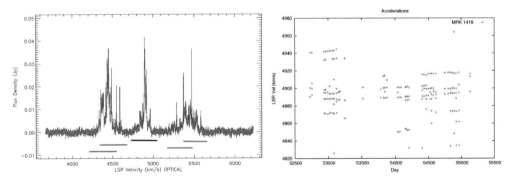

Figure 1. *Left:* GBT spectrum of Mrk 1419, taken on December 16, 2009. At the bottom, the position of the VLBI bands are marked for each of the spectral components. Note that for the red and systemic bands, each IF represents two polarizations. *Right:* Plot of the systemic maser velocity (on the left shown with the black bar underneath) as a function of time. The velocity of each maser spot is determined from the GBT observations. The slope in the fit gives the acceleration for each component.

All our observations were carried out in self-calibration mode. Figure 1 (left) shows a typical single dish spectrum of Mrk 1419, taken in December, 2009. The systemic masers are typically 40 mJy, and range over $150 \, \mathrm{km \, s^{-1}}$. Given the relative weakness of the masers in this source, and in order to improve the quality of our calibration, we self-calibrated the data using a clump of systemic masers spreading over $10 \, \mathrm{km \, s^{-1}}$. VLBI observations were carried out with four IF bands and two polarizations (RCP and LCP), each of 16 MHz. Two of the bands were centred on the galaxy systemic velocity, two further were centered on the blue-shifted part of the spectrum, and the last two were centered on the red-shifted part of the spectrum, offset from each other because of the larger spread in velocity in this part (Figure 1, left). "Geodetic" blocks were placed at the start and end of our observations, in order to solve for atmospheric and clock delay residuals for each antenna.

Calibration was performed using AIPS, and included an a priori phase and delay calibration, zenith and atmospheric delays and clock drifts (with the geodetic block data), flux density calibration, a manual phase calibration to remove delay and phase differences among all bands, and selecting a maser feature as the interferometer phase-reference. After calibrating each dataset separately, the data were "glued" together, and imaged in all spectral channels for each of the IF bands. The image from each spectral channel appeared to contain a single maser spot, which we fitted with a Gaussian brightness distribution in order to obtain positions and flux densities.

Single dish GBT monitoring of Mrk 1419 was performed with approximately one observation per month, except for the summer months when the humidity makes observations at 22 GHz inefficient. The GBT spectrometer was configured with two 200 MHz spectral window each with 8192 channels, one centered on the systemic velocity of the galaxy and the second offset by 180 MHz. Each observation was carried out for about 4 h. Finally, data calibration was performed in GBTIDL, with a low order-polynomial fit to the line-free channels to remove the spectral baseline.

3. Results and discussion

We present here preliminary results from three VLBI epochs, out of the six epochs observed overall, and from the GBT acceleration measurement performed around those epochs. The VLBI epochs, labeled as BB261N, P and R, were all observed between December 2009 and January 2010. The rms noise level for each of the VLBI maps is $\sim 0.8 \, \mathrm{mJy \, beam^{-1}}$. Figure 2 (left) shows the maser distribution on the sky, for the three epochs combined, with east-west, and north-south offsets (in mas) relative to the maser components at systemic (black symbols). The position angle of the maser disk is $-131°$ and the inclination with respect to the observer is $89°$. The inner and outer radii of the disk are 0.13 and 0.37 pc, respectively. The disk shows some warping, especially towards the lower, blue-shifted part. While on the outer part the disk flattens out, there is clear evidence for a significant bending in the inner side. Towards the red-shifted part of the disk, however, the larger vertical spread in the maser distribution may be due in part to the masers being fainter ($\sim 10 \, \mathrm{mJy}$), lowering our signal-to-noise, but may also indicate a true scatter in the maser position, due to a larger inclination in this part of the disk, or may reveal a thicker disk. Figure 2 (right) shows the position-velocity, PV, diagram for Mrk 1419. The high-velocity masers trace a Keplerian rotation curve and the systemic masers fall on a linear slope. This slope can be extended to the rotation curve traced by the high-velocity features, and this intersection determines the angular radius of the disk and magnitude of the rotation velocity traced by systemic masers. The precise fit to the high velocity masers demonstrate that the disk is dominated by the gravitational

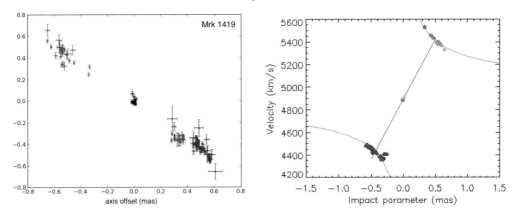

Figure 2. *Left:* Spatial distribution of the inner accretion disk traced by H_2O masers. *Right:* Position-velocity diagram for Mrk 1419.

potential of the supermassive black hole at the center. From this fit, we calculate the mass of the black hole to be $(1.16 \pm 0.05) \times 10^7$ M_{solar} (for a Hubble constant value of $H_0 = 73\,\mathrm{km\,s^{-1}\,Mpc}$; see Kuo *et al.* 2011).

From the parameters derived from both single dish and VLBI measurements, we measure the angular diameter distance to Mrk 1419 to be $81 \pm 10\,\mathrm{Mpc}$. Here, we calculate the distance using: $D = a^{-1}\,k^{2/3}\,\Omega^{4/3}$, where a is the acceleration from the GBT results (Figure 1, right), k is the Keplerian rotation constant, derived from the fit to the high velocity masers in the PV-diagram, and Ω is the slope velocity/impact parameter for the systemic masers (see Braatz *et al.* 2010). In our most simplified model of the disk, the maser emission originates in a thin, flat, edge-on disk and the dynamics are dominated by a central massive object, with all maser clouds in circular orbits. In this model, high-velocity masers trace gas near the tangential point at the edge of the disk. Systemic masers occupy part of a ring orbiting at a single radius and covering a small range velocities on the near side of the disk. The positive slopes seen in the maser velocities with time (Figure 1, right) indeed show clear evidence for the centripetal acceleration of masers, as they move across the line of sight in front of the central black hole. Using a "by-eye" method from similar plots, we measure two accelerations for the systemic masers, one for the masers with velocities $> 4940\,\mathrm{km\,s^{-1}}$, of $3.5\,\mathrm{km\,s^{-1}\,yr^{-1}}$, and one for the masers $< 4940\,\mathrm{km\,s^{-1}}$, of $2.1\,\mathrm{km\,s^{-1}\,yr^{-1}}$. Because the higher acceleration masers are fainter and not visible in our VLBI maps, we only take the lower acceleration into account for the distance estimation. While it is clear that with more than one acceleration the systemic masers likely originate from more than one radial distance from the black hole, as in our simplified assumption, more sensitive VLBI observations will be extremely important in the future to better constrain our models with the available information. Finally, we determined the slope for the systemic masers in the PV-diagram, Ω, by rotating the disk on the sky by $45°$ (counterclockwise in Figure 2, left) and measured the impact parameter for each maser component by its abscissa on the rotated axes. This method worked well, but has the caveat that the linear slope in the PV diagram of Mrk 1419 is best fitted when the systemic features are rotated by a smaller angle than the disk further out, thus giving further evidence for the presence of a significant warp in the part of the map that is not directly traced by the masers.

4. Summary and future work

We presented here VLBI images and single-dish GBT results of the water vapor masers in Mrk 1419. The spatial distribution of the masers in this source is nearly linear, with high-velocity masers on both sides of the masers at the galaxy systemic velocity. The water masers trace gas in Keplerian orbits at radii of $\sim 0.2\,\mathrm{pc}$, moving under the influence of a $\sim 1.16 \times 10^7\,\mathrm{M_{solar}}$ black hole. We model the rotation from the systemic masers assuming a narrow ring, and combine our results with the acceleration measurement from single dish observations to determine a distance to Mrk 1419 of $81 \pm 10\,\mathrm{Mpc}$. The main source of uncertainty in the distance comes from the measurement of the orbital curvature parameter Ω, and the uncertainty in the acceleration, while the contribution from the Keplerian rotation constant is negligible. However, the complex geometry in this source is evident from a significant warp in the disk, and the presence of more than one ring for the systemic masers. A more sophistcated modeling of the maser disk using a Bayesian fitting will therefore help solve these complications, and some first results using this method look promising (see Figure 3). Finally, the addition of more sensitive VLBI epochs to our analysis will improve the signal-to-noise ratio and can reduce the distance uncertainty to about 10%.

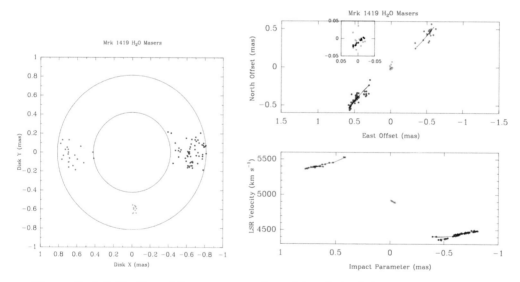

Figure 3. *Left:* Position of the masers seen from "above" the disk, determined from the output of the Bayesian fitting program. *Right:* Comparison between model and data.

References

Braatz, J. A., Reid, M. J., Humphreys, E. M. L., Henkel C., Condon, J. J., & Lo, K. Y. 2010, *ApJ*, 718, 657

Henkel, C., Braatz, J. A., Greenhill, L. J., & Wilson, A. S. 2002, *A&A*, 394, 23

Herrnstein, J. R., Moran, J. M., Greenhill, L. J., Diamond, P. J., Inoue, M., Nakai, N., Miyoshi, M, Henkel, C., & Riess, A. 1999, *Nature*, 400, 539

Kuo, C.-Y., Braatz, J. A., Condon, J. J., Impellizzeri, C. M. V., Lo, K. Y., Zaw, I., Schenker, M., Henkel, C., Reid, M., & Greene, J. 2011, *ApJ*, 727, 20

Lo, K. Y. 2005, *ARA&A*, 43, 625

Cosmic Masers - from OH to H_0
Proceedings IAU Symposium No. 287, 2012
R.S. Booth, E.M.L. Humphreys & W.H.T. Vlemmings, eds.

© International Astronomical Union 2012
doi:10.1017/S1743921312007247

Optical Properties of the Host Galaxies of Extragalactic Nuclear H_2O Masers

Ingyin Zaw[1,2], Guangtun Zhu[2], Michael Blanton[2], and Lincoln J. Greenhill[3]

[1]New York University Abu Dhabi
P.O. Box 129188, Abu Dhabi, UAE
email: ingyin.zaw@nyu.edu

[2]Center for Cosmology and Particle Physics, New York University,
4 Washington Place, New York, NY 10003, USA

[3]Harvard-Smithsonian Center for Astrophysics 60 Garden St.,
Cambridge, MA 02138, USA

Abstract. Although most nuclear 22GHz ($\lambda = 1.35$ cm) H_2O masers are in Seyfert 2 and LINER galaxies, only a small fraction of such galaxies host water masers. We systematically study the optical properties of the galaxies with and without nuclear H_2O maser emission to better understand the relationship between H_2O maser emission and properties of the central supermassive black hole and improve the detection rates in future surveys. To this end, we cross-matched the galaxies from H_2O maser surveys, both detections and non-detections, with the Sloan Digital Sky Survey (SDSS) low-redshift galaxy sample. We find that maser detection rates are higher at higher optical luminosity (M_B), larger velocity dispersion (σ), and higher [O III] $\lambda5007$ luminosity, with [O III] $\lambda5007$ being the dominant factor, and that the isotropic maser luminosity is correlated with these variables. These correlations are natural if maser emission depends on the host SMBH mass and AGN activity. We also find that the detection rate is higher for galaxies with higher extinction. These results indicate that, by pre-selecting galaxies with high extinction-corrected [O III] $\lambda5007$ flux, future maser surveys can increase detections efficiencies by a factor of ~3 to ~5.

Keywords. masers, galaxies: active, galaxies: nuclei, galaxies: Seyfert, radio lines: galaxies

---

1. Introduction

Water maser emission at 22 GHz ($\lambda = 1.35$ cm) has been mainly associated with active galactic nuclei (AGNs) and is currently the only resolvable tracer of warm, dense molecular gas in the inner parsec of AGNs. These masers are found in a variety of regions, such as jets/winds and nearly edge-on, rotating accretion disks. The latter, "disk masers", have been used to measure many properties of the central engine, including accurate SMBH masses (e.g. Kuo *et al.* 2011), accretion disk geometry (e.g. Greenhill *et al.* 2003), and geometric distances to host galaxies (e.g. Humphreys *et al.* 2008, Braatz *et al.* 2010). Unfortunately, these extragalactic nuclear H_2O masers are extremely rare. Discovering what sets maser host galaxies apart from other AGNs would lead not only to a better understanding of the connections between H_2O maser emissions and the properties of the AGN but also make surveys more efficient.

Earlier studies have found that most H_2O masers are found in Seyfert 2's and LINERS (instead of Seyfert 1's) (e.g. Braatz *et al.* 1997, Kondratko *et al.* 2006b) and that maser emission is more likely in systems with high X-ray obscuring columns (e.g. Madejski *et al.* 2006, Zhang *et al.* 2006, Greenhill *et al.* 2003, Zhang *et al.* 2010). Correlations have also been reported between H_2O maser isotropic luminosity and 2-10 keV X-ray luminosity (Kondratko *et al.* 2006a) and far IR luminosity (Henkel *et al.* 2005). However, there are currently no large samples of AGNs with measured 2-10 keV X-ray luminosities, and a high far IR luminosity does not guarantee that the galaxy will also be an AGN. Although most H_2O masers are in Seyfert 2's and LINERs, most Seyfert 2's and LINERs do not have maser emission. Since maser surveys may start with a sample of Seyfert 2's and LINERs, they are already targeting galaxies with optical spectra. In addition, there exist large optical surveys, such as the Sloan Digital Sky Survey (SDSS), 2dF Galaxy Redshift Survey (2dFGRS) and 6dF Galaxy Survey (6dFGS). Consequently, if masers can be shown to be in galaxies with certain optical properties, large maser surveys can select targets based on these properties. We systematically studied the optical properties of the maser hosts and known non-detections by cross-matching the SDSS low-redshift galaxy catalog with all galaxies surveyed for maser emission. The results are presented in later sections and further details can be found in Zhu *et al.* 2011.

2. Galaxy Samples and Optical Property Selection

For a full sample of maser detections and non-detections, we combine the catalogs (as of 1 Dec. 2010) on the Megamaser Cosmology Project (MCP†) and Hubble Constant Maser Experiment (HoME‡) web sites. In order to restrict ourselves to masers associated with AGNs, we exclude the galaxies associated with star formation regions, as noted in the MCP or HoME catalogs. We also construct a master list of non-detections listed by MCP and HoME and remove duplicates. This resulted in a total of 123 known masers, of which ~40 are noted as disk masers on the HoME web site based on very long baseline interferometry (VLBI) maps or single-dish spectra, and 3806 non-detections, for an overall detection rate of ~3%.

The SDSS Data Release 7 (DR7; Abazajian *et al.* 2009) catalog provides a complete sample of galaxies with uniform imaging and spectroscopy to systematically study optical properties related to maser emission. Since the masers are mostly nearby and the SDSS photometry and spectroscopy samples are most complete at low redshifts, we limit ourselves to galaxies with z < 0.05 (excluding only 4 maser galaxies). For photometry, we use the low-z catalog from the NYU Value Added Galaxy Catalog (NYU-VAGC; Blanton *et al.* 2005) and for spectroscopic parameters, we use the measurements of the MPA-JHU group¶ (e.g. Tremonti *et al.* 2004). We cross-match this SDSS low-z sample with maser detections and non-detections. We find 48 detections (15 disk masers) and 1588 non-detections which have SDSS photometry, of which 33 detections (10 disk masers) and 1030 non-detections have well measured spectral properties. The detection rates in both the photometric and spectroscopic samples are ~3%. This is similar to the total sample, and therefore, the cross-matching should not cause a bias for the analysis of optical properties of maser host galaxies.

SDSS provides a plethora of photometric and spectroscopic measurements. We choose to study the following properties based on the assumption that maser emission should be related to properties of the central SMBH: [O III] $\lambda5007$ luminosity ($L_{[OIII]\lambda5007}$, in erg s^{-1}), a well-known bolometric luminosity indicator (Heckman *et al.* 2005), and velocity dispersion (σ, in km s^{-1}), correlated with SMBH mass (Ferrarese *et al.* 2000). Since [O III] $\lambda5007$ can be obscured by the material in the host galaxy and we expect masers to be in systems with high obscuration, we also look at the Balmer decrement, $H\alpha/H\beta$, and correct the observed [O III] $\lambda5007$ luminosity, $L_{[OIII]\lambda5007,obs}$, to obtain the

---

† https://safe.nrao.edu/wiki/bin/view/Main/MegamaserCosmologyProject
‡ https://www.cfa.harvard.edu/~lincoln/demo/HoME/index.html
¶ http://www.mpa-garching.mpg.de/SDSS/DR7/

intrinsic [O III] $\lambda5007$ luminosity, $L_{[\mathrm{OIII}]\lambda5007,\mathrm{cor}}$. Finally, we study B band magnitude, M_B, as an indicator of optical luminosity.

3. Results

We find that maser detection rates depend on the optical properties of host galaxies. The strongest effect is that detection rates are higher for higher [O III] $\lambda5007$ luminosity. Detection rates are also higher for galaxies with higher $H\alpha/H\beta$, and consequently, the extinction-corrected [O III] $\lambda5007$ luminosity, $L_{[\mathrm{OIII}]\lambda5007,\mathrm{cor}}$, distribution for maser host galaxies and non-detections are even more discrepant than for observed [O III] $\lambda5007$ luminosity, $L_{[\mathrm{OIII}]\lambda5007,\mathrm{obs}}$. Figure 1 shows the distributions of $L_{[\mathrm{OIII}]\lambda5007,\mathrm{obs}}$, $L_{[\mathrm{OIII}]\lambda5007,\mathrm{cor}}$, and $H\alpha/H\beta$ for maser hosts and non-detections as well as the detection rates with respect to these variables.

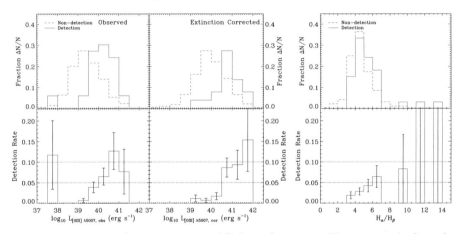

Figure 1. $L_{[\mathrm{OIII}]\lambda5007,\mathrm{cor}}$, $L_{[\mathrm{OIII}]\lambda5007,\mathrm{obs}}$, and Balmer decrement. The top panels show the distributions for maser detections (blue) and non-detections (red). The bottom panels show the detection rate as a function of these optical properties.

As shown in Figure 2, maser detection rates are also higher for higher B band magnitude, M_B, and velocity dispersion, σ. We denote the disk masers with green diamonds but, due to the smallness of the sample, we do not draw conclusions just based on the disk masers. Although weaker than the dependence on $L_{[\mathrm{OIII}]\lambda5007}$, the dependence of detection rates on M_B and σ persist even for galaxies with high $L_{[\mathrm{OIII}]\lambda5007}$ (Zhu *et al.* 2011).

We test whether the dependence of detection rate on $L_{[\mathrm{OIII}]\lambda5007}$, σ, and M_B is a consequence of an underlying correlation between these properties and maser luminosity. We collected 66 isotropic maser luminosities from literature (see table in Zhu *et al.* 2011). The left panel in Figure 4 shows the isotropic H_2O maser luminosity against redshift. The plot shows a flux-limited effect, with a maximum sensitivity of ~0.1 Jy km s^{-1}, similar to the sensitivities of surveys with the Green Bank Telescope (GBT) (e.g. Braatz *et al.* 2004). This is consistent with earlier findings that detection rates are higher for galaxies with smaller recession velocities (e.g. Braatz *et al.* 1997). We compare the isotropic luminosity with our optical properties of interest. We note that isotropic luminosity is not ideal for several reasons, e.g. maser emission is beamed and variable and that isotropic luminosities are from different observations with different sensitivities. In addition, we have supplemented our SDSS sample with measurements from literature to increase the sample size. All these effects will add scatter. Despite these shortcomings,

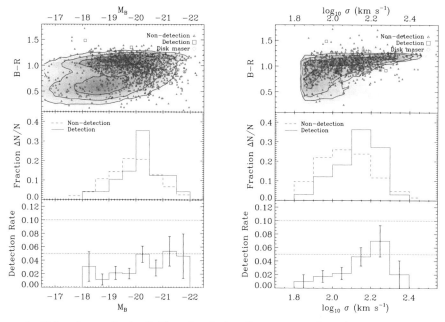

Figure 2. B band magnitude (left) and velocity dispersion, σ (right). Top panels show the color-magnitude diagrams with the full low-z photometric sample in grayscale, with contours representing 40%, 80%, and 90% of the sample. Middle panels show the distributions of these variables, and the bottom panels show the detection rate as a function of these variables.

we find that isotropic maser luminosity is correlated with $L_{[OIII]\lambda5007,cor}$, $L_{[OIII]\lambda5007,obs}$, σ, and M_B, albeit with large scatter, as shown in Figure 3. Larger samples are needed to further test these correlations and whether the correlations depend on maser environment (e.g. jets vs. disks).

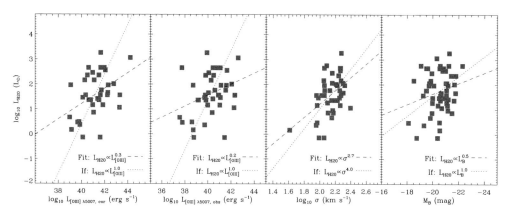

Figure 3. Relations between isotropic maser luminosity (L_{H2O}) and $L_{[OIII]\lambda5007,cor}$, $L_{[OIII]\lambda5007,obs}$, σ, and M_B of the host galaxies. The dashed lines are the linear least-squares fits assuming a uniform error of 0.5 dex in $\log_{10} L_{H2O}$. The dotted lines show the linear least-squares fits with the slopes fixed assuming that $L_{H2O} \propto L_{AGN} \propto M_{BH}$, and $M_{BH} \propto L_B \propto \sigma^4$, and $L_{AGN} \propto L_{[OIII]\lambda5007}$.

4. Discussion and Recommendations for Future Surveys

The correlations we see between isotropic maser luminosity with $L_{\text{[OIII]}\lambda5007,\text{cor}}$, $L_{\text{[OIII]}\lambda5007,\text{obs}}$, σ, and M_B would be natural if the strength of maser emission depends on AGN activity and SMBH mass. We fit these results to a simplified model in which we assume $L_{\text{H2O}} \propto L_{\text{AGN}} \propto M_{\text{BH}}$, $M_{\text{BH}} \propto L_B \propto \sigma^4$, and $L_{\text{AGN}} \propto L_{\text{[OIII]}\lambda5007}$ (see Zhu *et al.* 2011 for rationale). Using the intercepts from the fits, we translate the survey maser isotropic luminosity flux limit to corrected and observed [O III] $\lambda5007$ luminosity limits which are plotted against redshift in the right panel of Figure 4. If surveys are limited to the galaxies with high $L_{\text{[OIII]}\lambda5007}$, the detection rates could be improved by a factor of ~ 3 (for $L_{\text{[OIII]}\lambda5007,\text{obs}}$) to ~5 (for $L_{\text{[OIII]}\lambda5007,\text{cor}}$). Even if surveys do not cut on $L_{\text{[OIII]}\lambda5007}$, the galaxies should be ranked by $L_{\text{[OIII]}\lambda5007}$ so that discoveries are front-loaded. If $L_{\text{[OIII]}\lambda5007}$ is unavailable, surveys should rank galaxies by velocity dispersion or by optical luminosity.

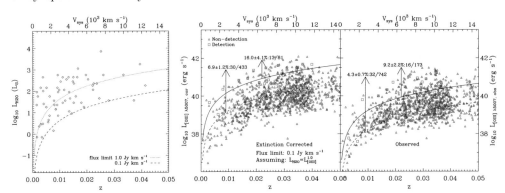

Figure 4. H_2O maser isotropic luminosity (left) and extinction-corrected [O III] $\lambda5007$ luminosity, $L_{\text{[OIII]}\lambda5007,\text{cor}}$, and observed [O III] $\lambda5007$ luminosity, $L_{\text{[OIII]}\lambda5007,\text{obs}}$ (right), as a function of redshift. The solid lines indicate a maser flux limit of 0.1 Jy km s^{-1} and the dashed lines are the flux limit shifted downward by -1.0 dex. The arrows and percentages are detection rates above the corresponding lines.

References

Abazajian K. N., Adelman-McCarthy, J. K., Agueros, M. A., *et al.* 2009, *ApJS*, 182, 543
Blanton, M. R., Schlegel, D. J., Strauss, M. A., *et al.* 2005, *AJ*, 129, 2562
Braatz, J. A., Wilson, A. S., & Henkel, C. 1997, *ApJS*, 110, 321
Braatz, J. A., Henkel, C., Greenhill, L. J., Moran, J. M., & Wilson, A. S. 2004, *ApJ*, 617, L29
Braatz, J. A., Reid, M. J., Humphreys, E. M. L., *et al.* 2010, *ApJ*, 718, 657
Ferrarese, L. & Merritt, D. 2000, *ApJ*, 539, L9
Greenhill, L. J., Booth, R. S., Ellingsen, S. P., *et al.* 2003, *ApJ*, 590, 162
Greenhill, L. J., Tilak, A., & Madejski, G. 2008, *ApJ*, 686, L13
Heckman, T. M., Ptak, A., Hornschemeier, A., Kauffmann, G., *et al.* 2005, *ApJ*, 643, 161
Henkel, C., Braatz, J. A., Tarchi, A., *et al.* 2005, *Ap&SS*, 295, 107
Humphreys, E. M. L., Reid, M. J., Greenhill, *et al.* 2008, *ApJ*, 672, 800
Kondratko, P. T., Greenhill, L. J., & Moran, J. M. 2006a, *ApJ*, 652, 138
Kondratko, P. T., Greenhill, L. J., Moran, J. M., *et al.* 2006b, *ApJ*, 638, 100
Kuo, C. Y., Braatz, J. A., Condon, J. J., *et al.* 2011, *ApJ*, 727, 20
Madejski, G., Done, C., Zycki, P. T., & Greenhill, L. 2006, *ApJ*, 636, 75
Tremonti, C. A., Heckman, T. M., Kauffmann, G., *et al.* 2004, *ApJ*, 613, 898
Zhang, J. S., Henkel, C., Kadler, M., *et al.* 2006, *A&A*, 450, 933
Zhang, J. S., Henkel, C., Guo, Q., Wang, H. G., & Fan, J. H. 2010, *ApJ*, 708, 1528
Zhu, G., Zaw, I., Blanton, M. R., & Greenhill, L. J. 2011, *ApJ*, 742, 73

T7:

AGN and megamasers *Chair: Lincoln Greenhill*

Cosmic Masers - from OH to H$_0$
Proceedings IAU Symposium No. 287, 2012
R.S. Booth, E.M.L. Humphreys & W.H.T. Vlemmings, eds.
© International Astronomical Union 2012
doi:10.1017/S1743921312007259

AGN and Megamasers

Andrea Tarchi

INAF - Osservatorio Astronomico di Cagliari, Capoterra (CA), Italy
email: `atarchi@oa-cagliari.inaf.it`

Abstract. Luminous extragalactic masers are traditionally referred to as the 'megamasers'. Those produced by water molecules are associated with accretion disks, radio jets, or outflows in the nuclear regions of active galactic nuclei (AGN). The majority of OH maser sources are instead driven by intense star formation in ultra-luminous infrared galaxies, although in a few cases the OH maser emission traces rotating (toroidal or disk) structures around the nuclear engines of AGN. Thus, detailed maser studies provide a fundamental contribution to our knowledge of the main nuclear components of AGN, constitute unique tools to measure geometric distances of host galaxies, and have a great impact on probing the, so far, paradigmatic Unified Model of AGN.

Keywords. Masers, Galaxies: active, Galaxies: nuclei, Radio lines: galaxies

1. Introduction

The widely accepted Unified Model of active galactic nuclei (AGN; e.g., Antonucci 1993, Urry & Padovani 1995) implies, in their very centres, the presence of a supermassive black hole surrounded by a parsec-scale accretion disk. The emission from the disk is particularly intense at ultraviolet (UV) and soft X-ray wavelengths. The accretion disk is then surrounded by a torus (or a thick disk) of atomic and molecular gas, with a size of 1-100 pc, that obscures the optical and UV emission along certain directions. Therefore, the object appears as either a type 1 or type 2 AGN depending on the line of sight. In type 1 AGN, the observer views the accretion disk and black hole through the hole in the torus, while in type 2 AGN the direct view of these nuclear components is obscured by the torus. The amount of radio loudness in each object (thus, if it is classified as a radio-quiet or radio-loud AGN) and its membership to an individual radio class of AGN (e.g., QSO, FR I, BL Lac, etc...) are instead more ascribable to the host galaxy type and/or to intrinsic properties of the nuclear components of the AGN (spin, mass, and accretion rate of the black hole, the relativistic jet power and orientation, etc...; see, e.g., Urry & Padovani 1995).

Studies of the central regions of AGN are complicated by the extremely small scales and complex structures of the nuclear components. In addition, particularly in type 2 AGN, the inner regions are often obscured at optical and UV wavelengths. Observations at infrared (IR, the band where most of the nuclear radiation absorbed by the torus is re-emitted), X-ray, and radio frequencies can, however, access these obscured regions. In particular, at radio wavelengths, water and OH maser studies are a unique tool for investigating the structure and kinematics of the gas close to and around the nuclear engines of AGN.

2. Masing molecules in extragalactic objects

The most common molecules found to produce maser emission in extragalactic environments are hydroxyl and water. The former molecule is detected at radio wavelengths from four hyperfine transitions of its ground rotational level with rest frequencies, in the radio band, of 1665, 1667, 1612, 1720 MHz. The emission from the first two transitions is the one typically observed in galaxies. The OH maser emission traces relatively warm $(100 < T_{kin} < 300$ K$)$ and dense $(10^4 < n(H_2) < 10^6$ cm$^{-3})$ gas. The water maser main line is instead produced by the transition between the rotational levels 6_{16} and 5_{23} at a rest radio frequency of 22.2 GHz. Extragalactic water maser emission at 183 GHz was also detected in two galaxies, NGC 3079 (Humphreys *et al.* 2005) and Arp 220 (Cernicharo, Pardo & Weiß 2006), that, however, will not be discussed in the following. The water maser emission traces much warmer $(T_{kin} > 300$ K$)$ and denser $(10^7 < n(H_2) < 10^{11}$ cm$^{-3})$ gas than OH and the emitting spots show extreme compactness and brightness temperatures.

While the association of water megamaser sources with AGN activity has been confidently assessed for a large number of galaxies, the association between OH megamasers and AGN processes is, so far, less understood. Hence, the remainder of this review will mostly focus on H_2O megamasers, although a brief description of OH megamasers and, in particular, on their association with the H_2O ones, will be reported in Sect. 4. More issues related to extragalactic OH masers will also be treated elsewhere in this volume.

2.1. *Disk masers*

When water masers are associated with accretion disks in AGN they often show a spectrum with a characteristic triple-peak pattern, with three distinct groups of features, one (the systemic lines) clustered around the systemic velocity of the galaxy and the other two (the high-velocity lines) almost symmetrically displaced from the first group toward the blue and the red sides by hundreds of km s^{-1}. Very Long baseline Interferometry (VLBI) and single-dish monitoring studies in the radio have enabled use of the water maser spots to map nuclear accretion disks and provide, for NGC 4258, a calibration of the cosmic distance scale (Miyoshi *et al.* 1995; Herrnstein *et al.* 1999). For many years,

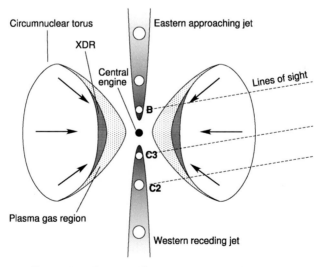

Figure 1. A cartoon illustrating the possible environment in the circumnuclear torus and jets in NGC 1052. An X-ray dissociation region (XDR) is formed on the inner layer of the torus and amplifies background continuum emission from the jet knots. On B and C3, we can see both the H_2O maser emission and free-free absorption (FFA). Only FFA appears, instead, on C2. The infalling of the gas inside the torus toward the central engine motivates the redshifted velocity of the water (Sawada-Satoh *et al.* 2008).

NGC 4258 has been the only galaxy where such detailed studies were possible. However, nowadays, the number of galaxies found to host water masers associated with accretion disks and with favourable characteristics (e.g., proximity, disk inclination, etc...) is rapidly rising thanks, in particular, to the Megamaser Cosmology Project (MCP). The description of this project and its main results achieved, so far, on the use of disk-masers to derive distances of the host galaxies, hence with a strong impact on improving the accuracy with which H_0 is known, are described by Henkel and Impellizzeri *et al.* (these proceedings) and will not be repeated here.

Another very relevant use of disk-masers is that, by modelling and analysing the Keplerian rotation curve of the accretion disks, as derived from the aforementioned maser studies, mass estimates of the nuclear engine can be obtained. Recently, Kuo *et al.* (2011) has estimated 6 new BH masses using this method, obtaining values from 0.75 to 6.5 $\times$ 10^7 $M_\odot$. For 3 of these galaxies, they also derived central densities large enough to rule out clusters of stars or stellar remnants as the central objects, thus reinforcing the 'standard' supermassive black hole scenario for AGN. Furthermore, using the BH mass estimates of Kuo *et al.* (2011), in addition to a few other BH masses estimated before using disk-masers, Greene *et al.* (2010) improved the low mass end of the $M_{BH} - \sigma^*$ relation, so far, almost uniquely derived for elliptical galaxies with larger BH masses. This way, they found a larger scatter in this relation than previously obtained, possibly hinting at a non-universal nature of the relation instead of that indicated before by the law derived from elliptical galaxies only.

2.2. Jet masers

Water maser emission can also be associated with radio jets, produced by either the interaction between the radio jet and an encroaching molecular cloud or by the amplification of the radio continuum from the jet from excited water molecules in a foreground cloud. Two examples of jet-maser sources (but more cases are reported in literature) can be briefly described to illustrate the two mechanisms: Mrk 348 and NGC 1052. In both cases, the maser spectrum is composed of a single broad (a few 100 km s^{-1}) line redshifted w.r.t. the systemic velocity of the host galaxy (for the single-dish detection spectra of Mrk 348 and NGC 1052, see Falcke *et al.* 2000 and Braatz *et al.* 1994, respectively), and the maser emission was confidently found to be located along the radio continuum of the jet, displaced from the position of the putative nucleus (for Mrk 348, see Peck *et al.* 2003; for NGC 1052, see Claussen *et al.* 1998).

Mrk 348 was extensively studied by Peck *et al.* (2003) using interferometric and single-dish multi-epoch observations. Their analysis led to a model for the origin of the emission related to a post-shock region at the interface between the energetic jet material and the molecular gas in the cloud where the jet is boring through. With the aid of the reverberation mapping technique, Peck *et al.* (2003) were also able to derive relevant physical quantities of the jet material, such as its velocity and density.

A somewhat different scenario is instead that brought about for NGC 1052 by Sawada-Satoh *et al.* (2008). According to their picture, the maser clouds are most likely located foreground to the jet in a circumnuclear torus (or disk), thus amplifying the continuum seed emission from the jet knots (Fig. 1). The maser gas is indeed found where the free-free opacity from foreground thermal plasma absorbing the jet synchrotron emission is large. Since the material in the torus/disk is also the putative source of accretion onto the nucleus, its contraction toward the central engine accounts for the redshifted velocity of all the maser features.

2.3. *Outflow masers*

A third class of AGN-associated water masers is named 'outflow-masers'. Presently, how-ever, there is only one case that has been thoroughly investigated, the nearby Seyfert 2 galaxy Circinus. VLBI maps of the water emission from this galaxy have shown that maser spots trace two different dynamical components, a warped edge-on accretion disk and a wide angle nuclear outflow up to ~ 1 pc from the central engine. Indeed, that in the Circinus galaxy represents the first direct evidence of dusty, high-density, molecular material in a nuclear outflow, at such small scales, thus allowing detailed study of the velocity and geometry of nuclear winds (Greenhill *et al.* 2003). At larger scales, more recently, the distribution of dust in the nuclear region of Circinus has been investigated by Tristram *et al.* (2007) using interferometric observations with the MID-infrared In-terferometric Instrument (MIDI) at the VLTI. They found that the dust is distributed in two components, a dense and warm disk component with a radius of 0.2 pc (where the disk-masers found by Greenhill *et al.* 2003 are located) and a less dense and slightly cooler geometrically-thick torus-like component up to pc scales. This dusty torus con-firms the presence of such structures in AGN, as expected from the Unified Model, and is seemingly the agent that collimates the AGN outflow traced by the water maser spots. However, evidence is also reported by Tristram *et al.* (2007) of a clumpy or filamentary dust distribution in the torus that, if confirmed for AGN tori in general, may have an impact on our understanding/classification of the different types of AGN.

3. Water masers in AGN: detection rates and host galaxies

So far, more than 3000 galaxies have been searched for water maser emission and detections have been obtained in about 150 of them[†], the majority being radio-quiet AGN classified as Seyfert 2 (Sy 2) or low-ionization nuclear emission-line regions (LINERs), in the local Universe (z < 0.05).

The overall detection rate for AGN is only of a few percent. This detection rate rises considerably (up to 20-25%) when considering only the nearest Sy 2 and LINER galaxies (e.g. those < 5000 km s^{-1}; Braatz *et al.* in prep.).

A question that naturally arises is: why some AGN, also among the same class, host H_2O maser sources while some others don't? One way to try answering this question is to investigate possible peculiarities of the masing galaxies (w.r.t. the non-masing ones).

Indeed, AGN hosting water maser emission tend to show a high column density ($N_H > 10^{23}$ cm^{-2}) or are even Compton-thick ($N_H > 10^{24}$ cm^{-2}; e.g. Zhang *et al.* 2006, Greenhill *et al.* 2008). Furthermore, a rough correlation has been found between maser isotropic luminosity and unabsorbed X-ray luminosity (Kondratko, Greenhill & Moran 2006). Although promising, these studies have been affected by a lack or poor quality of X-ray data for a large percentage ($\sim 50\%$) of the known maser galaxies. Positively, a survey of all known H_2O maser sources in AGN using the Swift satellite is ongoing (Castangia *et al.* 2011, Castangia *et al.* in prep.) that may help clarifying, on a firm statistical basis, the interplay between X-ray and maser emission.

A number of studies have been also recently led with the goal of finding correlations between the occurrence of maser emission and host galaxy characteristics at several wavelengths. The main results of these studies are: the confirmation of the fact that

---

† These values are compiled using information taken from the Megamaser Cosmology Project (MCP) and Hubble Constant Maser Experiment (HoME), 'https://safe.nrao.edu/wiki/bin/view/Main/MegamaserCosmologyProject' and 'https://www.cfa.harvard.edu/lincoln/demo/HoME/index.html', respectively.

H$_2$O masers reside primarily in X-ray absorbed sources (Zhang *et al.* 2010, Ramolla *et al.* 2011, Zhu *et al.* 2011), the larger radio luminosity found in H$_2$O maser hosts w.r.t. that in non-maser galaxies (Zhang *et al.* 2012), and the indication of the extinction-corrected [OIII]λ5007 flux as a good sample selection criterion to maximize the number of detections in maser searches (Zhu *et al.* 2011; see also Zaw *et al.*, these proceedings).

As already mentioned before, the highest occurrence of maser emission happens in type 2 Seyferts. Surely, identifying water maser emission associated with Sy 2 galaxies is not particularly surprising, given that the Unified Model for AGN requires an obscuring structure, which is probably an edge-on disk or torus, along the line of sight towards the nucleus of a type 2 AGN. This structure can provide a molecular reservoir and the amplification paths necessary for maser action. The relation between a type 1 Seyfert and the maser phenomenon is instead less obvious. According to the same paradigm, in type 1 objects, accretion disks and/or tori, if present at all, should be orientated face-on, making them less likely to produce detectable maser emission. Consistently, out of the 150 water maser sources detected so far, maser emission has been detected in only one 'pure' Sy 1, NGC 2782, apparently in strong agreement with the Unified Model. However, a recent result obtained by Tarchi *et al.* (2011b) seems to complicate the aforementioned conclusion.

Indeed, in the last years, a (sub)class of Seyfert 1 (Sy 1) galaxies, named Narrow-Line Seyfert 1 (NLS1s), has attracted particular attention among astronomers. NLS1s have the broad emission-line optical spectra of type 1 Seyfert galaxies, but with the narrowest Balmer lines from the broad line region and the strongest Fe II emission (e.g., Osterbrock 1989, Véron-Cetty *et al.* 2001). Extreme properties are also observed in X-rays (see Komossa 2008 for a review). By including all past surveys (Braatz *et al.* in prep) and the outcome of two new surveys, Tarchi *et al.* (2011b) has compiled a list of all NLS1s (71) searched, so far, in the water maser line. Five successful detections are reported, yielding a detection rate of $\sim$ 7%, which is comparable with the global rates of AGN surveys. While this result is surprising, the maser detection rate in NLS1s becomes even more impressive when we consider a volume-limited sample that somewhat minimizes the limitation in sensitivity of the survey(s). When considering only NLS1 galaxies at recessional velocities less than 10000 km s^{-1}, the detection rate goes up dramatically to $\sim$ 21 % (5/24). This value approaches the highest detection rates ever obtained for similarly volume-limited samples, in any class of AGN. Possible explanations for such a high water maser detection rate may reside in the NLS1s peculiar properties: accretion rates close to their Eddington limit, putative small black hole masses, strong radiation-pressure outflows, and viewing angles intermediate between type 1 and type 2 Seyferts.

3.1. *Masers in Elliptical and/or Radio-loud galaxies*

So far, (almost) no water masers have been found in elliptical and/or radio-loud galaxies despite a number of surveys that have been performed on different classes of AGN.

Among the AGN targeted in these surveys are:
- 50 Fanaroff-Riley I (FR I) with $z < 0.15$ (Henkel *et al.* 1998)
- 273 Type 2 QSOs with $0.3 < z < 0.83$ (Bennert *et al.* 2009)
- 79 radio galaxies, mostly Fanaroff-Riley II (FR II) with $z < 0.17$ (GBT project #AGBT02C_030, unpublished)
- 5 Grav. Lensed Quasars with $2.3 < z < 2.9$ (McKean *et al.* 2011)
- 17 Type 1 QSOs with $z < 0.06$ (König *et al.* 2012)

No new water maser detections were reported.

There are, however, a few exceptions:

— The radio galaxy NGC 1052 at z = 0.005 (Claussen *et al.* 1998; Sawada-Satoh *et al.* 2008). The maser source in this galaxy, discovered by Braatz *et al.* (1994), has been already described in Sect. 2.2.

— The FR II galaxy 3C 403 at z = 0.06. This galaxy displays a peculiar X-shaped radio morphology (Fig. 2). The maser emission showed two main broad lines experiencing extreme flux variability within about a year period. In view of linewidths and the lack of satellite lines, an interpretation in terms of an association of the maser emission with the radio jets is the most plausible one (Tarchi *et al.* 2003; Tarchi *et al.* 2007b)

— The type 2 QSO SDSS J0804+3607 at z = 0.66. The maser is composed of a single feature, redshifted w.r.t. to the systemic velocity of the galaxy, with an isotropic luminosity of 23000 $L_\odot$. No other maser features are reported. So far, the nature of the maser emission has not been assessed (Barvainis & Antonucci 2005; Bennert *et al.* 2009)

— The type 1 quasar MG J0414+0534 at z = 2.64. The most distant and luminous ($\sim$ 30000 $L_\odot$) maser source and, so far, the only one found in a gravitational lensed system (Impellizzeri *et al.* 2008; Castangia *et al.* 2011). For a description of this source, see also Castangia *et al.* (these proceedings).

— The NLS1 IGR 16385-2057 at z = 0.03. Indeed, while the nature of the galaxy hosting the maser emission has still to be confidently determined, IGR 16385-2057 is optically classified as an elliptical and displays a core+lobes radio morphology, resembling that of classical radio galaxies (Tarchi *et al.* 2011b; Castangia *et al.* in prep.).

Although the small number of detections does not allow us to derive any definite conclusion, we note that the detection of masers in E/radio-loud objects seem not to be very AGN-type dependent. Indeed, so far, there have been detections both in type 1 and

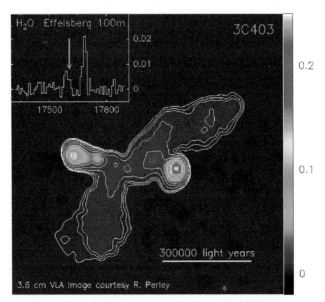

Figure 2. NRAO Very Large Array image of the radio galaxy 3C 403 at a wavelength of 3.6 cm. The intensity range of the colors (in Jansky, Jy, units) is indicated at the right hand side. The red arrow points at the galaxy's nucleus. The 22-GHz water maser spectrum shown in the upper left hand inset was taken with the Effelsberg 100m telescope. The green arrow points at the systemic radial velocity of the whole galaxy (Credits: National Radio Astronomy Observatory/Rick Perley (NRAO/AUI/NSF); the water maser spectrum is taken from Tarchi *et al.* 2003).

type 2 objects. Generally, masers in E/radio-loud objects are more seemingly associated with radio-jets or outflows, although the disk-maser scenario cannot be a priori ruled out. Obviously, VLBI observations are necessary to better assess the origin of the emission. However, these measurements have been successfully performed only for the nearest source, NGC 1052, due to the intrinsic weakness or flaring-down of the maser lines in the other cases.

Among the possible interpretations for the paucity of water maser detections in elliptical and/or radio-loud galaxies, we can report: i) the possible lack of molecular material (e.g. Henkel *et al.* 1998); ii) instability due to tidal disruption of molecular maser clouds in circumnuclear disks orbiting around black holes that are particularly massive, as those in powerful radio galaxies (e.g., Tarchi *et al.* 2007b); iii) a steep or non-evolving water maser luminosity function that would not provide us with maser sources luminous enough to be detectable at the large distances at which the elliptical/radio-loud galaxies are located (this option seems, however, to be ruled out by the recent work of McKean *et al.* 2011, where an indication for a slow - at least - evolution of the maser luminosity function with z is reported); iv) strong variability of the maser features, as seen, for example in 3C 403 (on the other hand, very high flux stability has been reported by Castangia *et al.* 2011 for the maser in MG J0414+0534); v) the insufficient sensitivity of the surveys led so far (in this framework, the enhanced sensitivity of some of the present and upcoming radio facilities will help to quantify the true relevance of this item).

4. H$_2$O and OH megamasers: brotherhood or mutual exclusion?

4.1. *OH megamasers in AGN*

OH megamasers (so far, about 100 sources are known) are almost uniquely associated with LIRGs ($L_{IR} > 10^{11}$ L$_\odot$) and ULIRGs ($L_{IR} > 10^{11}$ L$_\odot$), are IR radiatively pumped, and a relation between OH and FIR luminosities exists, $L_{OH} \propto L_{FIR}^{1.2}$ (e.g., Baan 1989; Darling & Giovanelli 2002).

The OH emission is typically the contribution of two components. One component is diffuse and was explained as low-gain, unsaturated amplification of background radio continuum by foreground clouds (e.g., Baan 1985). The second component is compact and produced by high-gain saturated emission (e.g., Lonsdale 2002). A model explaining both emissions by a single gas phase has been more recently described in Parra *et al.* (2005; see also the following text).

The region of maser emission has an extent of up to 100 pc and it is typically found in high concentration of molecular gas where intense/extreme starbursts are ongoing. It has been speculated that the association with starbursts may indicate that the OH megamaser occurrence in a galaxy is a short-lived phenomenon (e.g., Lo 2005, and references therein).

Although, as just mentioned, OH megamasers seem to have a star-formation origin, two cases have been reported where a connection between the OH megamaser and AGN activity is present, Mrk 231 and IIIZw 35. In the former source, the OH maser emission (mapped with the VLBI) traces a rotating, dusty, molecular torus (or thick disk) located between 30 and 100 pc from the central engine (Klöckner, Baan, & Garrett 2003; Fig. 3). Similarly, in IIIZw 35, the OH emission is produced by pc-size OH maser clouds (overlapping) in the tangent points of a nearly edge-on torus (Pihlström *et al.* 2001). For this source, both the diffuse and compact emission components have been successfully explained by a single phase of unsaturated clumpy gas in a ring structure amplifying background continuum (Parra *et al.* 2005).

Another OH megamaser source that has recently attracted a lot of attention is that in Arp 299, a merger system produced by the close interaction of two galaxies, IC 694 and NGC 3690. OH megamaser emission was detected from the nuclear region of IC 694 (Baan 1985). The OH emission was found coincident in position with the radio continuum peak and confined in a 100 pc rotating structure (Polatidis & Aalto 2001). In the same galaxy, water maser emission was also detected (Tarchi *et al.* 2007a), with a position slightly offset w.r.t. that of the OH maser, and associated with an expanding slab of material, seemingly a nuclear outflow (Tarchi *et al.* 2011a). While VLBI observations of the maser emission from both molecules are planned, interestingly, Pérez-Torres *et al.* (2010) has found strong evidence for the presence of the long-sought putative AGN in the system, associated with the nucleus of IC 694.

4.2. *On the OH and H_2O maser relation*

By compiling all sources where searches for maser emission from both H_2O and OH have been, so far, reported in the literature, it has been found that, out of the resulting 51 galaxies, 33 galaxies show only water maser emission, 13 galaxies show only OH maser emission, and 5 sources confidently show maser emission from both molecules (Tarchi *et al.* 2011a; their Fig. 11). In particular, the two well-known starburst galaxies NGC 253 and M 82 show weak maser emission from H_2O and OH, the two water megamaser galaxies NGC 3079 and NGC 1068 also host a weak OH maser, and Arp 299 shows megamaser emission from both molecules. Thus, only Arp 299 shows the contemporary presence, in IC 694 (see previous paragraph), of luminous OH and H_2O maser emission (Surcis *et al.* 2009; Tarchi *et al.* 2011a).

What is then possibly causing this apparent lack of contemporary detections in the same object of megamaser emission from both molecules (duration of the maser phenomena, merger stage in some systems, other)? To answer this question a more systematic approach (parallel surveys to detect H_2O maser emission in OH maser galaxies, and

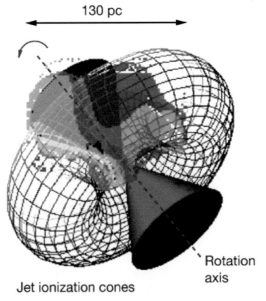

Figure 3. The inferred model of the nuclear torus in Mrk 231 is displayed as a wire diagram with symmetric ionization cones. This model takes into account all large–scale characteristics of the nuclear radio emission and the OH emission. The molecular material moves from top–right to bottom–left, i.e., northwest to southeast (Klöckner, Baan, & Garrett 2003).

viceversa) is necessary. Furthermore, a more detailed study of the, so far, unique case, IC 694, is auspicious since it offers the possibility to investigate different physical conditions and dynamic structures in the same AGN.

5. Concluding remarks

The results obtained, so far, strengthen the uniqueness of the contribution provided by megamaser studies to our understanding of the nuclear regions of AGN. H_2O megamasers are preferentially found in nearby (radio-quiet) type 2 Seyferts with some exceptions (e.g., NLS1s), somewhat consistently, with the AGN Unified Model. H_2O megamasers are, instead, rarely found in elliptical and radio-loud galaxies with only a few special cases. In this case, given the low statistics, a connection with the Unified Model is more difficult. In addition, OH megamasers are preferentially found in ULIRGs associated with extreme star formation rather than AGN activity. Nevertheless, in a few cases, OH emission traces the rotating disk/torus structures invoked by the Unified Model for AGN. A mutual exclusion between OH and H_2O megamasers seems also to be present, with Arp 299 as the only exception.

While a large number of OH and water maser sources is under detailed investigation, the overall understanding of the maser phenomenon in the framework of the standard Unified Model for AGN is still far from being assessed, mostly due to the lack of maser detections and/or detailed-enough observations in the radio-loud domain of AGN.

Furthermore, a number of open issues still needs to be clarified. Among these, the most relevant are the presently-unknown beaming of the maser radiation and the degree of clumpiness of the material in the disk/torus. These two elements may have a relevant influence on the true luminosity estimates and excitation requirements of the maser sources, and on the classification of the AGN type hosting the maser emission.

A final mention needs to be given to the existence of the alternative or revised versions of the standard Unified Model for AGN (e.g., Nenkova *et al.* 2008; Elitzur 2012). When trying to derive a comprehensive picture of the AGN 'zoo', observational and theoretical results should necessarily be compared with these models as well. Once again, in this framework, maser studies offer among the best tools to test both the standard and alternative scenarios.

Acknowledgements

The author would like to thank Paola Castangia for critically reading the manuscript and Jim Braatz for providing useful information prior to publication.

This review is dedicated to the memory of Albert Greve, a mentor and a friend.

References

Antonucci, R. 1993, *ARAA*, 31, 473
Baan, W. A. 1985, *Nature*, 315, 26
Baan, W. A. 1989, *ApJ*, 338, 804
Barvainis, R. & Antonucci, R. 2005, *ApJ*, 628, L89
Bennert, N., Barvainis, R., Henkel, C., & Antonucci, R. 2009, *ApJ*, 695, 276
Braatz, J. A., Wilson, A. S., & Henkel, C. 1994, *ApJ*, 437, L99
Castangia, P., Tilak, A., Kadler, M., Henkel, C., Greenhill, L., & Tueller, J. 2010, *X-ray Astronomy 2009* Proc. AIPC, Vol. 1248, p. 347
Castangia, P., Impellizzeri, C. M. V., McKean, J. P., Henkel, C., Brunthaler, A., Roy, A. L., Wucknitz, O., Ott, J., & Momjian, E. 2011 *A&A*, 529, 150
Cernicharo, J., Pardo, J. R., & Weiß, A., 2006, *ApJ*, 646, L49
Claussen, M. J., Diamond, P. J., Braatz, J. A., Wilson, A. S., & Henkel, C. 1998 *ApJ*, 500, L129

Darling, J. & Giovanelli, R. 2002, *ApJ*, 572, 810

Elitzur, M. 2012, *ApJ*, 747, L33

Falcke, H., Henkel, C., Peck, A. B., Hagiwara, Y., Prieto, M. A., & Gallimore, J. F. 2000 *A&A*, 358, 17

Greene, J. E., Peng, C. Y., Kim, M., Kuo, C. -Y., *et al.* 2010, *ApJ*, 721, 26

Greenhill, L. J., Booth, R. S., Ellingsen, S. P., *et al.* 2003, *ApJ*, 590, 162

Greenhill, L. J., Tilak, A., & Madejski, G. 2008, *ApJ*, 686, L13

Henkel, C., Wang, Y. P., Falcke, H., Wilson, A. S., & Braatz, J. A. 1998, *A&A*, 335, 463

Herrnstein, J. R., Moran, J. M., Greenhill, L. J., Diamond, P. J., Inoue, M., Nakai, N., Miyoshi, M., Henkel, C., & Riess, A. 1999, *Nature*, 400, 539

Humphreys, E. M. L., Greenhill, L. J., Reid, M. J., Beuther, H., Moran, J. M., Gurwell, M., Wilner, D. J., & Kondratko, P. T. 2005, *ApJ*, 634, 133

Impellizzeri, C. M. V., McKean, J. P., Castangia, P., Roy, A. L., Henkel, C., Brunthaler, A., & Wucknitz, O. 2008, *Nature*, 456, 927

Klöckner, H.-R., Baan, W. A., & Garrett, M. A. 2003, *Nature*, 421, 821

König, S., Eckart, A., Henkel, C., & García-Marín, M. 2012 *MNRAS*, 420, 2263

Komossa, S. 2008, *Rev. Mexicana AyA* (Conference Series), vol. 32, 86

Kondratko, P. T., Greenhill, L. J. & Moran, J. M. 2006, *ApJ*, 652, 136

Kuo, C. Y., Braatz, J. A., Condon, J. J., *et al.* 2011, *ApJ*, 727, 20

Lo, K. Y. 2005, *ARAA*, 43, 625

Lonsdale, C. J. 2002, *Cosmic Masers: From Proto-Stars to Black Holes*. Proc. IAU Symposium No. 206 (San Francisco: ASP), p. 413

McKean, J. P., Impellizzeri, C. M. V., Roy, A. L., Castangia, P., Samuel, F., Brunthaler, A., Henkel, C., & Wucknitz, O. 2011, *MNRAS*, 410, 2506

Miyoshi, M., Moran, J., Herrnstein, J., Greenhill, L., Nakai, N., Diamond, P., & Inoue, M. 1995, *Nature*, 373, 127

Nenkova, M., Sirocky, M. M., Nikutta, R., Ivezi, Z., & Elitzur, M., 2006, *ApJ*, 685, 160

Osterbrock, D. E. 1989, *Astrophysics of gaseous nebulae and active galactic nuclei*, ed. University Science Books

Parra, R., Conway, J. E., Elitzur, M., & Pihlström, Y. M. 2005, *A&A*, 443, 383

Peck, A. B., Henkel, C., Ulvestad, J. S., Brunthaler, A., Falcke, H., Elitzur, M., Menten, K. M., & Gallimore, J. F. 2003, *ApJ*, 590, 149

Pérez-Torres, M. A., Alberdi, A., Romero-Canizales, C., & Bondi, M. 2010, *A&A*, 519, L5

Pihlström, Y. M., Conway, J. E., Booth, R. S., Diamond, P. J., & Polatidis, A. G. 2001, *A&A*, 377, 413

Polatidis, A. G. & Aalto, S. 2001, *Galaxies and their Constituents at the Highest Angular Resolutions*. Proc. IAU Symposium No. 205 (San Francisco: ASP), p. 1

Ramolla, M., Haas, M., Bennert, V. N., & Chini, R. 2011, *A&A*, 530, 147

Sawada-Satoh, S., Kameno, S., Nakamura, K., Namikawa, D., Shibata, K. M., & Inoue, M. 2008, *ApJ*, 680, 191

Surcis, G., Tarchi, A., Henkel, C., Ott, J., Lovell, J., & Castangia, P. 2009, *A&A*, 502, 529

Tarchi, A., Henkel, C., Chiaberge, M., & Menten, K. M. 2003, *A&A*, 407, L33

Tarchi, A., Castangia, P., Henkel, C., & Menten, K. M. 2007a, *New Astron. Revs*, 51, 67

Tarchi, A., Brunthaler, A., Henkel, C., Menten, K. M., Braatz, J., & Weiß, A. 2007b, *A&A*, 475, 497

Tarchi, A., Castangia, P., Henkel, C., Surcis, G., & Menten, K. M. 2011a, *A&A*, 525, 91

Tarchi, A., Castangia, P., Columbano, A., Panessa, F., & Braatz, J. A. 2011b, *A&A*, 532, 125

Tristram, K. R. W., Meisenheimer, K., Jaffe, W., *et al.* 2007, *A&A*, 474, 837

Urry, C. M. & Padovani, P. 1995, *PASP*, 107, 803

Véron-Cetty, M., Véron, P., & Gonçalves, A. C. 2001, *A&A*, 372, 730

Zhang, J. S., Henkel, C., Kadler, M., Greenhill, L. J., Nagar, N., Wilson, A. S., & Braatz, J. A. 2006, *A&A*, 450, 933

Zhang, J. S., Henkel, C., Guo, Q., Wang, H. G., & Fan, J. H. 2010, *ApJ*, 708, 1528

Zhang, J. S., Henkel, C., Guo, Q., & Wang, J. 2012, *A&A*, 538, 152

Zhu, G., Zaw, I., Blanton, M. R., & Greenhill, L. J. 2011, *ApJ*, 742, 73

Cosmic Masers - from OH to H$_0$
Proceedings IAU Symposium No. 287, 2012
R.S. Booth, E.M.L. Humphreys & W.H.T. Vlemmings, eds.
© International Astronomical Union 2012
doi:10.1017/S1743921312007260

Masers in Starburst Galaxies

Jeremy Darling

Department of Astrophysical and Planetary Sciences
University of Colorado
Boulder, CO 80309, United States of America
email: jdarling@colorado.edu

Abstract. Masers in starburst galaxies are outstanding probes of a range of phenomena related to galaxy and black hole evolution, star formation, and magnetic fields. Here I briefly discuss five related topics: (1) Galactic analog water masers in nearby galaxies; (2) multiwavelength solutions to the OH megamaser puzzle in major galaxy mergers; (3) formaldehyde anti-inversion in starburst galaxies; (4) OH spoofing in HI surveys; and (5) new discovery space in radio line surveys. New insights into the physical conditions responsible for OH megamasers, including indications of a critical molecular gas density obtained from the formaldehyde "densitometer," will be applicable to future surveys, particularly surveys for redshifted hydrogen where OH lines arising in major galaxy mergers can "contaminate" the disk population identified by the HI 21 cm line. Blind radio spectral line surveys also offer the opportunity for unexpected discoveries of new nonthermal radio lines.

Keywords. masers, radiation mechanisms: nonthermal, radiative transfer, surveys, galaxies: evolution, galaxies: interactions, galaxies: ISM, galaxies: starburst, infrared: galaxies, radio lines: galaxies

1. Introduction

The state of observations of extragalactic masers is woefully inadequate, particularly compared to the wealth of Galactic maser data. It is important to remember that all masers seen in the Galaxy occur in other galaxies and that there are masers produced in other galaxies that are not seen in our own. The latter group include both the well-known water and hydroxyl megamasers as well as masers yet to be discovered (see Section 6).

The reason for the paucity of information about Galactic analog masers in other galaxies lies in the raw sensitivity needed to detect masers at Mpc distances. For example, moving from kpc to Mpc distances, resolution drops 1000-fold, and luminosity sensitivity is effectively diminished million-fold: parsecs become kpc, and 1 Jy becomes 1 μJy. There are thus only a few maser species detected in external galaxies, most often water and hydroxyl, and they are generally not Galactic analogs.

Here I present a brief discussion of five related topics: (1) Galactic analog water masers in nearby galaxies; (2) solutions to the OH megamaser puzzle in major galaxy mergers; (3) formaldehyde anti-inversion in starburst galaxies; (4) OH spoofing in HI surveys; and (5) new discovery space in radio line surveys.

2. Water Masers in Star-Forming Galaxies

Despite the difficulty of detecting Galactic analog masers in galaxies beyond the Magellanic Clouds, modern telescopic facilities now have the raw sensitivity to detect Galactic analog water masers to distances of ~ 20 Mpc. If we call a Galactic analog maser one that simply has a peak flux density or isotropic line luminosity equal to or less than the brightest Galactic masers, then we now have numerous examples of such extragalactic

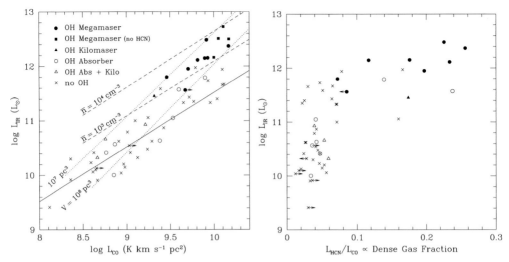

Figure 1. Global star formation properties of OH megamaser host galaxies (after Darling 2007). Left: IR luminosity versus CO line luminosity in HCN-detected galaxies with known OH properties from the Gao & Solomon (2004a) sample. The legend indicates symbols for OH megamasers, OH kilomasers, OH absorbers, and objects with no detected OH lines. The solid line is a linear fit by Gao & Solomon (2004b) to galaxies with $L_{\rm IR} < 10^{11} L_\odot$ ($L_{\rm IR} = 33 L_{\rm CO}$ in the units above). The dotted lines indicate a constant total volume of molecular material, and the dashed lines indicate the mean H_2 density derived from Krumholz & Thompson (2007). Right: IR luminosity versus $L_{\rm HCN}/L_{\rm CO}$, a proxy for the dense gas fraction. OH megamaser hosts form the majority of the high density population, suggesting an extreme stage of star formation and tidal concentration of the ISM.

masers (e.g., Castangia *et al.* 2008, Darling *et al.* 2008, Surcis *et al.* 2009, Tarchi *et al.* 2011). But in most cases the physical conditions, such as those found in dwarf starbursts or the Antennae Galaxies (a major merger), are not analogous to those found in the Galaxy (e.g., Brogan *et al.* 2010).

There have, however, recently been detections of true Galactic analog masers in the Andromeda Galaxy, M31: Sjouwerman *et al.* (2010) detected a 6.7 GHz methanol maser, and Darling (2011) detected five water maser complexes. All of these masers appear to be associated with star formation and lie within the molecular gas- and dust-rich part of the spiral disk of M31.

3. The OH Megamaser Puzzle

Lo (2005) posed a useful question about the hosts of OH megamasers that can be paraphrased as: suppose that one can identify two apparently identical major mergers/(U)LIRGs, but one shows OH megamaser (OHM) emission while the other does not. What is the cause of the OHM? What is physically different between these galaxies? It may be that one cannot find such a situation, but more broadly, it is the case that the global properties, such as IR luminosity, IR color, morphology, and optical spectrum do not distinguish between OHM hosts and non-masing mergers once the selection for the proper type of galaxy (e.g., late-stage merger, $L_{\rm IR} \gtrsim 10^{11.6} \, L_\odot$, warm IR colors, starburst-dominated spectrum) has been made. Could the difference be beaming? Small scale physical conditions? OH abundance? AGN influence? Or could it be some property not yet observed?

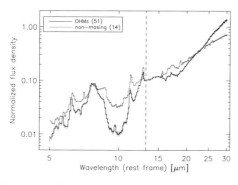 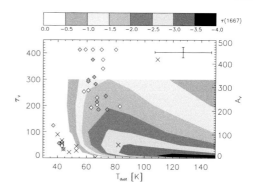

Figure 2. Comparison of IR properties of OH megamaser host galaxies to non-masing but otherwise similar galaxies (after Willett *et al.* 2011b). Left: The average mid-IR *Spitzer* spectrum of OHM hosts shows a deeper 10 μm silicate dust absorption feature as well as a steeper 20–30 μm dust slope when compared to non-masing galaxies with similar global properties. These differences are two sides of the same coin: a very geometrically and optically thick dust blanket ($\tau_V \sim 300$) in OHM hosts (Ivezić & Elitzur 1997, Levenson *et al.* 2007). Right: Comparison of the observed dust optical depth versus dust temperature to the theoretical predictions of Lockett & Elitzur (2008). Colors in contours indicate the predicted 1667 MHz OH line optical depth (negative indicates masing), and colors in the points indicate the observed apparent 1667 MHz OH line optical depth. Crosses indicate non-masing galaxies. While the OHM sample is well-separated from the non-masing sample in this space, and we can now *predict* the most likely galaxies to produce OHM activity, the theory cannot yet predict the observed maser emission based on global dust properties alone.

The root physical cause of OH megamaser activity has been observationally revealed to lie in the extreme concentration of gas and dust in major mergers, and OHMs appear to select the time of peak concentration during the merging process.

Darling (2007) show that OHMs lie in a distinct locus in plots of star formation rate (as traced by IR luminosity) versus CO line luminosity: the OHM hosts are the starburst galaxies that cause the IR-CO relation to break away from the linear relationship that holds for star-forming galaxies across many orders of magnitude (Gao & Solomon 2004b). This suggests that OHM hosts are undergoing a special stage of merger-triggered star formation. OH "kilomasers" and absorbers can be found throughout the linear part of the relation, suggesting that OH abundance is not a factor in producing OHMs (Figure 1). OHM hosts are also the majority of galaxies in the high dense gas fraction phase of the merging process, suggesting that while OHM emission is beamed, it is stochastic and fairly isotropic when conditions are favorable for OHM production.

Willett *et al.* (2011a, 2011b) show that OHM hosts have a deeper 10 μm silicate dust absorption feature and steeper 20–30 μm dust continuum slope than non-masing but otherwise similar galaxies (Figure 2). Ivezić & Elitzur (1997) demonstrate that for silicate dust, these are two aspects of the phenomenon: as dust opacity increases, the mid-IR silicate dust absorption features become so deep that they steepen the mid-IR slope of the dust emission and move the peak to longer wavelengths. In fact, for $\tau_V \sim 300$, the peak of the dust emission is shifted to maximize the IR pumping at 35 and 53 μm that is most likely responsible for OHM emission.

Willett *et al.* (2011b) directly compare observations of the OHM and non-masing galaxy host dust temperatures and apparent dust optical depths to the locus theoretically predicted by Lockett & Elitzur (2008) for OHM production and find a significant separation between OHM hosts and non-masing galaxies, verifying that non-masing galaxies do not lie in the expected dust opacity-temperature region required for masing. Willett *et al.* (2011b) also demonstrate that the theoretical dust optical depths may be lower than what is observed (Figure 2).

Finally, in the Mangum *et al.* (2008) study of formaldehyde (H_2CO) in nearby star-forming galaxies, OHM hosts show the 6 cm H_2CO line in emission while all non-masing host galaxies show this line in absorption. This line sign-flip occurs when the 6 cm line excitation temperature crosses the cosmic microwave background (CMB) temperature, which corresponds to a molecular hydrogen density threshold of $n(H_2) \sim 10^{5.6}$ cm^{-3} (the H_2CO cm lines are collisionally driven to low excitation temperatures; see Section 4). Again, this suggests a density threshold above which OHM activity can (or perhaps must) occur.

The pieces of the OHM puzzle are falling into place and are in good agreement: we have a picture of a stage during major galaxy mergers when there is extreme concentration of gas and dust in a period of peak triggered star formation. The combination of extreme concentration of a large fraction of a galaxy's ISM into the inner kpc with warm and geometrically and optically thick dust driven to peak emission near the OHM pumping lines produces the conditions for OHM production in the inner ~ 100 pc. These conditions are physically wide-spread and stochastic, producing many masing sightlines, making the OHM detectable from most viewing angles. This also explains why OHM emission is predictable based on global gas and dust properties of the host galaxies rather than being an amplification of small-scale conditions.

4. Formaldehyde Anti-Inversion

Centimeter-wave K-doublet rotational transitions of formaldehyde (H_2CO) can be collisionally "refrigerated" below the cosmic microwave background (CMB) temperature such that these transitions absorb CMB photons (Palmer *et al.* 1969, Townes & Cheung (1969), Evans *et al.* 1975, Garrison *et al.* 1975). The manner in which the cm lines of H_2CO are cooled is analogous to a maser pumping process: a pump drives an over-population of states, but in this case they are lower-energy states, and the level populations become "anti-inverted" compared to thermal. When the line excitation temperature drops below the local CMB temperature, CMB photons may be absorbed. This effect of "anomalous" H_2CO absorption was playfully called the dasar: darkness amplification by stimulated absorption of radiation (an inaccurate description since the "darkness" is in fact a 2.73 K blackbody and the absorption is not stimulated).

Since absorption lines are detectable independent of distance, and since the CMB illuminates all gas and provides an exceptionally uniform illumination that lies behind every galaxy, a Formaldehyde Deep Field can provide a *mass-limited* census of the dense molecular gas associated with star formation. If multiple lines are observed then physical molecular gas densities may also be obtained fairly independently of the gas kinetic temperature or other factors. The utility of the "formaldehyde densitometer" has been well-studied in the Galaxy (e.g., Mangum & Wootten 1993, Ginsburg *et al.* 2011), in nearby star-forming galaxies (e.g., Mangum *et al.* 2008), in the $z = 0.67$ gravitational lens and molecular absorption line system (Zeiger & Darling 2010), and theoretically for arbitrary redshift (Darling & Zeiger 2012, submitted).

5. OH Spoofing HI

Briggs (1998) predicted that 18 cm OH megamaser lines produced in major galaxy mergers will "contaminate" blind HI 21 cm surveys (which typically identify spiral disks), and the level of contamination will depend on the sensitivity and redshift coverage of the survey. In fact, for any given survey depth, there is a point in redshift space where the sky density of the "contaminating" lines surpasses the sky density of the survey object line and becomes the dominant population. This is due to the relatively fixed HI masses of massive disk galaxies (fixed to within one order of magnitude, even for early gas-rich galaxies) compared to the highly nonlinear maser action that can make lines disproportionately bright compared to the gas mass responsible for the maser emission.

The expected sky density of disks versus mergers, for a given survey sensitivity and frequency span (redshift range), depends primarily on the merging rate of galaxies as a function of cosmic time and on the OH luminosity function (Briggs 1998, Darling & Giovanelli 2002). The galaxy merging rate versus redshift is a topic of ongoing contention and is highly selection-biased. The OH luminosity function, while fairly well-measured given the small sample size, is not well-defined at the high luminosity end, which is precisely the luminosity that future surveys will detect first given the large volumes that will be sampled and the severe Malmquist bias of OHM samples in general.

To address these issues, we have begun a multi-wavelength data-mining approach to sifting HI survey samples for OH contamination using the Arecibo Legacy Fast Arecibo L-band Feed Array (ALFALFA) survey (Giovanelli *et al.* 2005) as a test case, training set, and proof of concept. While mergers are IR-luminous, lower redshift star-forming disk galaxies can show a similar flux density in the mid-IR bands, so IR brightness alone is an insufficient selection condition. However, lack of a bright IR counterpart can exclude OH, thus confirming HI as the line identification. We also employ optical counterpart identification (which is not always unique), optical morphology, mid-IR color, photometric redshift, and radio line profile to sift OH from HI signals. None of these criteria alone are sufficient, but taken together they can usually distinguish disks at $z = 0$–0.05 from mergers at $z = 0.17$–0.23. In some cases, however, the only certain discrimination comes from an optical spectroscopic redshift.

We anticipate that this process of sorting disks from mergers will be challenging in redshifted HI surveys planned for Square Kilometer Array prototypes such as MeerKAT, ASKAP, SKAMP, FAST, and Apertif on the WSRT, both due to noisy spatially unresolved signals toward faint optical/IR counterparts and due to the higher "contamination" rate expected in these surveys compared to ALFALFA. It is also likely that new lines will be detected that are neither HI nor OH, making the sifting process even more challenging.

6. New Discovery Space

Low energy (frequency) transitions are slow: the Einstein coefficient for spontaneous emission scales as $A \propto \nu^3$, so metastable states are more common at low energies. Many low-energy transitions connect energy levels split by hyperfine structure or lambda-doubling, and so are also forbidden. Moreover, slight imbalances in rates of higher energy (fast) transitions can produce large nonthermal populations in low energy states, driving inversion (or anti-inversion). Note that the Boltzmann factors for low-energy states are generally roughly unity: $1 - \epsilon$ thanks to the CMB $T_{CMB} = (1 + z) \times 2.73$ K radiation field ($kT_{CMB} \gg h\nu$).

Pumping cycles for masers are not intuitive and rely on the detailed structure of the many states and rates connected (and even not connected) to the states involved in the maser. It is difficult to explain — much less predict — maser (or dasar) activity. Examples include Lyα pumping of the 21 cm HI line, which is fairly straightforward, to H_2 collisional pumping of H_2CO, to IR radiative pumping of OH. Note that the latter two examples were not predicted before they were observed.

The point of this discussion is that we simply do not know what bright maser or other nonthermal lines will appear at low frequencies. It is likely that new masers will be discovered in the course of redshifted HI 21 cm line surveys with the Square Kilometer Array and its prototypes. Extensive and sensitive sub-GHz spectral surveys have not been pursued due in large part to RFI, but also due to lack of community interest. It is difficult to propose to "make discoveries" when one does not know what one is looking for or might find; no one wrote a proposal to detect "mysterium," the first astrophysical maser.

7. Conclusions

Current telescopic facilities now have the raw sensitivity to detect Galactic analog masers in other galaxies. Using Galactic analog masers as extragalactic probes will be a growing field of astrophysical inquiry.

In contrast, the exceptional OH megamasers detected in starbursts have no local analogs and have been a long-term puzzle, but by using multi-wavelength observations combined with theoretical models we now have a better notion of the physical conditions and host galaxies indicated by the bright OH megamaser signposts. With this new understanding, we can interpret the populations of mergers and exceptional starbursts that "contaminate" HI 21 cm surveys via OH megamaser "spoofing." We can also learn how to separate disks from mergers in blind high-redshift HI 21 cm surveys planned for Square Kilometer Array prototypes.

Anti-inverted formaldehyde lines provide a molecular gas "densitometer" and the absorption of CMB photons by H_2CO provides a probe of star formation at any redshift. We envision a Formaldehyde Deep Field that will make a dust extinction-free redshift-independent molecular gas mass-limited census of the history of cosmic star formation (Darling & Zeiger 2012, submitted).

Finally, it is likely that new nonthermal lines will be discovered in low frequency radio spectral surveys provided that observers don't pre-bias surveys and data pipelines to only detect what is already expected and known.

References

Briggs, F. H. 1998, *A&A*, 336, 815
Brogan, C., Johnson, K., & Darling, J. 2010, *ApJ*, 716, L51
Castangia, P., Tarchi, A., Henkel, C., & Menten, K. M. 2008, *A&A*, 479, 111
Darling, J. & Giovanelli, R. 2002, *ApJ* 572, 810
Darling, J. & Giovanelli, R. 2006, *AJ* 132, 2596
Darling, J. 2007, *ApJ*, 669, L9
Darling, J., Brogan, C., & Johnson, K. *ApJ*, 685, L39
Darling, J. 2011, *ApJ*, 732, L2
Evans, N. J. II, Zuckerman, B., Morris, G., & Sato, T. 1975, *ApJ*, 196, 433
Gao, Y. & Solomon, P. M. 2004a, *ApJS*, 152, 63
Gao, Y. & Solomon, P. M. 2004b, *ApJ* 606, 271
Garrison, B. J., Lester Jr., W. A., Miller, W. H., & Green, S. 1975, *ApJ*, 200, L175
Ginsburg, A., Darling, J., Battersby, C., Zeiger, B., & Bally, J. 2011, *ApJ*, 736, 149
Giovanelli, R., Haynes, M. P., Kent, B. R., et al. 2005, *AJ*, 130, 2598
Henkel, C., Walmsley, C. M., & Wilson, T. L. 1980, *A&A*, 82, 41
Ivezić, Ž. & Elitzur, M. 1997, *MNRAS*, 287, 799
Jethava, N., Henkel, C., Menten, K. M., Carilli, C. L., Reid, M. J., & Walmsley, C. M. 2007, *A&A*, 472, 435

Krumholz, M. R. & Thompson, T. A. 2007, *ApJ*, 669, 289

Levenson, N. A., Sirocky, M. M., Hao, L., Spoon, H. W. W., Marshall, J. A., Elitzur, M., & Houck, J. R. 2007, *ApJ*, 654, L45

Lo, K. Y. 2005, *ARAA* 43, 625

Lockett, P. & Elitzur, M. 2008, *ApJ*, 677, 985

Mangum, J. G. & Wootten, A. 1993, *ApJS*, 89, 123

Mangum, J. G., Darling, J., Menten, K. M., & Henkel, C. 2008, *ApJ*, 673, 832

Palmer, P., Zuckerman, B., Buhl, D., & Snyder, L. E. 1969, *ApJ*, 156, L147

Sjouwerman, L. O., Murray, C. E., Pihlstrom, Y. M., Fish, V. L., & Araya, E. D. 2010, *ApJ*, 724, L158

Surcis, G., Tarchi, A., Henkel, C., Ott, J., Lovell, J., & Castangia, P. 2009, *A&A*, 502, 529

Tarchi, A., Castangia, P., Henkel, C., Surcis, G., & Menten, K. M. 2011, *A&A*, 525, A91

Townes, C. H. & Cheung, A. C. 1969, *ApJ*, 157, L103

Willett, K. W., Darling, J., Spoon, H. W. W., Charmandaris, V., & Armus, L. 2011, *ApJS*, 193, 18

Willett, K. W., Darling, J., Spoon, H. W. W., Charmandaris, V., & Armus, L. 2011, *ApJ*, 730, 56

Zeiger, B. & Darling, J. 2010, *ApJ*, 709, 386

Cosmic Masers - from OH to H_0
Proceedings IAU Symposium No. 287, 2012
R.S. Booth, E.M.L. Humphreys & W.H.T. Vlemmings, eds.
© International Astronomical Union 2012
doi:10.1017/S1743921312007272

Long term Arecibo monitoring of the water megamaser in MG J0414+0534

Paola Castangia[1], C. M. Violette Impellizzeri[2]†, John P. McKean[3], Christian Henkel[4], Andreas Brunthaler[4], Alan L. Roy[4], and Olaf Wucknitz[5]

[1]INAF-Osservatorio Astronomico di Cagliari,
Loc. Poggio dei Pini, Strada 54, 09012 Capoterra (CA), Italy
email: pcastang@oa-cagliari.inaf.it

[2]ALMA, Chile
email: vimpelli@alma.cl

[3]ASTRON,
Oude Hoogeveensedijk 4, 7991 PD Dwingeloo, the Netherlands

[4]Max-Planck-Institut für Radioastronomie,
Auf dem Hügel 69, D-53121 Bonn, Germany

[5]Argelander-Institut für Astronomie,
Auf dem Hügel 71, D-53121 Bonn, Germany

Abstract. We monitored the 22 GHz maser line in the lensed quasar MG J0414+0534 at z = 2.64 with the 300-m Arecibo telescope for almost two years to detect possible additional maser components and to measure a potential velocity drift of the lines. The main maser line profile is complex and can be resolved into a number of broad features with line widths of 30-160 km s^{-1}. A new maser component was tentatively detected in October 2008 at a velocity of +470 km s^{-1}. After correcting for the estimated lens magnification, we find that the H_2O isotropic luminosity of the maser in MG J0414+0534 is ~26,000 solar luminosities, making this source the most luminous ever discovered. Both the main line peak and continuum flux densities are surprisingly stable throughout the period of the observations. An upper limit on the velocity drift of the main peak of the line has been estimated from our observations and is of the order of 2 km s^{-1} per year. We discuss the results of the monitoring in terms of the possible nature of the maser emission, associated with an accretion disk or a radio jet. This is the first time that such a study is performed in a water maser source at high redshift, potentially allowing us to study the parsec-scale environment around a powerful radio source at cosmological distances.

Keywords. Masers, galaxies: active, galaxies: nuclei.

1. Introduction

To date, most surveys searching for extragalactic water masers have targeted spiral galaxies and radio-quiet AGN in the local Universe. Likely, this is the reason why, with the exception of a type 2 quasar at $z = 0.66$ (Barvainis & Antonucci 2005), the majority of the known extragalactic water masers have been found in Seyfert 2 or LINER galaxies at low redshift ($z < 0.06$). Observations of objects at higher redshifts are limited in part by the range of frequencies available but mainly by the sensitivity of current radio telescopes. To overcome this limitation, we have been carrying out a survey of water masers in known gravitationally lensed quasars with the Effelsberg and Arecibo telescopes (McKean *et al.* 2011; McKean *et al.* in prep). Observing gravitationally lensed quasars allows us to use the magnification provided by the lensing galaxy to increase the

† Speaker

measured flux density of the background AGN, strongly reducing the integration time necessary to detect the signal originating from distant objects ($z > 1$). This potentially allows us to discover water masers at cosmological distances and to study the parsec-scale environment of AGN in the early Universe. Our first confirmed high-redshift water maser was found toward the lensed quasar MG J0414+0534 at $z = 2.64$ (Impellizzeri *et al.* 2008). The line was originally detected with the Effelsberg radio telescope and subsequently confirmed with the EVLA, that found the emission to be coincident with the lensed images of the quasar (A1 and A2). The apparent isotropic luminosity of the line ($\sim$10,000 $L_\odot$, after correcting for the estimated lens magnification) makes the water maser in MG J0414+0534 not only the most distant but also one of the most luminous water masers ever detected and suggests that the emission is associated with the AGN. Based on the large line width of the Effelsberg and EVLA spectra, our initial hypothesis on the origin of the maser was that the emission is associated with the jet(s) of the radio loud quasar. In order to reveal possible variations in the maser flux density, typical of the masers produced by the interactions between a molecular cloud and a radio jet, and determine if a correlation exists between the maser and the continuum emission, we monitored the line and the radio continuum in MG J0414+0534 with the 300-m Arecibo telescope during a time interval of 15 months. Here we report the results of this monitoring campaign and discuss them within the framework of the two main scenarios for the origin of the maser, i.e. jet-maser and disk-maser emission (for more details on this work, see Castangia *et al.* 2011). We adopt a cosmology with $\Omega_M = 0.3$, $\Omega_\Lambda = 0.7$ and $H_0 = 70\,\mathrm{km\,s^{-1}\,Mpc^{-1}}$. In the following, the quoted line velocities are defined w.r.t. the optical redshift of MG J0414+0534 ($z = 2.639$; Lawrence *et al.* 1995), using the optical velocity definition in the heliocentric frame.

2. Results

We have monitored the redshifted (rest frequency: 22 GHz) radio continuum and maser emission in MG J0414+0534 for $\sim$15 months at $\sim$6 week intervals and found that both are surprisingly stable. Absolute deviations of the continuum flux from the mean are on average comparable with the flux calibration uncertainty (7%). The 6 GHz (observed frequency) continuum flux density of MG J0414+0534 thus remained nearly constant for the duration of the entire monitoring period, with an average flux density of 0.71 ± 0.02 Jy, the error being the standard deviation of the mean. The line peak flux density is also surprisingly stable throughout the period of the observations. Small fluctuations do not exceed the limits of uncertainty (between 10% and 50%). From the analysis of the 11 epochs of the monitoring, we can place an upper limit on the velocity drift of the line peak of $2\,\mathrm{km\,s^{-1}\,yr^{-1}}$.

A significant change in the line profile seems to have occurred between the Effelsberg and EVLA observations and the first epoch of the Arecibo monitoring campaign. Indeed, the line appears to be much broader in the first Arecibo spectrum (taken in October 2008) w.r.t. the previous observations. The full width at half maximum (FWHM) of the Gaussian profile fitted to the line is $174 \pm 5\,\mathrm{km\,s^{-1}}$, i.e. more than a factor of two larger than the FWHM of the Effelsberg and EVLA spectra (Impellizzeri *et al.* 2008). As a consequence, we measure an unlensed isotropic luminosity of 26,000 $L_\odot$, that makes the maser in MG J0414+0534 the most luminous that is currently known. Furthermore, in October 2008 we tentatively detected a weaker satellite line at $+470\,\mathrm{km\,s^{-1}}$ (Fig. 1) that, however, was not confirmed by the spectra of the other epochs. This second feature, detected with a signal-to-noise ratio of three, is displaced by about $800\,\mathrm{km\,s^{-1}}$ from the main line and is five time less luminous. In February 2009 we performed deeper observations aimed at confirming the presence of this feature. No emission line other than the main one at the velocity of about $-300\,\mathrm{km\,s^{-1}}$ was detected above a 3σ noise level of $0.3\,\mathrm{mJy}$ per $19.2\,\mathrm{km\,s^{-1}}$ channel. However, a weak feature is seen in the spectrum at the velocity of about $+490\,\mathrm{km\,s^{-1}}$ (see Fig. 2, lower panel). The satellite line remains undetected also in the spectrum produced by averaging all of the epochs with the same weights (Fig. 2, upper panel). Nonetheless, we note that the range between 200 and $500\,\mathrm{km\,s^{-1}}$ looks spiky and that, interestingly, one of these spikes is at the position of the satellite line. Averaging the spectra using different weights (e.g. $1/\mathrm{r.m.s.}^2$ or the integration time) does not change the shape of the resulting spectrum significantly. This may indicate that many weak lines are present in the range $200\text{--}500\,\mathrm{km\,s^{-1}}$ and that in October 2008 we saw one of these lines flaring.

The high SNR of the February 2009 spectrum (~13; see Fig. 2, lower panel) reveals that the main line has a complex profile that is likely the result of the blending of at least four components with line widths between 30 and $160\,\mathrm{km\,s^{-1}}$. In order to inspect the variability of the individual velocity features, we produced a spectrum by averaging with equal weights the last three epochs of the monitoring campaign (September and November 2009 and January 2010). The resulting spectrum (Fig. 2, middle panel) has an r.m.s comparable with that of the February 2009 observation. Comparing the Gaussian peak velocities, we find that the velocities of components I and II did not change, while the velocities of components III and IV have marginally increased by $+15 \pm 3\,\mathrm{km\,s^{-1}}$ and $+10 \pm 3\,\mathrm{km\,s^{-1}}$, respectively. Since velocity drifts have been observed only in very narrow lines (FWHM $\sim 1\text{--}4\,\mathrm{km\,s^{-1}}$), we think that the change in the peak velocities of these features maybe due to variations of a large number of sub-components, simulating a change in the radial velocity as observed most notably in NGC 1052 (Braatz *et al.* 1996).

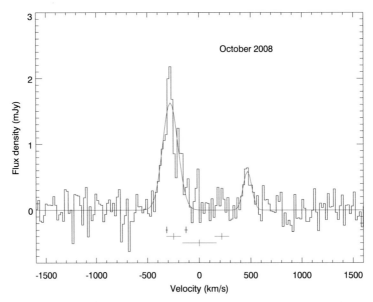

Figure 1. Water maser spectrum observed towards MG J0414+0534 in October 2008 (black histogram). The fitted Gaussian profiles are overlaid (blue line). The channel spacing is $19.2\,\mathrm{km\,s^{-1}}$. The root-mean-square (r.m.s.) noise level of the spectrum is $0.2\,\mathrm{mJy}$ per channel. The velocity scale is relative to redshift 2.639 using the optical velocity definition in the heliocentric frame. The red cross marks the systemic velocity and the associated uncertainty. The blue and the black crosses indicate the peaks of the CO emission (Barvainis *et al.* 1998) and the H I absorption components (Moore *et al.* 1999), respectively, with their errors.

3. Discussion and concluding remarks

Our monitoring data are partially consistent with our initial hypothesis that the emission is associated with the prominent relativistic jets of the quasar (Impellizzeri *et al.* 2008). First of all, even when the maser line profile is resolved into multiple velocity components, individual emission features have line widths between 30 and 160 km s^{-1} that resemble those of known H_2O masers associated with radio jets (e.g. Mrk 348; Peck *et al.* 2003). Our non-detection of a radial acceleration of the main maser peak is also compatible with the jet-maser scenario. The extreme stability of the main line peak and the continuum flux density in MG J0414+0534 resulting from our study, seems to exclude a jet-maser scenario similar to that in Mrk 348 (Peck *et al.* 2003) and NGC 1068 (Gallimore *et al.* 2001), while the reported significant variations in the line profile of our target may hint at similarities with the case of NGC 1052 (Braatz *et al.* 1996 and 2003).

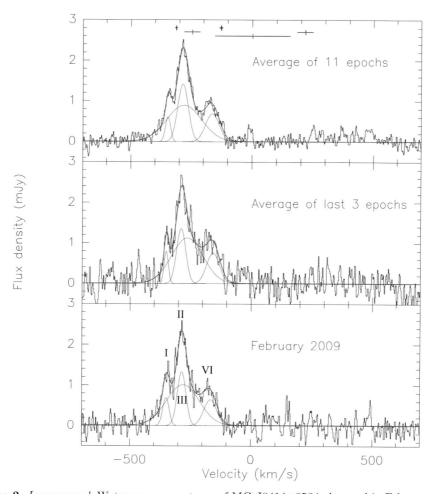

Figure 2. *Lower panel*: Water maser spectrum of MG J0414+0534 observed in February 2009. The labels I to IV indicate the best fit Gaussian components. *Middle panel*: Average of the last three epochs (September and November 2009 and January 2010) obtained using equal weights. *Upper panel*: Final spectrum produced by averaging all the epochs with the same weight. Individual Gaussian profiles fitted to the spectra are overlaid in blue together with the resulting profile. The channel spacing is 2.4 km s^{-1}. The r.m.s noise level is 0.2 mJy per channel in the spectra of the lower and middle panels and 0.1 mJy per channel in the upper panel.

We note, however, that the number of sources in which the maser emission is confidently associated with the jet(s) is very low and that more of these masers should be studied in detail in order to investigate the properties of these kind of sources. Furthermore, the tentative detection of the redshifted feature in the October 2008 spectrum is compatible with the disk-maser hypothesis. If the main maser line and the satellite line at $+470\,\mathrm{km\,s^{-1}}$ are considered as the blueshifted and redshifted lines from the tangentially seen part of an edge-on accretion disk in Keplerian rotation, then the radius at which the emission originates is given by $R = GM_{\mathrm{BH}}V_{\mathrm{R}}^{-2}$, where G is the gravitational constant, M_{BH} is the black hole mass, and V_{R} is the rotational velocity at radius R. From the difference between the line of sight velocities of the main and satellite maser lines (V_{obs}), we obtain $V_{\mathrm{R}} = V_{\mathrm{obs}} \cdot \sin(i)^{-1} \sim 370 \cdot \sin(i)^{-1}\,\mathrm{km\,s^{-1}}$. Adopting the black hole mass of $M_{\mathrm{BH}} = 10^{9.0}\,\mathrm{M_\odot}$ calculated by Pooley *et al.* (2007) for MG J0414+0534, and assuming an edge-on orientation ($i = 90°$) for the accretion disk, we get a radius of $R \sim 30\,\mathrm{pc}$. This value is fairly large compared to the radii at which maser emission is found in the accretion disks of nearby radio quiet AGN (typically, 0.1 to 1 pc). We should keep in mind however, that MG J0414+0534 is a radio loud quasar, while known disk-maser hosts are mainly radio quiet Seyfert or LINER galaxies with a mass of the nuclear engine that is two orders of magnitude lower ($\sim 10^7\,\mathrm{M_\odot}$; Kuo *et al.* 2011).

In conclusion, although we have been able to provide useful elements to determine the nature of the maser in MG J0414+0534, our current data are presently insufficient to confidently rule out whether the maser emission is due to a jet-cloud interaction or a rotating circumnuclear disk. VLBI observations and longer time-scale single-dish monitoring will be essential to shed light on the origin of the H_2O maser in this distant quasar.

References

Barvainis, R. & Antonucci, R. 2005, *ApJ*, 628, 89

Braatz, J. A., Wilson, A. S., & Henkel, C. 1996, *ApJS*, 106, 51

Braatz, J. A., Wilson, A. S., Henkel, C., Gough, R., & Sinclair, M. 2003, *ApJS*, 146, 249

Castangia, P., Impellizzeri, C. M. V., McKean, J. P., Henkel, C., Brunthaler, A., Roy, A. L., Wucknitz, O., Ott, J., & Momjian, E. 2011, *A&A*, 529, 150

Gallimore, J. F., Henkel, C., Baum, S. A., *et al.* 2001, *ApJ*, 556, 694

Kuo, C. Y., Braatz, J. A., Condon, J. J., *et al.* 2011, *ApJ*, 727, 20

Impellizzeri, C. M. V., McKean, J. P., Castangia, P., Roy, A. L., Henkel, C., Brunthaler, A., & Wucknitz, O. 2008, *Nature*, 456, 927

Lawrence, C. R., Elston, R., Januzzi, B. T., & Turner, E. L. 1995, *AJ* 110, 2570

McKean, J. P., Impellizzeri, C. M. V., Roy, A. L., Castangia, P., Samuel, F., Brunthaler, A., Henkel, C., & Wucknitz, O. 2011, *MNRAS*, 410, 2506

Moore, C. B., Carilli, C. L., & Menten, K. M. 1999, *ApJ*, 510, L87

Peck, A. B., Henkel, C., Ulvestad, J. S., *et al.* 2003, *ApJ*, 590, 149

Pooley, D., Blackburne, J. A., Rappaport, S., & Schechter, P. L. 2007, *ApJ*, 661,19

Cosmic Masers - from OH to H$_0$
Proceedings IAU Symposium No. 287, 2012
R.S. Booth, E.M.L. Humphreys & W.H.T. Vlemmings, eds.

© International Astronomical Union 2012
doi:10.1017/S1743921312007284

Searching for new OH megamasers out to redshifts $z > 1$

Kyle W. Willett[1,2]

[1] Center for Astrophysics and Space Astronomy, University of Colorado
UCB 391, Boulder, CO 80304, United States

[2] School of Physics and Astronomy, University of Minnesota
116 Church St. SE, Minneapolis, MN 55455, United States
email: willett@physics.umn.edu

Abstract. We have carried out a search for 18-cm OH megamaser (OHM) emission with the Green Bank Telescope. The targeted galaxies comprise a sample of 121 ULIRGs at $0.09 < z < 1.5$, making this is the first large, systematic search for OHMs at $z > 0.25$. Nine new detections of OHMs are reported, all at redshifts $z < 0.25$. For the remainder of the galaxies, observations constrain the upper limit on OH emission; this rules out OHMs of moderate brightness ($L_{OH} > 10^3\ L\odot$) for 26% of the sample, and extremely bright OHM emission ($L_{OH} > 10^4\ L\odot$) for 73% of the sample. Losses from RFI result in the OHM detection fraction being significantly lower than expected for galaxies with $L_{IR} > 10^{12}\ L_\odot$. The new OHM detections are used to calculate an updated OH luminosity function, with $\Phi \propto L_{OH}^{-0.66}$; this slope is in agreement with previous results. Non-detections of OHMs in the COSMOS field constrain the predicted sky density of OHMs; the results are consistent with a galaxy merger rate evolving as $(1 + z)^m$, where $m \leqslant 6$.

Keywords. masers; radio lines: galaxies; infrared: galaxies; galaxies: interactions

1. Introduction

OH megamasers (OHMs) trace of some of the most extreme physical conditions in the universe - in particular, the presence of an OHM signals specific stages in the merger process of gas-rich galaxies. OHMs can thus be used as probes of their environments, both directly and indirectly. Characteristics of the maser emission itself can be used to measure extragalactic magnetic fields (via Zeeman splitting) and gas kinematics, while the *presence* of an OHM is a signpost for phenomena associated with galaxy mergers, including extreme star formation and merging black holes. OHMs are a unique tool in this respect due to their extreme luminosities and ability to be seen at cosmic distances.

The total number of OHMs detected to date is still low. As of 2012, there are ~ 113 OHMs published in the literature, with roughly 50% discovered in the Arecibo survey of Darling & Giovanelli (2000, 2001, 2002a). No OHMs at a distance of greater than 1300 Mpc ($z = 0.265$) have been detected, and no large, systematic searches for high-z OHMs have been carried out. The association of OHMs with IR-bright merging galaxies, however, means that the density of OHMs is expected to be much higher at $z \simeq 1 - 2$, coinciding with an increase in both merging rate and cosmic star formation. We have conducted a search for high-redshift OHMs using the Green Bank Telescope (GBT).

2. Sample selection and observations

We constructed three samples of galaxies to search for high-redshift OHMs. None of the samples are fully complete or flux-limited, but draw on the catalogs of IR-luminous galaxies with well-defined redshifts that were available at the time.

Table 1. Properties of new OHM detections from the GBT survey

Galaxy	z_{hel}	S^{peak}_{1667} [mJy]	W_{1667} [MHz]	$\log L_{OH}$ $[L_\odot]$	$\log L^{pred}_{OH}$ $[L_\odot]$
IRAS 00461−0728	0.2427	8.93	0.937	3.58	2.94−3.30
IRAS 01298−0744	0.136181	118.28	0.785	3.78	3.10
IRAS 01355−1814	0.192	6.99	0.344	2.81	3.25
IRAS 01569−2939	0.1402	40.67	1.821	3.93	2.95
IRAS 10597+5926	0.196	17.56	0.915	3.85	3.13
IRAS 12071−0444	0.128355	4.27	0.598	2.45	3.07
IRAS 16090−0139	0.13358	9.00	3.110	3.53	3.35
FF 0758+2851	0.126	4.67	0.396	2.67	2.39
FF 2216+0058	0.212	18.40	0.347	3.53	2.58

IRAS PSCz galaxies not visible from Arecibo: The first sample of galaxies consisted of IRAS sources included in the redshift catalog of the PSCz survey. Galaxies were selected according to similar criteria as in the flux-limited Arecibo sample, but included objects lying outside the declination limits of Arecibo ($-1° < \delta < 38°$). Galaxies selected for GBT observations had: a declination range of $-40° < \delta < 0°$ or $\delta > 37°$, a redshift range of $0.10 < z < 0.25$, and a lower-luminosity threshold of $L_{60\mu m} > 10^{11.4} L_\odot$. 153 galaxies in the PSCz met these criteria, of which 47 of the brightest candidates were observed according to the LST windows during the early commissioning phase of the GBT in 2002.

Sub-mm and ULIRG galaxies from the field: The second sample of potential OHM hosts was assembled from flux-limited catalogs of ULIRGs at higher redshifts. We began with 35 galaxies in the FSC-FIRST catalog (Stanford *et al.* 2000), which consists of targets detected in both the IRAS Faint Source Catalog and the 20-cm VLA FIRST survey. This was supplemented with 5 IR-bright galaxy pairs and 26 sub-millimetre galaxies. All galaxies have $L_{IR} > 10^{11} L_\odot$, with more than half having $L_{IR} > 10^{12} L_\odot$. The highest redshift in this sample is at $z = 1.55$.

Starburst galaxies from COSMOS: The third group of 19 OHM candidates was the last observed in our program, and made explicit use of the results from the first two samples. Targets were selected from the COSMOS field, a 2-deg^2 survey with deep spectral coverage from X-ray through radio wavelengths. Recent infrared (Willett *et al.* 2011) studies show that while OHMs are found in infrared-bright galaxies, the OHM fraction is much higher for starburst-dominated galaxies vs. AGN. Selection of OHM candidates began with COSMOS galaxies detected by *Spitzer* at 70 μm, and then eliminating all targets except the LIRGs, ULIRGs, and HyLIRGs identified by Kartaltepe *et al.* (2010). We removed all galaxies identified as AGN or with $L_{IR} < 10^{12} L_\odot$. Finally, we culled the target list based on the expected RFI conditions near the observed frequency bands. By limiting the observed OH frequencies to cleaner regions, we have a broader margin for error on the galaxy redshift. The two windows used are at $\nu_{obs} = 825 - 830$ MHz and $960 - 1005$ MHz, equivalent to redshifted OH at $0.97 < z < 1.01$ and $0.67 < z < 0.73$.

Observations: We observed the OHM candidates in several sessions at the GBT from 2002–2010, totaling approximately 150 hours. The majority of observations used the maximum available bandwidth of 50 MHz and 8192 channels. Integration times were selected with the goal of achieving ~ 1 mJy rms per channel for each galaxy. The data were reduced using standard routines in GBTIDL, including extensive flagging for RFI. After flagging, we fit the radio continuum around the expected line center with a polynomial function of order $n = 5$. This removed both intrinsic continuum structure from the target itself and any baseline structure not removed by the position-switching technique. After stacking, the spectra were smoothed to a rest-frame velocity resolution of 10 km s^{-1}.

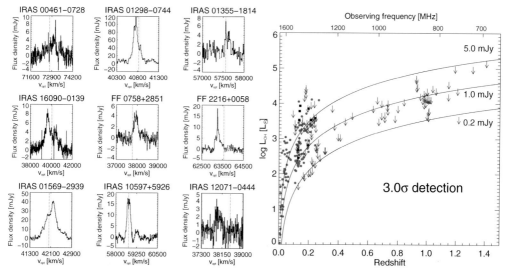

Figure 1. Left: 18-cm spectra of the nine new OHMs discovered with the Green Bank Telescope. Right: Integrated OH luminosities all known OHMs, including data from the literature (red) and the new GBT detections (black). Upper limits from the GBT survey are indicated with the green arrows. The black curves show detection thresholds for a $\Delta v = 150$ km s^{-1} OHM at a range of telescope rms sensitivities as a function of redshift.

3. Survey results

OH megamasers: Out of 128 galaxies observed for OH, we detected new OH megamaser emission in nine objects (Figure 1). Seven of the detections were PSCz galaxies from the first sample, while the other two were from the FSC-FIRST catalog in the second sample. All nine OHMs have redshifts near the lower end of the sample distribution, with the most distant lying at a redshift of $z = 0.2427$. The observed frequencies are in the range of $1300 - 1500$ MHz, which is covered by the L-band receiver and has relatively little RFI compared to the GBT prime focus bands. The OH emission has been confirmed with GBT follow-up observations for five of the galaxies. We also confirmed the detection of the previously-discovered OHM IRAS 09539+0857 Darling & Giovanelli (2001).

Table 1 lists the 18-cm radio properties of the new OHM detections. We give the galaxy's optical redshift (z_{hel}), peak flux density of the OHM (S_{1667}^{peak}), the ratio of the integrated 1667 MHz emission to its peak flux density (W_{1667}), measured OH luminosity (log L_{OH}), and predicted OH luminosity (L_{OH}^{pred}) based on the $L_{OH} - L_{FIR}$ relationship in Darling & Giovanelli (2002a).

OH non-detections: 112 galaxies showed no confirmed detections of OH. The L_{OH}^{max} for each galaxy was (conservatively) derived assuming a boxcar line profile with a linewidth $\Delta v = 150$ km s^{-1} and a 1.5σ detection. The rms was measured from baseline-subtracted continuum centered on the optical redshift of the galaxy and in a frequency range sufficient to cover the uncertainty in the optical redshift ($\Delta\nu_{obs} = \Delta z \times \nu_{rest}/[1 + z]$).

4. Updating the OH luminosity function

One of the goals of performing a search for OHMs at higher redshifts was to improve the measurements of the OH megamaser luminosity function (LF). Darling & Giovanelli (2002b) used the results of the flux-limited Arecibo survey to construct a well-sampled LF between $10^{2.2}$ $L_\odot < L_{OH} < 10^{3.8} L_\odot$, which followed a power law in integrated line luminosity of $\Phi[L] \propto L_{OH}^{-0.64}$ Mpc^{-3} dex^{-1}. This measurement was limited to a narrow redshift range, spanning $0.1 < z < 0.23$. We constructed a new OH LF by combining the GBT and Arecibo OHM detections, using the $1/V_a$ method and combining limits on both spectral line and continuum emission. We fit a power-law to all bins with more than one detection, yielding:

$$\log \Phi[L] = (-0.66 \pm 0.14) \log L_{OH} - (4.91 \pm 0.41), \qquad (4.1)$$

where Φ is measured in Mpc^{-3} dex^{-1} and L_{OH} in $L_\odot$. The new detections only change the slope measured by Darling & Giovanelli (2002b) by -0.02 and the offset by $+0.10$. Both values are well within the uncertainties of the combined LF, as well as that of the original Arecibo LF.

Assuming the Malmquist-corrected relationship of $L_{OH} \propto L_{IR}^{1.2}$ from Darling & Giovanelli (2002a), this gives $\Phi[L_{IR}] \propto L_{IR}^{-0.83 \pm 0.18}$. We compare this to the LF of ULIRGs in the local Universe using the AKARI measurements of Goto et al. (2011). Folding in the OHM fraction derived from the combined Arecibo and GBT samples results in a slope of (-0.6 ± 0.2), a much shallower value than that measured from the AKARI galaxies (-2.6 ± 0.1). This inconsistency may suggest either that OHMs are highly saturated or that the maser strength is only weakly correlated with global properties such as L_{IR} (Darling & Giovanelli 2002a). The OHM LF of Briggs (2000) assumed a quadratic OH-IR relation corresponding to unsaturated masing; a decrease in saturation could potentially steepen the OHM LF up to a slope of -1.5.

5. Constraining the evolution of the cosmic merger rate

One of the ultimate goals of high-redshift OHM surveys is to use megamasers as tracers of the populations of merging galaxies as a function of redshift. Models of the merger rate as a function of redshift are typically parameterized with an evolutionary factor of $(1 + z)^m$. The value of m, however, is not well-constrained (e.g., Kim & Sanders 1998; Bridge et al. 2010). Sufficiently deep surveys of OHMs can provide an independent measurement of the parametrization of the merging rate. We calculated the predicted sky density of OHMs as a function of redshift:

$$\frac{dN}{d\Omega d\nu}[z] = \frac{cD_L^2}{H_0 \nu_0 \sqrt{(1+z)^3 \Omega_M + \Omega_\Lambda}} \left(\frac{b}{a \ln 10}\right) \times \left((L_{OH,max})^a - (L_{OH,min})^a\right), \quad (5.1)$$

where a and b are parameters of the OHM LF from Equation 4.1 ($\Phi[L_{OH}] = bL_{OH}^a$), $L_{OH,min}$ is the minimum OH luminosity that can be observed at a given sensitivity level, and $L_{OH,max}$ is the upper physical limit on OHM luminosity (Darling & Giovanelli 2002b).

Using the upper limit of zero OHMs detected in the COSMOS field, we place an upper limit on the merger rate of $m \lesssim 6$. While this is still within uncertainties for the highest estimated values of m, the COSMOS limit is an important first step in using OHMs as an independent tracer. Further measurements of OH deep fields at higher redshifts will be crucial for a more accurate constraint.

6. Future OHM searches

The final results for the GBT OHM survey yielded a much lower detection rate $(9/121 = 7\%)$ of megamasers than expected. Based on the high detection fraction for galaxies with $L_{IR} > 10^{12} L_\odot$ from the Arecibo survey, in addition to our careful selection of starburst-dominated galaxies, we had predicted a success rate of 20–30%. We attribute one of the primary causes of the low OHM fraction to be the sensitivity of the observations. Figure 1 shows the upper limits for OHM candidates from our survey, along with the necessary rms sensitivity to detect OH lines as a function of redshift. To detect median-luminosity OHMs at $z = 1$ will require rms levels of 100 μJy, and perhaps significantly more time devoted to individual targets.

It must also be mentioned that RFI is a major culprit at $\nu_{obs} < 1$ GHz, significantly restricting the redshift path and increasing the noise in each receiver band. Future OHM searches may benefit from interferometric observations (for which celestial RFI is uncorrelated from dish to dish), or from observations in more radio-quiet environments. The GMRT (India) and ASKAP (Australia) are promising instruments for interferometry, while the upcoming 64-m Sardinia Radio Telescope (Italy) and 500-m FAST (China) will be options to continue the search for OHMs at high redshifts.

7. Acknowledgments

This work was part of the Ph.D. thesis of KWW at the University of Colorado, and will be submitted in early 2012 as Willett, Darling, Kent, & Braatz (2012). Observations were funded in part by student support grants from NRAO. We are indebted to the staff of the Green Bank Telescope, whose expertise and assistance made these observations possible. The National Radio Astronomy Observatory is a facility of the National Science Foundation operated under cooperative agreement by Associated Universities, Inc.

References

Bridge, C. R., Carlberg, R. G., & Sullivan, M. 2010, *ApJ*, 709, 1067
Briggs, F. H. 1998, *A&A*, 336, 815
Darling, J. & Giovanelli, R. 2000, *AJ*, 119, 3003
Darling, J. & Giovanelli, R. 2001, *AJ*, 121, 1278
Darling, J. & Giovanelli, R. 2002a, *AJ*, 124, 100
Darling, J. & Giovanelli, R. 2002b, *ApJ*, 572, 810
Goto, T. *et al.* 2011, *MNRAS*, 410, 573
Kartaltepe, J. *et al.* 2010, *ApJ*, 709, 572
Kim, D. & Sanders, D. B. 1998, *ApJS*, 119, 41
Stanford, S. A., Stern, D., van Breugel, W., & De Breuck, C. 2000, *ApJS*, 131, 185
Willett, K. W., Darling, J., Spoon, H. W. W., Charmandaris, V., & Armus, L. 2011, *ApJ*, 730, 56

Cosmic Masers - from OH to H$_0$
Proceedings IAU Symposium No. 287, 2012
R.S. Booth, E.M.L. Humphreys & W.H.T. Vlemmings, eds.
© International Astronomical Union 2012
doi:10.1017/S1743921312007296

Expectations of maser studies with FAST

Jiang Shui Zhang[1], Di Li[2] and Jun Zhi Wang[3]

[1] Center for astrophysics, Guangzhou university,
Guangzhou 510006,
email: jszhang@gzhu.edu.cn

[2] National Astronomical Observatories, Chinese Academy of Sciences
A20 Datun Road, Chaoyang District, Beijing 100012
email: dili@nao.cas.cn

[3] Department of Astronomy, Nanjing University,
22 Hankou Road, Nanjing 210093, China
email: junzhiwang@nju.edu.cn

Abstract. The Five-hundred-meter Aperture Spherical radio Telescope (FAST) is being built by the Chinese and will be the largest single dish radio telescope in the world. FAST, with much increase in sensitivity, will give astronomers good opportunities to answer many fundamental questions in astronomy. Here we give a brief introduction of FAST and its enormous potential for studying Galactic and extragalactic masers.

Keywords. telescopes, masers

1. Introduction

The Five-hundred-meter Aperture Spherical radio Telescope (FAST) will be the largest single dish telescope in the world upon its finish in five years. Its location (25°.647222N and 106°.85583E) is about 170 km from Guiyang, the capital of Guizhou province in the southwest of China. Its construction began in March 2011 and its first light is expected at the end of 2016. With respect to existing radio telescopes, FAST has many unprecedented advantages: largest filled aperture, large sky coverage, extremely radio quiet site, etc. This will provide astronomers a good opportunity to explore many important science goals and the potential to obtain some big discoveries, such as, H I galaxy surveys, pulsar-black hole systems, extragalactic pulsars, radio signals from exoplanets (Nan *et al.* 2011).

One fundamental science proposal (Frontiers in radio astronomy and FAST Early Science) has been granted now, which will be carried out before the scheduled first light in 2016. The group of the proposal includes five science teams and one receiver group. The main goals of the proposal focus on defining key FAST programs and early science. In addition, tens of graduate students and young radio astronomers will be trained as part of the proposal.

FAST frequency coverage ranges from 70 MHz to 3 GHz and nine sets of receivers are planned according to the requirements of the science goals. For maser-related areas, OH masers ($\lambda \sim 18$čm) could be the main object of FAST. Among those nine receivers, the L-wide single beam receiver (No. 7) can be used to detect both Galactic and extragalactic OH masers. The No. 3 (0.28–0.56 GHz) and No. 4 (0.56–1.02 GHz) receivers can be used to detect high redshift OH maser sources (OH megamasers or gigamasers).

Here we present our investigation of existing OH maser observations and our proposed studies of both Galactic and extragalactic OH maser studies with the future FAST.

2. Investigations of OH maser studies

Galactic OH masers: To date, more than 3000 OH maser sources have been detected in our galaxy since their first discovery in interstellar space (Weaver *et al.* 1965). They can be basically classified as interstellar and circumstellar masers. Interstellar masers were considered to originate in dense molecular gas in star formation regions and circumstellar masers in the molecular circumstellar envelopes of evolved giant and supergiant stars (Reid 2002). About 80% of all Galactic OH masers belong to the latter (e.g., Lo 2005, Mu *et al.* 2010) and their catalogue and detailed information can be found in the Hamburg University database of circumstellar OH masers (http://www.hs.uni-hamburg.de/ st2b102/maserdb/index.html, also from Dr. Engels, these proceedings). In addition, OH masers with only 1720 MHz emission are believed to be associated with supernova remnants (SNRs) and have proven to be excellent tracers of SNR-molecular cloud interactions. About 10% of SNRs ($\sim$20) were detected with this type of OH maser emission (e.g., Frail *et al.* 1996, Yusef-Zadeh *et al.* 1999, Hewitt *et al.* 2008, also Dr. Gray and Dr. Wardle, these proceeedings).

Searching for Galactic OH masers mainly consists of blind sky surveys and sample-selection surveys. Blind sky surveys were mostly performed long ago (in the 1970s and 1980s) with low sensitivity and the telescopes used include Parkes 64 m, Onsala 25.6 m, NRAO 42 m, Effelsberg 100m etc. Many Galactic OH masers were found from these blind surveys (e.g., Caswell *et al.* 1980, Caswell & Haynes 1983, Caswell & Haynes 1987). Meanwhile, choosing different samples for searching for Galactic OH masers is also successful. Selected-samples mainly include color-selected IRAS samples (e.g., Eder *et al.* 1988, Lewis *et al.* 1990, te Lintel Hekkert 1991), star-forming-region samples (e.g., Szymczak *et al.* 2004, Wouterloot *et al.* 1993) and high mass protostellar object samples (e.g., Edris *et al.* 2007) etc.

Extragalactic OH masers: There are about 125 OH megamasers so far ($\lambda \sim 18$ cm, Darling & Giovanelli 2000, Chen *et al.* 2007, Fernandez *et al.* 2010, ten or so unpublished sources from Dr. Darling, also Dr. Willett, this symposium) since the first detection in galaxy IC4553 (Baan *et al.* 1982). The most luminous OH megamaser is IRAS14070+0525 with a redshift of 0.265, detected by Arecibo (Baan *et al.* 1992). Host galaxies of those 125 OH megamasers are (ultra)luminous infrared galaxies, which are believed to be the product of galaxy mergers (e.g., Clements *et al.* 1996). The rate of galaxy mergers was measured to be proportional to $(1+z)^m$ (m: $3 \sim 8$, e.g, Le Fevre *et al.* 2000, Kim & Sanders 1998). Therefore, we can expect more OH megamasers at high redshift, given that OH megamasers trace galaxy mergers.

Existing surveys of OH megamasers came from Arecibo 305 m, NRAO 91 m, JB MkIA 76 m, Nancay 300m and Parkes 64m etc. The most successful survey so far is the upgraded Arecibo survey with a detection rate of $\sim$17% (Darling & Giovanelli 2000). They chose a sample from the IRAS Point Source Catalog Redshift Survey ($f_{60\,um} > 0.6$ Jy) (Saunders *et al.* 2000), with a declination range of $0°$–$37°$ (Arecibo sky coverage) and a redshift range of 0.1–0.3. Statistical results show a relation between OH and FIR luminosity of $L_{OH} \propto L_{FIR}^{1.2\pm0.1}$ (Darling & Giovanelli 2002a) and a luminosity function of $\phi \propto L_{OH}^{-0.6}\, Mpc^{-3} dex^{-1}$ (Darling & Giovanelli 2002b, also Dr. Willett, this symposium).

3. Expectations for maser studies

Galactic OH masers: FAST has a sensitivity at least one order more than those telescopes used in the 1970s & 1980s Galactic surveys. To achieve the same sensitivity limit, the FAST surveying speed would be 2 orders of magnitude better. This brings high efficiency for both sample selection surveys and Galactic plane blind sky surveys. As mentioned above, the L-wide single beam receiver could be used for these surveys. Certainly, an upgraded multi-beam receiver (L-wide receiver) will be better for wide area blind surveys. In addition, we believe that many more 1720 MHz OH masers in SNRs could be detected with the high sensitivity FAST, which could help us understand the interaction between SNRs and adjacent molecular clouds, and further related sciences, such as the acceleration of relativistic particles, and the structure and physical conditions of the Galactic interstellar medium (Green *et al.* 1997).

Extragalactic OH masers: As discussed above, detections of OH megamasers come mainly from the Arecibo 305 m telescope. Compared with Arecibo, FAST has three times better raw sensitivity and about 10 times higher surveying speed. In addition, FAST has a zenith angle of 40 degrees, thanks to the deep karst depression and the innovative design of the active primary surface. It makes the declination range of FAST about -15 to 65 degrees (its latitude is $25°.647222$N), which is 2–3 times Arecibo's sky coverage (Nan *et al.* 2011). Based on the FAST advantages, the following advances are expected:

1) Numbers of OH megamasers: three times better sensitivity can extend the detection limit by a factor of $\sim\sqrt{3}$ in terms of distance, thus increasing the sample size of detectable sources by a factor of $\sim 3\sqrt{3}$ times. Taking into account its 2–3 times sky coverage increase, we can expect the number of OH megamasers to increase 10–20 times, assuming a similar detection rate and luminosity function. Thus more than 1000 OH megamasers could be detected by FAST.

2) Higher sensitivity and larger sky coverage provide more opportunities to detect high redshift OH megamasers or Gigamasers. The Arecibo survey gives a typical rms flux density of 0.65 mJy or so for a 12 minute integration. Given FAST with 3 times better sensitivity, the rms value of a FAST similar survey is ~ 0.22 mJy for the same integration time. According to a plot of sensitivity thresholds of OH megamaser detections (Darling & Giovanelli 2002b), FAST could detect most OH megamsers with $L_{OH} > 10^3 L_\odot$ out to $z \sim 1$ and could detect OH gigamasers out to roughly $z \sim 2$ in a 12 minute integration.

3) The most distant OH megamaser (1720 MHz) was detected by the Green Bank Telescope in a gravitationally lensed source PMN J0134-0931 (Kanekar *et al.* 2005). FAST detections of OH maser emission could be expected in similar, but more distant (e.g., z>1) sources.

Acknowledgements

This work is supported by the China Ministry of Science and Technology under the State Key Development Program for Basic Research (2012CB821800) and the Natural Science Foundation of China (No. 11043012, 11178009). We made use of the NASA Astrophysics Data System Bibliographic Services (ADS) and the NASA/IPAC extragalactic Database (NED), which is operated by the Jet Propulsion Laboratory, California Institute of Technology, under contract with NASA.

References

Baan, W. A., Wood, P. A. D., & Haschick, A. D. 1982, *ApJ*, 260, L49
Baan, W. A., Rhoads, J., Fisher, K., Altschuler, D. R., & Haschick, A. D. 1992, *ApJ*, 396, L99
Caswell, J. L., Haynes, R. F., & Goss, W. M. 1980, *Australian Journal of Physics*, 33, 639
Caswell, J. L. & Haynes, R. F. 1983, *Australian Journal of Physics*, 36, 361
Caswell, J. L. & Haynes, R. F. 1987, *Australian Journal of Physics*, 40, 215
Chen, P. S., Shan, H. G., & Gao, Y. F. 2007, *AJ*, 133, 496

Clements, D. L., Sutherland, W. J., McMahon, R. G., & Saunders, W. 1996, *MNRAS*, 279, 477

Darling, J. & Giovanelli, R. 2000, *AJ*, 119, 3003

Darling, J. & Giovanelli, R. 2002a, *AJ*, 124, 100

Darling, J. & Giovanelli, R. 2002b, *ApJ*, 572, 810

Eder, J., Lewis, B. M., & Terzian, Y. 1988, *ApJS*, 66, 183

Edris, K. A., Fuller, G. A., & Cohen, R. J. 2007, *A&A*, 465, 865

Fernandez, M. X., Momjian, E., Salter, C. J., & Ghosh, T. 2010, *AJ*, 139, 2066

Frail, D. A., Goss, W. M., Reynoso, E. M., Giacani, E. B., Green, A. J., & Otrupcek, R. 1996, *AJ*, 111, 1651

Green, A. J., Frail, D. A., Goss, W. M., & Otrupcek, R. 1997, *AJ*, 114, 205

Hewitt, J. W., Yusef-Zadeh, F., & Wardle, M. 2008, *ApJ*, 683, 189

Kanekar, N., Carilli, C. L., & Langston, G. I. *et al.* 2005, *PhRvL*, 95, 1301

Kim, D.-C. & Sanders, D. B. 1998, *ApJS*, 119, 41

Le Fvre, O., Abraham, R., & Lilly, S. J. 2000, *MNRAS*, 311, L565

Lewis, B. M., Eder, J., & Terzian, Y. 1990, *ApJ*, 362, 634

Lo, K. Y. 2005, *ARA&A*, 43, 625

Mu, J. M., Esimbek, J., Zhou, J. J., & Zhang, H. J. 2010, *RAA*, 2, 166

Nan, R. D., Li, D., & Jin, C. J. *et al.* 2011, *IJMPD*, 20, 989

Reid, M. 2002, *IAU Symp.*, 206, 506

Saunders, W., Sutherland, W. J., & Maddox, S. J. *et al.* 2000, *MNRAS*, 317, 55

Szymczak, M. & Grard, E. 2004, *A&A*, 423, 209

te Lintel Hekkert, P. 1991, *A&A*, 248, 209

Weaver, H., Williams, D. R. W., Dieter, N. H., & Lum, W. T. 1965, *Nature*, 208, 29

Wouterloot, J. G. A., Brand, J., & Fiegle, K. 1993, *A&AS*, 98, 589

Yusef-Zadeh, F., Goss, W. M., Roberts, D. A., Robinson, B., & Frail, D. A. 1999, *ApJ*, 527, 172

Cosmic Masers - from OH to H$_0$
Proceedings IAU Symposium No. 287, 2012
R.S. Booth, E.M.L. Humphreys & W.H.T. Vlemmings, eds.

© International Astronomical Union 2012
doi:10.1017/S1743921312007302

The origin of Keplerian megamaser disks

Mark Wardle[1] and Farhad Yusef-Zadeh[2]

[1] Department of Physics & Astronomy and Research Centre for Astronomy, Astrophysics
& Astrophotonics, Macquarie University, Sydney NSW 2109, Australia
email: mark.wardle@mq.edu.au

[2] Department of Physics & Astronomy, Northwestern University, Evanston IL 60208, USA
email: zadeh@northwestern.edu

Abstract. Several examples of thin, Keplerian, sub-parsec megamaser disks have been discovered in the nuclei of active galaxies and used to precisely determine the mass of their host black holes. We show that there is an empirical linear correlation between the disk radius and black hole mass and that such disks are naturally formed as molecular clouds pass through the galactic nucleus and temporarily engulf the central supermassive black hole. For initial cloud column densities below about $10^{23.5}$ cm^{-2} the disk is non-self gravitating, but for higher cloud columns the disk would fragment and produce a compact stellar disk similar to that observed around Sgr A* at the galactic centre.

Keywords. accretion, accretion disks, masers, Galaxy: center, galaxies: Seyfert

---

Fourteen Seyfert 2 nuclei are known to host powerful 22 GHz water masers located in a circumnuclear disk within a parsec of the central massive black hole. In eight edge-on systems, the maser kinematics trace the rotation curve of a thin, Keplerian disk, enabling accurate determination of the black hole mass (Herrnstein *et al.* 2005; Kuo *et al.* 2010). 22 GHz water masers occur where the disk surface density exceeds $1\,\mathrm{g\,cm^{-2}}$ and warping exposes the surface to X-ray irradiation from the centre (Neufeld *et al.* 1994; Maloney 2002). We propose instead that Keplerian megamaser disks are created through the partial capture of molecular clouds (Wardle & Yusef-Zadeh 2012), a model we have previously used to explain the compact disk of stars orbiting within 0.1 pc of Sgr A* in our own Galactic centre (Wardle & Yusef-Zadeh 2008). In this scenario a dense molecular cloud sweeps through the central few parsecs, temporarily engulfing the massive black hole. Streams of material passing on opposite sides of the hole are gravitationally deflected towards each other and collide, dissipating kinetic energy and partly cancelling angular momentum. The shocked gas cools efficiently, circularises, and forms a compact, thin molecular disk.

The disk size is determined by the black hole mass M, the initial speed v of the incoming cloud and the effectiveness of angular momentum cancellation. We make 2 simple assumptions: cloud material with initial impact parameter less than $r_{\mathrm{acc}} = 2GM/v^2$ is captured; and the angular momentum of this material is reduced to a fraction λ of its initial magnitude. The radius R of the disk is estimated by equating the specific orbital angular momentum at the disk edge, $\sqrt{GMR}$, to that of the material that is just barely captured, i.e. $\lambda r_{\mathrm{acc}} v$; this yields $R = 4G\lambda^2 M/v^2$. For $v = 225\,\mathrm{km\,s^{-1}}$ and $\lambda = 0.3$ the predicted disk radius is 0.3 pc for a 10^7 M$_\odot$ black hole and matches the empirical linear relationship of maser disk size with black hole mass (see left panel of Fig. 1).

The disk surface density Σ_D can be estimated from the captured mass $\sim \pi r_{\mathrm{acc}}^2 \Sigma_{\mathrm{cloud}}$, where Σ_{cloud} is the cloud surface density. Σ_D must meet two conditions to host 22 GHz masers: $\Sigma_D \gtrsim 1\,\mathrm{g\,cm^{-2}}$ so that x-ray irradiation produces a warm, dense, H$_2$O-rich layer; and $\Sigma_D \lesssim c_s\Omega/\pi G$, where $c_s \approx 1\,\mathrm{km\,s^{-1}}$ is the sound speed in the layer, so that the disk is

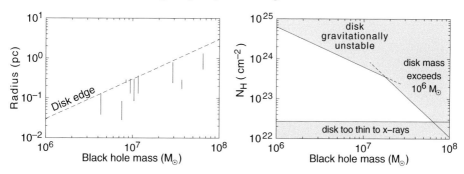

Figure 1. *(Left)* Vertical bars: observed radial extent of Keplerian rotation curves traced by megamasers in Seyfert 2 nuclei. Dashed curve: predicted disk edge in the cloud capture scenario. From left to right the systems are Sgr A* (compact stellar disk, Lu *et al.* 2009); NGC 2273, NGC 4388, NGC 6323, UGC 3789, NGC 2960, and NGC 6264 (Kuo *et al.* 2011), NGC 4258 (Herrnstein *et al.* 2005), and NGC 1194 (Kuo *et al.* 2011). *(Right)* Constraints on cloud column density as a function of black hole mass. The column density must lie in the unshaded region to produce a thin, Keplerian disk hosting 22 GHz megamasers.

not self-gravitating. In addition the inferred disk mass should neither exceed the notional cloud mass nor be sufficient to distort the rotation curve. The corresponding constraints on the column density of the incoming cloud are shown in Fig. 2.

This partial capture scenario naturally produces megamaser disks on sub-parsec scales, with the observed linear correlation between disc radius and black hole mass. The outer edge of the maser discs is due to physical truncation rather than a breakdown in masing conditions within an extended disk. This picture relies on dense clouds occasionally sweeping through the inner few parsecs of galactic nuclei. The circumnuclear molecular ring at 1.7 pc from Sgr A* (e.g. Christopher *et al.* 2005), and the 10 pc scale circumnuclear rings in numerous Seyfert galaxies suggest that there is an ample supply of material from larger radii, perhaps controlled by angular momentum transfer between massive clouds during gravitational encounters (Namekata & Habe 2011). The formation of gravitationally *unstable* disks may be a more common event because molecular clouds in the inner regions of galaxies tend to have column densities $\gtrsim 10^{24}$ cm^{-2}. These transient disks may also host megamasers (Milosavljević & Loeb 2004), so that only a small fraction of megamaser AGN may have the Keplerian disks necessary for accurate black hole mass determinations.

References

Christopher, M. H. *et al.* 2005, *ApJ* 622, 346

Herrnstein, J. R. *et al.* 2005, *ApJ* 629 719

Kuo, C. Y., *et al.* 2011, *ApJ* 727, 20

Lu, J. R., *et al.* 2009, *ApJ* 690, 1463

Maloney, P. R. 2002, *PASA* 19, 401

Milosavljević, M. & Loeb, A. 2004, *ApJ* 604, L45

Namekata, D. & Habe, A. 2011, *ApJ* 731, 57

Neufeld, D. A., Maloney, P. R., & Conger, S. 1994, *ApJ* 436, L127

Wardle, M. & Yusef-Zadeh, F. 2008 *ApJ* 683, L37

Wardle, M. & Yusef-Zadeh, F. 2012, *ApJ* (Letters) in press (arXiv:1112.3279)

A memento presented to Shuji Deguchi in honor of his retirement.

T8:

Maser Astrometry *Chair: Huib Jan van Langevelde*

Cosmic Masers - from OH to H$_0$
Proceedings IAU Symposium No. 287, 2012
R.S. Booth, E.M.L. Humphreys & W.H.T. Vlemmings, eds.
© International Astronomical Union 2012
doi:10.1017/S1743921312007314

Maser Astrometry: from Galactic Structure to Local Group Cosmology

Mark J. Reid

Harvard-Smithsonian Center for Astrophysics, 60 Garden Street, Cambridge, MA 02138, USA
email: reid@cfa.harvard.edu

Abstract. This review summarizes current advances in astrometry of masers as they pertain to large-scale Galactic structure and dynamics and Local Group cosmology. Parallaxes and proper motions have now been measured for more than 60 massive star forming regions using the Japanese VERA array, the EVN and the VLBA. These results provide "gold standard" distances and 3-dimensional velocities for sources across the Milky Way, revealing its spiral structure. Modeling these data tightly constrains the fundamental parameters of the Milky Way: R$_0$ and Θ$_0$. Proper motions of Local Group galaxies have been measured, improving our understanding of the history and fate of the Group.

Keywords. astrometry, masers, stars: formation, Galaxy: fundamental parameters, Galaxy: structure, Galaxy: kinematics and dynamics, galaxies: kinematics and dynamics, Local Group

1. Introduction

Figure 1 displays an image of NGC 1232, a galaxy classified SAB(rs)c that might resemble the Milky Way. It displays a multi-armed spiral with regular and bifurcated arms and a central bar, which may be weaker than in the Milky Way. Using NGC 1232 as a model for Milky Way, the Sun would be about two-thirds of the way from the nucleus to the edge of the stellar disk. The Hipparcos satellite measured $\sim 10^5$ stellar parallax with ≈ 1 mas accuracy and mapped the Solar Neighborhood. However, the Solar Neighborhood would cover an area not much larger than one of the H II regions that dot its spiral arms. Far better astrometric accuracy is needed to map the Milky Way.

The GAIA mission, scheduled for launch in 2013, hopes to achieve parallaxes to $\sim 10^9$ stars with accuracies of ≈ 20 μas. With this accuracy, one can measure a distance of 5 kpc with 10% uncertainty and begin to map large portions of the Milky Way. However, GAIA observes at optical wavelengths and will be limited to a small number of lines of sight in the Galactic plane owing to dust extinction. In order to map the spiral structure of the Milky Way, one needs observations at longer wavelengths. Very Long Baseline Interferometry (VLBI) at cm-wavelengths can accomplish this.

At cm-wavelengths dust in the Milky Way is transparent. VLBI parallax accuracy routinely is ~ 20 μas, and some measurements have achieved accuracies of 5 μas! Thus, one can measure distances to compact radio sources throughout the Galactic plane. Astronomical H$_2$O and CH$_3$OH masers are good astrometric targets and are found in regions of massive star formation. Such star forming regions are home to OB-type stars that ionize their placental material, and it is these H II regions that best define spiral structure in other galaxies. Thus, parallaxes of masers are an excellent method to map spiral structure in the Milky Way.

Figure 1. NGC 1232: a Milky Way look-alike?

2. VLBI Parallax Accuracy

While a thorough discussion of sources of error in phase-referenced VLBI astrometry is beyond the scope of this review, one can gain an appreciation of what is involved with $\sim \mu$as astrometry from a few key points. For a more thorough discussion see, eg, Honma, Tamura & Reid (2008) or Reid *et al.* (2009a).

The fringe spacing of a two-element interferometer is given by $\theta_f \approx \lambda/B$, where λ is the observing wavelength and B is the baseline length. For $B = 8000$ km and $\lambda = 1$ cm, $\theta_f = 250$ μas. The centroid position of a point-source imaged with a multi-baseline array with a FWHM beam size of $\approx \theta_f$ can be measured with a precision of $\approx 0.5\theta_f/$SNR, where SNR is the peak signal-to-noise ratio in the image. Thus, even for modest SNR values of ~ 10, one can obtain a centroid positional precision of ~ 10 μas.

However, in most cases, systematic sources of error dominate and one's precision is far better than one's true accuracy. For observations at short cm-wavelengths, where *un-modeled* ionospheric delays are smaller than tropospheric delays, one can often achieve residual zenith path delay errors of ~ 1 cm. These un-modeled path delays are usually associated with large-scale irregularities in water vapor that change by ~ 0.5 cm per hour in a quasi-random-walk fashion. This level of zenith path delay noise can be achieved with regular measurement (via "geodetic blocks" or GPS data) and correction of the interferometer data. At a "typical" source zenith angle of $30°$, this corresponds to ~ 2 cm of total path delay error. For an observing wavelength of 1 cm, this causes a $\sim 2\lambda$ shift in the optical path between antenna pairs, which can lead to a systematic shift of two fringes or about ~ 500 μas in the absolute position of a target. But, following a suggestion originally attributed to Galileo for parallax measurement, by measuring a *relative* position between a target and calibrator nearby in angle on the sky, this systematic position uncertainty can be largely canceled. For a typical target-calibrator separation of $1°$ (ie, 0.02 radians), one can achieve $\sim 500 \times 0.02 \sim 10$ μas accuracy.

3. Some Individual Parallax Results

One of the first high-accuracy parallax results was for W3OH. This source had a long-standing distance discrepancy: an optical photometric distance (assuming an association with O-type stars separated by $\sim 1°$ on the sky) gave 2.2 kpc, while a kinematic distance gave 4.3 kpc. Xu *et al.* (2006) observed 12 GHz (Class II) CH$_3$OH masers with the

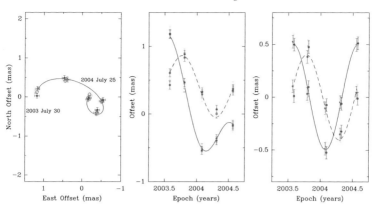

Figure 2. Parallax data for 12 GHz CH$_3$OH masers in W3OH from Xu *et al.* (2006). Left panel: positions on the sky of 1 maser spot measured relative to 3 background sources. Middle panel: eastward (solid line) and northward (dashed line) offsets versus time. Right panel: same as middle panel, but with proper motion removed, showing only the parallax signature.

Very Long Baseline Array (VLBA) and obtained a parallax of 0.512 ± 0.010 mas (see Figure 2). This corresponds to a distance of 1.96 kpc and indicates the photometric distance was accurate to about ±10%. The measured proper motion indicates a peculiar (non-circular) velocity of 22 km s^{-1} largely toward the Sun, explaining the anomalous kinematic distance.

Honma *et al.* (2007) using the Japanese VERA array measured the first parallax for a distant, outer Galaxy source, S269 (see Figure 3). Their results confirmed the flatness of earlier rotation curve data well beyond the Solar Circle and, therefore, the need for substantial dark matter in the halo of the Milky Way. It is important to note that previous results, mostly using H I data were based on less reliable kinematic distances and only 1-dimension (radial) of velocity information. The VERA result was the first time that full 3-dimensional location and velocity information had been obtained for a distant outer Galaxy source.

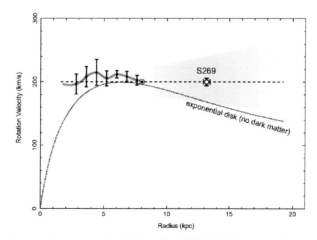

Figure 3. Galactic rotation curve including the VERA S 269 parallax and proper motion, indicating the need, from 3-dimensional data, for dark matter to sustain the rotation speed in the outer galaxy.

The 3-dimensional structure of the Cygnus-X region has been explored with European VLBI Network (EVN) observations by Rygl *et al.* (2011), who measured parallaxes for

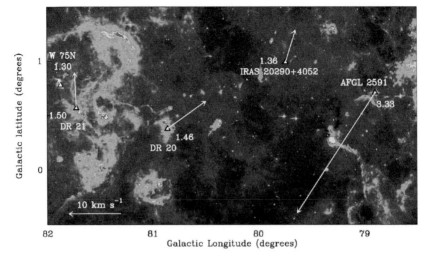

Figure 4. The Cygnus-X region from Rygl *et al.* (2011). Four sources are at a distance of
≈ 1.4 kpc; however AFGL 2591 is at 3.3 kpc and not associated with the "Cygnus-X" complex.

five massive star forming regions (see Figure 4). Four of these (W 75N, DR 20, DR 21 &
IRAS 20290 + 4052) are consistent with being at a distance of 1.40 ± 0.08 kpc and are
clearly associated with the same giant molecular cloud. However, one source, AFGL 2591,
was found to be at a significantly greater distance (3.33 ± 0.11 kpc) and is not associated
with the others. Thus, the Cygnus-X complex is not a single physical association. Toward
this region one is looking down the Local Arm where multiple star forming regions are
projected together.

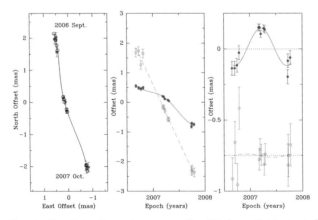

Figure 5. Parallax to the Galactic center source Sgr B2 from Reid *et al.* (2009c), yielding
$R_0 = 7.9 \pm 0.8$ kpc. See Fig. 2 for a description of the plot.

Sgr B2 is located in the Galactic center region and is one of the most massive star
forming regions in the Milky Way. Reid *et al.* (2009c) observed the H_2O masers in Sgr B2N
and Sgr B2M and find a parallax of 0.129 ± 0.012 mas (see Figure 3), corresponding to a
distance of 7.8 ± 0.8 kpc. While the line-of-sight distance from the Galactic Center (and
its supermassive black hole Sgr A*) is unknown, Sgr B2 is probably ≈ 0.1 kpc nearer
than Sgr A*. Evidence for this comes from i) Sgr B2's small projected distance of ≈ 90 pc
from Sgr A* and ii) its 3-dimensional velocity (from the measured proper motion) which
suggests a line-of-sight offset of 130 ± 60 pc (assuming a low eccentricity Galactic orbit)

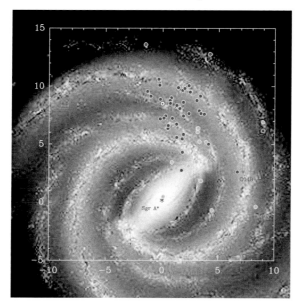

Figure 6. Locations of massive star forming regions with maser parallaxes measured by VERA, the EVN, and the VLBA overlaid on a schematic image of the Milky Way's spiral structure by R. Hurt. Blue circles are used when distance uncertainty is < 0.5 kpc, and green circles are used if the uncertainty is larger (typically three times the size of the circles). The spiral arm pitch angle plot (Fig. 7) suggests that G048 is associated with the Perseus arm and W49N with the Outer arm; hence the schematic image underlying the plot needs adjusting.

toward the Sun. Correcting for this small 0.1 kpc offset from the Galactic center gives a direct measurement of $R_0 = 7.9 \pm 0.8$ kpc. On going observations should yield a more accurate result.

4. Spiral Structure

Figure 6 shows the locations of 62 sources with parallaxes measured by VERA, the EVN, or the VLBA, overlaid on an artist's conception of the spiral structure of the Milky Way. Five sources (including W49; see the discussion of spiral pitch angles for justification for the arm assignment) now trace the Outer arm. Eleven sources (including G048) outline the Perseus arm, which is the arm nearest the Sun outside the solar circle. Moving inward, the Sun is embedded in the Local (or Orion) "spur." Inside the solar circle the definition of arms is less clear, especially as the sources approach the end of the bar near (3,3) kpc.

The Milky Way spiral arms might be approximated by a log-periodic function,

$$\log_{10}(R/R_{ref}) = -(\beta - \beta_{ref}) \tan \psi,$$

where R and β are the Galactocentric radius and azimuth and the subscript "ref" indicates reference values (eg, for the start of the arm). Such a spiral would map to a straight line on a plot of $\log_{10}(R/R_{ref})$ versus β, with slope given by $- \tan \psi$. The spiral pitch angle ψ controls how tightly wound is the spiral; values close to 16° (or 8°) would have arms wrap once (or twice) around the Galaxy before reaching the end of the stellar disk.

Figure 7 displays such a pitch-angle plot for sources with uncertainties in $\log_{10}(R/\text{kpc}) < 0.1$ and uncertainties in $\beta < 10°$. Reid *et al.* (2009b), based on 5 sources in the Perseus arm, found $\beta \approx 16°$. Using this pitch angle (ie, holding the slopes constant)

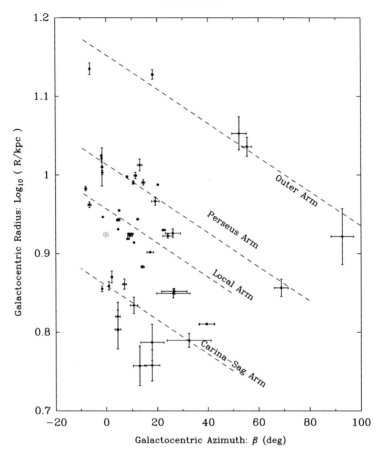

Figure 7. Spiral pitch angle plot. Plotted is the $\log_{10}$ of Galactocentric radius vs. Galactocentric azimuth; on such a plot, log-periodic spirals map to straight lines. The slope of the lines was fixed for a pitch angle of 16° from an analysis of 5 sources in the Perseus arm by Reid *et al.* (2009b); only a vertical offset was adjusted to better fit the data.

and adjusting only the vertical offsets for the lines in Figure 7, one can fit the data for the Outer, Perseus and Local arms reasonably well. (Interior to the Local arm, more data will be needed.) This leads to several interesting conclusions. Firstly, W49 at $(93°, 0.92)$ belongs to the Outer arm and G048 at $(69°, 0.86)$ belongs to the Perseus arm. Thus, the artistic conception of the Milky Way in Figure 6 needs to be adjusted. Secondly, the Local "spur" has a similar pitch angle as other arms, suggesting it is not a spur, which generally protrudes perpendicularly from an arm, and instead is a arm-segment. Thirdly, a globally averaged pitch angle value near 16° is generally characteristic of galaxies of Hubble type Sbc to Sc (Kennicutt (1981)).

5. Galactic Dynamics

With 3-dimensional position and velocity vectors for each source, one can construct an azimuthally averaged rotation curve for the Milky Way. For each source one must convert from the Heliocentric frame, in which the measurements are made, to a Galactocentric frame. This involves adding the full vector motion of the Sun in its Galactic orbit to each measured source vector and then calculating the projection of that vector in the direction of Galactic rotation. Figure 8 shows rotation curves generated for two values of Θ_0 (220

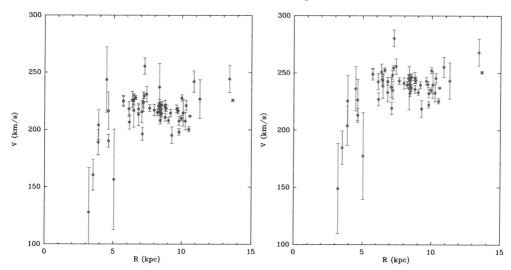

Figure 8. Azimuthally averaged rotation curves from the 3-dimensional data afforded by parallax and proper motion measurements. Left panel: used $\Theta_0 = 220$ km s^{-1} when converting from Heliocentric to Galactocentric (R,V) coordinates. Right panel: same as middle panel except $\Theta_0 = 245$ km s^{-1} was used.

and 245 km s^{-1}). Note the two curves resemble each other except for an offset of 25 km s^{-1}, corresponding to the difference in the two values of Θ_0. Thus, the average rotation speed seen in a rotation curve is essentially determined by the value of Θ_0 assumed when converting frames. Also, it is worth pointing out that all previous rotation curves were based on less reliable distances and only the radial component of the velocity vectors. Inferring Θ_0 from parallax and proper motions must come from modeling the entire data set, and not from the construction of an azimuthally averaged rotation curve.

Reid *et al.* (2009b) modeled the Milky Way using 18 maser parallax/proper motions and concluded that $R_0 = 8.4 \pm 0.6$ kpc and $\Theta_0 = 254 \pm 16$ km s^{-1} (adopting $V_\odot = 5$ km s^{-1} from Dehnen & Binney (1998)). A similar modeling of the 62 measurements can now be done. Using a "conservative formulation" of a Bayesian analysis described in section 8.3.1 of Sivia & Skilling (2006) that allows for outlying data, we find $R_0 = 8.5 \pm 0.3$ kpc and $\Theta_0 = 246 \pm 9$ km s^{-1}(adopting a new $V_\odot = 12$ km s^{-1} from Schoenrich, Binney & Dehnen (2010); using the older $V_\odot = 5$ km s^{-1} from Dehnen & Binney (1998) would give $\Theta_0 = 253 \pm 9$ km s^{-1}). Note that one expects the motions of some star forming regions to deviate from a simple model of Galactic rotation. The Galactic bar is known to cause large non-circular motions; also some sources may be affected by super bubbles from multiple supernovae and have sizeable peculiar velocities.

In addition, the new modeling confirms the conclusion of Reid *et al.* (2009b) that high-mass star forming regions lag circular orbits in the Galaxy, but by a smaller value of 6 ± 2 km s^{-1} if one adopts the updated solar motion.

6. Local Group Dynamics

In the 1920's, Adrian van Maanen reported measurement of proper motions of stars in the galaxy M 33. His motions were ~ 10 mas yr^{-1}, requiring that the "spiral nebula" M 33 was a Galactic object (to avoid $V > c$). van Maanen's motions were nearly a factor of 1000 too large and to this day the reason for the error lacks a clear explanation. A decade ago, Brunthaler (2004) for his Ph.D. thesis attempted to repeat the "van Maanen"

Figure 9. Image of M 33 with the locations and expected orbital motions of two H_2O masers. Credit: A. Brunthaler.

observations, using H_2O masers as astrometric targets. He succeeded both in measuring the internal rotation of M33 and its absolute motion across the sky relative to background quasars.

Brunthaler *et al.* (2005) compared the angular rotation of the H_2O masers with that predicted by the rotation speed and inclination of M 33 (determined from H I maps) and estimated the distance to the galaxy to be 730 ± 168 kpc. This was the first successful extragalactic "rotational parallax." With improvements in the observations (which are on-going) and in the rotation curve measurements, this geometric distance estimate can improve to better than 10% accuracy.

The absolute motion of M 33, a satellite of the Andromeda galaxy, also provides valuable information about the dynamics of the Local Group of galaxies. The Local Group is anchored by the Andromeda and Milky Way galaxies. We now know the 3-dimensional velocity of M 33 (relative to the Milky Way) and the radial velocity of Andromeda. Were we to know Andromeda's proper motion, we could integrate backwards in time and determine the history of the dominant galaxies of the Local Group. Constraints on the proper motion of Andromeda can be obtained from the measured velocity vector of M 33 and the fact that it hasn't been tidally heated by close encounters with Andromeda in the past. Loeb *et al.* (2005) rule out a null proper motion for Andromeda and suggest that its motion is probably near 100 km s^{-1}.

In the near future it will be possible to measure Andromeda's proper motion directly, owing to two recent developments. Firstly, after decades of searches for H_2O masers in Andromeda by several teams, Darling (2011), using the Green Bank Telescope (GBT), recently discovered 5 sites of H_2O masers in spiral arms that ring the galaxy. This will allow proper motion measurements similar to those done for M33, yielding both a rotational parallax and measurement of Andromeda's proper motion. Interestingly, Darling has noted that the "ring" of masers will *appear* to grow measurably in diameter over a decade as Andromeda approaches the Milky Way! Secondly, the recording rate will soon be 2 Gb/s on the VLBA. This increased bandwidth, coupled with the additional collecting area of the High Sensitivity Array (which adds 100-m class telescopes like the GBT, Effelsberg, and the phased-EVLA to the VLBA), will make it possible to directly measure the proper motion of M31* (the AGN of Andromeda). Indeed, a demonstration detection has already been accomplished (Brunthaler, private communication).

References

Brunthaler, A. 2004, *Ph.D. Thesis, Univ. Bonn*
Brunthaler, A., Reid, M., Falcke, H., Greenhill, L. J., & Henkel, C. 2005, *Science*, 307, 5714
Darling, J. 2011, *ApJ*, 732, L2
Dehnen, W. & Binney, J. J. 1998, *MNRAS*, 298, 387
Honma, M. *et al.* 2007, *PASJ*, 59, 889
Honma, M., Tamura, Y., & Reid, M. J. 2008, *PASJ*, 60, 951
van Maanen, A. 1923, *ApJ*, 57, 264
Kennicutt, R. C. Jr. 1981, *AJ*, 86, 1847
Loeb, A., Reid, M. J., Brunthaler, A., & Falcke, H. 2005, *ApJ*, 633, 894
Reid, M. J., Menten, K. M., Brunthaler, A., Zheng, X. W., Moscadelli, L., & Xu, Y. 2009a, *ApJ*, 693, 397
Reid, M. J. *et al.* 2009b, *ApJ*, 700, 137
Reid, M. J., Menten, K. M., Zheng, X. W., Brunthaler, A., & Xu, Y. 2009c, *ApJ*, 705, 1548
Rygl, K. *et al.* 2011, arXiv:1111.7023
Schönrich, R., Binney, J., & Dehnen, W. 2010, *MNRAS*, 403, 1829
&Sivia, D. S. with Skilling, J. 2006, *Data Analysis: A Bayesian Tutorial*, 2nd *edition*, (Oxford University Press, Oxford UK), p. 168
Xu, Y., Reid, M. J., Zheng, X. W., & Menten, K. M. 2006, *Science*, 311, 53

Cosmic Masers - from OH to H$_0$
Proceedings IAU Symposium No. 287, 2012
R.S. Booth, E.M.L. Humphreys & W.H.T. Vlemmings, eds.
© International Astronomical Union 2012
doi:10.1017/S1743921312007326

Methanol Maser Parallaxes and Proper Motions

Y. Xu[1], M. J. Reid[2], L. Moscadelli[3], K. M. Menten[4], X. W. Zheng[5], A. Brunthaler[4], B. Zhang[4], K. L. J. Rygl[6], J. J. Li[1], and A. Sanna[4]

[1] Purple Mountain Observatory, China
email: xuye@pmo.ac.cn

[2] Harvard-Smithsonian Center for Astrophysics, USA

[3] Arcetri Obs., Italy

[4] Max-Planck-Institute für Radioastronomie, Germany

[5] Nanjing University, China

[6] INAF-IAPS, Italy

Abstract. Due to their compactness, persistence and slow motion, Class II CH$_3$OH masers are excellent targets for parallax and proper motion measurements for massive star-forming regions in the Galactic Disk. These measurements can be used to improve our understanding of the spiral structure and dynamics of the Milky Way. At the same time, Class II CH$_3$OH masers can also be used to study gas kinematics close to the exciting star, tracing rotation, infall and/or outflow motions.

Keywords. CH$_3$OH maser, parallaxes and proper motions, Galactic structure

1. Introduction

Class II CH$_3$OH masers at 6.7 and 12.2 GHz are strong (10-100 Jy), compact (typical spot size of a few milli-arcseconds (mas)), persistent (over many years), and are generally associated with slowly moving gas (velocities $\leqslant$ 10 km s^{-1}). Thus, they are excellent astrometric targets. About a thousand 6.7 GHz and a hundred 12.2 GHz CH$_3$OH masers have been detected. Large numbers are located in the "molecular ring" and share the same spatial distribution as CO gas. Since the 6.7 and 12.2 GHz masers are always found in regions of on-going massive star-formation, they are good tracers of the spiral structure in the Galactic disk. In addition, they can also be used to infer the three-dimensional (3-D) gas kinematics around the massive (proto)stars at radii of tens to hundreds of astronomical unit (AU), which is not accessible via interferometric observations of thermal tracers.

2. Source of astrometric error

The spiral structure of the Milky Way is still poorly known. The classical model is mainly based on the kinematic distances (Georgelin & Georgelin 1976), which have been shown to be inaccurate (Xu *et al.* 2006) by up to 30% or more. Masers allow us a direct measurement of distances via trigonometric parallaxes; however, a (relative) positional accuracy of about 10 μas is necessary to determine accurate distances throughout the Milky Way. Achieving such an accuracy with Very Long Baseline Interferometry (VLBI) observations at centimeter wavelengths is not an easy task. Many factors can degrade the measurement accuracy. Systematic errors are related to the separation angle between target and calibrator, motivating one to find background calibrators close to the target.

Either the (maser) target or the background sources can be used as a phase-reference calibrator. If the maser target is strong enough to be detected within the interferometer coherence time, a quasar (QSO) as weak as a few mJy/beam can be revealed by imaging after applying the phase corrections derived from the maser. Statistically, the weaker the QSO, the higher the chance for it to be found near the maser. Using several interferometers (VLA, ATCA, MERLIN and VLBA), we have searched for weak QSOs and determined their positions accurately. Typically we find 3 suitable QSOs within 1 to 2 degrees of any CH_3OH maser source. Using multiple background calibrators helps in reducing systematic errors due to time variation in the atmosphere and in calibrator structure.

Poor knowledge of the vertical atmospheric delay for each station of the VLBI array is often the major source of systematic error in position. At high elevation angles, the VLBA correlator model is typically in error by about 10 cm of excess path length at 12 GHz, while at 6.7 GHz the model inaccuracy can be larger than 20 cm, dominated by ionospheric path delay. To achieve an astrometric accuracy $\leqslant 50$ μas, it is crucial to remove the residual tropospheric and ionospheric effects, in particular for low-frequency observations. One way to measure and remove the tropospheric delays involves geodetic-like observing blocks. By observing about 15 QSOs, whose positions are known to better than 1 mas, the residual atmospheric delays above each antenna can be estimated accurately. Ionospheric (dispersive) delays can be partially removed by applying a global ionospheric model derived from GPS measurements. Our VLBA observations always include geodetic-like observing blocks and correct for the GPS ionospheric model during the data calibration.

One other source of observational error comes from the absolute position of the phase-reference calibrator. If the error in the correlated position of the calibrator is $\geqslant 10$ mas, the phase-referenced image can be distorted and affected by second-order positional errors. To be able to detect weak (a few mJy/beam) signals in the phase-reference map, it is therefore important to know the calibrator position with high accuracy. Note that most of the VLBA calibrators have a position accuracy of ~ 1 mas.

3. Parallaxes and proper motions

After calibrating and removing the sources of error discussed in the previous section, one can measure parallaxes and proper motions of the CH_3OH masers with high accuracy. Figure 1 shows the results for the 12 GHz masers towards the UC HII region W3(OH). It displays the sky-projected trajectory of a single 12 GHz CH_3OH maser spot relative to three QSOs, which results from the sum of the annual parallax, Solar Motion, differential Galactic rotation, peculiar motion (of the target star with respect to *its* Local standard of rest (LSR)) and internal motion (of the maser spot with respect to the star. The projection of the maser spot path on each equatorial coordinate (RA and Dec) can be fitted with a linear plus a sinusoidal motion, the latter reflecting the annual parallax. For W3(OH), we obtain a parallax of 512 ± 10 μas and spot proper motions are derived with an accuracy of ~ 1 km s^{-1}(Xu *et al.* 2006). Such a parallax accuracy is enough to measure distances to ~ 10 kpc, comparable with the size of the Milky Way disk, with 10% uncertainty.

In the last years, the 6.7 GHz methanol masers were successfully used for parallax measurements with the European VLBI Network, achieving accuracies approaching 20 μas (Rygl *et al.* 2010). The obtained distances were found to be in agreement with VERA water masers parallaxes (Sato *et al.* 2008, Nagayama *et al.* 2011) for two star-forming regions. Recently, the distance to the Cygnus X region was measured using 6.7

GHz methanol masers. The parallaxes toward four star-forming regions, W 75N, DR 21, DR 20, and IRAS 20290+4052 in Cygnus X North, were found to be consistent with a single distance of 1.40 ± 0.08 kpc for the Cygnus X complex (Rygl *et al.* 2012). Figure 2 shows the parallax fit of W 75N based on 10 masers spots observed in eight EVN epochs. The resulting parallax was 0.772 ± 0.042 mas, which corresponds to a distance of 1.30 ± 0.07 kpc.

To date, the CH_3OH masers have been used to determine, parallaxes and proper motions for 26 massive star-forming regions (Bartkiewicz *et al.* 2008, Reid *et al.* 2009a, Moscadelli *et al.* 2009, Xu *et al.* 2009, Xu *et al.* 2011, Zhang *et al.* 2009, Brunthaler *et al.* 2009, Sanna *et al.* 2009, Rygl *et al.* 2010, Rygl *et al.* 2012), tabulated in Table 1. Most of them have parallax accuracy of better than 10%. Reid *et al.* (2009b) modeled 18 parallaxes available in 2009 and found that massive star-forming regions lag circular rotation by about 15 km s^{-1}. This provided motivation for a re-analysis of the Solar Motion by Schönrich *et al.* (2010), resulting in an increase in the component in the direction of Galactic rotation by 7 km s^{-1}, and a corresponding reduction in the star forming region velocity lag.

We can update these model results based on the 26 parallax/proper motions. In the frame rotating with the Galaxy, taking the distance to the Galactic center of 8.3 kpc and circular rotation speed of 239 km s^{-1} Brunthaler *et al.* (2011), and the updated Solar Motion values from Schönrich *et al.* (2010), we find an average lag of 12 km s^{-1}in the direction of Galactic rotation and an average motion of 11 km s^{-1} toward Galactic cen-

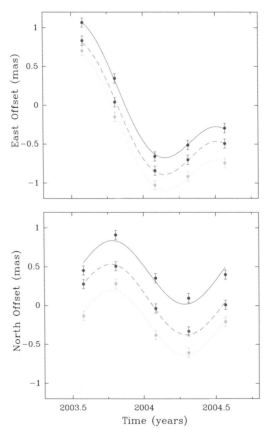

Figure 1. Position versus time for a maser spot relative to each QSO in W3OH.

ter (see Figure 3) However, one can see that G9.62+0.20 & G23.6-0.1 have large peculiar motions, probably caused by the gravitational potential of the bar making them unsuitable for studying average systematic motions. If we drop them and use the remaining 24 sources we find, similar to Reid *et al.* (2009b), on average a systematic motion of 5 km s^{-1} toward the Galactic center and 7 km s^{-1} of counter rotation. Since the CH$_3$OH masers are only associated with massive star-forming regions, this indicates that newly formed massive stars orbit the Milky Way $\sim$7 km s^{-1} slower than for circular orbits and $\sim$5 km s^{-1} towards the Galactic center. The systematic motions might be caused by a wrong estimate of the Solar Motion and an inadequate Galaxy rotation curve, while the accurate values would be obtained by the BeSSeL project.

4. Gas kinematics close to the massive (proto)star

Presently massive star-formation is a very active field of research but several theoretical issues associated with the ionization action and the dynamical effects of the radiation from the massive (proto)star on the surrounding gas are still unclear. However, the observation of massive star-forming regions is severely limited by the large distances (typically of several kpc) and the clustered formation mode, which prevent detailed studies of single (proto)stellar-systems, with all but VLBI observations of their associated maser emission.

For several years, we have been conducting a campaign of multi-epoch VLBI observations of the strongest interstellar molecular masers (OH at 1.6 GHz, CH$_3$OH at 6.7 GHz, and H$_2$O at 22 GHz) towards a small sample of massive star-forming regions (Moscadelli *et al.* 2007, 2011; Sanna *et al.* 2010; Goddi *et al.* 2011). One of the major results of our observations is the result that the (VLBI measurable) 3-D velocity distribution of

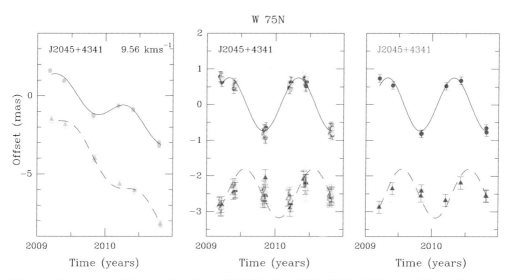

Figure 2. Results of the parallax fit for W 75N using QSO J2045+4341 as position reference. *Left panel:* Results of the parallax fit for the maser spot at $V_{\rm LSR} = 9.56\,{\rm km\,s}^{-1}$ without the removal of the proper motions. The right ascension and declination data have been offset for clarity. *Middle and right panel:* Results of the parallax fit for W 75N after removing proper motions and positional offsets; combined fit on 10 maser spots (*middle panel*) and fit to the averaged data of the 10 maser spots (*right panel*). The declination data have been offset for clarity. The dots mark the data points in right ascension, while the filled triangles mark the declination data points. The solid lines show the resulting parallax fit in right ascension and the dashed lines show the fit in declination. The scale of the Y-axis of the *middle* and *right panels* is different from the scale of the *left panel*.

Table 1. Parallaxes and Proper Motions of High-mass Star-Forming Regions.

Source	l (deg)	b (deg)	Parallax (mas)	μ_x (mas yr^{-1})	μ_y (mas yr^{-1})	v_{LSR} (km s^{-1})	Maser (GHz)
G9.62+0.20	9.62	+0.20	0.194 ± 0.023	−0.580± 0.054	−2.49 ± 0.27	+2 ± 5	12
G12.89+0.49	12.89	+0.49	0.452 ± 0.029	+0.20 ± 0.04	−1.74 ± 0.76	+34 ± 3	12
G15.03−0.68	15.03	−0.68	0.505 ± 0.033	+0.68 ± 0.05	−1.42 ± 0.09	+20 ± 3	12
G 23.0−0.4	23.01	−0.41	0.218 ± 0.017	−1.72 ± 0.04	−4.12 ± 0.30	+81 ± 3	12
G 23.4−0.2	23.44	−0.18	0.170 ± 0.032	−1.93 ± 0.10	−4.11 ± 0.07	+97 ± 3	12
G 23.6−0.1	23.66	−0.13	0.313 ± 0.039	−1.32 ± 0.02	−2.96 ± 0.03	+83 ± 3	12
G27.36−0.16	27.36	−0.16	0.125 ± 0.042	−1.81 ± 0.08	−4.11 ± 0.26	+92 ± 3	12
G 35.2−0.7	35.20	−0.74	0.456 ± 0.045	−0.18 ± 0.06	−3.63 ± 0.11	+28 ± 3	12
G 35.2−1.7	35.20	−1.74	0.306 ± 0.045	−0.71 ± 0.05	−3.61 ± 0.17	+42 ± 3	12
W 51 IRS 2	49.49	−0.37	0.195 ± 0.071	−2.49 ± 0.08	−5.51 ± 0.11	+56 ± 3	12
G 59.7+0.1	59.78	+0.06	0.463 ± 0.020	−1.65 ± 0.03	−5.12 ± 0.08	+27 ± 3	12
ON 1	69.54	−0.98	0.389 ± 0.045	−3.24 ± 0.89	−5.42 ± 0.46	+12 ± 1	6.7
IRAS 20126+4104	78.12	+3.63	0.61 ± 0.02	−4.14 ± 0.13	−4.14 ± 0.13	−3.5± 3	6.7
IRAS 20290+4052	79.74	+0.99	0.737 ± 0.062	−2.84 ± 0.09	−4.14 ± 0.54	−1.4±3	6.7
DR 20	80.86	+0.38	0.687 ± 0.038	−3.29 ± 0.13	−4.83 ± 0.26	−3.0±3	6.7
DR 21	81.74	+0.59	0.666 ± 0.035	−2.84 ± 0.15	−3.80 ± 0.22	−3.0±3	6.7
W 75N	81.87	+0.78	0.772 ± 0.042	−1.97 ± 0.10	−4.16 ± 0.15	+9.0± 3	6.7
L 1206	108.18	+5.52	1.289 ± 0.153	+0.27 ± 0.23	−1.40 ± 1.95	−10 ± 3	6.7
Cep A	109.87	+2.11	1.430 ± 0.080	+0.50 ± 1.10	−3.70 ± 0.20	−10 ± 5	12
NGC 7538	111.54	+0.78	0.378 ± 0.017	−2.45 ± 0.03	−2.44 ± 0.06	−57 ± 3	12
L 1287	121.30	+0.66	1.077 ± 0.039	−0.86 ± 0.11	−2.29 ± 0.56	−18 ± 3	6.7
NGC 281-W	123.07	−6.31	0.421 ± 0.022	−2.69 ± 0.16	−1.77 ± 0.11	−32 ± 3	6.7
W3(OH)	133.95	+1.06	0.512 ± 0.010	−1.20 ± 0.20	−0.15 ± 0.20	−45 ± 3	12
S 252	188.95	+0.89	0.476 ± 0.006	+0.02 ± 0.01	−2.02 ± 0.04	+11 ± 3	12
S 255	192.60	−0.05	0.628 ± 0.027	−0.14 ± 0.54	−0.84 ± 1.76	+8 ± 3	6.7
G 232.6+1.0	232.62	+1.00	0.596 ± 0.035	−2.17 ± 0.06	+2.09 ± 0.46	+23 ± 3	12

water masers, emerging from shocked molecular gas at the interface between the fast
flowing and the ambient material, efficiently trace the structure of (proto)stellar outflows close to (within 10–100 AU) the (proto)star. Our study also provides evidence that
the methanol masers are found at (relatively) large separations from the protostellar jet

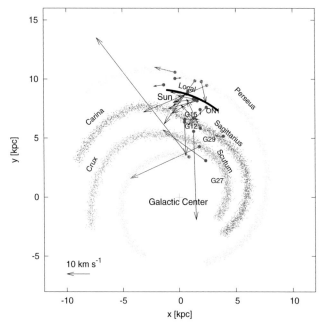

Figure 3. Peculiar motions of massive star forming regions. The amplitude scale is given at
the bottom.

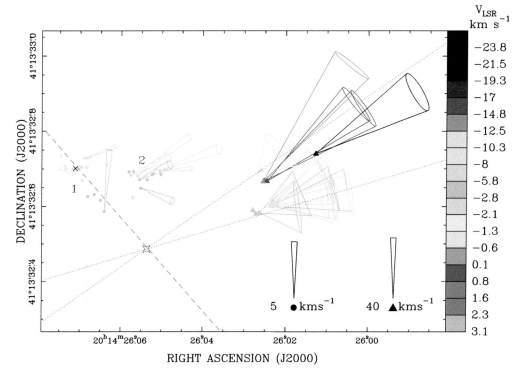

Figure 4. Water (triangles) and 6.7 GHz methanol (circles) masers in IRAS 20126+4104. The cones indicate the 3D velocities of the maser features with respect to the star, for the H_2O masers, or relative to a chosen feature (marked with a cross), for the CH_3OH masers. Points without an associated cone have been detected only at one or two epochs and the associated proper motion cannot be computed or is considered unreliable. The cone opening angle corresponds to 1σ motion uncertainties. The length of the cones is proportional to the speed, with different scales used for the two maser types as indicated in the figure. The colours correspond to the line of sight velocities, as shown in the colour scale to the right. The starred polygon marks the location of the star, obtained from the model fit to the H_2O maser positions and 3D velocities. Labels "1" and "2" denote two elongated groups of CH_3OH maser features. The dotted line outline the projection of the bipolar H_2O maser jet onto the plane of the sky, while the dashed line is a linear fit to the CH_3OH maser features of group 1.

and trace more quiescient gas than the water masers. In the following we describe the results obtained towards the sources IRAS 20126+4104 and AFGL5142. In the former, the 6.7 GHz CH_3OH masers might originate a the Keplerian disk, partly rotating with the disk material and partly being entrained into a slow disk-wind. In the latter, we found evidence that the CH_3OH masers emerge from an infalling envelope at a radius of $\approx$300 AU from the protostar.

4.1. *IRAS 20126+4104*

Figure 4 shows positions and velocities of the 22 GHz H_2O and 6.7 GHz CH_3OH masers towards the intermediate-massive protostar IRAS 20126+4104. The H_2O maser velocities trace a bipolar, collimated, NE–SW oriented outflow at a distance of a few hundred AU from the star. Moscadelli *et al.*(2011), by fitting a simple model of a conical jet, derive the main parameters of the water maser outflow: an outflow axis PA of 115°, an inclination with respect to the line-of-sight (LOS) of 80°, an opening angle of 18°, and the position of the central star (i.e., the cone vertex) with an accuracy of $\approx$100 mas.

The 6.7 GHz methanol masers are clustered to the NE of the star and their positions and LOS velocities are compatible with their origin from the northeastern side of the Keplerian disk detected in several high-density thermal tracers about the star (Cesaroni *et al.* 1997). Looking at Fig. 4, one sees that the CH_3OH maser features are roughly outlining two linear structures, one oriented NE–SW (group 1) and the other SE–NW (group 2). Moscadelli *et al.* (2011) have calculated the *relative* proper motions (see Fig. 4) of the CH_3OH maser features of both groups with respect to a feature belonging to group 1, which appears to be the most reliable for the persistency of the internal structure over the observing epochs.

It is worth noting two facts. First, the masers in group 2 move towards NW, i.e. in the same direction as the blue lobe of the H_2O maser jet. Despite the different speeds (up to ~ 100 km s^{-1} for H_2O and only ~ 5 km s^{-1} for CH_3OH), this suggests a relationship between the two. Second, the mean velocity of group 1 features along the line of sight is -6.9 km s^{-1}, which differs by -3.4 km s^{-1} with respect to the systemic LSR velocity of -3.5 km s^{-1}. The mean distance of the same features from the nominal position of the star (starred symbol in Fig. 4) is $0\farcs27$ or ~ 460 AU. If the masers are undergoing Keplerian rotation about the 7 $M_\odot$ star, the expected velocity is ~ -3.7 km s^{-1}, consistent with the value of -3.4 km s^{-1} quoted above.

Based on this, Moscadelli *et al.* (2011) proposed that group 1 masers trace the Keplerian disk, while for group 2 masers it is suggested that they also lie in the disk, but very close to the transition region between the disk surface and the outer part of the jet. This hypothesis may explain why the relative proper motions of group 2 features (see Fig. 4) are roughly perpendicular to the plane of the disk. Group 2 features could be tracing material lifted from the disk surface and accelerated along the rotation axis. This scenario is reminiscent of the MHD "disk wind", driven by the toroidal magnetic pressure (see, e.g., Banerjee & Pudritz 2005 and references therein), invoked to explain the formation of large scale outflows in Young Stellar Objects.

4.2. *AFGL5142*

Figure 5 shows the proper motions of the 22 GHz water masers (left panel) and 6.7 GHz methanol masers (right panel) measured towards the millimeter core AFGL 5142 MM-1 by Goddi *et al.* (2011). The high-angular resolution maps show a collimated bipolar outflow, at radii 140 to 400 AU, traced by radio continuum emission from its ionized component and by water masers from its molecular component. The two peaks of radio continuum emission are associated with the two clusters of water masers which are expanding, suggesting that the radio continuum probably comes from an ionized jet.

Looking at the right panel of Figure 5, the 6.7 GHz CH_3OH yellow spots have relatively large proper motions, directed towards the geometric center of the red features, which move mostly along the LOS, indicating motions towards the protostar. Since all methanol masers are seen in the foreground of the 22 GHz continuum emission (which is optically thick at 6.7 GHz) and have red-shifted LOS velocities, they most likely trace infall toward the protostar. Goddi *et al.* (2011) have reproduced the methanol maser velocity distribution with a simple model of spherical infall, deriving a central mass of 4 ± 1 $M_\odot$, a radius for the infalling envelope of 290 ± 70 AU, an infall velocity of 5 ± 1 km s^{-1}, and a stellar position (i.e., the sphere center) coincident (within the fit error of 20 mas) with the geometric center of the red features. The derived mass infall rate of 6×10^{-4} $M_\odot$ yr^{-1} implies a protostellar age $\lesssim 10^{-4}$ yr. Note that methanol masers, for the first time, show evidence of infall of material on to the protostar from the 3-D velocity field of molecular gas.

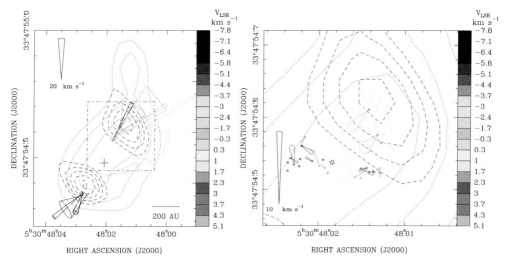

Figure 5. Proper motions of H_2O (*Left panel*) and CH_3OH (*Right panel*) masers measured towards AFGL 5142 MM-1. The dot-dashed rectangle in the left panel shows the area plotted in the right panel. The colored cones indicate proper motion orientation and uncertainties for individual maser features (the amplitude scale is given at the left of each panel). Colors denote line-of-sight velocities and contour maps show the 22 GHz (dotted line) and 8.4 GHz (dashed line) continuum emissions observed with the VLA. The black star in the right panel identifies the expected location of the protostar, corresponding to the geometric mean of positions of the red methanol masers. Water masers and radio continuum together identify a collimated bipolar (ionized/molecular) outflow.

5. Summary

1. Class II CH_3OH masers are closely associated with massive star formation and are very good targets for astrometry because they are compact, long lived and slow moving.
2. Parallax accuracies of better than 10 μms have been obtained for CH_3OH masers, indicating that they can be used to construct the spiral structure throughout the Milky Way's disk.
3. They can be used to study the Solar Motion and the Galaxy's rotation curve.
4. They can also be used to study the gas kinematics close to the exciting star, which might trace rotation, infall and/or outflowing motion.

Acknowledgements

This work was supported by the Chinese NSF through grants NSF 11133008, NSF 11073054, NSF 10621303, NSF 10733030 and the Key Laboratory for Radio Astronomy, CAS. K.L.J.R is funded by an ASI fellowship under contract number I/005/07/1.

References

Banerjee, R. & Pudritz, R. E. 2005, "Outflows and Jets from Collapsing Magnetized Cloud Cores", Protostars and Planets V, 2005.

Bartkiewicz, A., Brunthaler, A., Szymczak, M. van Langevelde, H. J & Reid, M. J. 2008, *A&A*, 490, 787

Brunthaler, A., Reid, M. J, Menten, K. M. *et al.* 2011, *AN*, 332, 461

Brunthaler, A., Reid, M. J., Menten, K. M., Zheng, X. W., Moscadelli, L., & Xu, Y. 2009, *ApJ*, 693, 424

Cesaroni, R., Felli, M., Testi, L., Walmsley, C. M., & Olmi, L. 1997, *A&A*, 325, 725

Dehnen, W. & Binney, J. J., *MNRAS*, 1998, 298,387

Georgelin, Y. M. & Georgelin, Y. P., 1976, *A&A*, 49, 57

Goddi, C., Moscadelli, L., & Sanna, A. 2011, *A&A*, 535, L8.

Moscadelli, L., Reid, M. J., Menten, K. M., Brunthaler, A., Zheng, X. W., & Xu, Y. 2009, *ApJ*, 693, 406

Moscadelli, L., Cesaroni, R., Rioja, M. J., Dodson, R., & Reid, M. J. 2011, *A&A*, 526, 66

Nagayama, T., Omodaka, T., Nakagawa, A., *et al.* 2011, *PASJ*, 63, 23

Reid, M. J., Menten, K. M., Brunthaler, A., Zheng, X. W., Moscadelli, L., & Xu, Y. 2009, *ApJ*, 693, 397

Reid, M. J., Menten, K. M., Zheng, X. W. *et al.* 2009, *ApJ*, 700, 137

Rygl, K. L. J., Brunthaler, A., Reid, M. J., Menten, K. M., van Langevelde, H. J., & Xu, Y. 2010, *A&A*, 511, 2

Rygl, K. L. J., Brunthaler, A., Sanna, A. *et al.* 2012, *A&A*, 539, 79

Sanna, A., Reid, M. J., Moscadelli, L., Dame, T. M., Menten, K. M., Brunthaler, A., Zheng, X. W., & Xu, Y. 2009, *ApJ*, 706, 464

Sanna, A., Moscadelli, L., Cesaroni, R., Tarchi, A., Furuya, R. S., & Goddi, C. 2010, *A&A*, 517, A78

Sato, M., Hirota, T., Honma, M., *et al.* 2008, *PASJ*, 60, 975

Schönrich, R., Binney, J. J., & Dehnen, W., 2010, *MNRAS*, 403, 1829

Xu, Y., Reid, M. J., Zheng, W. W., & Menten, K. M. 2006, *Science*, 311, 54

Xu, Y., Reid, M. J., Menten, K. M., Brunthaler, A., Zheng, X. W., & Moscadelli, L. 2009, *ApJ*, 693, 413

Xu, Y., Moscadelli, L., Reid, M. J. *et al.*, 2011, *ApJ*, 733, 25

Zhang, B., Zheng, X. W., Reid, M. J., Menten, K. M., Xu, Y., Moscadelli, L., & Brunthaler, A. 2009, *ApJ*, 693, 419

Cosmic Masers - from OH to H_0
Proceedings IAU Symposium No. 287, 2012
R.S. Booth, E.M.L. Humphreys & W.H.T. Vlemmings, eds.

© International Astronomical Union 2012
doi:10.1017/S1743921312007338

VLBI multi-epoch water maser observations toward massive protostars

José M. Torrelles[1], José F. Gómez[2], Nimesh A. Patel[3], Salvador Curiel[4], Guillem Anglada[2], and Robert Estalella[5]

[1]ICE(CSIC)-UB/IEEC, Barcelona (Spain)

[2]IAA(CSIC), Granada (Spain)

[3]Harvard-Smithsonian, CfA, Cambridge (USA)

[4]IAUNAM, México D.F. (México)

[5]UB/IEEC, Barcelona (Spain)

Abstract. VLBI multi-epoch water maser observations are a powerful tool to study the gas very close to the central engine responsible for the phenomena associated with the early evolution of massive protostars. In this paper we present a summary of the main observational results obtained toward the massive star-forming regions of Cepheus A and W75N. These observations revealed unexpected phenomena in the earliest stages of evolution of massive objects (e.g., non-collimated "short-lived" pulsed ejections in different massive protostars), and provided new insights in the study of the dynamic scenario of the formation of high-mass stars (e.g., simultaneous presence of a jet and wide-angle outflow in the massive object Cep A HW2, similar to what is observed in low-mass protostars). In addition, with these observations it has been possible to identify new, previously unseen centers of high-mass star formation through outflow activity.

Keywords. ISM: general, stars: formation, ISM: jets and outflows

1. Introduction

It is well-established that low-mass stars form via an accretion process. A natural consequence of this process is the formation, during the early stages of evolution of a young stellar object (YSO), of a system (with typical scales of $\simeq 100$ AU) that comprises a central protostar, surrounded by a circumstellar (protoplanetary) disk, and a collimated outflow, ejected perpendicular to the disk (e.g., Anglada 1996). The disk is the reservoir of material from which the central protostar accretes further matter, while the collimated outflow releases the necessary angular momentum and magnetic flux for this accretion to proceed. This accretion scenario seems to be generally applicable in the formation of stars up to $\simeq 20$ $M_\odot$ (e.g., Garay & Lizano 1999). However, very few massive protostar-disk-outflow systems have been identified and studied in detail, at scales $\leqslant 3000$ AU (Patel *et al.* 2005, Jiménez-Serra *et al.* 2007, Torrelles *et al.* 2007, Zapata *et al.* 2009, Davies *et al.* 2010, Carrasco-González *et al.* 2010a, 2011, Fernández-López *et al.* 2011). This scarcity of studies is probably due to observational limitations (sensitivity and angular resolution), given that high-mass stars are more rare, are typically located at larger distances, and form in a more clustered environment than their low-mass counterparts.

Observations of water maser emission ($\lambda = 1.35$ cm) towards massive YSOs can overcome these observational limitations (sensitivity and angular resolution). Water maser emission is compact ($\leqslant 1$ mas) and strong (brightness temperatures can reach 10^{10} K), which make it an ideal tool for observations with Very Long Baseline Interferometry (VLBI), i.e., with angular resolutions < 1 mas. Therefore, we can study shocked, warm

(~ 500 K), and dense ($\simeq 10^8$-10^9 cm^{-3}) gas, at scales of only 10-1000 AU from massive protostars (Reid & Moran 1981), with the possibility of accurately tracing their kinematics, by measuring proper motions of a few km s^{-1} in timescales of a few weeks. In this paper we summarize the main results of our studies obtained toward the high-mass star forming regions of Cepheus A and W75N, using VLBI multi-epoch water maser observations with an angular resolution of ~ 0.5 mas. These results were reported by Torrelles *et al.* (2001a,b, 2003, 2011) and summarized in Torrelles *et al.* (2012). Other very interesting examples (not included in this paper) can be found in the studies on IRAS 06061+2151 (Motogi *et al.* 2008), AFGL 5142 (Goddi, Moscadelli, & Sanna 2011), AFGL 2591 (Sanna *et al.* 2012, Trinidad *et al.* 2003), and Cep A HW3d (Chibueze *et al.* 2012) (see also talks in this conference by A. Bartkiewicz-H. van Langevelde, J. Chibueze, C. Goddi, T. Hirota, and A. Sanna).

2. Cepheus A

Cepheus A is an active high-mass star-forming region, located at only 700 pc (Moscadelli *et al.* 2009), which contains a cluster of at least 16 sources within $\simeq 30''$ (Hughes & Wouterloot 1984, Garay *et al.* 1996). Its star-formation activity is dominated by the source HW2. The radio continuum emission in HW2 seems to trace an ionized jet, powered by a massive protostar ($\simeq 15 - 20$ M$_\odot$). Extremely broad millimeter wavelength hydrogen recombination lines were recently detected toward this source with terminal velocity > 500 km s^{-1} for the ionized gas (Jiménez-Serra *et al.* 2011). This source was the first identified case of a disk-protostar-jet system in a massive YSO, at scales of $\simeq 1000$ AU (Rodríguez *et al.* 1994, Patel *et al.* 2005, Curiel *et al.* 2006, Jiménez-Serra *et al.* 2007, Torrelles *et al.* 2007, Vlemmings *et al.* 2010). This suggests that this high-mass object formed by an accretion mechanism, in a similar way to low-mass stars.

HW2 is associated with strong water maser emission, which has been studied in detail with the Very Large Array (VLA, beam size $\simeq 80$ mas) and the Very Long Baseline Array (VLBA, beam size $\simeq 0.5$ mas), for a total of 9 different epochs (Torrelles *et al.* 2001a,b, 2011). Fig. 1 shows the location of the water masers, overlaid on the 1.3 cm continuum map of the HW2 radio jet and the nearby HW3c and HW3d objects (located $\sim 3''$ south from HW2; see Chibueze *et al.* 2012 and Chibueze's talk in this conference for a detailed study of the distribution and kinematics of the water masers associated with HW3d). The VLBA data showed that individual maser features tend to be organized both spatially and kinematically, forming linear microstructures of a few mas in size. The flattened appearance of these water maser linear "microstructures" and their proper motions indicate that they are originated through shock excitation by outflows, as expected from theory (e.g., Elitzur *et al.* 1992). These linear microstructures are the building blocks of larger, coherent linear/arcuate structures of $\simeq 40 - 100$ mas. The identified structures in the neighborhood of HW2 are in the regions labelled R1 to R8 in Figure 1.

In particular, the masers in the R4 subregion trace a section of a nearly elliptical ring of $\simeq 70$ mas size (50 AU), with expanding motions of ~ 15-30 km s^{-1} (Figure 2), from which an extremely short dynamical age of 4-8 year can be derived. This expansion must be driven by a yet undetected probably massive YSO (to explain the high luminosity of the water masers), located at the geometrical center of the ring, ~ 130 AU from HW2. A detailed inspection of Fig. 2 also shows some internal structure within the ring in individual epochs, with several "shells" that may have been created by multiple successive ejections. Future sensitive cm and (sub)mm observations may help us to identify the massive object powering this structure.

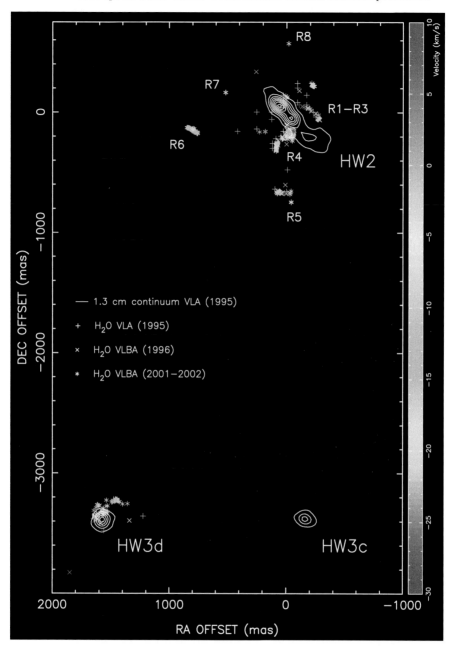

Figure 1. (Published online) Positions and radial velocities (color code) of the H_2O masers overlaid onto the 1.3 cm continuum maps (contours) of Cep A HW2, HW3c, and HW3d. Sub-regions "R" discussed in this paper are numbered. (Figure from Torrelles *et al.* 2011).

Another remarkable subregion is R5, located $\sim 6''$ south of HW2. In our 1996 observations, the masers in this region traced part of a nearly perfect circle (to an accuracy of 1/1000), and expanding at 9 km s^{-1}. We interpreted this as a short-lived (dynamical age $\simeq 30$ yr) spherical wind, powered by a high-mass object at its center (later detected in radio continuum observations by Curiel *et al.* 2002). Only five years after its discovery, this maser structure has already lost its spherical symmetry, probably due to its interaction

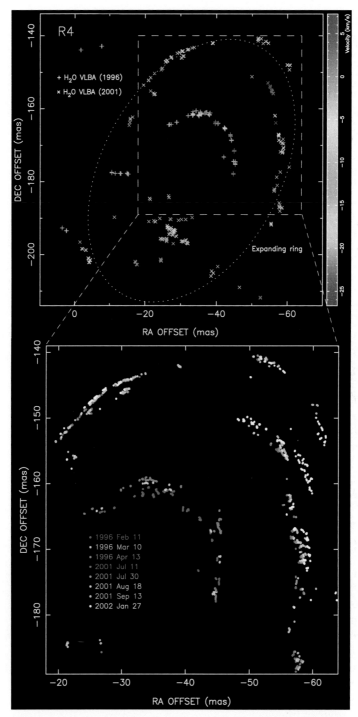

Figure 2. (Published online) *Upper panel:* Water maser positions measured in sub-region R4 of Cepheus A (see also Figure 1). Color code indicates the LSR radial velocity (km/s) of the masers. *Lower panel:* Zoom showing the evolution of the expanding motions in the sky for all the observed VLBA epochs. Color code indicates the epoch. (Figure from Torrelles *et al.* 2011).

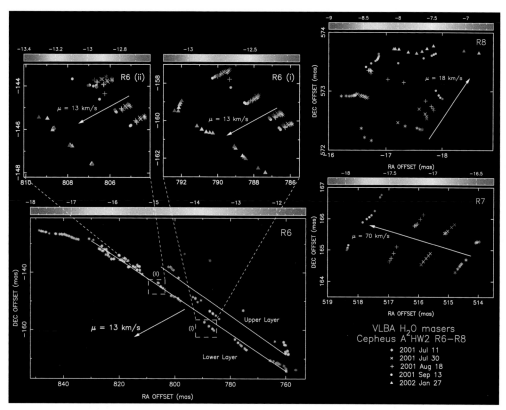

Figure 3. (Published online) Positions and proper motions of the H_2 masers in sub-regions R6, R7, and R8 of Cepheus A (see also Figure 1). Color code indicates the LSR radial velocity (km s^{-1}) of the masers. (Figure from Torrelles *et al.* 2011).

with the surrounding interstellar medium. The discovery of a nearly spherical ejection is especially relevant, since this phenomenon is difficult to explain in current models of star-formation, which assume that mass loss in YSOs is produced by the transformation of rotational energy in the disk into collimated outflows via magnetohydrodynamic mechanisms. A few other cases of nearly-isotropic ejections have been reported later in other star-forming regions (e.g., W75N, see §3). This indicates that uncollimated, episodic ejections may occur during the earliest stages of evolution of massive YSOs. The mechanism powering those ejections is still unclear.

While the maser structures in R4 and R5 are believed to be associated with different sources, the remaining ones (subregions R1 to R3, and R6 to R8) could trace an outflow powered by HW2. Significant differences in the magnitude and direction of the proper motions in these structures (Fig 3.), suggest that we can be witnessing the simultaneous presence of a collimated jet and a wide-angle outflow. In this scenario (Fig. 4), R6 (moving to the southeast at ~ 13 km s^{-1}) and R8 (to the northeast at ~ 18 km s^{-1}), would be tracing shock fronts at the walls of expanding cavities created by a wide-angle outflow from HW2, with an opening angle of $\simeq 100°$. On the other hand, there is clearly a collimated jet traced by the radio continuum emission. The R7 masers, moving closer to the axis of this jet ($\simeq 30°$ from the axis), show expansion velocities of ~ 70 km s^{-1}, which are intermediate between those observed in R6/R8 ($\sim 13 - 18$ km s^{-1}) and in the jet (~ 500 km s^{-1}). The R1-R3 masers, located on the opposite side from the central

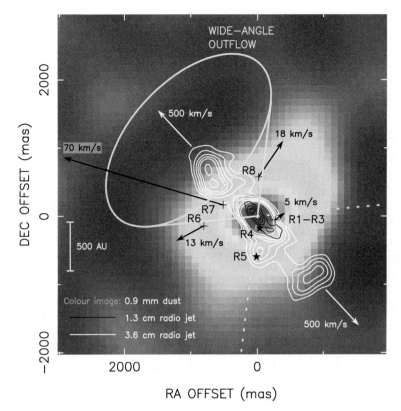

Figure 4. (Published online) Wide-angle outflow and jet in Cepheus A HW2. The radio jet (opening angle of ∼ 18°) exhibits ejections in opposite directions, moving away at ∼ 500 km s^{-1} from the central source, and is surrounded by a dust/molecular disk (Patel *et al.* 2005). R6, R8, and R1-3 trace emission fronts from the shocked walls of expanding cavities, created by the wide-angle wind of HW2 (opening angle of ∼ 100°). The R7 masers, with motions along an axis at an angle of ∼ 30° with respect to the radio jet axis, are excited inside the cavity by the wide-angle wind. They exhibit higher velocity than R6, R8, and R1-3 (which are located at the expanding cavity walls) but lower than the velocity of the jet. The R6, R7, and R8 masers (observed towards the blue-shifted lobe of the 1 arcmin, large-scale bipolar molecular outflow; Gómez *et al.* 1999) are blue-shifted with respect to the systemic velocity of the circumstellar disk, while R1-3 (observed towards the red-shifted lobe of the large-scale molecular outflow) are red-shifted. The position of the two massive YSOs required to excite the R4 and R5 maser structures are indicated by star symbols (see text). The star associated with R4 is not yet detected. (Figure and caption adapted from Torrelles *et al.* 2011, 2012).

source, moving at ∼ 5 km s^{-1} and with a difference in PA of ≃ −80° with respect to the radio jet direction, would represent the corresponding shocked walls of the southwestern cavities created by the wide-angle outflow.

The simultaneous presence of a collimated and a wide-angle outflow has been previously found in low-mass YSOs (e.g., see Velusamy *et al.* 2011 and references therein). Different theoretical models have been proposed to explain these two kind of outflows in low-mass YSOs (e.g., "X-wind", "Disk-wind" models; see Machida *et al.* 2008). The evidence that they are also found in a high-mass object, further reinforces that similar processes could be at work in both low- and high-mass star formation. Moreover, these observations in HW2 should provide important observational constraints for future models trying to reproduce the presence of outflows with different opening angles in high-mass YSOs.

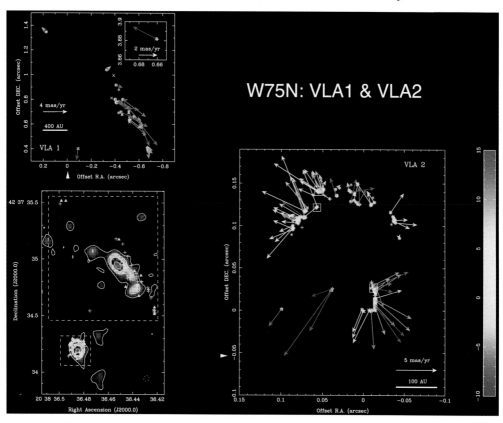

Figure 5. (Published online) *Bottom left:* 1.3 cm continuum contour map of VLA 1 and VLA 2. The positions of the H_2O masers detected with the VLA (triangles) and VLBA (plus symbols) are indicated. *Top left and bottom right:* Proper motions (arrows) of the H_2O masers in VLA 1 and VLA 2. Color code indicates the LSR velocity of the masers in km s^{-1}. (Figure from Torrelles *et al.* 2003, 2012).

3. W75N

We have focused our studies in three YSOs (VLA 1, VLA 2, VLA 3) in this region of high-mass star formation (see, e.g., Hunter *et al.* 1994, Torrelles *et al.* 1997, Persi *et al.* 2006, and Carrasco-González *et al.* 2010b). Despite being located within a region of only 1.5″ (3000 AU at a distance of 2 kpc), they are believed to be in different evolutionary stages. In particular, VLA 1 and VLA 2 show distinctive mass-loss characteristics (see Figure 5): while in VLA1 the water masers observed with the VLBA, and the radio continuum emission trace a collimated jet of $\sim$ 2000 AU size, the radio continuum emission in VLA 2 is compact, and the water masers seem to expand with no preferential direction, tracing a shell-like outflow of $\sim$ 160 AU radius, moving at $\sim$ 30 km s^{-1} (dynamical age of $\simeq$ 13 yr). Given the close proximity between these sources (projected distance 1400 AU), they are likely to share a common molecular gas environment. This led Torrelles *et al.* (2003) to suggest that their different degree of outflow collimation is probably not due to ambient conditions, but to the different evolutionary stages of these sources, with VLA2 being younger. In this scenario, it is expected that the VLA 2 outflow would become more collimated as the source evolves. In fact, recent VLBA water maser observations (Surcis *et al.* 2011) indeed suggest that a jet is being formed in VLA 2.

The case of VLA 2 is reminiscent of that of the R5 subregion in Cepheus A, suggesting the presence of non-collimated outflows in very early stages of high-mass YSOs. It is still unclear whether the observed non-collimated outflows are non-standard phenomena in particular types of sources, or if all massive YSOs undergo short-lived pulsed ejection phases with these type of processes. Nevertheless, their presence poses new challenges in our knowledge of the earliest stages of stellar evolution.

Acknowledgement

GA, RE, JFG, and JMT acknowledge support from MICINN (Spain) AYA2008-06189-C03 and AYA2011-30228-C03 grants, co-funded with FEDER funds. SC acknowledges support from CONACyT (Mexico) 60581 and 168251 grants.

References

Anglada, G. 1996, *ASPC*, 93, 3
Carrasco-González, C., Rodríguez, L. F., Anglada, G., Martí, J., Torrelles, J. M., & Osorio, M. 2010a, *Science*, 330, 1209
Carrasco-González, C. *et al.* 2011, *RMxAC*, 40, 229
Carrasco-González, C. *et al.* 2010b, *AJ*, 139, 2433
Chibueze, J. M. *et al.* 2012, *ApJ*, in press
Curiel, S. *et al.* 2006, *ApJ*, 638, 878
Curiel, S. *et al.* 2002, *ApJ*, 564, L35
Davies, B., Lumsden, S. L., Hoare, M. G., Oudmaijer, R. D., & de Wit, W-J. 2010, *MNRAS*, 402, 1504
Elitzur, M., Hollenbach, D. J., & McKee, C. F. 1992, *ApJ*, 394, 221
Fernández-López, M., Girart, J. M., Curiel, S., Gómez, Y., Ho, P. T. P., & Patel, N. 2011, *AJ*, 142, 97
Garay, G. & Lizano, S. 1999, *PASP*, 111, 1049
Garay, G., Ramírez, S., Rodríguez, L. F., Curiel, S., & Torrelles, J. M. 1996, 1996, *ApJ*, 459, 193
Goddi, C., Moscadelli, L., & Sanna, A. 2011, *A&A*, 535, L8
Gómez, J. F. *et al.* 1999, *ApJ*, 514, 287
Hughes, V. A. & Wouterloot, J. G. A. 1984, *ApJ*, 276, 204
Hunter, T. R., Taylor, G. B., Felli, M., & Tofani, G. 1994, *A&A*, 284, 215
Jiménez-Serra, I. *et al.* 2007, *ApJ*, 661, L187
Jiménez-Serra, I. *et al.* 2011, *ApJ*, 732, L27
Machida, M. N., Inutsuka, S.-i., & Matsumoto, T. 2008, *ApJ*, 676, 1088
Moscadelli, L. *et al.* 2009, *ApJ*, 693, 406
Motogi, K. *et al.* 2008, *MNRAS*, 390, 523
Patel, N. A., Curiel, S., Sridharan, T. K., Zhang, Q., Hunter, T. R., Ho, P. T. P.., Torrelles, J. M., Moran, J. M., Gómez, J. F. G., & Anglada, G. 2005, *Nature*, 437, 109
Persi, P., Tapia, M., & Smith, H. A. 2006, *A&A*, 445, 971
Reid, M. J. & Moran, J. M. 1981, *ARAA*, 19, 231
Rodríguez, L. F., Garay, G., Curiel, S., Ramírez, S., Torrelles, J. M., Gómez, Y., & Velázquez, A. 1994, *ApJ*, 430, L65
Sanna, A., Reid, M. J., Carrasco-González, C., Menten, K. M., Brunthaler, A., Moscadelli, L., & Rygl, K. L. J. 2012, *ApJ*, 745, 191
Surcis, G.,Vlemmings, W. H. T., Curiel, S., Hutawarakorn Kramer, B., Torrelles, J. M., & Sarma, A. P. 2011, *A&A*, 527, 48
Torrelles, J. M., Gómez, J. F., Rodríguez, L. F., Ho, P. T. P., Curiel, S., & Vázquez, R. 1997, *ApJ*, 489, 744
Torrelles, J. M., Patel, N., Anglada G., Gómez, J. F., Ho, P. T. P., Cantó, J., Curiel, S., Lara, L., Alberdi, A., Garay, G., & Rodríguez, L. F. 2003, *ApJ*, 598, L115

Torrelles, J. M., Patel, N. A., Curiel, S., Estalella, R., Gómez, J. F., Rodríguez, L. F., Cantó, J., Anglada, G., Vlemmings, W., Garay, G., Raga, A. C., & Ho, P. T. P. 2011, *MNRAS*, 410, 627

Torrelles, J. M., Patel, N. A., Curiel, S., Ho, P. T. P., Garay, G., & Rodríguez, L. F. 2007, *ApJ*, 666, L37

Torrelles, J. M., Patel, N., Gómez, J. F., Ho, P. T. P., Rodríguez, L. F., Anglada, G., Garay, G., Greenhill, L., Curiel, S., & Cantó, J. 2001a, *Nature*, 411, 277

Torrelles, J. M., Patel, N., Gómez, J. F., Ho, P. T. P., Rodríguez, L. F., Anglada, G., Garay, G., Greenhill, L., Curiel, S., & Cantó, J. 2001b, *ApJ*, 560, 853

Torrelles, J. M., Patel, N., Curiel, S., Gómez, J. F., Anglada, G., & Estalella, R. 2012, Boletín Asoc. Argentina de Astronomía, Eds. J.J. Clarià, A.E. Piatti, R. Barbá, P. Benaglia and F. Bareilles, No. 54, in press

Trinidad, M. *et al.* 2003, *ApJ*, 589, 386

Velusamy, T., Langer, W. D., Kumar, M. S. N., & Grave, J. M. C. 2011, *ApJ*, 741, 60

Vlemmings, W. H. T. *et al.* 2010, *MNRAS*, 404, 134

Zapata, L. A. *et al.* 2009, *ApJ*, 698, 1422

Cosmic Masers - from OH to H_0
Proceedings IAU Symposium No. 287, 2012
R.S. Booth, E.M.L. Humphreys & W.H.T. Vlemmings, eds.

© International Astronomical Union 2012
doi:10.1017/S174392131200734X

Maser astrometry with VERA
and Galactic structure

**Mareki Honma[1,2], Takumi Nagayama[1], Tomoya Hirota[1,2],
Naoko Matsumoto[1], Nobuyuki Sakai[1,2], Noriyuki Kawaguchi[1,2]
and VERA project members**

[1]Mizusawa VLBI observatory, NAOJ, Osawa, Mitaka, Tokyo, 181-8588, Japan
email: mareki.honma@nao.ac.jp

[2]Department of Astronomical Science, School of Physical Sciences,
The Graduate University of Advanced Studies (Sokendai),
Osawa, Mitaka, Tokyo, 181-8588, Japan

Abstract. Since 2007 VERA (VLBI Exploration of Radio Astrometry) has been producing astrometric results (distances and/or proper motions) for Galactic maser sources. Nearly 30 parallaxes have been obtained for star-forming regions and late-type stars. By using VERA's astrometric results for star-forming regions, combined with those obtained with VLBA and EVN, fundamental Galactic parameters and Galactic structure may be derived. Our results show that $R_0 = 8.4 \pm 0.4$ kpc and $\Omega_\odot \equiv \Omega_0 + V_\odot/R_0 = 30.7 \pm 0.8$ km s^{-1} kpc^{-1}, and also show that the rotation curve of the Galaxy is nearly flat. The determinations of Galactic parameters and structures demonstrate that the maser astrometry can not only contribute significantly to research of individual maser sources, but also to studies of the structure of the Galaxy.

Keywords. maser, astrometry, VERA, the Galaxy

1. Introduction

Maser astrometry with phase-referencing VLBI is a powerful tool to measure parallaxes and proper motions of sources at kpc-scale distances. VERA (VLBI Exploration of Radio Astrometry) is a VLBI array dedicated to phase-referencing maser astrometry (Honma *et al.* 2000). With its unique dual-beam system with which one can effectively compensate tropospheric fluctuations, we have achieved accurate parallax and proper motion measurements of Galactic maser sources with distances up to ~ 5 kpc (e.g., Honma *et al.* 2007, Nagayama *et al.* 2011a). Since 2007, VERA has been regularly producing astrometric results of Galactic masers, and so far close to 30 parallax measurements have been obtained (e.g, see two VERA special issues in PASJ in 2008 and 2011). Here we summarize the recent output of VERA, focusing on astrometric results for individual sources as well as Galactic parameter determinations.

2. Summary of recent results for individual sources

Since the first measurements of parallaxes (Honma *et al.* 2007; Hirota et al. 2007), we have contiued to conduct astrometrc measurements of Galactic maser sources. In fact, 28 parallax measurements have already been published, and there are several more sources for which preliminary parallax data is avalable.

Figure 1 summarizes the current status of VERA's results: it shows distributions of the maser sources for which astrometric measurements are obtained. As seen figure 1, reliable parallaxes are available for sources within 5 kpc, while beyond 5 kpc it is still challenging

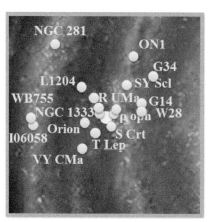 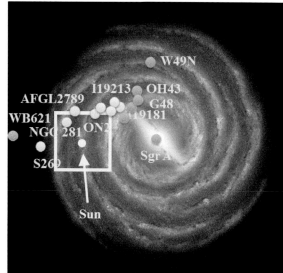

● Parallax + proper motion ○ Proper motion

Figure 1. Galactic distributions of maser sources with VERA's astrometric results. Right panel shows a face-on view of the whole Galaxy, and left panel shows a zoom-up of the area around the Sun, which is indicated by a square in the right panel (Background image: NASA/JPL-Caltech/R. Hurt)

to obtain accurate distances. However, for these distant sources accurate proper motions can readily be measured.

Before discussing Galactic structures revealed by such kpc-scale astrometry, we would like to present a brief summary of the highlights of our observational results for individual maser sources. Most-recent parallax measurements are summarized in the PASJ special VERA issue (PASJ, 2011, Vol. 63, no. 2), where new parallaxes for 9 sources are presented. These include L1448 (Hirota *et al.* 2011a), IRAS 06061+2151 (Niinuma *et al.* 2011), ON1 (Nagayama *et al.* 2011b), ON2 (Ando *et al.* 2011) etc. These results are of great importance in the derivation of accurate physical properties of these star-forming regions e.g. physical size, mass, luminosity and etc.

Since VERA regularly monitors maser sources, our data is also useful for studies of the maser sources themselves, through measurements of flux variations and internal maser proper motions. One of such example is the study of the maser burst in Orion KL (Hirota *et al.* 2011b; Hirota *et al.* 2012), in which the location and motion of the bursting maser component has been identified with mas-precision for the first time. Based on its location and motion, Hirota *et al.*(2011b) suggested that the burst was probably caused by the interaction of an outflow and the molecular gas in the Orion Compact Ridge. Another example is H_2O maser monitoring of G353.273+0.641 conducted by Motogi *et al.*(2011), which revealed a possible acceleration in the maser motions associated with high velocity outflow.

Installation of new C-band receivers to VERA added another window for maser research through observations of 6.7 GHz methanol masers. In collaborations with other radio telescopes in Japan and China, we have been conducting a massive imaging survey for methanol maser sources (Fujisawa *et al.* 2011). This methanol maser data will be useful not only for VLBI astrometry, but also for understanding the massive star formation process. In fact, sources with mas-scale maser distributions will be potential targets for

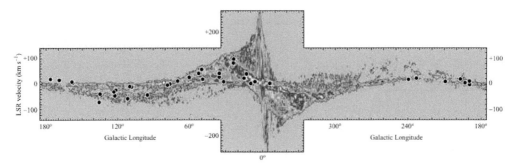

Figure 2. Distributions in *l-v* diagram of 44 maser sources used for the analyses of Galactic structures. The sources include those observed with VERA, VLBA and EVN.

future observations with ALMA. Measurements of thermal emissions from dust combined with molecular data on gas with maser emission will provide for a deeper understanding of star-formation processes.

3. Galactic Structure

Astrometric measurements can also be used to explore fundamental details of Galactic structure such as Galactic constants and the shape of Galactic rotation curve. To obtain the best estimates of such parameters, we compiled a list of astrometric results obtained with VERA, VLBA and EVN. The list consists of 44 star-forming regions with precise astrometry, from which we determine the basic structure of the Galaxy (Honma *et al.* 2012). In figure 2, we show distributions of 44 star-forming regions in the *l-v* diagram of the Galaxy, overlaid on the CO *l-v* diagram (Dame et al. 2001). As seen in figure 2, the sources are basically distributed in the northern part of the Galaxy, with a hole around the region between $l = 240°$ to $350°$ due to the bias in array location. However, except for the hole in the southern hemisphere, the sources are well scattered over a wide range of Galactic longitude l (2/3 of the whole Galaxy), and are thus useful for tracing Galactic structure.

In order to explore the basic structure of the Galaxy, we introduce thos parameters to be determined from the data of maser astrometry: the Galactic constants R_0 and Ω_0, which are the distance from the Sun to the Galactic center, and the angular Galactic rotation velocity at Local Standard of Rest (note that $\Omega_0 \equiv \Theta_0/R_0$, where Θ_0 is the Galactic rotation velocity at the LSR), the power-law index of the rotation curve α (which describes the shape of rotation curve as $\Theta = \Theta_0(R/R_0)^\alpha$), and the mean peculiar motion of the star-forming regions ($U_{\mathrm{SFR}}, V_{\mathrm{SFR}}, W_{\mathrm{SFR}}$). With these sample sources and parameters, we have run MCMC (Markov-Chain Monte Carlo) simulations to evaluate the best estimate of the Galactic parameters. The 44 sources in our sample could include some outliers which have significant deviation from circular Galactic rotation, and could affect the parameter determinations. To handle the effect of outliers with care, we have conducted MCMC simulations with different samples by removing possible outliers that have large deviations. By doing such a careful analysis for 36 of the 44 sources (eliminating up to 8 outliers), we determined the best values of the parameters with estimates of the systematic errors caused by possible outliers (Honme et al. 2012).

The determined parameters are summarized in table 1. Here we adopted a solar motion of $V_\odot = 5.25 \, \mathrm{km \, s^{-1}}$ by Dehnen & Binney (1998), but note that recently there are claims of upward modifications up to $\sim 12 \, \mathrm{km \, s^{-1}}$ (Schönrich *et al.* 2010). We adopted $V_\odot = 5.25 \, \mathrm{km \, s^{-1}}$ just for simplicity in comparison with previous studies, which were mostly

Table 1. Summary of Galactic parameter determinations based on 44 star-forming regions.

parameter	value	unit	note
R_0	8.4 ± 0.4	kpc	distance to Galactic center
Ω_0	30.1 ± 0.8	km s^{-1} kpc^{-1}	angular rotation velocity at LSR
α	0.01 ± 0.03		power-low index of rotation curve
$U_{\rm SFR}$	3.6 ± 1.2	km s^{-1}	mean peculiar motion of sources toward Galactic center
$V_{\rm SFR}$	-13.4 ± 1.5	km s^{-1}	mean peculiar motion of sources toward Galactic rotation
$W_{\rm SFR}$	-1.3 ± 1.3	km s^{-1}	mean peculiar motion of sources toward north Galactic pole

All the results shown here are based on the solar motion $V_\odot = 5.25$ km s^{-1} (Dehnen & Binney 1998). The results could be affected by adopting a different value of $V_\odot$, especially the parameters related to Galactic rotations such Ω_0 and $V_{\rm SFR}$.

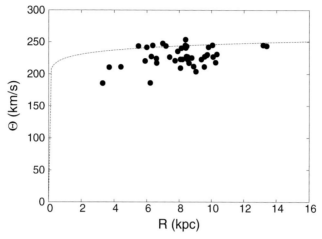

Figure 3. Galactic rotation curve determined in this study. The curve shows the power low fit, and filled circle are observational data. In addition to scatter which appears rather random, there is systematic offset between the curve and observed data, showing possible lag of star-forming regions. Note that this curve is for $V_\odot = 5.25$ km s^{-1}, and the lag strongly depends on adopted value of $V_\odot$ as is described in text.

based on Dehnen & Binney (1998). Although the number of sources is still around ~ 40, as seen in table 1, Galactic parameters are constrained reasonably well. For instance, R_0 is determined at 5% level, and is consistent with recent determinations of R_0 which are independent of our results (e.g., Ghez *et al.* 2008; Gillessen *et al.* 2009). Ω_0 is also determined at 3% level, which can be converted into $\Theta_0 = 254 \pm 14$ km s^{-1} by adopting $R_0 = 8.4 \pm 0.4$ kpc. However, we point out that Ω_0 and Θ_0 are dependent of the adopted value of $V_\odot$, and hence the values presented here are not decisive ones. Instead, we can define another parameter which is less dependent on $V_\odot$ as $\Omega_\odot \equiv \Omega_0 + V_\odot/R_0$. We obtained $\Omega_\odot = 30.7 \pm 0.8$ km s^{-1} kpc^{-1}, with little dependence of choice of $V_\odot$.

The rotation curve index is found to be $\alpha = 0.01 \pm 0.03$, indicating that the Galactic rotation curve is basically flat. This result is consistent with previous studies of rotation curves (e.g., Sofue & Rubin 2001). Regarding $U_{\rm SFR}$ and $W_{\rm SFR}$ components, mean peculiar motion of star-forming regions are not prominent. In contrast, in $V_{\rm SFR}$ component, there is a notable lag behind the Galactic rotation ($V_{\rm SFR} = -13.4 \pm 1.5$ km s^{-1}), which confirms the suggestion originally made by Reid *et al.* (2009). However, we also note that this lag is dependent of the adopted value of $V_\odot$, and we found a clear relation between $V_{\rm SFR}$ and $V_\odot$ as $V_{\rm SFR} = V_\odot - 19(\pm 2)$ km s^{-1}. Therefore, the lag of star-forming regions could be an artifact caused by an inappropriate value of V_{SFR}, but we still have to wait for precise determination of $V_\odot$ to obtain final conclusion.

4. Future prospect

In the next decade, VERA will continue astrometric monitoring of maser sources, and hopefully obtain astrometric results for 300–400 sources by ~2020. To accelerate the observations, we are in collaboration with KVN (Korean VLBI Network), which has three of 21-m telescopes in Korea, to combine KVN and VERA. The combined array of KVN and VERA will be powerful for Galactic maser astrometry in terms of better baseline coverage as well as better sensitivity. Targets are H_2O masers at 22 GHz and SiO masers at 43 GHz, with a possible extension to CH_3OH masers at 6.7 GHz (VERA already has receivers for this band, and there is a plan to install new receivers for this band on KVN). Another possibly-new target is CH_3OH maser at 44 GHz, originally discovered Morimoto *et al.*(1985). Recent test observations with KVN show that some sources are detectable with VLBI (with relatively short baselines), and so they may be new target sources for maser astrometry in near future.

References

Ando *et al.* 2011, *PASJ*, 63, 45
Dame *et al.* 2001 *ApJ*, 547, 792
Dehnen & Binney 1998, *MNRAS*, 298, 387
Fujisawa *et al.* 2011, in this volume
Ghez, A., *et al.* 2008, *ApJ*, 689, 1044
Gillessen, S. *et al.* 2009, *ApJ*, 692, 1075
Hirota T. *et al.* 2007, *PASJ*, 59, 897
Hirota T. *et al.* 2011a, *PASJ*, 63, 1
Hirota T. *et al.* 2011b, *ApJL*, 739, L59
Hirota T. *et al.* 2012, in this volume
Honma, M. *et al.* 2000, *Proceedings of SPIE*, 4015, 624
Honma, M. *et al.* 2007, *PASJ*, 59, 889
Honma, M. *et al.* 2011, *PASJ*, 63, 17
Honma, M. *et al.* 2012, *PASJ*, submitted
Nagayama, M. *et al.* 2011, *PASJ*, 63, 23
Morimoto *et al.* 1985 *ApJ*, 288, L11
Motogi, *et al.* 2011, *MNRAS*, 417, 238
Reid M. *et al.* 2009, *ApJ*, 693, 397
Schönrich *et al.* 2010, *MNRAS*, 403, 1829
Sofue Y., Rubin V. 2001, *ARA&A*, 39, 137

Cosmic Masers - from OH to H_0
Proceedings IAU Symposium No. 287, 2012
R.S. Booth, E.M.L. Humphreys & W.H.T. Vlemmings, eds.

© International Astronomical Union 2012
doi:10.1017/S1743921312007351

Astrometry of Galactic Star-Forming Regions ON1 and ON2N with VERA

Takumi Nagayama[1] and VERA project members

[1] Mizusawa VLBI Observatory, National Astronomical Observatory of Japan, 2-21-1 Osawa, Mitaka, Tokyo, 181-8588, Japan
email: takumi.nagayama@nao.ac.jp

Abstract. We conducted the astrometry of H_2O masers in the Galactic star-forming regions ON1 and ON2N with the VLBI Exploration of Radio Astrometry (VERA). The measured distances to ON1 and ON2N are 2.47 ± 0.11 kpc and 3.83 ± 0.13 kpc, respectively. In the case that ON1 and ON2N are on a perfect circular rotation, we estimate the angular rotation velocity of the Galactic rotation at the Sun (the ratio of the Galactic constants) to be 28 ± 2 km s^{-1} kpc^{-1} using the measured distances and three-dimensional velocity components of ON1 and ON2N. This value is larger than the IAU recommended value of 25.9 km s^{-1} kpc^{-1}, but consistent with other results recently obtained with the VLBI technique.

Keywords. astrometry, Galaxy: fundamental parameters, Galaxy: kinematics and dynamics

1. Introduction

The Galactic constants, the distance from Sun to the Galactic center (R_0) and the Galactic rotation velocity at Sun (Θ_0) are major parameters to study the structure of the Milky Way Galaxy (MWG). Although IAU has recommended the values of $R_0 = 8.5$ kpc and $\Theta_0 = 220$ km s^{-1} since 1985, Recent studies report the values different from them (e.g. Reid *et al.* 2009). However, observational estimation of the Galactic constants is difficult. This is because the observational estimation of Galactic constants is affected by several independent assumptions; the peculiar motion of the source, systemic non-circular motions of both the source and the LSR due to the spiral arm and the non-axisymmetric potential of the MWG, and relative motion of Sun to the LSR (Reid *et al.* 2009; McMillan & Binney 2010). To minimize these effects, we should observe many sources located at various positions in the MWG.

The tangent point and the Solar circle are kinematically unique positions in the MWG. We can estimate R_0 and Θ_0 from the source proper motion and distance. This is because that in the case of the source located at tangent point, the proper motion depends only on the Galactic rotation of Sun, Θ_0. The tangent point, Sun, and the Galactic center make the right triangle. In the case of the source on the solar circle, the proper motion depends only on Θ_0. The source, Sun, and the Galactic center make an isosceles triangle. Even if the source is not exactly located at tangent point or on the Solar circle, but near there, we can estimate the ratio of the Galactic constants, $\Omega_0 = \Theta_0/R_0$, as described in section 3. The radial velocity of ON1 is 12 ± 1 km s^{-1} (e.g. Bronfman *et al.* 1996). It is close to the terminal velocity at $l = 69.54°$ of 15 ± 5 km s^{-1} Dame *et al.* 2001. The radial velocity of ON2N is 0 ± 1 km s^{-1} (e.g. Olmi & Cesaroni 1999). These suggest that ON1 is located near the tangent point, and ON2N is located near the Solar circle. In the present study, we report about the parallax measurements of ON1 and ON2N with VERA. This is a first step to estimate the Galactic constants using VERA. This study is published in Nagayama *et al.* 2011 and Ando *et al.* 2011.

2. Observations

We observed H_2O masers in the star-forming regions ON1 and ON2N with VERA at 11 epochs in 2006–2008. The position reference sources of ON1 and ON2N are J2010+3322 and J2015+3710, respectively. Their separation angles are 1.85° and 1.27°, respectively. The maser sources and the position reference sources were simultaneously observed in a VERA dual-beam mode for about 10 hours. The data were recorded onto magnetic tapes at a rate of 1024 Mbps, providing a total bandwidth of 256 MHz, which consists of 16×16 MHz IF channels. One IF channel was assigned to the maser source, and the other 15 IF channels were assigned to the position reference source, respectively. Correlation processing was carried out on the Mitaka FX correlator. Data reduction was conducted using the NRAO Astronomical Image Processing System (AIPS).

3. Results and Discussion

Figure 1 and 2 show Parallaxes the H_2O masers in ON1 and ON2N, respectively. The positional variations show systematic sinusoidal modulation with a period of one year caused by the parallax. We conducted a combined parallax fit, in which the positions of ten features are fitted simultaneously with one common parallax but different proper motions and position offsets for each spot. The resulting parallaxes of ON1 and ON2N are 0.404 ± 0.019 mas (2.47 ± 0.11 kpc) and 0.261 ± 0.009 mas (3.83 ± 0.13 kpc), respectively. The obtained distance of ON1 is consistent with the 6.7 GHz CH_3OH maser parallax corresponding to $2.57^{+0.34}_{-0.27}$ kpc measured by Rygl et al. 2010. Figure 3 shows the position of ON1 and ON2N in the MWG. ON1 and ON2N are appeared to be located near the tangent point at $l = 69.54°$ and the Solar cirlce, respectively.

The systemic proper motions of ON1 and ON2N are derived to be $(\mu_\alpha \cos \delta, \mu_\delta) = (-3.10 \pm 0.18, -4.70 \pm 0.24)$ and $(-2.79 \pm 0.13, -4.66 \pm 0.17)$ mas yr^{-1} from the averages of the proper motions of maser features. We convert to the proper motions with respect to LSR using the Solar motion in the traditional definition of $(U_\odot, V_\odot, W_\odot) = (10.3, 15.3, 7.7)$ km s^{-1}. The proper motion of ON1 and ON2N with respect to LSR in the direction of l and b are calculated to be $(\mu_l, \mu_b) = (-6.00 \pm 0.22, 0.69 \pm 0.20)$ and $(-5.42 \pm 0.16, -0.36 \pm 0.14)$ mas yr^{-1}, respectively. This proper motions correspond to a velocity of $(v_l, v_b) = (-70.2 \pm 2.6, 8.1 \pm 2.3)$ and $(-98.4 \pm 2.9, -6.6 \pm 2.6)$ km s^{-1}.

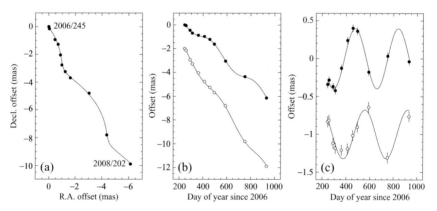

Figure 1. Parallax and proper motion data and fits of the H_2O maser in ON1. (a): Positions on the sky with first and last epochs labeled. (b): x (filled circles) and y (open circles) position offsets versus time. (c): same as (b) panel, expect the proper motion fit has been removed, allowing the effects of only the parallax to be seen.

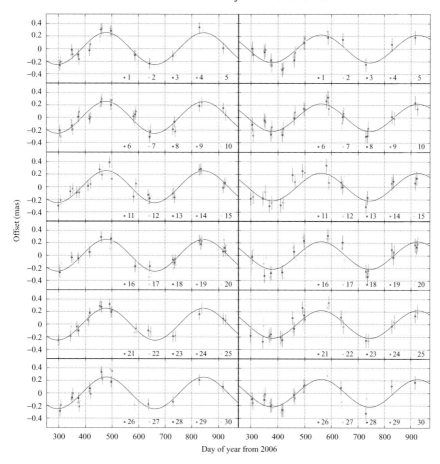

Figure 2. Parallax of the H$_2$O masers in ON2N. Individual proper motions and position offsets are removed. The left panel shows the parallax in right ascension. The right panel shows the parallax in declination.

In the case that the source is located exactly at the tangent point, and it is on pure circular rotation, the source, Sun, and the Galactic center make the right triangle, and the source proper motion on the sky depends only on Θ_0 This geometry is shown in Figure 4(a). Therefore, R_0 and Θ_0 are determined from the observed distance D and the proper motion along the Galactic plane, v_l, as

$$R_0 = D/\cos l \tag{3.1}$$

$$\Theta_0 = -v_l/\cos l. \tag{3.2}$$

The Galactic constants are estimated to be $R_0 = 7.1 \pm 0.3$ kpc and $\Theta_0 = 201 \pm 7$ km s^{-1}, respectively, from $D = 2.47 \pm 0.11$ kpc and $v_l = -70.2 \pm 2.6$ km s^{-1}. These values are approximately 10–20% smaller than the IAU recommended values of $R_0 = 8.5$ kpc and $\Theta_0 = 220$ km s^{-1}, the recently estimated value of $R_0 = 8.4 \pm 0.6$ kpc and $\Theta_0 = 254 \pm 16$ km s^{-1} (Reid *et al.* 2009). However, our estimated values would not be inconsistent with the previous estimates. This is because our estimation is affected by the ambiguities of the two assumptions that ON1 is on the pure circular rotation, and located exactly at the tangent point. If we consider the effects of this amibiguities, the errors of our estimated values increase to approximately 70% (Nagayama *et al.* 2011).

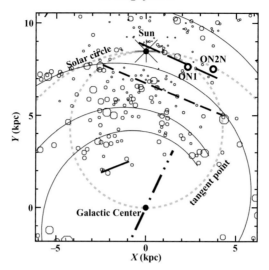

Figure 3. Positions of ON1 and ON2N in the MWG. The background is the four spiral arm structure of the MWG (Russeil 2003).

Although the estimated values of the Galactic constants are strongly affected by the assumption of the source location in the MWG, we found that the ratio of Galactic constants, Θ_0/R_0, can be estimated with small ambiguity. In the case that the source is on pure circular rotation at any position in the Galactic disk, the radial and tangential velocities of the source can be written as

$$v_r = \left(\frac{\Theta}{R} - \frac{\Theta_0}{R_0}\right) R_0 \sin l, \tag{3.3}$$

$$v_l = \left(\frac{\Theta}{R} - \frac{\Theta_0}{R_0}\right) R_0 \cos l - \frac{\Theta}{R} D. \tag{3.4}$$

From these equations, the relation between the Θ_0 and R_0 is obtained to be

$$\Theta_0 = \left[-\frac{v_l}{D} + v_r \left(\frac{1}{D \tan l} - \frac{1}{R_0 \sin l}\right)\right] R_0 = \left[-a_0 \mu_l + v_r \left(\frac{1}{D \tan l} - \frac{1}{R_0 \sin l}\right)\right] R_0, \tag{3.5}$$

where a_0 is a conversion constant from a proper motion to a linear velocity (4.74 km s^{-1} mas^{-1} yr kpc^{-1}). The equation (3.5) is graphed in Figure 5(a) using the observed values of $D = 2.47 \pm 0.11$ kpc, $\mu_l = -6.00 \pm 0.22$ mas yr^{-1}, and $v_r = 12 \pm 1$ km s^{-1}. We found that the slope in Figure 5(a) is a nearby constant at $7 \leqslant R_0 \leqslant 9$ kpc. The slope yields the ratio Θ_0/R_0, which is described as

$$\frac{\Theta_0}{R_0} = -\frac{v_l}{D} + v_r \left(\frac{1}{D \tan l} - \frac{1}{R_0 \sin l}\right) = -a_0 \mu_l + v_r \left(\frac{1}{D \tan l} - \frac{1}{R_0 \sin l}\right), \tag{3.6}$$

The equation (3.6) is graphed in Figure 5(b). The ratio is estimated to be $\Theta_0/R_0 = 28.7 \pm 1.3$ km s^{-1} kpc^{-1} using the above observed values, and $7 \leqslant R_0 \leqslant 9$ kpc. The error of Θ_0/R_0 mainly depends on that of μ_l. The errors of Θ_0/R_0 depend on those of v_r and D are ± 0.02 and ± 0.05 km s^{-1} kpc^{-1}, respectively, and they can be neglected in this estimation. This is because that $D \tan l \simeq R_0 \sin l$ in the case that the source is located near the tangent point (see Figure 4(b)).

The ratio can be also estimated from the observed distance, proper motion, radial velocity of the source near the Solar circle. The radial velocity of the source near the solar circle is close to zero. Therefore, the ratio is also free from R_0. The ratio is obtained to be $\Theta_0/R_0 = 27.3\pm0.8$ km s^{-1} kpc^{-1} using the observed ON2N parameters of $D = 3.83\pm0.13$ kpc, $\mu_l = -5.76 \pm 0.16$ mas yr^{-1}, and $v_r = 0 \pm 1$ km s^{-1}.

We estimated the ratio $\Theta_0/R_0 = 28 \pm 2$ km s^{-1} kpc^{-1} from the average of obtained values of ON1 and ON2N. This value is consistent with the value of $\Theta_0/R_0 = 28.6\pm0.2$ km s^{-1} kpc^{-1} obtained from the proper motion measurement of Sgr A* (Reid & Brunthaler 2004), which is revised using the traditional definition of the solar motion by us. However, this value is inconsistent to that derived from the IAU recommended values 220 km s^{-1}/8.5 kpc=25.9 km s^{-1} kpc^{-1}.

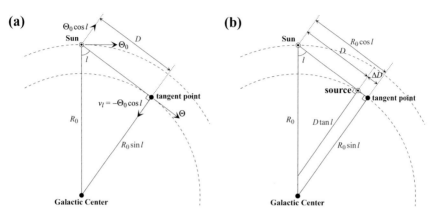

Figure 4. The geometry of the Galactic center, Sun, the tangent point, and the source. (a): The geometry in the case that the source is located at the tangent point. (b): The geometry in the case that there is a offset between the source and the tangent point.

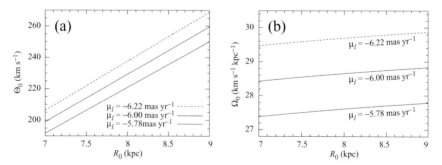

Figure 5. (a): The relation of R_0 and Θ_0 shown in the equation (3.5). (b): The relation of R_0 and Θ_0/R_0 shown in the equation (3.6).

References

Ando, K., Nagayama, T., & Omodaka, T., *et al.* 2011, *PASJ*, 63, 45
Bronfman, L., Nyman, L.-A., & May, J. 1996, *A&AS*, 115, 81
Dame, T. M., Hartmann, D., & Thaddeus, P. 2001, *ApJ*, 547, 792
McMillan, P. J. & Binney, J. J. 2010, *MNRAS*, 402, 934
Nagayama, T., Omodaka, T., Nakagawa, A., *et al.* 2011, *PASJ*, 63, 23
Olmi, L. & Cesaroni, R. 1999, *A&A*, 352, 266
Reid, M. J. & Brunthaler, A. 2004, *ApJ*, 616, 872
Reid, M. J., *et al.* 2009, *ApJ*, 700, 137
Russeil, D. 2003, *A&A*, 397, 133
Rygl, K. L. J., Brunthaler, A., Reid, M. J., Menten, K. M., van Langevelde, H. J., & Xu, Y. 2010, *A&A*, 511, A2

Cosmic Masers - from OH to H$_0$
Proceedings IAU Symposium No. 287, 2012
R.S. Booth, E.M.L. Humphreys & W.H.T. Vlemmings, eds.

© International Astronomical Union 2012
doi:10.1017/S1743921312007363

VLBI maser kinematics in high-mass SFRs: G23.01–0.41

Alberto Sanna[1], Luca Moscadelli[2], Riccardo Cesaroni[2] and Ciriaco Goddi[3]

[1] Max-Planck-Institut für Radioastronomie, Auf dem Hügel 69, 53121 Bonn, Germany
email: asanna@mpifr-bonn.mpg.de

[2] INAF, Osservatorio Astrofisico di Arcetri, Largo E. Fermi 5, 50125 Firenze, Italy
email: mosca@arcetri.astro.it; cesa@arcetri.astro.it

[3] European Southern Observatory, Karl-Schwarzschild-Strasse 2, D-85748 Garching bei München, Germany
email: cgoddi@eso.org

Abstract. Very Long Baseline Interferometry studies of different maser species observed at multiple epochs allow complementary measurements of the 3-dimensional velocity field of gas close ($\lesssim 10^3$ AU) to massive young stellar objects. Here, we review our recent results toward the high-mass star-forming region G23.01–0.41, where all the strongest molecular maser transitions known to date cluster within 2000 AU from the center of an hot molecular core and are associated with a so called extended green object. The overall maser kinematics reveals a common outflowing motion from a central object; the details of the spatial distribution and velocity field of each maser species hint at the presence of different dynamical structures: a collimated jet, a wide-angle wind, and a flattened rotating core. We further compare the simultaneous presence of maser emission from different molecular species with a recent evolutionary sequence for masers associated with massive young stellar objects.

Keywords. masers, techniques: high angular resolution, stars: formation, stars: individual (G23.01–0.41).

1. Introduction

How do high-mass stars (M$_{star} > 8$ M$_\odot$) form? And what are the physical properties and dynamics of their environments locally? Maser emission in high-mass star-forming regions (HMSFRs) is a signpost of the earliest evolutionary phases of massive young stellar objects (MYSOs). The spatial compactness and the narrow linewidth (owing to the required high degree of velocity coherence) make maser emission an ideal tool for testing ordered velocity fields from the inner protostellar cocoon ($\lesssim 10^3$ AU). At typical distances of a few kpc for MYSOs, mas-resolution (i.e., Very Long Baseline Interferometry, VLBI) observations of the spatial distribution of each masing cloudlets (as small as a few AU) monitored over time allows us to measure the full-space kinematics of the masing gas. In particular, those sites exhibiting multiple molecular masers gain greater interest since different molecular species can trace distinct and complementary physical environments, thus providing us with information useful to distinguish among different dynamical processes such as rotation, expansion, or contraction.

Almost a decade ago, we started a campaign of multi-epoch, VLBI observations of the three most powerful maser transitions, of water (H$_2$O) at 22.2 GHz, methanol (CH$_3$OH) at 6.7 GHz, and hydroxyl (OH) at 1.665 GHz toward HMSFRs already well studied with interferometers in thermal, continuum and line emission (e.g., Moscadelli *et al.* 2007,2011; Goddi *et al.* 2007,2011; Sanna *et al.* 2010a,b). Our idea was to complement

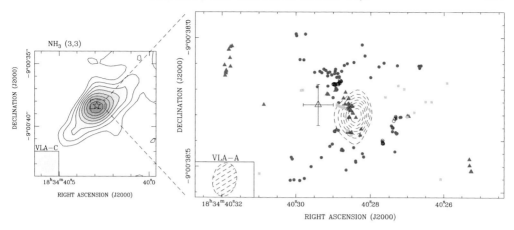

Figure 1. Collection of the subarcsec observations toward G23.01–0.41. *Left panel:* map of the NH$_3$ (3,3) line from Codella *et al.* (1997) with contour levels at multiples of 10% of the peak emission (46 mJy beam^{-1}). The star marks the peak position of the 3 mm continuum emission as determined by Furuya *et al.* (2008). *Right panel:* zoom in on the maser emitting region. Red triangles, blue dots, and green squares represent respectively H$_2$O, CH$_3$OH, and OH maser positions from our VLBI measurements (Sanna *et al.* 2010b). The empty circles mark the position of the 12.2 GHz CH$_3$OH maser cloudlets (Sanna *et al.* in prep.). The empty triangle with the error bars indicates the position (and the associated uncertainty) of the 4.8 GHz H$_2$CO maser feature derived by Araya *et al.* (2008). Dashed contours show the 1.3 cm continuum emission with contour levels at multiples of 10% of the peak emission (0.72 mJy beam^{-1}).

the information from thermal tracers of outflows (e.g., CO and SiO) and hot molecular cores (HMCs) e.g., dust continuum, NH$_3$, and CH$_3$CN), at typical scales from about 0.1 pc to several 1000 AU, with that on the inner few 1000 AU from the (proto)star, provided by maser emission and radio continuum from shock-induced ionization (i.e., thermal jets/winds) or photoionization by the central MYSO (i.e., young H II regions).

2. An overview of G23.01–0.41

G23.01–0.41 is an active site of massive star formation (L$_{bol}$ ∼ 10^{5}L$_\odot$) with a copious amount of maser emission from different molecular species (see Tab. 1). This star-forming region has an accurate distance measurement of 4.6 ± 0.4 kpc, determined through the trigonometric parallax of the associated 12 GHz methanol masers (Brunthaler *et al.* 2009; see also the chapter by Reid *et al.*), and belongs to a larger star-forming complex possibly triggered by the close supernova remnant W41 (Leahy & Tian 2008). On a pc scale, the 4.5 µm excess of the Spitzer IRAC GLIMPSE observations (Benjamin *et al.* 2003; see also Cyganowski *et al.* 2009, their Figure 1) show a shocked molecular clump powered by a massive, ^{12}CO bipolar outflow detected with the IRAM interferometer by Furuya *et al.* (2008). Because of the small width-to-length ratio of both the blueshifted and redshifted outflow lobes (to the NE and SW, respectively), such a geometry indicates that the inclination angle of the outflow axis with respect to the plane of the sky must be rather small (particularly evident in our new SMA observations; Sanna *et al.* in prep.). Following the statistical analysis by Cyganowski *et al.* (these proceedings), the association of G23.01–0.41 with such an extended green object (EGO) suggests an enhanced outflow activity at an early stage of massive star formation, and thus (presumably) an actively accreting MYSO. At ten-times smaller scale, CH$_3$CN and NH$_3$ thermal emissions reveal an HMC of 70 M$_\odot$ with a flattened structure and a velocity gradient (∼ 1 km s^{-1} over 0.1 pc) along its major axis (Furuya et al. 2008; Codella *et al.* 1997). Since the HMC is

Table 1. Overall maser detections/upper limits toward G23.01–0.41

Molecule	ν (GHz)	Telescope	F_{peak} (Jy)	V_{LSR} (km s^{-1})	Ref.[1]
H_2O	22.2	VLBA	320	72.6	1
OH	1.612	Nançay Radio tel.	0.2	78	2
	1.665	VLBA	2.6	67.9	1
	1.667	Nançay Radio tel.	9	75	2
CH_3OH	6.7	EVN	440	74.7	1
	12.2	VLBA	20	75.0	3
	19.9	Tidbinbilla 70-m	< 0.26		4
	23.1	Parkes 64-m	< 0.9		5
	37.7	Mopra 22-m	< 0.9		6
	38.3	Mopra 22-m	< 0.9		6
	38.5	Mopra 22-m	< 0.9		6
	44.1	Parkes 64-m	40	77.1	7
	85.5	SEST	< 3.8		8
	95.2	Mopra 22-m	3	77.4	9
	107.0	SEST	5.2	75.9	10
	108.8	Mopra 22-m	< 5.1		11
	156.6	SEST	< 2		10
H_2CO	4.8	VLA-B	0.048	73.6	12

Notes:
F_{peak} and V_{LSR} are the peak flux density and its Local Standard of Rest velocity for each maser transition, respectively.
[1] Ref.: (1) Sanna *et al.* (2010b); (2) Szymczak & Gérard (2004); (3) Brunthaler *et al.* (2009); (4) Ellingsen *et al.* (2004); (5) Cragg *et al.* (2004); (6) Ellingsen *et al.* (2011); (7) Slysh *et al.* (1994); (8) Cragg *et al.* (2001); (9) Val'tts *et al.* (2000); (10) Caswell *et al.* (2000); (11) Val'tts *et al.* (1999); (12) Araya *et al.* (2008).

close to the center of the bipolar outflow and is elongated perpendicular to the outflow axis, it has been interpreted as a candidate rotating toroid, i.e. a massive, non-equilibrium structure accreting onto the embedded (proto)star(s) over a timescale shorter than its rotation period.

All the strongest maser species known to date are observed to cluster at the center of the HMC together with faint continuum emission at centimeter wavelengths (Sanna *et al.* 2010b; Brunthaler *et al.* 2009; Araya *et al.* 2008). The spectral index of the radio continuum is consistent with shock excited emission from a thermal jet, whereas its spatial morphology at the longest VLA baselines clearly draws a bright knot along the NE–SW direction of the molecular outflow (Sanna *et al.* 2010b). Between 2005 and 2007, we conducted phase-referencing, VLBA and EVN observations of the H_2O, CH_3OH (both at several epochs), and of the OH masers (at a single epoch) in the K, C, and L bands, respectively (see Tab. 1). Recently, we have also complemented this data set with archival VLBA observations in the U band, in order to study the details of the methanol maser kinematics at 12 GHz with respect to the 6.7 GHz transition (Sanna *et al.* in prep.). These observations reveal that all the maser emissions in the region participate in an overall expansion from a common center, which possibly denotes the position of the YSO (Sanna *et al.* 2010b). Furthermore, these maser species sample different positions in the gas around G23.01–0.41 and their radial and transverse velocities pinpoint 3 distinct kinematical structures (see Fig. 1). The H_2O masers trace both: I) shocked gas along the route of the same bipolar jet exciting the radio continuum and II) a wide-angle, faster wind projected close to the putative position of the YSO (where H_2O masers are also flaring; see Figure 5b of Sanna *et al.* 2010b). Similarly, the OH masers projected close to the fast-moving water masers are strongly blueshifted with respect to the systemic velocity of the region (Figure 3c of Sanna *et al.* 2010b). We have hypothesized that: I) this emission may be tracing an expanding layer of gas at a larger distance from the YSO than the H$_2$O gas and II) the same wide-angle wind responsible for the fast water maser shocks could power the expansion of the OH gas. Finally, the spatial distribution and velocity field of the CH$_3$OH cloudlets sketch out a funnel-like structure that probably results from a combination of both rotation and expansion inside the NH$_3$ toroid (Figure 6 of Sanna *et al.* 2010b).

3. Details on the Maser Emission

Water masers. It is interesting to note that, the momentum rate of the water maser jet can be directly compared with that of the large-scale ^{12}CO outflow. This calculation shows consistent values of a few 0.1 M$_\odot$ yr^{-1} km s^{-1} and has two implications: I) it brings further support to the hypothesis that water masers more detached from the radio continuum trace the primary wind (i.e., the jet) driving the large-scale molecular outflow; II) provided that molecular outflows are momentum driven and their momentum rate is correlated with the mass and luminosity of the YSO (e.g., Figure 4 of Beuther *et al.* 2002), the high value measured above implies indeed a ZAMS star of early B/late O spectral type. Furthermore, the detection of a wide-angle wind from water masers close to the MYSO can explain the observation of a width-to-length ratio for the molecular outflow of $\sim 0.4 - 0.3$, much higher than in a pure jet-driven outflow.

Methanol masers. Focusing on the 6.7 GHz maser transition, by observing 3 different EVN epochs spanning 2 years, *we have measured for the first time the internal proper motions of individual, methanol maser cloudlets with relative uncertainties less than 30%* (together with the source G16.59–0.05 from Sanna *et al.* 2010a). The line-of-sight velocity distribution is clearly bipolar, showing two distinct groups of methanol masers, one moving away from the observer (southern redshifted cluster) and one toward the observer (northern blueshifted cluster). By studying the distribution of the 3-dimensional methanol maser velocities with respect to the jet direction on the sky and its inclination with respect to the l.o.s., we have proposed that the methanol gas undergoes rotation about the jet axis and is simultaneously dragged into the outflow motion along this axis (Figure 8 of Sanna *et al.* 2010b). If we assume centrifugal equilibrium, the rotational component inferred from the velocity field of the methanol gas implies again a central, massive ZAMS star with an early B/late O spectral type. A further, detailed comparison of the spatial distribution and kinematics of the strong maser emission at 12 GHz and 6.7 GHz suggests that both transitions are excited from the same methanol gas (Sanna *et al.* in prep.). According to Cragg *et al.* (2005), this evidence constrains the gas temperature and density to values of T$_k$ = 30–50 K and n$_{H_2}$ = 10^6–10^7 cm^{-3}, well in agreement with the physical parameters of the ammonia core. This analysis strongly supports the hypothesis that the Class II CH$_3$OH masers may emerge from the inner part of the massive toroid traced on a larger scale with the CH$_3$CN and NH$_3$ molecules. It is also worth noting that G23.01–0.41 is one of the rare cases where H$_2$CO maser emission at 4.8 GHz was recently found by Araya *et al.* (2008). This emission arising from a region consistent both in position and l.o.s. velocity with the blueshifted methanol masers may, when refined models for H$_2$CO pumping mechanism become available (e.g., Araya *et al.* 2007), help to further constrain the physical conditions in the region.

Evolutionary sequence in G23.01–0.41. Finally, we want to use a recent evolutionary sequence proposed to explain the appearance and relative lifetimes of different maser species in HMSFRs (cf. Figure 6 of Breen *et al.* 2010; see also Breen *et al.*, these proceedings) to speculate on the relative phase of evolution for G23.01–0.41. First of all, the star-forming region is associated with strong 4.5 μm emission (i.e., an EGO) and shows also bright 44 GHz methanol masers along the direction of the outflow (e.g., Cyganowski *et al.* 2009). The presence of Class I methanol masers and strong outflow activity, together with our findings of faint radio continuum emission mostly due to shock-induced ionization, rule out the far side of the Breen's scale, where photoionization from the central MYSO should produce a bright UCH II region. On the other hand, very bright 6.7 GHz and 12 GHz methanol masers (as found in G23.01–0.41) seem to be associated with a somewhat late stage of evolution for MYSO (Breen *et al.* 2011), that rules out the first half of the Breen's scale as well. The evidence for an intermediate evolutionary phase of G23.01–0.41 is further supported by the simultaneous presence of bright, mainline, OH masers, that are usually found in the vicinity of a detectable UCH II region. Following this evolutionary sequence, we can speculate that: I) G23.01–0.41 should develop in the near future a visible UCH II region; II) since high accretion rates are the main mechanism proposed for quenching of UCH II regions, G23.01–0.41 should be still actively accreting material from its protostellar envelope, that would be detectable with ALMA observations of thermal, dense gas tracers.

Acknowledgements

This work has been supported by the ERC Advanced Grant GLOSTAR under grant agreement no. 247078.

References

Araya, E., Hofner, P., & Goss, W. M. 2007, IAU Symposium, 242, 110
Araya, E. D., Hofner, P., Goss, W. M., *et al.* 2008, *ApJS*, 178, 330
Benjamin, R. A., Churchwell, E., Babler, B. L., *et al.* 2003, *PASP*, 115, 953
Beuther, H., Schilke, P., Sridharan, T. K., *et al.* 2002, *A&A*, 383, 892
Breen, S. L., Ellingsen, S. P., Caswell, J. L., & Lewis, B. E. 2010, *MNRAS*, 401, 2219
Breen, S. L., Ellingsen, S. P., Caswell, J. L., *et al.* 2011, *ApJ*, 733, 80
Brunthaler, A., Reid, M. J., Menten, K. M., *et al.* 2009, *ApJ*, 693, 424
Caswell, J. L., Yi, J., Booth, R. S., & Cragg, D. M. 2000, *MNRAS*, 313, 599
Codella, C., Testi, L., & Cesaroni, R. 1997, *A&A*, 325, 282
Cragg, D. M., Sobolev, A. M., Ellingsen, S. P., *et al.* 2001, *MNRAS*, 323, 939
Cragg, D. M., Sobolev, A. M., Caswell, J. L., Ellingsen, S. P., & Godfrey, P. D. 2004, *MNRAS*, 351, 1327
Cragg, D. M., Sobolev, A. M., & Godfrey, P. D. 2005, *MNRAS*, 360, 533
Cyganowski, C. J., Brogan, C. L., Hunter, T. R., & Churchwell, E. 2009, *ApJ*, 702, 1615
Ellingsen, S. P., Cragg, D. M., Lovell, J. E. J., *et al.* 2004, *MNRAS*, 354, 401
Ellingsen, S. P., Breen, S. L., Sobolev, A. M., *et al.* 2011, *ApJ*, 742, 109
Furuya, R. S., Cesaroni, R., Takahashi, S., *et al.* 2008, *ApJ*, 673, 363
Goddi, C., Moscadelli, L., Sanna, A., Cesaroni, R., & Minier, V. 2007, *A&A*, 461, 1027
Goddi, C., Moscadelli, L., & Sanna, A. 2011, *A&A*, 535, L8
Leahy, D. A. & Tian, W. W. 2008, *AJ*, 135, 167
Moscadelli, L., Goddi, C., Cesaroni, R., Beltrán, M. T., & Furuya, R. S. 2007, *A&A*, 472, 867
Moscadelli, L., Cesaroni, R., Rioja, M. J., Dodson, R., & Reid, M. J. 2011, *A&A*, 526, A66
Sanna, A., Moscadelli, L., Cesaroni, R., *et al.* 2010a, *A&A*, 517, A71
Sanna, A., Moscadelli, L., Cesaroni, R., *et al.* 2010b, *A&A*, 517, A78
Slysh, V. I., Kalenskii, S. V., Valtts, I. E., & Otrupcek, R. 1994, *MNRAS*, 268, 464
Szymczak, M. & Gérard, E. 2004, *A&A*, 414, 235
Val'tts, I. E., Ellingsen, S. P., Slysh, V. I., *et al.* 1999, *MNRAS*, 310, 1077
Val'tts, I. E., Ellingsen, S. P., Slysh, V. I., *et al.* 2000, *MNRAS*, 317, 315

Cosmic Masers - from OH to H_0
Proceedings IAU Symposium No. 287, 2012
R.S. Booth, E.M.L. Humphreys & W.H.T. Vlemmings, eds.

© International Astronomical Union 2012
doi:10.1017/S1743921312007375

3D velocity fields from methanol and water masers in an intermediate-mass protostar

C. Goddi[1], L. Moscadelli[2] and A. Sanna[3]

[1] European Southern Observatory, Karl-Schwarzschild-Strasse 2, D-85748 Garching, Germany
email: cgoddi@eso.org

[2] INAF, Osservatorio Astrofisico di Arcetri, Largo E. Fermi 5, 50125 Firenze, Italy

[3] Max-Planck-Institut für Radioastronomie, Auf dem Hügel 69, 53121 Bonn, Germany

Abstract. We report multi-epoch VLBI observations of molecular masers towards the high-mass star forming region AFGL 5142, leading to the determination of the 3D velocity field of circumstellar molecular gas at radii $<0''.23$ (or 400 AU) from the protostar MM–1. Our observations of CH_3OH maser emission enabled, for the first time, a direct measurement of infall of a molecular envelope on to an intermediate-mass protostar (radius of 300 AU, velocity of 5 km s^{-1}, and infall rate of 6×10^{-4} n_8 $M_\odot$ yr^{-1}, where n_8 is the ambient volume density in units of 10^8 cm^{-3}). In addition, our measurements of H_2O maser (and radio continuum) emission revealed a collimated bipolar molecular outflow (and ionized jet) from MM–1. The evidence of simultaneous accretion and outflow at small spatial scales, makes AFGL 5142 an extremely compelling target for high-angular resolution studies of high-mass star formation.

1. Introduction

Observational signatures of infalling envelopes and outflowing material in early stages of protostellar evolution, at small radii from the protostar, are essential to further the understanding of the mass-accretion and mass-loss processes, and constrain theoretical models of star formation. To date, bulk infall motions have been detected, mostly by observing inverse P-Cygni profiles of molecular lines (with redshifted absorption and blueshifted emission) along the line-of-sight (l.o.s.) to the infalling gas, against a bright background (dust or H II emission), in both low-mass (e.g., Lee *et al.* 2009) and high-mass (e.g., Beltrán *et al.* 2011) protostars. This method is, however, prone to ambiguous interpretation because line asymmetries may be associated with competing processes such as rotation and outflows, or even arise from superposition of intervening clouds along the l.o.s. Confusion is more severe in the case of high-mass protostars, which are on average more distant (>1 kpc) and form embedded in rich protoclusters. Evidence of infall has been found in only a handful of high-mass star forming regions and on scales of protoclusters rather than individual protostars (>1000 AU: Beltrán *et al.* 2011, and references therein). In order to establish definitely whether the infalling material at large scales eventually accretes on to individual protostars as well as to resolve individual outflows in the protocluster, infall and outflow motions should be measured in the proximity (≪1000 AU) to the protostar.

We report here a convincing signature of infall in a circumstellar molecular envelope with a radius of only 300 AU, associated with the protostar MM–1 in the high-mass star-forming region AFGL 5142 (Zhang *et al.* 2007). MM–1 shows hot-core chemistry (Palau *et al.* 2011), exhibits radio continuum emission from ionized gas, and powers strong water and methanol masers (Goddi *et al.* 2007). Multi-epoch VLBI observations of CH_3OH masers enabled us to measure the 3D velocity field of molecular gas, providing the

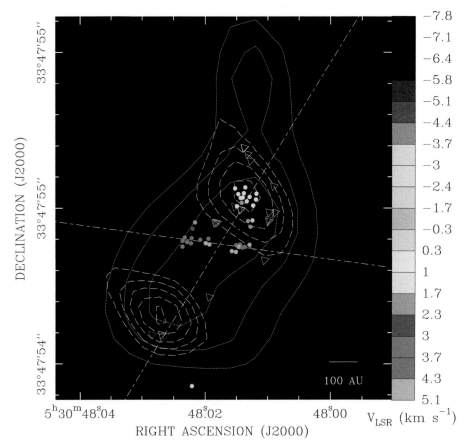

Figure 1. Molecular masers and radio continuum in AFGL 5142 MM–1. Positions and l.o.s. velocities of H_2O masers (observed with the VLBA) are indicated by triangles and CH_3OH masers (observed with the EVN) are indicated by circles. Colors code l.o.s. velocities, according to the wedge to the right (the systemic velocity, –1.1 km s^{-1}, is in green). The contours show the continuum emissions (observed with the VLA) at 22 GHz (dotted contours, representing 3, 4, and 5 times the 46 μJy beam^{-1} rms) and 8.4 GHz (dashed contours, representing 3, 4, 5, 6 and 7 times the 23 μJy beam^{-1} rms). The two small-dashed lines show the linear fits to the positions of all water masers (PA –32°) and of only the red methanol masers (PA 82°).

most direct, unbiased measurement, to date, of infall on to an intermediate- to high-mass protostar. We were also able to characterise the outflow structure using the kinematics of H_2O masers and the physical properties of radio continuum emission, thus obtaining a complete picture of star formation on scales <400 AU.

2. Observations and Results

We observed the 6.7 GHz CH_3OH masers with the EVN at three distinct epochs (years 2004–2009), the 22.2 GHz H_2O with the VLBA at four distinct epochs (years 2003–2004), and the radio continuum emission with the VLA at 22 GHz and 8.4 GHz. The 3D kinematics of H_2O and CH_3OH masers is discussed in detail in Goddi *et al.* (2006, 2007, 2011). Figure 1 shows positions and l.o.s. velocities of the 22 GHz water and the 6.7 GHz methanol masers, overplotted on the contour maps of the continuum emission at 8.4 GHz and 22 GHz. The 22 GHz continuum emission (0.″24 beamwidth) appears elongated NW-SE, while the 8.4 GHz emission (0.″16 beamwidth) is resolved into two components separated by 300 mas (or 540 AU) on the plane of the sky. H_2O masers are concentrated in two clusters, associated with the two components of the 8.4 GHz continuum: that towards the SE has l.o.s. velocities blue-shifted with respect to the systemic velocity of the region (-1.1 km s^{-1}; Zhang *et al.* 2007); the second, located to the NW has red-shifted l.o.s. velocities. CH_3OH masers are distributed across an area similar to the H_2O masers, and consist mainly of two clusters: one with moderately red-shifted l.o.s. velocities is associated with the NW continuum peak ("yellow" features); the other cluster with the most red-shifted l.o.s. velocities lies in between the two 8.4 GHz peaks ("red" features). The dashed lines in Fig. 1 provide the best (least-squares fit) to the positions of all the water masers (position angle or PA of $-32°$) and to only the red methanol masers (PA of $82°$), respectively.

Figure 2 shows the proper motions of H_2O masers (left panel) and CH_3OH masers (right panel). Since measurements of *absolute* proper motions are affected by the combined uncertainty of the Solar motion and Galactic rotation curve (up to 15 km s^{-1}; Reid *et al.* 2009), we prefer to base our analysis on *relative* velocities for both water and methanol masers. This requires us to adopt a suitable reference frame, ideally centered on the protostar. For water, we calculated the geometric mean of positions at individual epochs, for all the masers persistent over four epochs (hereafter "center of motion") and we referred the proper motions to this point (this is equivalent to subtracting from all masers, the average proper motion of the selected features). The proper motions indicate that the two clusters are moving away from each other, NW-SE, with velocities ~15 km s^{-1}. We used positions and 3D velocities of water masers to define geometric parameters of the outflow (for simplicity, supposed of conical shape): PA of the sky-projected axis ($-40°$), inclination angle of the axis with the plane of the sky ($25°$), and opening angle ($25°$).

Since the amplitude of methanol proper motions is much lower than that of water (with mean values ~3 km s^{-1} and ~15 km s^{-1}, respectively), the choice of a suitable reference for methanol velocities is more critical. The red masers have *internal* proper motions (calculated using the center of motion of only red masers; 1–2 km s^{-1}) much smaller than the internal proper motions of the yellow masers (relative to the center of motion of only yellow features; 1–10 km s^{-1}); the opposite is true for their l.o.s. velocities (3–6 km s^{-1} vs. 0–2 km s^{-1}). Since red masers are projected in the plane of the sky closer to the protostellar position and move mostly along the l.o.s., we assume that their average proper motion gives an estimate of the protostar proper motion. Hence, we calculated the center of motion of methanol using only red features with a stable spatial and spectral structure. The resulting proper motions (Fig. 2, right), should represent the sky-projected velocities as measured by an observer co-moving with the star. The yellow masers have proper motions with larger amplitudes, directed towards the the red maser centroid, which instead move mostly along the l.o.s.; both aspects indicate infall towards the protostar.

3. Discussion

Our VLBI observations show that water and methanol masers, despite being excited at similar radii from MM–1, trace different kinematics: outflow and infall, respectively.

Collimated outflow from MM–1. We have identified a collimated bipolar outflow, at radii 140 to 400 AU from the driving protostar, traced by radio continuum emission in its ionized component and by water masers in its molecular component. For optically thin emission, $F\,d^2 = 10^{3.5}\,(\Omega/4\pi)\,\dot{P}_{jet}$ (Sanna *et al.* 2010), where F is the continuum flux density in mJy, $\dot{P}_{jet}$ is the jet momentum rate in $M_\odot$ yr^{-1} km s^{-1}, Ω is the jet solid angle in

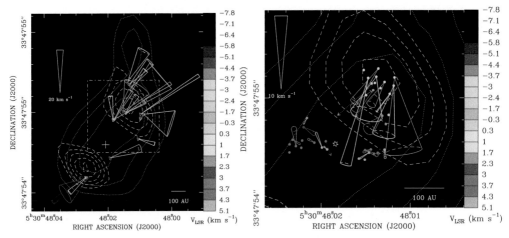

Figure 2. Proper motions of H_2O (*left panel*) and CH_3OH (*right panel*) masers in AFGL 5142 MM–1, as measured relative to their centers of motion, independently calculated for water (*cross*) and methanol (*star*). The rectangle in the left panel shows the area plotted in the right panel. The cones indicate orientation and uncertainties of measured proper motions (the amplitude scale is given in each panel) and colors denote l.o.s. velocities. Contour maps show the VLA 22 GHz (dotted line) and 8.4 GHz (dashed line) continuum.

sr, and d is the source distance in kpc. For d =1.8 kpc and F=0.24 (0.17) mJy for the NW (SE) component at 8.4 GHz, we derive: $\dot{P}_{jet} = 2.5(1.7) \times 10^{-4} \, (\Omega/4\pi)^{-1} \, M_\odot \, \mathrm{yr}^{-1} \, \mathrm{km \, s}^{-1}$. Based on our measurements of positions and 3D velocities of water masers and under reasonable assumptions for gas densities, we can estimate the mass-loss rate, $\dot{M}_{out} = 4\pi R^2 n_{H_2} m_{H_2} V \, (\Omega/4\pi)$, and the momentum rate, $\dot{P}_{out} = \dot{M}_{out} V$, of the molecular outflow. For an average projected distance of water masers from the protostar of 290 AU and an average velocity of 15 km s^{-1}, we derive $\dot{M}_{out} = 1.9 \times 10^{-3} \, (\Omega/4\pi) \, n_8 \, M_\odot \, \mathrm{yr}^{-1}$, where Ω is the (conical) flow solid angle and n_8 is the ambient volume density in units of 10^8 cm^{-3}. If the water masers originate in shocks produced by the interaction of the ionized jet with the ambient medium, then Ω can be determined by requiring that the momentum rate in the maser outflow equals that from the continuum emission. We derive Ω=1–1.2 sr, corresponding to an outflow semi-opening angle of 32–36°, which provides $\dot{M}_{out} = 1.6 \times 10^{-4} \, n_8 \, M_\odot \, \mathrm{yr}^{-1}$ and $\dot{P}_{out} = 2.4 \times 10^{-3} \, n_8 \, M_\odot \, \mathrm{yr}^{-1} \, \mathrm{km \, s}^{-1}$, indicating strong outflow activity from MM–1.

Gas infall on to MM–1 from a molecular envelope. Infall onto MM–1 was first claimed by Goddi *et al.* (2007), based solely on positions and l.o.s. velocities of methanol masers. The measurement of proper motions has provided the missing kinematic observable to confirm the infall hypothesis.

We adopted a simple spherical model: $\mathbf{V}(\mathbf{R}) = (2 \, G \, M/R^3)^{1/2} \, \mathbf{R}$, where $\mathbf{V}$ is the velocity field, $\mathbf{R}$ is the position vector from the protostar, M is the total gas mass within a sphere of radius R centered on the protostar. The best-fit parameters we derived from the model are: R=290 AU and M=4 $M_\odot$, corresponding to V_{inf}=5 km s^{-1}. Figure 3 shows the measured and best-fit model 3D velocity vectors of methanol masers projected onto the plane containing the protostar and perpendicular to the outflow axis (the equatorial plane; left panel) and the plane containing the l.o.s. and the outflow axis (right panel). Most of the measured 3D velocities are well reproduced by the infall model. The two projections show also that methanol masers do not sample the whole spherical envelope but concentrate into preferred areas. We interpret this behavior in terms of amplification of a background emission. The yellow features lie on top of the NW 8.4 GHz continuum peak,

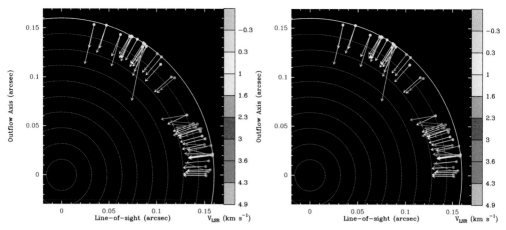

Figure 3. Measured (*color arrows*) and best-fit model (*black arrows*) 3D velocity vectors of CH$_3$OH masers. We show projections onto the equatorial plane of the envelope (*left panel*) and a plane containing the l.o.s. and the outflow axis (*right panel*). We assume an edge-on equatorial plane. The dotted lines indicate concentric circles at steps of 10% of the envelope radius (0.''16) around the protostar at the (0,0) position.

which naturally provides the background radiation being amplified. The red features are projected closer to the protostellar position and background source may be a compact HII region and/or an ionized wind with a size smaller than the extent of methanol masers (0.''14). Alternately, the elongation of red features could be explained by the larger densities expected in the equatorial plane (a disk?), providing longer amplification path for the maser radiation.

We can now use the best-fit parameters for the sphere radius and infall velocity derived from our model, to estimate a mass infall rate, $\dot{M}_{\rm inf} = 4\pi R^2 n_{H_2} m_{H_2} V_{\rm inf}$, and an infall momentum rate, $\dot{P}_{\rm inf} = \dot{M}_{\rm inf} V_{\rm inf}$. For $R = 290$ AU and $V_{\rm inf} = 5$ km s^{-1}, we obtain: $\dot{M}_{\rm inf} = 6 \times 10^{-4} \, n_8 \, M_\odot$ yr^{-1} and $\dot{P}_{\rm inf} = 3 \times 10^{-3} \, n_8 \, M_\odot$ km s^{-1} yr^{-1}. On the one hand, the high infall rate might indicate that the protostar is actively accreting. On the other hand, a ratio of 0.27 for the mass-loss to the infall rate indicates that the outflow can efficiently remove mass (and angular momentum) from the system, as expected from magneto-centrifugal ejection. We caution however that the estimated mass rates are accurate only to within 1–2 orders of magnitude, owing to the wide range of densities required for maser excitation (Kaufman & Neufeld 1996; Cragg *et al.* 2005). Nevertheless, our measurements show that the two maser species used to derive the mass-loss (H$_2$O) and the mass-infall (CH$_3$OH) rates are distributed across a similar area, making plausible the hypothesis that they may be excited in molecular gas with similar densities, (10^8 cm^{-3}). Hence, the dependence of the outflow/infall rate ratio on gas density should be less critical.

We estimated 4 $M_\odot$ gas mass for the infalling envelope, consistent with the value estimated from dust continuum emission (3 $M_\odot$; Zhang *et al.* 2007). Hence, AFGL 5142 MM–1 may be the lowest mass protostar known to be associated with Class II methanol masers. We argue that accretion onto a 4 $M_\odot$ protostar may explain the bolometric luminosity of the region and possibly provide the strong IR background to pump the methanol masers. We define the accretion luminosity as $L_{\rm acc} = G\dot{M}_{\rm inf} M_*/R_*$, where $\dot{M}_{\rm inf}$ is the mass infall rate, and M_* and R_* are the mass and radius of the protostar. Our infall model provides an estimate for $\dot{M}_{\rm inf}$ and M_*, while R_* can be derived from the mass-radius relation in protostellar models: $R_* = 8 \, R_\odot$ for $M_* = 4 \, M_\odot$ and $\dot{M}_{\rm inf} = 10^4 \, M_\odot$ yr^{-1} (Palla & Stahler 1993). We then derive: $L_{\rm acc} \sim 10^4 \, n_8 \, L_\odot$, comparable with the bolometric luminosity of the region (Zhang *et al.* 2007). Alternatively, our analysis may provide only a lower limit to the protostellar mass, as suggested by the powerful outflow activity, evidenced by strong water maser and radio continuum emission, and the large mass of the CO outflow ($3\ M_\odot$). In fact, our infall model considers only gravity but likely non-gravitational (e.g., centrifugal and magnetic) forces can influence gas dynamics and possibly lead to underestimates of the central object mass. Future interferometric observations of thermal dense gas tracers at sub-arcsecond resolution (with ALMA) may help to resolve the accretion disk around MM–1 and to derive a rotation curve, which would give a robust estimate of the protostellar mass.

References

Beltrán, M. T., Cesaroni, R., Neri, R., & Codella, C. 2011, *A&A*, 525, A151

Cragg, D. M., Sobolev, A. M., & Godfrey, P. D. 2005, *MNRAS*, 360, 533

Goddi, C. & Moscadelli, L. 2006, *A&A*, 447, 577

Goddi, C., Moscadelli, L., Sanna, A., Cesaroni, R., & Minier, V. 2007, *A&A*, 461, 1027

Goddi, C., Moscadelli, L., & Sanna, A. 2011, *A&A* (Letters), 535, L8

Kaufman, M. J. & Neufeld, D. A. 1996, *ApJ*, 456, 250

Lee, C.-F., Mao, Y.-Y., & Reipurth, B. 2009, *ApJ*, 694, 1395

Palau, A., Fuente, A., Girart, J. M., *et al.* 2011, *ApJ* (Letters), 743, L32

Palla, F. & Stahler, S. W. 1993, *ApJ*, 418, 414

Reid, M. J., Menten, K. M., Zheng, X. W., *et al.* 2009, *ApJ*, 700, 137

Sanna, A., Moscadelli, L., Cesaroni, R., *et al.* 2010, *A&A*, 517, A71

Zhang, Q., Hunter, T. R., Beuther, H., *et al.* 2007, *ApJ*, 658, 1152

Cosmic Masers - from OH to H_0
Proceedings IAU Symposium No. 287, 2012
R.S. Booth, E.M.L. Humphreys & W.H.T. Vlemmings, eds.

© International Astronomical Union 2012
doi:10.1017/S1743921312007387

Trigonometric Parallax of the Protoplanetary Nebula OH 231.8+4.2

Y. K. Choi[1]*, A. Brunthaler[1], K. M. Menten[1] and M. J. Reid[2]

[1] Max-Planck-Institut für Radioastronomie, Auf dem Hügel 69, 53121 Bonn, Germany
*email: ykchoi@mpifr-bonn.mpg.de
[2] Harvard-Smithsonian Center for Astrophysics, 60 Garden Street, Cambridge, MA 02138, USA

Abstract. We report a trigonometric parallax measurement for the H_2O masers around the protoplanetary nebula OH 231.8+4.2 carried out with the Very Long Baseline Array (VLBA). Based on astrometric monitoring for 1 year, we measured a parallax of 0.65 ± 0.01 mas, corresponding to a distance of 1.54 $^{+0.02}_{-0.01}$ kpc. The spatial distribution of H_2O masers is consistent with that found in the previous studies. After removing the average proper motion of 1.4 mas yr^{-1}, corresponding to 10 km s^{-1}, the internal motions of the H_2O maser spots indicate a bipolar outflow.

Keywords. masers, astrometry, stars: late-type, stars: distances, stars: individual (OH 231.8+4.2)

1. Introduction

A protoplanetary nebula (PPN) is in a transition phase from an asymptotic giant branch (AGB) star to a planetary nebula (PN). Compared with the spherically symmetric circumstellar envelopes around AGB stars, PPNe and PNe show asymmetries, frequently clear bipolarity, and collimated winds. The evolution of post-AGB objects is still not well understood. Several models have postulated the presence of rings or disks close to the post-AGB stars. The accretion of material from such structures might create a post-AGB jet and the interaction between such a jet and the remnant AGB envelope produces shocks in the PPN lobes. Since H_2O and SiO masers probe the structure and kinematics of the inner nebular region, they are good tools to understand the formation of bipolar PPNe.

OH 231.8+4.2 is a well studied PPN, showing H_2O and SiO maser emission. The central source is a binary system, composed of an M9-10 III Mira variable (i.e. an AGB star) and an A0 main-sequence companion (Sánchez Contreras *et al.* 2004). This bipolar nebula shows signs of post-AGB evolution: fast bipolar outflows with velocities 200–400 km s^{-1}, shock-excited gas and shock-induced chemistry (Morris *et al.* 1987).

Previous VLBI phase referencing observations (Desmurs *et al.* 2007) successfully determined the absolute positions of the H_2O and SiO ($v = 2, J = 1 - 0$) masers. The H_2O maser clumps are detected in two regions separated by $\sim$ 60 mas nearly in the north-south direction, suggestive of a bipolar outflow. The northern region has blue-shifted emission and the southern red-shifted emission, relative to the central star's velocity. The SiO masers, most likely indicating the position of the Mira component of the binary system, are located between the two H_2O maser regions. The distribution of the SiO masers in OH 231.8 is elongated perpendicular to the nebular axis, suggesting the presence of an equatorial torus or disk around the central star.

Measuring trigonometric parallaxes of PPNe is crucial for determining their fundamental properties of luminosity, mass-loss rate, age, and initial mass. Thanks to the VLBI technique, distance measurements via trigonometric parallaxes have become possible even beyond several kpc. Here, we present the results from our multi-epoch phase-referencing VLBI observations of the H_2O masers toward OH231.8+4.2.

2. Observations

We have conducted Very Long Baseline Array (VLBA) observations to study the H_2O masers at a rest frequency of 22.235080 GHz toward the PPN OH 231.8+4.2. In order to measure a trigonometric parallax, we used phase-referencing observations by fast switching between the maser target and the extragalactic continuum source, J0746-1555, separated by 1.55 degrees from OH 231.8+4.2. The VLBA observations were scheduled under program BC188 at four epochs: 2009 May 01, Oct 19, Nov 09 and 2010 May 01. These dates were optimized to get the maximum and minimum of the parallax signal in right ascension.

We placed a strong source, J0530+1331, near the beginning, middle, and end of the observations to monitor delay and clock offsets. In order to calibrate atmospheric delays, we placed geodetic blocks before and after our phase-referencing observations (Reid *et al.* 2009).

The data were correlated in two passes with the VLBA correlator in Socorro, NM. The four dual-polarized frequency bands of 8 MHz bandwidth were processed with 16 spectral channels for each frequency band. The one (dual-polarized) frequency band containing maser emission was processed with 1024 channels, giving a velocity resolution of 0.10 km s^{-1}. The data reduction was performed with the NRAO AIPS package and ParselTongue scripts.

3. Results

3.1. *Spatial distribution of maser spots*

Fig. 1 (left) shows the 22 GHz H_2O maser spectrum toward OH231.8+4.2 on 2010 May 01 produced by scalar averaging the data over all times and baselines. The H_2O masers span LSR velocities ranging from 24 to 45 km s^{-1}. Though there are variations in the

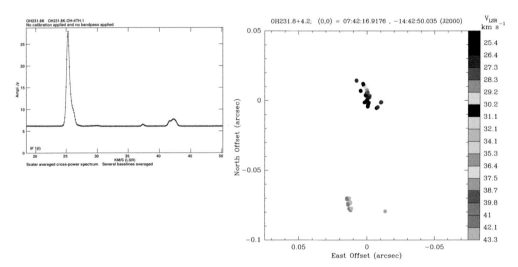

Figure 1. Spectrum and spatial distribution of the H_2O masers toward OH231.8+4.2. Left panel: the spectrum on 2010 May 01 produced by scalar averaging the data over all times and baselines. Right panel: the spatial distribution on 2010 May 01. The LSR velocity of the maser spots is indicated by the color bar to the right.

flux density, the overall structures of the maser spectrum are common to all epochs, indicating that the maser spots survive over the observing period of 1 year. We selected a stable and strong maser spot at an LSR velocity of 25.42 km s^{-1} as phase-reference.

The spatial distribution of H_2O masers toward OH231.8+4.2 relative to the reference maser spot at the LSR velocity of 25.42 km s^{-1} on 2010 May 01 is shown in the Fig. 1 (right). The H_2O maser spots are located within 40 mas in right ascension and 100 mas in declination. Compared with the previous study by Desmurs *et al.* (2007), the red-shifted components in south and the blue-shifted components in north indicate an outflow along the north-south direction.

3.2. *Parallax measurements*

Fig. 2 shows position measurements of the H_2O maser component at the LSR velocity of 25.42 km s^{-1} relative to the background continuum source J0746-1555 over 1 year. The position offsets are with respect to α(J2000.0)= 07^{h}42^{m}16^{s}9176 and δ(J2000.0)= $-14°42'50''035$. Assuming that the movements of the maser features are composed of a linear motion and the annual parallax, we obtain a proper motion and an annual parallax by least-squares fitting. We measured the parallax of OH 231.8+4.2 to be 0.65 ± 0.01 mas, corresponding to a distance of 1.54 $^{+0.02}_{-0.01}$ kpc. This is the first high-accuracy distance that has been determined for OH 231.8+4.2. The absolute proper motion for this maser spot is μ_x = -4.35 ± 0.02 mas yr^{-1} toward the east and μ_y = 0.71 ± 0.59 mas yr^{-1} toward the north.

There are two more spots detected at all four epochs after phase-referencing. We estimated parallaxes and proper motions for those maser spots and list these in table 1. The results are consistent with each other. While the parallaxes should be identical within measurement uncertainties, the proper motions are expected to vary among the spots because of their internal motions.

3.3. *Internal motion of maser spots*

We also calculated the internal motions of the maser spots in right ascension and declination by a least-square fit for linear motion with respect to the reference spot. After

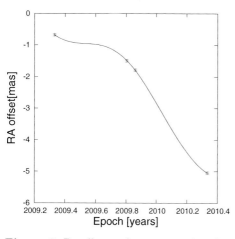 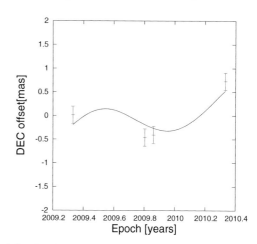

Figure 2. Parallax and proper motion data and fits for OH 231.8+4.2. Points with error bars are position measurements of the H_2O maser spot at an LSR velocity of 25.42 km s^{-1} relative to background quasar J0746-1555. Solid lines represent the best-fit model with an annual parallax and linear proper motions. Left panel: the change of position in right ascension as a function of time (year). Right panel: the same as the left panel in declination.

Table 1. Parallax and proper motions fits

LSR velocity (km s^{-1})	parallax (mas)	μ_x (mas yr^{-1})	μ_y (mas yr^{-1})
25.42	0.65 ± 0.01	-4.35 ± 0.02	0.71 ± 0.59
37.32	0.65 ± 0.02	-4.94 ± 0.07	-1.65 ± 0.02
41.85	0.66 ± 0.03	-4.05 ± 0.11	-1.75 ± 0.07
Combined	0.65 ± 0.01	-4.45 ± 0.14	-0.89 ± 0.45

removing the average relative motion of all maser spots of -0.136 mas yr^{-1} eastward and -1.43 mas yr^{-1} northward, the remaining internal motion vectors are shown in Fig. 3 and suggest a bipolar outflow. The blue-shifted components show an average motion of about 11 km s^{-1} to the north, and the red-shifted components show an average motion of about 7–13 km s^{-1} to the south for a distance of 1.5 kpc. Thus, the separation velocity between two components is about 20 km s^{-1}, which is consistent with the result reported by Leal-Ferreira *et al.* (2012).

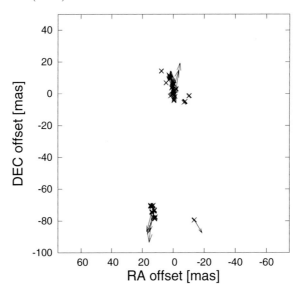

Figure 3. The internal motion of the H$_2$O maser spots toward OH 231.8+4.2 after removing the average relative motion with respect to the reference maser spot. Absolute coordinates of the map origin are α(J2000.0)= $07^{\rm h}42^{\rm m}16^{\rm s}9176$ and δ(J2000.0)= $-14°42'50''035$.

Acknowledgements

This work has been supported by the ERC Advanced Grant GLOSTAR under grant agreement no. 247078.

References

Desmurs *et al.* 2007, *A&A*, 468, 189
Leal-Ferreira *et al.* 2012, *A&A*, 540, A42
Morris *et al.* 1987, *ApJ*, 321, 888
Reid *et al.* 2009, *ApJ*, 693, 397
Sánchez Contreras *et al.* 2004, *ApJ*, 616, 519

Cosmic Masers - from OH to H_0
Proceedings IAU Symposium No. 287, 2012
R.S. Booth, E.M.L. Humphreys & W.H.T. Vlemmings, eds.

© International Astronomical Union 2012
doi:10.1017/S1743921312007399

Astrometry of water masers in post-AGB stars

Hiroshi Imai[1] and VERA collaboration[1,2]

[1]Department of Physics and Astronomy, Graduate School of Science and Engineering,
Kagoshima University, 1-21-35 Korimoto, Kagoshima 890-0065, Japan
email: hiroimai@sci.kagoshima-u.ac.jp

[2]Mizusawa VLBI Observatory, National Astronomical Observatory of Japan,
2-12 Hoshigaoka, Mizusawa-Ku, Oshu-shi, Iwate 023-0861, Japan

Abstract. Measurements of trigonometric parallaxes and secular motions of evolved stars, especially post-AGB stars including central objects of planetary nebulae and water fountain sources as well as peculiar or unclassified stars, provide unambiguous source distance scales and information on their orbits in the Milky Way Galaxy. True source luminosities and kinematical properties should lead us to elucidate the true characteristics and evolutional tracks of these stars. Here we present the recent results of astrometry towards H_2O maser sources with the VLBI Exploration of Radio Astrometry (VERA). The target sources include a planetary nebula (K3−35), a pre-PN (IRAS 19312+1950), a water fountain (IRAS 18286−0959) and a K-type star (IRAS 22480+6002). We have demonstrated that parental stars of the former three sources should be intermediate-mass stars from their luminosities and orbits in the Milky Way. It is suggested that IRAS 22480+6002 should be a K-type supergiant previously suggested rather than an RV Tau variable star.

Keywords. masers, stars:distances, individuals (K3−35, IRAS 19312+1950, IRAS 18286−0959, IRAS 22480+6002)

The dual-beam receiver system of the VERA telescopes enables us to simultaneously track a Galactic H_2O maser source and an position-reference quasar, yielding sub-milliarcsecond-level astrometry between the two sources separated by $0.3°$–$2.2°$. In spite of high time variability of the H_2O masers, a few maser spots survive for one year or longer, enabling us to monitor their motions to measure their trigonometric parallaxes and linear proper motions.

We have conducted astrometric VLBI observations of H_2O masers in four post-AGB stars with VERA. Some of the results were published by Imai et al. (2011) and Tafoya et al. (2011)). Fig. 1 shows one of the results of the astrometric observations. Table 1 gives astrometric and Galactic kinematical parameters of these sources derived from the astrometric results. These sources harbor bipolar morphology of PNe or "water fountains" (WFs). From the VERA astrometry, we have learnt the following.

1. Stellar luminosities derived from the spectral energy distributions and the trigonometric parallax distances are higher than 10000 $L_\odot$ for IRAS 18286−0959 and IRAS 19312+1950 and 35000 $L_\odot$ for IRAS 22480+6002. Taking into account the size of the H_2O maser distribution (>100 AU) as well, IRAS 22480+6002 should be a K-type supergiant rather than an RV Tau variable star.

2. Large deviations from the Galactic circular rotation, >50 km s^{-1} for IRAS 18286−0959, implies kinematical property different from that of massive young stellar objects in the Galactic thin disk. These imply that the central stars may be intermediate-mass post AGB stars. Our recent sub-mm CO line observations also support this mass estimation (c.f. IRAS 16342−3814, Imai et al. 2012).

Table 1. Source parameters derived from the VERA astrometry.

Object	Type	D (kpc)	$R_{\rm Gal}$† (kpc)	z‡ (pc)	V_R¶ (km s^{-1})	V_θ¶ (km s^{-1})	V_z¶ (km s^{-1})	Reference
K3$-$35	PN	$3.9^{+0.7}_{-0.5}$	$7.11^{+0.08}_{-0.06}$	140^{+25}_{-18}	33 ± 16	233 ± 11	11 ± 2	Tafoya et $al.$ (2011)
IRAS 19312+1950	Pre-PN	$3.8^{+0.8}_{-0.6}$	7.07 ± 0.12	28 ± 3	33 ± 28	214 ± 4	-14 ± 8	Imai et $al.$ (2011)
IRAS 18286$-$0959	WF	$3.9^{+1.1}_{-0.7}$	4.93 ± 0.69	22 ± 2	64 ± 30	133 ± 36	-17 ± 31	Imai et $al.$ in prep.
IRAS 22480+6002	K-type	$2.50^{+0.28}_{-0.23}$	9.10 ± 0.14	55 ± 4	203 ± 5	55 ± 4	4 ± 1	Imai et $al.$ in prep.

†Galactocentric distance to the object.   ‡Height from the Galactic midplane.
¶V_R, V_θ, and V_z are the secular motion vector components in the Galactic radial, azimuthal, and rotation axis directions, respectively.

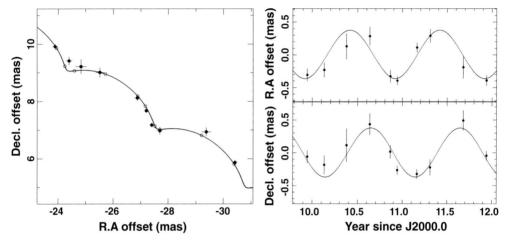

Figure 1. H$_2$O maser spot position in IRAS 22480+6002 and the kinematical model for the spot's motion. *Left* R.A. and decl. offsets of the maser spot observed (a filled circle) and model-predicted (an opened circle). A solid curve shows the modeled motion including an annual parallax and a constant velocity proper motion. *Right* R.A. and decl. variations of the spot position with time. The estimated linear proper motion is subtracted from the observed spot position.

3. Accuracy of kinematical parameters of the maser sources depend on accuracy of the relative motion of the position reference maser feature, whose annual parallax and secular motion is measured, with respect to the systemic motion of the maser source ($\sigma \sim$10 km s^{-1}), except for water fountain sources whose jet velocities are higher than several kilometers per second.

References

Imai, H., Chong, S.-N., He, J.-H., Nakashima, J., Hsia, C.-H., Sakai, T., Deguchi, S., & Koning, N. 2012, *PASJ*, submitted
Imai, H., Tafoya, D., Honma, M., Hirota, T., & Miyaji, T. 2011, *PASJ*, 63, 87
Tafoya, D., Imai, H., Gómez, Y. Torrelles, J. M., Patel, N. A., Anglada, G., Miranda, L. F., Honma M., Hirota, T., & Miyaji, T. 2011, *PASJ*, 63, 71

Cosmic Masers - from OH to H$_0$
Proceedings IAU Symposium No. 287, 2012
R.S. Booth, E.M.L. Humphreys & W.H.T. Vlemmings, eds.

© International Astronomical Union 2012
doi:10.1017/S1743921312007405

Relative parallaxes in the massive star forming region W33

K. Immer[1,2], M. J. Reid[2] and K. M. Menten[1]

[1] Max-Planck-Institut für Radioastronomie, Auf dem Hügel 69, 53121 Bonn, Germany
email: kimmer@mpifr-bonn.mpg.de

[2] Harvard-Smithsonian Center for Astrophysics, 60 Garden Street, 02138 Cambridge, MA, USA

Abstract. The massive star forming complex W33 contains several molecular clouds at different stages of star formation activity, ranging from quiescent to highly active clouds. Our trigonometric parallax observations of water masers in this complex, conducted with the VLBA at 22.2 GHz, show that all water masers have the same distance of 2.4 kpc, locating the W33 complex in the Sagittarius spiral arm.

Keywords. masers, astrometry, stars: distances, stars: formation, Galaxy: structure

1. Background

W33 is a massive star forming complex, containing objects in various stages of evolution, from quiescent infrared-dark clouds (G012.86−0.27, G012.85−0.23) to highly active, infrared-bright regions (G012.81−0.20) (see Fig. 1). The peculiar kinematic structure of W33 (two different velocity groups at ∼36 and ∼60 km/s spread over distinct parts of the region) was discovered via spectral line observations (Bieging *et al.* 1978, Goss *et al.* 1978, Bieging *et al.* 1982). This was explained by either one connected star forming region with large internal motions at a near kinematic distance of 4 kpc, or a superposition of several independent star forming regions arranged along the line of sight.

2. Observations

As part of the BeSSeL project (Brunthaler *et al.* 2011), we performed trigonometric parallax observations of the four water masers G012.68−0.18, G012.81−0.20, G012.90−

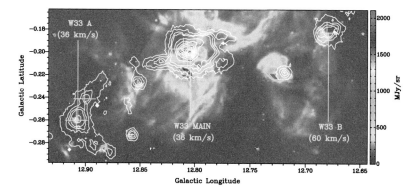

Figure 1: The W33 complex. Background image: 8 μm emission from the GLIMPSE survey. Contours: 870 μm dust emission from the ATLASGAL survey (contour levels: 4% to 10% in steps of 2% and 10% to 100% in steps of 10%). Crosses: Water masers.

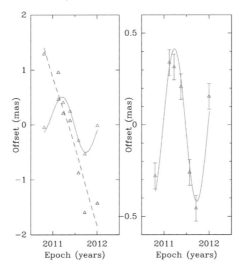

Figure 2: Position measurements of G012.68−0.18 relative to the background quasar J1825−1718. Left panel: East (continuous line) and North (dashed line) position offsets with parallax and proper motion fits versus time. Right Panel: Right ascension parallax fit of 0.416 ± 0.028 mas with the best-fit proper motions removed.

Maser	V_{peak} (km/s)	Relative Parallax (mas)	Distance (kpc)	
G012.68−0.18	61	+0.416* ± 0.028	2.4 ± 0.2	
G012.81−0.20	35, −3	+0.054 ± 0.031	2.1 ± 0.2	* Absolute parallax.
G012.90−0.24	35	−0.055 ± 0.018	2.8 ± 0.2	
G012.91−0.26	35	+0.006 ± 0.030	2.4 ± 0.2	

Table 1: Relative parallaxes and distances of the water masers in W33.

0.24, G012.91−0.26 with the Very Large Baseline Array (VLBA) at 22.2 GHz to determine their distances and proper motions. The observations were split in eight epochs from 2010 October to 2012 January to well-sample the right ascension parallax signatures in time. The position measurements of G012.68−0.18 are made relative to the background quasars J1825−1718 and J1808−1822 (see Fig. 2). The other three water masers are too weak to be used as phase reference sources and were phase referenced to G012.68−0.18, yielding relative parallaxes.

3. Results

The relative parallaxes of G012.81−0.20, G012.90−0.24 and G012.91−0.26 are consistent with zero (± 0.06 mas) (see Table 1). Thus, these water masers are located at the same distance as G012.68−0.18 and W33 is one connected star forming complex. Furthermore, we determined the distance to W33 to be 2.4 kpc, almost half the kinematic distance, locating W33 in the Sagittarius spiral arm and not in the Scutum-Crux arm.

References

Bieging, J. H., Pankonin, V., & Smith, L. F. 1978, *A&A*, 64, 341
Bieging, J. H., Wilson, T. L., & Downes, D. 1982, *A&AS*, 49, 607
Brunthaler, A., *et al.* 2011, *Astronomische Nachrichten*, 332, 461
Goss, W. M., Matthews, H. E., & Winnberg, A. 1978, *A&A*, 65, 307

Cosmic Masers - from OH to H_0
Proceedings IAU Symposium No. 287, 2012
R.S. Booth, E.M.L. Humphreys & W.H.T. Vlemmings, eds.
© International Astronomical Union 2012
doi:10.1017/S1743921312007417

3-Dimensional kinematics of water/SiO masers in Orion-KL

Mi Kyoung KIM, Tomoya Hirota, Kobayashi Hideyuki and VERA project team members

National Astronomical Observatory of Japan,
2-21-1, Osawa, Mitaka, Tokyo, Japan

Abstract. Results of multi-epoch VLBI observations toward water/SiO masers in Orion-KL are presented. We conducted high-resolution VLBI observations of water/SiO masers with VERA to probe the structure and the kinematics of the disk/outflow in Orion-KL. The VERA observations provide the positions and proper motions of masers features in Orion-KL with the highest accuracy ever observed. The results of water and SiO maser observations suggest that Source I is a massive YSO with an accretion disk and a collimated outflow.

Keywords. masers, stars:formation

1. Introduction

It is known that there are two kind of outflows in Orion-KL: a wide-angle, northwest-southeast high-velocity outflow, and a northeast-southwest low-velocity outflow. Although the explosive outflows have been studied by a number of observations (see Genzel & Stutzki (1989) for a review), the origins of outflows and the properties of the driving sources remain debated until now. To investigate the nature of the outflows and its driving source, it is essential to study the 3-dimensional motion of gas associated with the outflows in Orion-KL.

The bright water and SiO masers are detected in Orion-KL, and the masers are a good tracers of an outflow and a disk surrounding YSOs. Thus, we conducted high-resolution VLBI observations of water and SiO masers in Orion-KL with VERA to probe the structure and the kinematics of the disk/outflow in this region. 22.2 GHz water masers are observed 6 times during 2005-2006, and simultaneous observations for $v=1, 2$ $j=1-0$ SiO masers were done 11 times during 2006-2008. In all observations, Orion-KL and J0541-0541 were observed simultaneously in dual-beam mode for phase-referencing observation. Typical beam size was 1.7 ×0.9 mas for water maser, and 0.8 ×0.4 mas for SiO masers. The velocity resolution for both maser line was 0.21 km/s.

2. Results and Discussion

For both water and SiO $v=1, 2$ $J=1-0$ masers, we made multi-epoch maps and measured the absolute proper motions of maser features. SiO masers are confined to four arms of ~200 × 200 mas in an X-shape distribution, and water masers are distributed in a ~20 × 20 arcsec region along the low-velocity outflow in Orion-KL. To obtain the absolute proper motions, we measured the proper motions relative to the reference spot, then we added the absolute proper motion of the reference spot to all measured relative proper motions. For the absolute proper motions of the reference spots, we referred to the values from Hirota *et al.* (2007) for the water maser and Kim *et al.* (2008) For the SiO $v=2$ $J=1-0$ maser. For the SiO $v=1$ $J=1-0$ maser, we measured an absolute proper

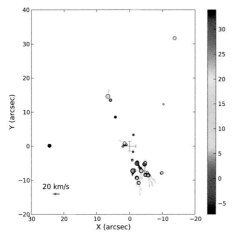

Figure 1. The positions and the absolute proper motions of water masers. The blue cross represents the position of the center from fitting result. The positions of YSOs, I, n and BN are marked by blue dots.

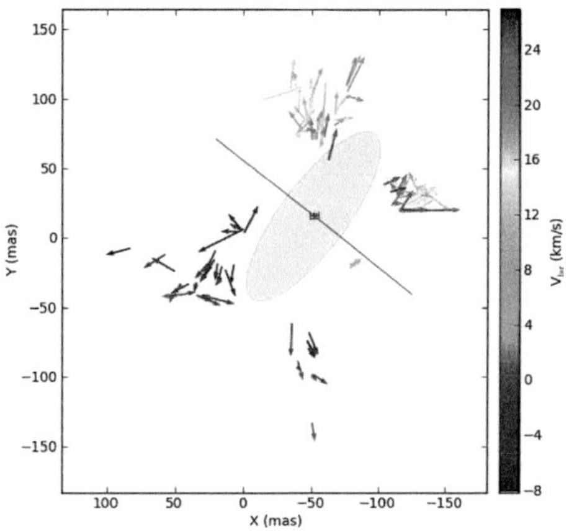

Figure 2. The positions and the internal motions of SiO masers. The proper motions of Source I is subtracted from the absolute proper motions of each features. The results of fitting are marked: The position of center is marked by the blue cross, the grey line represents the axis of rotation. The aspect ratio of the ellipse is consistent with the inclination of 19.39°.

motion with the annual parallax of 2.39 mas (Kim *et al.* 2008). The results are shown in Fig. 1, 2.

Then, we fitted the observed data $(V_{x,i}, V_{y,i}, V_{z,i})$, (X_i, Y_i) to the kinematic models by the least-square fitting method. For the SiO masers, we supposed an expanding, rotating velocity field V_{obs}=(Keplerian rotation)+(expansion)+(systemic motion), because the V_{los} gradient of ~-0.5 km/s/AU along arms and the existence of system velocity features between arms can be explained by rotating motions (Kim *et al.* 2008, Matthews *et al.* 2010). For water masers, we simply suppose that the water masers are moving outward from a center of expansion, V_{obs}=(expansion)+(systemic motion), since their motions show no tendency of acceleration or rotation.

The fitting results show that: 1) the SiO masers arise from rotating, expanding material around Source I, possibly on the surface of an accretion disk. Its rotating velocity is ~24 km/s at 25 AU from Source I and the estimated central mass is M ~14 ± 1 M$_\odot$. 2) the estimated position and proper motion of center of expansion of water masers are consistent with the position and the proper motion of Source I. Water masers trace the outflow from Source I with the expanding velocity of 22 km/s, which is perpendicular to the disk traced by SiO masers.

References

Genzel, R. & Stutzki, J. 1989, *ARAA*, 27, 41–85
Hirota, T., *et al.* 2007, *PASJ*, 59, 897
Kim, M., *et al.* 2008, *PASJ*, 60, 991
Matthews, L. D., *et al.* 2010, *ApJ*, 708, 80

Cosmic Masers - from OH to H$_0$
Proceedings IAU Symposium No. 287, 2012
R.S. Booth, E.M.L. Humphreys & W.H.T. Vlemmings, eds.

© International Astronomical Union 2012
doi:10.1017/S1743921312007429

Annual Parallax Measurements of an Infrared Dark Cloud MSXDC G034.43+00.24

Tomoharu Kurayama

Department of Physics and Astronomy, Graduate School of Science and Engineering,
Kagoshima University, 1-21-35 Korimoto, Kagoshima 890-0065, Japan
email: kurayama@sci.kagoshima-u.ac.jp

Abstract. We have measured the annual parallax of the H$_2$O maser associated with an infrared dark cloud, MSXDC G034.43+00.24, with VERA. The parallax is 0.643±0.049 mas, corresponding to a distance of $1.56^{+0.12}_{-0.11}$ kpc. This value is less than half of the previous kinematic distance of 3.7 kpc. We revised the core-mass estimates of MSXDC G034.43+00.24 from the previous estimations of $1000 M_\odot$ to hundreds of $M_\odot$. The spectral type is still consistent with that of the massive star.

Keywords. astrometry, IRDC, star formation

1. Introduction

MSXDC G034.43+00.24 is one of the Infrared Dark Clouds (IRDCs), which are known as sites of massive star formation. Various parameters are derived for IRDCs (e.g., Sanhueza *et al.* (2010)). MSXDC G034.43+00.24 has four millimeter cores from 1.2 mm continuum observations (Rathborne *et al.* (2005)). An IRAS point source and a compact H II region are associated with the brightest millimeter core MM2. Three millimeter cores MM1, MM3 and MM4 have H$_2$O maser sources revealed from VLA observations (Wang *et al.* (2006)). The kinematic distance of MSXDC G034.43+00.24 is 3.7 kpc (Simon *et al.* (2006)), which is the only one distance estimation so far.

Various parameters, such as masses, luminosities and spectral types are deived from various observations (e.g., Sanhueza *et al.* (2010)), but many of them depend on the distance. However, some kinematic distances are far from the actual distances, derived from the annual parallaxes (e.g., Sato *et al.* (2010), Motogi *et al.* (2011)). It is not good to use kinematic distance for each source. Therefore, we have measured the annual parallax of this source with VERA.

2. Observations, Results and Discussion

Observations were carried out with VERA four stations. They were phase-referencing VLBI observations at 22 GHz band. The phase-reference source is GPSR5 35.946+0.379 (= VCS2 J1855+0251), which is separated by 1°.6 from MSXDC G034.43+00.24.

The results of position measurements are shown in Figure 1. We can trace the movements on the sky for five maser features in MM1. Declination data have large scatters caused by the strong sidelobes in the synthesized beam because the declination of this source ($\sim +1°.5$) is very close to 0°.

We carried out the least square fitting for the annual parallax, using the data points of right ascensions only. The resultant annual parallax is 0.643 ± 0.049 mas, which corresponds to the distance of $1.56^{+0.12}_{-0.11}$ kpc. This value is less than the half of the kinematic distance of 3.7 kpc. The details of this fitting is described in Kurayama *et al.* (2011).

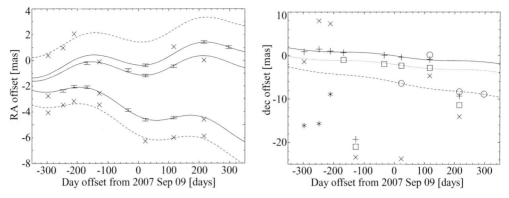

Figure 1. (Left) Plot of right ascensions versus time. Each line shows the different maser features. Crosses denotes the data which are not used for the least square fitting to derive the annual parallax. (Right) Plot of declinations versus time. Different signs denote different maser features. Lines are drawn from the least square fitting of the proper motion and the initial position by using the annual parallax of 0.643 ± 0.049 mas.

Millimeter core	Kinematic distance (3.7 kpc)				Distance from annual parallax (1.56 kpc)			
	MM1	MM2	MM3	MM4	MM1	MM2	MM3	MM4
Virial mass	1130	1510	1370	. . .	476	637	578	. . .
LTE mass	330	1460	. . .	. . .	59	260	. . .	. . .
Dust mass	1187	1284	301	253	211	228	54	45
Bolometric luminosity	32000	. . .	9000	12000	5700	. . .	1600	2100
Spectral type	O9.5	. . .	B0.5	B0.5	B1	. . .	B3	B2

Table 1. Modification of physical parameters of millimeter cores in the infrared dark cloud, MSXDC G034.43+00.24. The units of masses and luminosities are solar masses and solar luminosities. Data from Sanhueza *et al.* (2010), Rathborne, Jackson & Simon (2006), and Rathborne *et al.* (2005).

Distance is very fundamental parameter, so variours parameters have changed by the change of distances. Table 1 shows examples of this change. Masses are reduced from $\sim 1000 M_{\odot}$ to hundreds $M_{\odot}$. Spectral types still shows that they will be massive stars.

References

Kurayama, T., Nakagawa, A., Sawada-Satoh, S., Sato, K., Honma, M., Sunada, K., Hirota, T., & Imai, H. 2011, *PASJ*, 63, 513

Motogi, K., Sorai, K., Habe, A., Honma, M., Kobayashi, H., & Sato, K. 2011, *PASJ*, 62, 101

Rathborne, J. M., Jackson, J. M., & Simon, R. 2006, *ApJ*, 641, 389

Rathborne, J. M., Jackson, J. M., Chambers, E. T., Simon, R., Shipman, R., & Frieswijk, W. 2005, *ApJ*, 630, L181

Sanhueza, P., Garay, G., Bronfman, L., Mardones, D., May, J., & Saito, M. 2010, *ApJ*, 715, 18

Sato, M., Hirota, T., Reid, M. J., Honma, M., Kobayashi, H., Iwadate, K., Miyaji, T., & Shibata, K. M. 2010, *PASJ*, 62, 287

Simon, R., Rathborne, J. M., Shah, R. Y., Jackson, J. M., & Chambers, E. T. 2006, *ApJ*, 653, 1325

Wang, Y., Zhang, Q., Rathborne, J. M., Jackson, J., & Wu, Y. 2006, *ApJ*, 651, L125

Cosmic Masers - from OH to H_0
Proceedings IAU Symposium No. 287, 2012
R.S. Booth, E.M.L. Humphreys & W.H.T. Vlemmings, eds.

© International Astronomical Union 2012
doi:10.1017/S1743921312007430

The bar effect in the galactic gas motions traced by 6.7 GHz methanol maser sources with VERA

Naoko Matsumoto[1], Mareki Honma[1,2] and VERA project members

[1]Mizusawa VLBI Observatory, National Astronomical Observatory of Japan,
2-21-1 Osawa, Mitaka, Tokyo, Japan
email: naoko.matsumoto@nao.ac.jp

[2]Department of Astronomical Science, The Graduate University of Advanced studies
(SOKENDAI), 2-21-1 Osawa, Mitaka, Tokyo, Japan
email: mareki.honma@nao.ac.jp

Abstract. To establish the existence of the galactic bar, ten methanol maser sources around the starting points of the spiral arms were observed with VERA (VLBI Exploration of radio astrometry) using the phase-referencing technique at 6.7 GHz band. For six out of ten sources, absolute proper motions were obtained with better than 3σ accuracy. Using VLBI 3-D data of eight sources, including our five sources, we compared the observed data with three galactic models and found that the model including the bar effect is better to explain the 3-D data, than a flat circular rotation model. A non-flat circular rotation model is also consistent with the VLBI data. Based on a dynamical model with a bar, we estimate an inclination angle of the bar around $35°$, which is consistent with previous studies.

Keywords. masers, Galaxy: kinematics and dynamics, radio lines: ISM

1. Introduction

VLBI astrometry is one a powerful approach to obtain 3-D gas kinematics. For such astrometric observation, 6.7 GHz methanol maser sources have some advantages as the galactic gas tracer (e.g., large source numbers, long lifetime, and small internal proper motions), excepting the larger angular resolution than H_2O/SiO maser sources. Methanol has the special advantage for galactic bar studies, in that most of the sources concentrate around the molecular ring on the position-velocity map (Pestalozzi *et al.* 2005).

In May 2009, new 6.7 GHz receivers were installed in the VERA stations, and we confirmed a capability of 6.7 GHz astrometric observation with VERA (Matsumoto *et al.* 2011). It shows that parallax measurements using 6.7 GHz methanol maser are currently reached at 2 kpc with VERA. Thus, for far sources around the galactic bar, we need to present proper motions on the sky plane in units of $\mathrm{mas\,yr^{-1}}$, instead of $\mathrm{km\,s^{-1}}$ because the parallax could not be detected.

2. Observations and Result

In Nov. 2009, we started astrometric observations of ten methanol maser sources with VERA. The frequency of the maser line is 6.668518 GHz ($CH_3OH\,5_1 \rightarrow 6_0A^+$). The ten maser sources were selected from Pestalozzi et al. (2005) with following criteria: (1) the galactic longitude of $40°$ or less, (2) declination larger than $-37°$, (3) the kinematic distance from the galactic center is less than 5 kpc, (4) the flux is brighter than 15 Jy or more in Pestalozzi et al. (2005) and detected with VERA, (5) existence of detectable

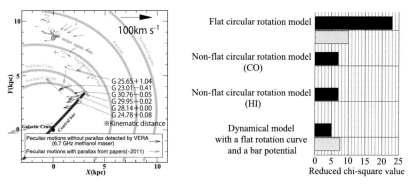

Figure 1. Left: Absolute proper motions of SFRs on the galactic plane at $(R_0, \Theta_0, U_{\odot HIP}, V_{\odot HIP}, W_{\odot HIP}) = (8.5$ kpc, 220 km s^{-1}, 10 km s^{-1}, 5.25 km s^{-1}, 7.17 km s$^{-1})$. Right: Reduced chi-square values of each model via VLBI 3-D data (black), and each model via H I terminal velocity (gray).

phase-reference sources with VERA within $4°$ separation from the target maser sources. We observed the ten sources every few months until May 2011. The typical size of synthesized beam is $\sim 3 \times 6$ - 5×10 mas^2.

In figure 1, the accurate proper motions obtained by our observations are plotted using kinematic distance with other parallax measurement data. Except G 25.65+1.04, all our sources have a similar trend in proper motion, much slower than 220 km s^{-1} galactic rotation

3. Discussion

We compared our observed 3-D data with some kinematic models (the galactic longitude [deg], proper motions for the direction of the galactic longitude [mas yr^{-1}] and the systemic velocities [km s^{-1}]). In this comparison, we also used data of G 23.01−0.41, G 23.44−0.18 and G 23.65−0.127 from Brunthaler *et al.* (2009) and Bartkiewicz *et al.* (2008). These three sources also satisfy our selection criteria.

We found our 3-D data sample cannot be reproduced by a flat rotation model with $\Theta_0 = 220$ km s^{-1}. However, our data are consistent with both existing rotation curves with H I/CO and the dynamical model with a bar (Wada 1994; Sakamoto *et al.* 1999). From the dynamical model and VLBI data, an acceptable parameter of the inclination angle of the Galactic bar was derived as $\sim 35°$. This is consistent with previous studies. Thus, the observed VLBI proper motions can be explained better with the Galactic bar. In time, if more sources can be observed, the H I/CO rotation curve and the dynamical model may be discriminated better.

This work was supported by Grant-in-Aid for JSPS Fellows for Young Scientists (No. 22-171).

References

Pestalozzi, M. R., Minier, V., & Booth, R. S. 2005, *A&A*, 432, 737
Matsumoto, *et al.* 2011, *PASJ*, 63, 1345
Wada, K. 1994, *PASJ*, 46, 165
Sakamoto, K., Okumura, S. K., Ishizuki, S., & Scoville, N. Z. 1999, *ApJS*, 124, 403
Brunthaler, A., *et al.* 2009, *ApJ*, 693, 424
Bartkiewicz, A., *et al.* 2008, *A&A*, 490, 797

Cosmic Masers - from OH to H_0
Proceedings IAU Symposium No. 287, 2012
R.S. Booth, E.M.L. Humphreys & W.H.T. Vlemmings, eds.

© International Astronomical Union 2012
doi:10.1017/S1743921312007442

Mass distribution of the Galaxy with VERA

Nobuyuki Sakai[1], Mareki Honma[1,2], Hiroyuki Nakanishi[3], Hirofumi Sakanoue[3], Tomoharu Kurayama[3] and VERA project members

[1] The Graduate University for Advanced Studies (Sokendai), Mitaka, Tokyo 181-8588, Japan

[2] Mizusawa VLBI Observatory, NAOJ, Mitaka, Tokyo 181-8588

[3] Faculty of Science, Kagoshima University, 1-21-35 Korimoto, Kagoshima, Kagoshima 890-0065
email: nobuyuki.sakai@nao.ac.jp

Abstract. We aim to reveal the mass distribution of the Galaxy based on a precise rotation curve constructed using VERA observations. We have been observing Galactic H_2O masers with VERA. We here report one of the results of VERA for IRAS 05168+3634. The parallax is 0.532 ± 0.053 mas which corresponds to a distance of $1.88^{+0.21}_{-0.17}$ kpc, and the proper motions are ($\mu_\alpha \cos\delta$, μ_δ) = $(0.23 \pm 1.07, -3.14 \pm 0.28)$ mas yr^{-1}. The distance is significantly smaller than the previous distance estimate of 6 kpc based on a kinematic distance. This drastic change places the source in the Perseus arm rather than in the Outer arm. Combination of the distance and the proper motions with the systemic velocity provides a rotation velocity of 227^{+9}_{-11} km s^{-1} at the source assuming $\Theta_0 = 240$ km s^{-1}. The result is marginally slower than the rotation velocity at LSR with $\sim$ 1-σ significance, but consistent with previous VLBI results for six sources in the Perseus arm. We also show the averaged disk peculiar motion over the seven sources in the Perseus arm as $(U_{\mathrm{mean}}, V_{\mathrm{mean}}) = (11 \pm 3, -17 \pm 3)$ km s^{-1}. It suggests that the seven sources in the Perseus arm are systematically moving toward the Galactic center, and lag behind the Galactic rotation with more than 3-σ significance.

Keywords. astrometry — **ISM**:individual (IRAS 05168+3634) — techniques: interferometric — **VERA**

1. Introduction

The rotation curve can be used to estimate mass for spiral galaxies. It has been revealed that there exist plenty of flat rotation curves beyond optical disks in external spiral galaxies. This indicates the existence of large quantities of dark matter in the outer region of galaxies. Today, a well-developed interferometer technique at radio wavelengths can be used to conduct galactic astrometry over a kilo-parsec scale (e.g. VERA, VLBA). To construct a rotation curve of the Milky Way with high accuracy, we have been using VERA to observe Galactic objects with H_2O maser emission in the Galactic-outer region. In this paper, we report the observational results for IRAS 05168+3634.

2. Results & Discussion

Eleven VLBI observations with VERA between October 2009 and May 2011 yielded the parallax and the proper motions for IRAS 05168+3634. The parallax is 0.532 ± 0.053 mas, corresponding to a distance of $1.88 \pm^{0.21}_{0.17}$ kpc. The proper motions are ($\mu_\alpha \cos\delta$, μ_δ) = $(0.23 \pm 1.07, -3.14 \pm 0.28)$ mas yr^{-1}. The distance is significantly smaller than the previous estimate of the kinematic distance, being 6 kpc (Molinari *et al.* 1999). Our result places the source in the Perseus arm rather than in the Outer arm. Combination of the distance and the proper motions with the systemic velocity provides a rotation velocity of 227^{+9}_{-11} km s^{-1} at the source assuming $\Theta_0 = 240$ km s^{-1}. The result shows

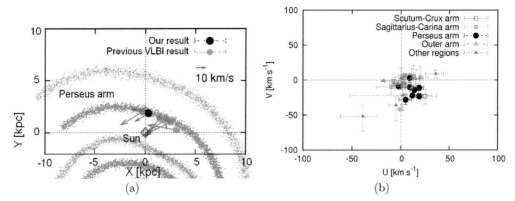

Figure 1. (a) Position of IRAS 05168+3634 based on our observational result (black circle), while the green circles represent previous VLBI results in the Perseus arm. They are superposed on the Galactic-face on image (Georgelin & Georgelin 1976). The solar position is (X, Y) = (0, 0) kpc. The red arrows show the peculiar motions. (b) Each source of Scutum-Crux (open square), Sagittarius-Carina (filled square), Perseus (circle), Outer arms (open triangle), and other regions (filled triangle) is plotted on the peculiar motions plane of U and V for 33 sources based on VLBI observations.

marginally slower rotation with respect to flat Galactic rotation ($\Theta(R) = \Theta_0$). In addition, the slower rotation is almost consistent with previous VLBI results in the Perseus arm. Figure 1a shows the sources in the Perseus arm superposed on the Galactic face-on image (Georgelin & Georgelin 1976). The black circle shows our result, while the green circles represent previous VLBI results. Figure 1b also shows peculiar motions plane of U and V on which previous VLBI results are plotted. U is directed toward the Galactic center, and V is also directed toward the Galactic rotation. Obviously, sources in the Perseus arm (black circle) are systematically located in the lower right region of the U-V plane. It means that the sources in the Perseus arm are systematically moving toward the Galactic center, and lag behind the Galactic rotation. We determine the averaged peculiar motions for seven sources in the Perseus arm as $(U_{\mathrm{mean}}, V_{\mathrm{mean}}) = (11 \pm 3, -17 \pm 3)$ km s^{-1}. The result reveals the peculiar motions in the Perseus arm with more than 3-σ significance. The peculiar motions may be caused at the inner edge of the Perseus arm where a shock front predicted by density-wave theory occurs (Mel'Nik *et al.* 1999).

Acknowledgements

NS acknowledges financial support from the SOKENDAI travel grant for overseas.

References

Georgelin, Y. M. & Georgelin, Y. P. 1976, *A&A*, 49, 57
Mel'Nik, A. M., Dambis, A. K., & Rastorguev, A. S. 1999, *Astron. Lett.*, 25, 518
Molinari, S., Brand, J., Cesaroni, R., & Palla, F. 1996, *A&A*, 308, 573

Cosmic Masers - from OH to H_0
Proceedings IAU Symposium No. 287, 2012
R.S. Booth, E.M.L. Humphreys & W.H.T. Vlemmings, eds.

© International Astronomical Union 2012
doi:10.1017/S1743921312007454

Distance and Maser Outflows of the Galactic Star-forming Region W51 Main/South

Mayumi Sato[1], Mark J. Reid[2], Andreas Brunthaler[1] and Karl M. Menten[1]

[1] Max-Planck-Institut für Radioastronomie,
Auf dem Hügel 69, 53121 Bonn, Germany
email: msato@mpifr-bonn.mpg.de

[2] Harvard-Smithsonian Center for Astrophysics,
60 Garden Street, Cambridge, MA 02138, USA

Abstract. We report on high-resolution astrometry of 22 GHz H_2O maser emission in the Galactic massive star-forming region W51 Main/South using the Very Long Baseline Array. We measured the trigonometric parallax of W51 Main/South to be 0.185 ± 0.010 mas, corresponding to a distance of $5.41^{+0.31}_{-0.28}$ kpc. The H_2O maser emission in W51 Main/South traces four powerful bipolar outflows within a 0.4 pc size region, three of which are associated with dusty molecular hot cores and/or hyper- or ultra-compact HII regions. The maser outflows in W51 Main/South have a relatively small range of internal 3D speeds, suggesting that multiple speed maser outflows in other Galactic massive star-forming regions may come from separate young stellar objects closely spaced on the sky.

Keywords. astrometry, masers, stars: formation, ISM: jets and outflows

1. Observations and Results

W51 Main/South is a well-studied massive star-forming region that hosts one of the strongest H_2O maser sources in the Galaxy. We conducted phase-referencing observations of the 22 GHz H_2O maser line in W51 Main/South using the Very Long Baseline Array (VLBA) at four epochs on 2008 October 22, 2009 April 27 and 30, and 2009 October 20. Two background quasars J1922+1530 and J1924+1540 were observed as phase references to measure the absolute position of the maser source in W51 Main/South.

We measured the parallax of W51 Main/South to be $\pi = 0.183 \pm 0.006$ mas using J1922+1530 and $\pi = 0.187 \pm 0.009$ mas using J1924+1540. The uncertainty of each parallax fit was obtained from the formal fitting uncertainty. By combining the two fits and estimating the partially correlated errors for the two quasars, we obtained $\pi = 0.185 \pm 0.010$ mas as the best estimate, which corresponds to a source distance of $5.41^{+0.31}_{-0.28}$ kpc. Our parallax distance is in good agreement with previous distance measurements of W51 (e.g., $5.1^{+2.9}_{-1.4}$ kpc by Xu *et al.* 2009 from a 12 GHz methanol maser parallax), but has a much higher accuracy of $\approx 5\%$ and thus gives a strong constraint on the distance to W51.

We searched for maser spots in W51 Main/South by mapping a field of $\approx 20'' \times 20''$ and, in the first epoch's data, detected 1362 spots making up 280 unique "features" (i.e., groups of spots at the same position but in adjacent velocity channels). We detected 37 H_2O maser features that we could identify at all four epochs. Figure 1 gives a map of positions and internal motions of the detected maser features, after subtracting the mean motion of all four regions. The positions of the hyper- and ultra-compact HII regions (e.g., W51e2-E, e2-NW) are also plotted by crosses with sizes marking the approximate beam sizes from previous observations. We classified all 37 features into four separate groups according to their likely exciting sources: W51e2-NW, W51e2-E and W51e8, and a region

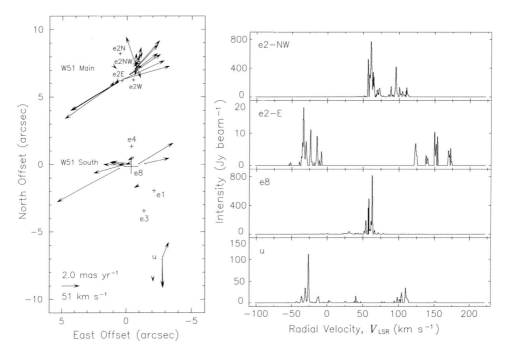

Figure 1. Left: map of the positions and internal motions of H_2O maser features in W51 Main/South, after subtracting the mean motion of all regions. Absolute coordinates of the map origin are R.A. (J2000)= $19^h 23^m 43^s.93427$ and Dec. (J2000)= $14°30'28''.3498$. The hyper- and ultra-compact H_{II} regions are marked by crosses with positions and beam sizes from previous observations (Gaume, Johnston & Wilson 1993; Zhang & Ho 1997; Shi, Zhao & Han 2010). Right: H_2O maser spectra of the four regions with intensities summed over all detected spots.

we call "u" without detected millimeter or centimeter continuum emission. For the W51e2 region, we identified two outflows: a high-velocity outflow and a low-velocity outflow that arise separately from the e2-E and e2-NW cores, respectively. Each H_2O maser outflow in W51 Main/South shows a relatively small range of internal 3D speeds. This is contrary to previous observations of other Galactic star-forming regions, e.g., Orion-KL (Genzel & Downes 1977) and W49N (Gwinn et al. 1992), which suggested that high and low velocity outflows might originate from the same young stellar object. In W49, a strong acceleration region at a radius of 0.1 pc is required if all masers are associated with a single young stellar object. However, these conclusions were based on observations with less precise maser motions and much lower resolution dust emission maps. Our results for W51 suggests that multiple speed outflows may come from separate young stellar objects that can be very closely spaced on the sky (see Sato et al. 2010 for details of the measurements and for further discussion).

References

Gaume, R. A., Johnston, K. J., & Wilson, T. L. 1993, *ApJ*, 417, 645

Genzel, R. & Downes, D. 1977, *A&A*, 61, 117

Gwinn, C. R., Moran, J. M., & Reid, M. J. 1992, *ApJ*, 393, 149

Sato, M., Reid, M. J., Brunthaler, A., & Menten, K. M. 2010, *ApJ*, 720, 1055

Shi, H., Zhao, J.-H., & Han, J. L. 2010, *ApJ*, 710, 843

Xu, Y., Reid, M. J., Menten, K. M., Brunthaler, A., Zheng, X. W., & Moscadelli, L. 2009, *ApJ*, 693, 413

Zhang, Q. & Ho, P. T. P. 1997, *ApJ*, 488, 241

Cosmic Masers - from OH to H_0
Proceedings IAU Symposium No. 287, 2012
R.S. Booth, E.M.L. Humphreys & W.H.T. Vlemmings, eds.

© International Astronomical Union 2012
doi:10.1017/S1743921312007466

Trigonometric Parallax of RCW 122

Y. W. Wu[1,2], Y. Xu[1], K. M. Menten[2], X. W. Zheng[3] and M.J. Reid[4]

[1]Purple Mountain Observatory, Chinese Academy of Sciences, Nanjing 210008,
email: ywwu@pmo.ac.cn

[2]Max-Planck-Institut für Radioastronomie, Auf dem Hügel 69, 53121 Bonn, Germany
[3]School of Astronomy and Space sience, Nanjing University, Nanjing 210093, China
[4]Harvard-Smithsonian Center for Astrophysics, 60 Garden Street, Cambridge, MA 02138, USA

Abstract. As a part of the BeSSeL (Bar and Spiral Structure Legacy) survey, we report a trigonometric parallax for the massive star-forming region G348.70−1.04. Its distance is $3.38^{+0.33}_{-0.27}$ kpc, indicating that it is in the Scutum-Centaurus arm. Its proper motion is -0.73 ± 0.04 mas yr^{-1} toward the east and -2.83 ± 0.50 mas yr^{-1} toward the north.

Keywords. astrometry, masers, stars: formation, techniques: high angular resolution

1. Introduction

The BeSSeL survey is a VLBA Key Science Project that will determine proper motions and parallaxes for methanol and water masers in several hundred massive star forming regions in the Galaxy. This project will allow us to locate spiral arms and determine the 3D motions of massive star-forming regions in our Galaxy. Currently, we are observing water and methanol masers towards 9 star-forming regions that are expected to trace the Carina-Sagittarius arm. Here we present results of G348.70−1.04.

G348.70−1.04 is located in the southern star-forming complex RCW 122. Earlier studies indicated very active star formation, e.g., containing a number of O-type stars (Arnal, *et al.* 2008). Approximately 20′ north of RCW 122 there is an open cluster Havlen-Moffat 1(HM1) (Havlen & Moffat 1977), harbouring a group of Of stars (Sanduleak 1974) and two Wolf-Rayet (WR) stars WR87 and WR89. In addition, another WR star, WR91, is located on the south-east edge of RCW 122. Dust emission from this field shows a ring-like structure (Fig. 1), with the open cluster HM1 at the ring's center and RCW 122 projected toward the southern part of the ring.

The spatial arrangement of these young objects is consistent with a triggered star-forming scenario between HM1 and RCW 122. However, whether or not they are truly associated is still under debate (Arnal *et al.* 2008), due to an inconsistency of their distances. For HM1, Havlen & Moffat (1977) found a spectrosopic distance of 2.9 ± 0.4 kpc, and Vázquez & Baume (2001), using deep UBVRI photomtry, got a distance of 3.3 kpc. The molecular cloud associated with the 12 GHz methanol masers we have observed has an LSR velocity near 13 km s^{-1}, corresponding a near kinematic distance of 1.7 kpc, while 21 cm wavelength H I spectra toward its H II region(s) show absorption features with LSR velocities out to -44 km s^{-1}, suggests a distance of 5 kpc (Radhakrishnan, *et al.* 1972). Thus the distance to RCW 122 and its relationship to other sources nearby in angle is not well established.

2. Observations and Results

We report 8-epoch observations with the VLBA toward the 12 GHz methanol maser source G348.70−1.04 from September 10, 2010 to February 11, 2012. The observing

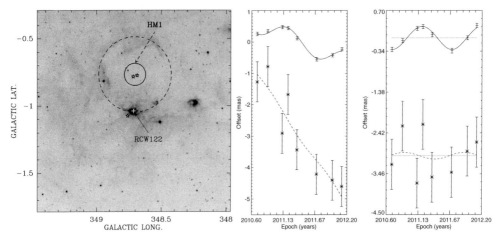

Figure 1. Left panel: MSX 8μ m dust emission. The plus-sign denotes the maser, the solid circle denotes the open cluster HM1, and penagrams denote WR stars. The dust emissions shows a ring-like structure (dashed line), with the open cluster HM1 at its center. Parallax fit results: the middle panel shows the eastward and northward position offsets and best fitting model; the right panel is the same as the middle panel, but with the best fit proper motion removed.

dates were chosen to sample the peaks of the parallax signature in right ascension only, as the amplitude of the signature in declination is considerably smaller. Three VLBA calibrators, J1723-3936, J1712-3736 and J1733-3722 (from Immer, Brunthaler, Reid, *et al.* 2011), were used as background position references.

After calibration, including phase referencing to a maser spot, we imaged the maser and the continuum sources. Only the continuum source nearest to the maser (J1723−3936) was detected, possibly owing to poor phase transfer to the other sources. We estimated the positions of the maser and the background source by fitting elliptical Gaussian brightness distributions to the images. The positions of the maser spots relative to the background source were then used to fit the proper motion and parallax (see the middle and right panel of Fig. 1). Best fit parallax is 0.296 ± 0.026 mas, giving a distance of $3.38^{+0.33}_{-0.27}$ kpc, and the proper motion is -0.73 ± 0.04 mas yr^{-1} toward the east and -2.83 ± 0.50 mas yr^{-1} toward the north.

The trigonometric parallax places the star formation region RCW 122, not in the Carina-Sagittarius arm, but in the Scutum-Centaurus arm. Also, it confirms the physical connection between the star-forming region RCW 122 and open cluster HM1.

This work was supported by the Chinese NSF through grants NSF 11133008, NSF 11073054, NSF 10621303, and the Key Laboratory for Radio Astronomy, CAS.

References

Arnal, E. M., Duronea, N. U., & Tesori, J. C. 2008, *A&A*, 486, 807

Havlen, R. J. & Moffat, A. F. J. 1977, *A&A*, 58, 351

Immer, K., Brunthaler, A., & Reid, M. J., *et al.* 2011, *ApJS*, 194, 25

Radhakrishnan, V., Goss, W. M., Murray, J. D., & Brooks, J. W. 1972, *ApJS*, 24, 49

Sanduleak, N 1974, *PASP*, 86, 461

Vázquez, R. A., & Baume, G. *A& A*, 371, 908

Cosmic Masers - from OH to H_0
Proceedings IAU Symposium No. 287, 2012
R.S. Booth, E.M.L. Humphreys & W.H.T. Vlemmings, eds.

© International Astronomical Union 2012
doi:10.1017/S1743921312007478

Distance and Size of the Red Hypergiant NML Cyg

B. Zhang[1,2], M. J. Reid[3], K. M. Menten[1], X. W. Zheng[4] and A. Brunthaler[1]

[1] Max-Plank-Institut für Radioastronomie, Auf dem Hügel 69, 53121 Bonn, Germany
email:bzhang@mpifr.de

[2] Shanghai Astronomical Observatory, Chinese Academy of Sciences, 80 Nandan Road, Shanghai 200030, China

[3] Harvard-Smithsonian Center for Astrophysics, 60 Garden Street, Cambridge, MA 02138, USA

[4] Department of Astronomy, Nanjing University, 22 Hankou Road, Nanjing 210093, China

Abstract. We present astrometric results of phase-referencing VLBI observations of 22 GHz H_2O maser and 43 GHz SiO maser emission towards the red hypergiant NML Cyg using VLBA. We obtained an annual parallax of 0.62 ± 0.04 mas, corresponding to a distance of $1.61^{+0.13}_{-0.11}$ kpc. With a VLA observation in its largest (A) configuration at 43 GHz, we barely resolve the radio photosphere of NML Cyg, and find a uniform-disk diameter of 44 ± 16 mas.

Keywords. astrometry — masers — instrumentation: interferometric — stars: distance — stars: individual (NML Cyg) — supergiants

1. Introduction

NML Cyg has been assumed to be at the same distance as the Cyg OB2 association (Morris & Jura 1983). The luminosity of NML Cyg derived using the previous estimated distance of 1.74 ± 0.2 kpc by Massey & Thompson (1991) places it near the empirical upper luminosity boundary on the H-R diagram (Schuster *et al.* 2006). However, a luminosity estimate of a star is dependent on the square of its distance. Thus, a trustworthy distance is crucial to derive a reliable luminosity of NML Cyg. In addition to the distance, another fundamental stellar parameter is the size. NML Cyg's high mass-loss rate results in a dense circumstellar envelope, thus the star is hardly observable at visual wavelengths due to high extinction, while it is extremely luminous in the infrared. Blöcker *et al.* (2001) obtained a central star luminosity by integrating the spectral energy distribution from 2 to 50 μm, and derived a stellar diameter of 16.2 mas assuming $T_{\mathrm{eff}} = 2500$ K. However, a direct detection of the star would be superior to indirect methods. Radio continuum emission from the evolved star's photosphere can be imaged by VLA using SiO masers as a phase reference (Reid & Menten 2007; Zhang *et al.* 2012), this allows us to directly determine the size of NML Cyg's radio photosphere.

2. Parallax of NML Cyg

We conducted VLBI phase-referencing observations of circumstellar 22 GHz H_2O and 43 GHz SiO masers toward NML Cyg and several extragalactic radio sources with the VLBA at five epochs in one year. We identified about 20 H_2O maser features with a V_{LSR} range of $-25.4 - 6$ km s^{-1}, while the SiO masers were very weak during our observation and only detectable at the last epoch. Using the positions of 18 H_2O maser spots relative to three background sources J2044+4005, J2046+4106 and J2049+4118 within $2°$ of

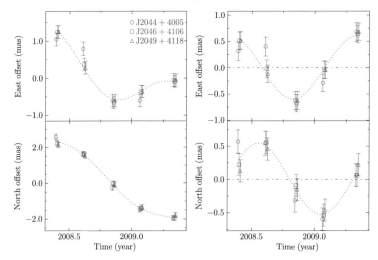

Figure 1. Parallax and proper motion data (*markers*) and best-fitting models (*dotted lines*) for the maser spot at the V_{LSR} of 6.48 km s^{-1}. Plotted are positions of the maser spot relative to the extragalactic radio sources J2044+4005 (*circles*), J2046+4106 (*squares*) and J2049+4118 (*triangles*). *Left panel:* Eastward (*solid lines*) and northward (*dotted lines*) offsets and best-fitting models versus time. Data of the eastward (*upper panel*) and northward (*bottom panel*) positions are offset horizontally for clarity. *Right panel:* Same as the *left panel*, except the best-fitting proper motion has been removed, displaying only the parallax signature.

NML Cyg, we determined a trigonometric parallax of 0.62 ± 0.04 mas, corresponding to a distance of $1.61^{+0.13}_{-0.11}$ kpc. Fig. 1 shows the parallax fit of one of the maser spots as an example.

3. Radio Photosphere of NML Cyg

We detected weak continuum emission at 43 GHz from the radio photosphere of NML Cyg using a calibration scheme in which circumstellar masers provide a phase reference for the continuum data with VLA. Taking into consideration possible biases introduced by the deconvolution process of images at low SNR, we fitted a round disk model to the uv-data directly and obtained a diameter of 44 ± 16 mas, which is comparable with two times that derived from the Stefan-Boltzmann law with a revised luminosity according to our distance and T_{eff} = 2500 K.

Acknowledgements

B. Zhang is supported by the National Science Foundation of China under grant 11073046, 11133008.

References

Blöcker, T., Balega, Y., Hofmann, K.-H., & Weigelt, G. 2001, *A&A*, 369, 142
Massey, P. & Thompson, A. B. 1991, *AJ*, 101, 1408
Morris, M. & Jura, M. 1983, *ApJ*, 267, 179
Reid, M. J. & Menten, K. M. 2007, *ApJ*, 671, 2068
Schuster, M. T., Humphreys, R. M., & Marengo, M. 2006, *AJ*, 131, 603
Schuster, M. T., *et al.* 2009, *ApJ*, 699, 1423
Zhang, B., Reid, M. J., Menten, K. M., & Zheng, X. W. 2012, *ApJ*, 744, 23

A memorable before dinner speech by Phil Diamond.

T9:

New masers and further developments in maser physics *Chair: Mark Reid*

Cosmic Masers - from OH to H₀
Proceedings IAU Symposium No. 287, 2012
R.S. Booth, E.M.L. Humphreys & W.H.T. Vlemmings, eds.

© International Astronomical Union 2012
doi:10.1017/S174392131200748X

New class I methanol masers

M. A. Voronkov[1,2], J. L. Caswell[1], S. P. Ellingsen[3], S. L. Breen[1], T. R. Britton[4,1], J. A. Green[1], A. M. Sobolev[5] and A. J. Walsh[6]

[1]CSIRO Astronomy and Space Science, PO Box 76, Epping NSW 1710, Australia

[2]Astro Space Centre, Profsouznaya st. 84/32, 117997 Moscow, Russia

[3]School of Mathematics and Physics, University of Tasmania, GPO Box 252-37, Hobart, Tasmania 7000, Australia

[4]Macquarie University, Department of Physics and Engineering, NSW 2109, Australia

[5]Ural State University, Lenin ave. 51, 620083 Ekaterinburg, Russia

[6]Centre for Astronomy, School of Engineering and Physical Sciences, James Cook University, Townsville, QLD 4814, Australia

Abstract. We review properties of all known collisionally pumped (class I) methanol maser series based on observations with the Australia Telescope Compact Array (ATCA) and the Mopra radio telescope. Masers at 36, 84, 44 and 95 GHz are most widespread, while 9.9, 25, 23.4 and 104 GHz masers are much rarer, tracing the most energetic shocks. A survey of many southern masers at 36 and 44 GHz suggests that these two transitions are highly complementary. The 23.4 GHz maser is a new type of rare class I methanol maser, detected only in two high-mass star-forming regions, G357.97-0.16 and G343.12-0.06, and showing a behaviour similar to 9.9, 25 and 104 GHz masers. Interferometric positions suggest that shocks responsible for class I masers could arise from a range of phenomena, not merely an outflow scenario. For example, some masers might be caused by interaction of an expanding HII region with its surrounding molecular cloud. This has implications for evolutionary sequences incorporating class I methanol masers if they appear more than once during the evolution of the star-forming region. We also make predictions for candidate maser transitions in the ALMA frequency range.

Keywords. masers – ISM: molecules – ISM: jets and outflows

1. Introduction

Methanol masers are associated with regions of active star formation, with more than 20 different cm- and mm-wavelength masing transitions discovered to date. The whole range of methanol maser transitions does not share the same behaviour, being loosely grouped in two classes. The division stems from early empirical distinctions (e.g. Batrla *et al.* 1987). Class II methanol masers (e.g. the most famous 6.7-GHz transition), along with OH and H₂O masers, occur in the immediate environment of young stellar objects (YSOs) recognisable from their characteristic infrared emission. The class II methanol masers are exclusive tracers of high-mass star-formation (e.g. Minier *et al.* 2003; Green *et al.* 2012). In contrast, the class I masers (e.g. at 36 and 44 GHz), which are the subject of this paper, are usually found offset from the presumed origin of excitation (e.g. Kurtz *et al.* 2004; Voronkov *et al.* 2006), and are found in regions of both high- and low-mass star formation (Kalenskii *et al.* 2010 and their paper in this volume).

Theoretical calculations can explain this empirical classification, with the pumping process of class I masers dominated by collisions (with molecular hydrogen), in contrast to class II masers which are pumped by radiative excitation (e.g. Cragg *et al.* 1992; Voronkov 1999; Voronkov *et al.* 2005a). The two mechanisms are competitive (see Voronkov *et al.* 2005a for illustration): strong radiation from a nearby infrared source quenches class I

masers and strengthens class II masers. The transitions of different classes occur in
opposite directions between two given ladders of energy levels (Fig. 1). The equilibrium
breaks first between the ladders giving rise to either class I or class II masers depending on
whether radiational or collisional excitation dominates (e.g. Voronkov 1999). Therefore,
bright masers of different classes residing in the same volume of gas are widely accepted
as mutually exclusive (with potential exceptions for weak masers). However, on larger
scales, they are often observed to coexist in the same star forming region within less than
a parsec of each other (while a few archetypal sources exist, displaying only one class of
methanol maser).

In addition to the gross classification, there are finer distinctions within the same class
of methanol maser transitions. At sensitivity levels typically attained in surveys, the
range of transitions can be further categorised into widespread masers (e.g. at 44 GHz)
and rare or weak masers (e.g. at 9.9 GHz). Models seem to suggest that the formation
of rare masers requires higher temperatures and densities (Sobolev *et al.* 2005). The
maser transitions of methanol tend to form series (individual transitions have different J
quantum numbers as is evident from Fig. 1). Observational properties such as whether
the individual transitions give rise or widespread masers are qualitatively similar within
the same series. Superposed are trends with J caused by the changes of excitation energy
and the efficiency of the sink process due to a different number of energy levels below.
Interestingly, all class II methanol maser series (with the exception of J_2-(J-1)$_3$ A$^\pm$ series
based on the 38-GHz maser) are going downwards (with J decreasing while frequency
increases) and eventually terminate. In contrast, all class I maser series extend upwards
(see Fig. 1). Therefore, the majority of candidate maser transitions searchable with the
Atacama Large Millimetre Array (ALMA) in the millimetre and sub-millimetre bands
belong to class I. In the following sections we review observational properties of all known

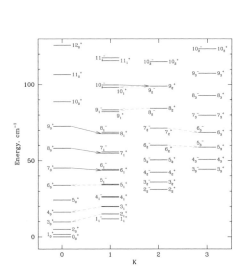

Figure 1. Energy level diagram for
A-methanol (energies are given with respect to the lowest level of A-methanol).
Solid (red) arrows represent known class I
maser transitions, dashed (green) arrows
show known class II masers.

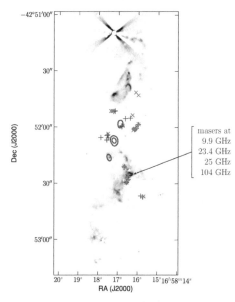

Figure 2. Distribution of the 36 (crosses)
and 44 GHz (pluses) maser spots on top
of the outflow image traced by 2.12μm H$_2$
emission in G343.12-0.06, contours show
the 12mm continuum emission (see also
Voronkov *et al.* 2006)

class I methanol maser series before summarising predictions for ALMA in bands 6 and 7. For simplicity, we refer to the maser series by the lowest frequency transition.

2. Widespread class I masers (series based on 36 and 44 GHz masers)

The J_0-$(J-1)_1$ A^+ methanol series includes the most studied 44 and 95-GHz class I methanol masers. A few hundred such masers are currently known, but the majority have only single dish data (e.g. Haschick *et al.* 1990; Slysh *et al.* 1994; Val'tts *et al.* 2000; Ellingsen 2005; Fontani *et al.* 2010; Chen *et al.* 2011; unpublished Mopra data from our group). The major published interferometric surveys are those of Kurtz *et al.* (2004) and Cyganowski *et al.* (2009). The second class I maser series in the widespread category is J_{-1}-$(J-1)_0$ E which is renowned for masers at 36 and 84 GHz. These two maser transitions are considerably less studied than the 44 and 95-GHz pair. As before, most observational data are obtained with single dish facilities (e.g. Haschick & Baan 1989; Kalenskii *et al.* 2001). The reported interferometric observations are scarce and confined to single source papers only (e.g. Voronkov *et al.* 2006; Voronkov *et al.* 2010; Sjouwerman *et al.* 2010; Fish *et al.* 2011). The typical spread of maser spots is comparable to or exceeds the beam size of a 20-m class single dish at the frequencies of these transitions (Kurtz *et al.* 2004; Voronkov *et al.* 2006). Therefore, interferometric observations, which allow us to measure positions of each maser sport accurately, are crucial even to get meaningful detection statistics.

To increase the number of class I masers studied at high angular resolution and to compare the morphologies observed in different maser transitions we carried out in 2007 a quasi-simultaneous interferometric survey at 36 and 44 GHz of all class I masers reported in the literature at the time of the observations and located south of declination -35° (a single source from the project was presented in Voronkov *et al.* 2010a). Fig. 2 shows the results of this survey for G343.12-0.06, which has been studied in detail in other transitions by Voronkov *et al.* (2006). The distribution of 36 and 44 GHz maser spots resembles that of 84- and 95-GHz maser spots from Voronkov *et al.* (2006), which is a good example of outflow association, but also has a few new spots found due to the larger primary beam and higher signal-to-noise ratio of the new observations. Note, that all rare masers in this source are located at the same position near the brightest knot of the 2.12 μm molecular hydrogen emission, which is a well known shock tracer (Voronkov *et al.* 2006). With the caveat about extinction variations, this supports the idea that the formation of rare class I masers requires higher temperatures and densities than the widespread masers (Sobolev *et al.* 2005).

In G343.12-0.06, the majority of 44-GHz maser spots have some 36-GHz emission and vice versa (Fig. 2). However, in many cases these two transitions were found to be highly complementary. Fig. 3 shows the maser spot distribution in G333.466-0.164, the best example of such a scenario that we currently have. The 44-GHz maser spots are distributed roughly along the line traced by the source of extended infrared emission with 4.5-μm excess (often referred to as an Extended Green Object or EGO). Without the 36-GHz data, this EGO would most likely be interpreted as tracing an outflow emanating from the location of the YSO marked by the 6.7-GHz maser (shown by square in Fig. 3). The chain of 36-GHz maser spots completes the second half of a bow-shock structure suggesting a different direction of the outflow. Another good example is the high-velocity feature blue-shifted by about 30 km s^{-1} from the systemic velocity which was found in G309.38$-$0.13 at 36-GHz only (Voronkov *et al.* 2010a). It is worth noting, that Sobolev *et al.* (2005) suggested that the 36 to 44-GHz flux density ratio is very sensitive to the orientation of the maser region.

3. Rare 9.9 and 104 GHz masers

These masers belong to the J_{-1}-$(J$-$1)_{-2}$ E methanol series, J=9 and 11, respectively. The first search for 9.9-GHz masers was carried out by Slysh *et al.* (1993) who reported a single maser detection towards W33-Met (G12.80−0.19). Recently, Voronkov *et al.* (2010b) carried out a sensitive (1σ limits as low as 100 mJy) survey at 9.9-GHz with the ATCA and found 2 new detections out of 46 targets observed. Two additional 9.9-GHz masers in G343.12−0.06 and G357.97−0.16 were found serendipitously (Voronkov *et al.* 2006, 2011). The latter maser is the strongest, with peak flux density around 70 Jy and the only one for which the absolute position has not been measured (although the position is expected to be the same as for the 23.4-GHz maser found in this source).

With the exception of the 104-GHz maser in G343.12−0.06 which had ATCA observations (Voronkov *et al.* 2006), all other currently known 104-GHz masers were found using single dish facilities (Voronkov *et al.* 2005b, 2007). In addition to the sources of 9.9-GHz maser emission described above, these observations brought only one new maser in G305.21+0.21. It is worth noting that a weak maser at 9.9 GHz was seen towards this source during test ATCA observations, but happened to be below the detection threshold of the regular survey (Voronkov *et al.* 2010b). This brings the total number of known masers in this series to 6, in contrast to more than 200 hundred known widespread masers.

Detailed investigations of these masers suggest that some class I masers (in all classes of transitions, not just 9.9-GHz) may be caused by expanding HII regions (see e.g. Fig. 4 and Voronkov *et al.* 2010b). This is an additional scenario to the commonly accepted mechanism for the formation of class I masers in the outflow shocks.

4. Evolutionary stage of star-formation with class I masers

The question whether different masers trace distinct evolutionary stages of high-mass star formation has recently become a hot topic (see e.g. Breen *et al.* 2010 and references therein), although the place of class I masers in this picture is still poorly understood. Ellingsen (2006) investigated the infrared colours of GLIMPSE (Galactic Legacy Infrared

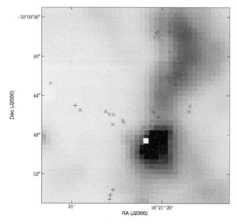

Figure 3. Distribution of the 36 (crosses) and 44 GHz (pluses) maser spots in G333.466-0.164. The position of the 6.7-GHz maser is shown by filled square. The background is GLIMPSE 4.5-μm image.

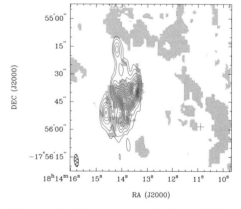

Figure 4. The position of the 9.9-GHz methanol maser in W33-Met. The contours represent 8.4-GHz continuum image, grayscale shows the distribution of the thermal NH_3 emission (for details see Voronkov *et al.* 2010b and references therein).

Mid-Plane Survey Extraordinaire) catalogue point sources associated with methanol masers and suggested that class I methanol masers may signpost an earlier stage of high-mass star formation than the class II masers. These and other considerations laid the foundation for a qualitative evolutionary scheme for different maser species proposed by Ellingsen *et al.* (2007). The scheme was further refined by Breen *et al.* (2010) in their Figure 6, but with no new survey data available on class I masers the conclusion about the evolutionary stage when these masers are present remained essentially the same. Voronkov *et al.* (2006, 2010b) pointed out that this statement is inconsistent with detailed studies of class I maser sources which do overlap with OH masers. Moreover, we carried out a search for the 44-GHz class I methanol masers towards known OH masers which were not detected at 6.7-GHz (class II) in the unbiased Methanol Multi-Beam (MMB) survey (Green *et al.* 2012 and references therein). Despite the inadequate spectral resolution (about 7 km s^{-1}) achieved in these test observations, which makes the survey insensitive to weak (<15 Jy) masers, the detection rate exceeded 50%. Therefore, it seems more appropriate to place the class I methanol masers as partly overlapping, but largely post-dating the evolutionary phase associated with class II methanol masers. The suggested association with expanding HII regions also implies that some of the sources are quite evolved. Whether there is a population of class I masers pre-dating the phase with the class II masers is unclear at present (see Voronkov *et al.* 2010b and Chen *et al.* 2011 for detailed discussion of these issues).

It is also important to keep in mind the major assumptions which underlie the evolutionary timeline suggested by Ellingsen *et al.* (2007) and Breen *et al.* (2010). In particular, that each major maser species arises only once during the evolution of a particular star formation region, and that all the maser species are associated with a single astrophysical object. The large spatial (and angular) spread of class I maser spots makes such identifications much less clear given a generally crowded environment of high-mass star formation where sequential or triggered star formation may take place. It is possible that one YSO in the cluster may have an associated 6.7-GHz maser, while another could be associated with OH masers and a well developed HII region. At this stage the number of sufficiently simple sources studied in detail is quite limited, hindering the statistical analysis and even in relatively simple sources some controversy may exist (see e.g. the case of G357.97−0.16 discussed by Britton *et al.*, this volume). It is worth noting that the survey of Chen *et al.* (2011) supports the hypothesis that class I masers may arise at more than one evolutionary phase.

5. Weak 25-GHz series

Historically, the first methanol masers found in space were from the J$_2$-J$_1$ E class I methanol maser series near 25-GHz towards Orion (Barrett *et al.* 1971). Very few additional sources were found in the following 3 decades, so these masers were widely believed to be rare. Voronkov *et al.* (2007) searched for J=5 transition of this maser series towards 102 targets with ATCA (although no absolute position was measured during this experiment). This search yielded 66 detections, but mostly weaker than 1 Jy. Other recent studies (e.g. Brogan *et al.* 2011; paper in this volume) also suggest that these masers are common, but typically weak. The new ATCA backend enabled simultaneous observations of up to 9 (limited by the receiver frequency range) transitions of this series (e.g. Wilson *et al.* (2011). Britton & Voronkov (paper in this volume) have recently followed up the majority of known southern 25-GHz masers in the J=2 to J=9 transitions.

6. New rare class I methanol maser at 23.4-GHz

The H_2O southern Galactic Plane Survey (HOPS; Walsh *et al.*, 2011 and their paper in this volume) brought an unexpected discovery of a new class I methanol maser at 23.4-GHz in G357.97−0.16 (see Voronkov *et al.* 2011 for details). This is the first transition in the J_1-$(J$-1$)_2$ A^- class I maser series (corresponding to J=10). A weaker 23.4-GHz maser was found in G343.12−0.06. Observational properties of this new maser are similar to other rare masers like that at 9.9-GHz. The source of strongest 23.4 and 9.9-GHz maser emission, G357.97−0.16, is discussed in detail by Britton *et al.* (this volume).

7. Class I methanol masers at high spatial resolution

In contrast to class II methanol masers, Very Long Baseline Interferometry (VLBI) observations of class I masers were effectively abandoned following a few unsuccessful attempts in early 1990s (which remained unpublished). The understanding of class I masers has certainly progressed since then and observations with connected element interferometers have become available (e.g. Kurtz *et al.* 2004; Cyganowski *et al.* 2009; see also earlier sections). In hindsight, the reasons for failure to detect class I masers in early VLBI experiments were not limited to larger intrinsic sizes of maser spots and a general difficulty doing high-frequency VLBI. First, only single dish positions (accurate up to arcmin) were available for most if not all of the targets. It is shown in the previous sections that images of class I masers typically contain several spots distributed over a large area often covering the whole primary beam of the telescope (see also Voronkov *et al.* 2006; Kurtz *et al.* 2004). Therefore, there are high chances to miss a spot of emission given a typical narrow field of view achieved in a VLBI experiment. In addition, the lack of images at arcsecond resolution made the selection of best most compact targets difficult. Recent VLBI observations (Tarchi, priv. comm.) at 44-GHz revealed fringes at the shortest baselines (similar observations were also carried out by Kim *et al.* (priv. comm.) with the Korean VLBI Network). The second issue is the selection of maser transition. The early VLBI attempts targeted the strongest and most widespread maser transitions at 44 and 95-GHz transitions. However, these are likely to have larger intrinsic source sizes being easier to excite than, for example, masers at 9.9 and 23.4 GHz along with the masers in the series near 25 GHz. It is worth noting that VLBI observations of 10 strongest 25-GHz masers have recently been carried out with the Long Baseline Array (LBA).

8. Millimetre and sub-millimetre masers with ALMA

One can extend the maser series reviewed in the previous sections to higher frequencies. ALMA bands 6 and 7 (already implemented) encompass the following candidate maser transitions. The series based on the 36-GHz maser giving 8_{-1}-7_0 E at 229 GHz (a known maser: Slysh *et al.* 2002; Fish *et al.* 2011), 9_{-1}-8_0 E at 278 GHz recently found in S255N with the SMA (Salii, Sobolev, Zinchenko, Liu & Su, priv. comm) and 10_{-1}-9_0 E at 327 GHz, although the system performance is poor for the latter transition. The series based on the 44-GHz maser giving 11_0-10_1 A^+ at 250 GHz, 12_0-11_1 A^+ at 303 GHz and 13_0-12_1 A^+ at 356 GHz. The series based on the 9.9-GHz maser giving 14_{-1}-13_{-2} E at 242 GHz, 15_{-1}-14_{-2} E at 287 GHz, 16_{-1}-15_{-2} E at 331 GHz. The new series based on the 23.4-GHz maser giving 14_1-13_2 A^- at 237 GHz, 15_1-14_2 A^- at 291 GHz and 16_1-15_2 A^- at 346 GHz.

The field of high-frequency class I methanol masers is essentially uncharted territory. For possible masers at even higher frequencies (e.g. ALMA band 9), it is hard to make sensible predictions. The transitions listed above correspond to excitation energies of about 300-500 K. The 25-GHz transitions are a cm-wavelength series but, between the same two ladders of levels, is the J_2-(J-1)$_1$ E series which gives mm-wavelength transitions including a 218-GHz maser (J=4) detected in the SMA observations mentioned above. It is worth noting that maser models predict population inversion for all these transitions (see also Sobolev *et al.*, this volume).

9. Conclusions

(*a*) Studies of different maser transitions are very complementary (filling the dots in morphology, high-velocity features and modelling).

(*b*) Rare/weak class I methanol masers (9.9, 23.4, 25 and 104 GHz) trace stronger shocks and higher temperatures and densities.

(*c*) Some class I masers may be caused by expanding HII regions, a scenario additional to their formation in outflows. An implication for a maser-based evolutionary sequence is that class I methanol masers may appear more than once during YSO evolution, and thus some regions with class I masers are probably quite evolved.

(*d*) The evolutionary stage with class I masers probably outlasts the stage when the 6.7-GHz methanol masers (class II) are present, overlapping with OH maser activity.

(*e*) Promising ALMA maser targets are G343.12-0.06 and G357.97-0.16.

Acknowledgments

The Australia Telescope Compact Array and the Mopra telescope are parts of the Australia Telescope National Facility which is funded by the Commonwealth of Australia for operation as a National Facility managed by CSIRO. The University of New South Wales Digital Filter Bank used for the observations with the Mopra Telescope was provided with support from the Australian Research Council. The research has made use of the NASA/IPAC Infrared Science Archive, which is operated by the Jet Propulsion Laboratory, California Institute of Technology, under contract with the National Aeronautics and Space Administration. AMS would like to thank the Russian Foundation for Basic Research (grant 10-02-00589-a) and the Russian federal task program "Research and operations on priority directions of development of the science and technology complex of Russia for 2007-2012" (state contract 16.518.11.7074).

References

Barrett, A. H., Schwartz, P. R., & Waters, J. W., 1971, *ApJ*, 168, 101

Batrla, W., Mathews, H. E., Menten, K. M., & Walmsley, C. M., 1987, *Nature*, 326, 49.

Breen, S. L., Ellingsen, S. P., Caswell, J. L., & Lewis, B. E., 2010, *MNRAS*, 401, 2219

Brogan, C. L., Hunter, T. R., Cyganowski, C. J., Friesen, R. K., Chandler, C. J., & Indebetouw, R., 2011, *ApJ*, 739, 16

Chen, X., Ellingsen, S. P., Shen, Z.-Q., Titmarsh, A., & Gan, C.-G., 2011, *ApJS*, 196, 9

Cragg, D. M., Johns, K. P., Godfrey, P. D., & Brown, R. D., 1992, *MNRAS*, 259, 203

Cyganowski, C. J., Brogan, C. L., Hunter, T. R., & Churchwell, E., 2009, *ApJ*, 702, 1615

Ellingsen, S. P., Voronkov, M. A., Cragg, D. M., Sobolev, A. M., Breen, S. L., & Godfrey, P. D., 2007, in: J. M. Chapman, W. A. Baan (eds.), *Astrophysical Masers and their Environments*, Proc. IAU Symposium No. 242 (Cambridge: CUP), p. 213 (arXiv:0705.2906)

Ellingsen, S. P., 2006, *ApJ*, 638, 241

Ellingsen, S. P., 2005, *MNRAS*, 359, 1498

Fish, V. L., Muehlbrad, T. C., Pratap, P., Sjouwerman, L. O., Strelnitski, V., Pihlström, Y. M., & Bourke, T. L., 2011, *ApJ*, 729, 14

Fontani, F., Cesaroni, R., & Furuya, R. S., 2010, *A&A*, 517, A56

Green, J. A., Caswell, J. L., Fuller, G. A., Avison, A., Breen, S. L., Ellingsen, S. P., Gray, M. D. Pestalozzi, M., Quinn, L., Thompson, M. A., & Voronkov, M. A., 2012, *MNRAS*, in press

Haschick, A. D., Menten, K. M., & Baan, W. A., 1990, *ApJ*, 354, 556

Haschick, A. D. & Baan, W. A., 1989, *ApJ*, 339, 949

Kalenskii, S. V., Slysh, V. I., Val'tts, I. E., Winnberg, A., & Johansson, L. E., 2001, *Astron. Rep.*, 45, 26

Kalenskii, S. V., Johansson, L. E. B., Bergman, P., Kurtz, S., Hofner, P., Walmsley, C. M., & Slysh, V. I., 2010, *MNRAS*, 405, 613

Kurtz, S., Hofner, P., & Álvarez, C. V., 2004, *ApJS*, 155, 149

Minier, V., Ellingsen, S. P., Norris, R. P., & Booth, R. S., 2003, *A&A*, 403, 1095

Sjouwerman, L. O., Pihlström, Y. M., & Fish, V. L., 2010, *ApJ*, 710, 111

Slysh, V. I., Kalenskii, S. V., & Val'tts, I. E., 2002, *Astron. Rep.*, 46, 49

Slysh, V. I., Kalenskii, S. V., Val'tts, I. E., & Otrupcek, R., 1994, *MNRAS*, 268, 464

Slysh, V. I., Kalenskii, S. V., & Val'tts, I. E., 1993, *ApJ*, 413, L133

Sobolev, A. M., Ostrovskii, A. B., Kirsanova, M. S., Shelemei, O. V., Voronkov, M. A., & Malyshev, A. V., 2005, in: E.Churchwell, P.Conti & M.Felli (eds.), *Massive star birth: A crossroads of Astrophysics*, Proc. IAU Symposium No. 227 (Cambridge: CUP), p.174 (astro-ph/0601260)

Val'tts, I. E., Ellingsen, S. P., Slysh, V. I., Kalenskii, S. V., Otrupcek, R., & Larionov, G. M., 2000, *MNRAS*, 317, 315

Voronkov, M. A., Walsh, A. J., Caswell, J. L., Ellingsen, S. P., Breen, S. L., Longmore, S. N., Purcell, C. R., & Urquhart, J. S., 2011, *MNRAS*, 413, 2339

Voronkov, M. A., Caswell, J. L., Britton, T. R., Green, J. A., Sobolev, A. M., & Ellingsen, S. P., 2010a, *MNRAS*, 408, 133

Voronkov, M. A., Caswell, J. L., Ellingsen, S. P., & Sobolev, A. M., 2010b, *MNRAS*, 405, 2471

Voronkov, M. A., Brooks, K. J., Sobolev, A. M., Ellingsen, S. P., Ostrovskii, A. B., & Caswell, J. L., 2007, in: J. M. Chapman, W. A. Baan (eds.), *Astrophysical Masers and their Environments*, Proc. IAU Symposium No. 242 (Cambridge: CUP), p. 182 (arXiv:0705.0355)

Voronkov, M. A., Brooks, K. J., Sobolev, A. M., Ellingsen, S. P., Ostrovskii, A. B., & Caswell, J. L., 2006, *MNRAS*, 373, 411

Voronkov, M. A., Sobolev, A. M., Ellingsen, S. P., & Ostrovskii, A. B., 2005a, *MNRAS*, 362, 995

Voronkov, M. A., Sobolev, A. M., Ellingsen, S. P., Ostrovskii, A. B., & Alakoz, A. V., 2005b, *Ap&SS*, 295, 217

Voronkov, M. A., 1999, *Astron. Lett.*, 25, 149 (astro-ph/0008476)

Walsh, A. J., Breen, S. L., Britton, T., Brooks, K. J., Burton, M. G., Cunningham, M. R., Green, J. A., Harvey-Smith, L., Hindson, L., Hoare, M. G., Indermuehle, B., Jones, P. A., Lo, N., Longmore, S. N., Lowe, V., Phillips, C.,J., Purcell, C. R., Thompson, M. A., Urquhart, J. S., Voronkov, M. A., White, G. L., & Whiting, M. T., 2011, *MNRAS*, 416, 1764

Wilson, W. E., Ferris, R. H., Axtens, P., Brown, A., Davis, E., Hampson, G., Leach, M., Roberts, P., Saunders, S., Koribalski, B. S., Caswell, J. L., Lenc, E., Stevens, J., Voronkov, M. A., Wieringa, M. H., Brooks, K., Edwards, P. G., Ekers, R. D., Emonts, B., Hindson, L., Johnston, S., Maddison, S. T., Mahony, E. K., Malu, S. S., Massardi, M., Mao, M. Y., McConnell, D., Norris, R., Schnitzeler, D., Subrahmanyan, R., Urquhart, J. S., Thompson, M. A., & Wark, R. M., *MNRAS*, 416, 832

Cosmic Masers - from OH to H$_0$
Proceedings IAU Symposium No. 287, 2012
R.S. Booth, E.M.L. Humphreys & W.H.T. Vlemmings, eds.

© International Astronomical Union 2012
doi:10.1017/S1743921312007491

OH Masers and Supernova Remnants

Mark Wardle and Korinne McDonnell

Department of Physics & Astronomy and Research Centre for Astronomy, Astrophysics &
Astrophotonics, Macquarie University, Sydney NSW 2109, Australia
email: mark.wardle@mq.edu.au, korinne.mcdonnell@mq.edu.au

Abstract. OH(1720 MHz) masers are created by the interaction of supernova remnants with molecular clouds. These masers are pumped by collisions in warm, shocked molecular gas with OH column densities in the range 10^{16}–10^{17} cm^{-2}. Excitation calculations suggest that inversion of the 6049 MHz OH line may occur at the higher column densities that have been inferred from main-line absorption studies of supernova remnants with the Green Bank Telescope. OH(6049 MHz) masers have therefore been proposed as a complementary indicator of remnant-cloud interaction.

This motivated searches for 6049 MHz maser emission from supernova remnants using the Parkes 63 m and Effelsberg 100 m telescopes, and the Australia Telescope Compact Array. A total of forty-one remnants have been examined by one or more of these surveys, but without success. To check the accuracy of the OH column densities inferred from the single-dish observations we modelled OH absorption at 1667 MHz observed with the Very Large Array towards three supernova remnants, IC 443, W44 and 3C 391. The results are mixed – the OH column is revised upwards in IC443, downwards in 3C391, and is somewhat reduced in W44. We conclude that OH columns exceeding 10^{17} cm^{-2} are indeed present in some supernova remnants and so the lack of any detections is not explained by low OH column density. We discuss the possibility that non-local line overlap is responsible for suppressing the inversion of the 6049 MHz line.

Keywords. masers, line: formation, molecular processes, shock waves, ISM: molecules, supernova remnants, radio continuum: ISM, radio lines: ISM

1. Introduction

OH (1720 MHz) masers associated with supernova remnants are unambiguous indicators of interaction with adjacent molecular clouds (Frail, Goss & Slysh 1994). The relative ease of identifying supernova remnant - molecular cloud interaction sites relative to other methods, such as searching for molecular gas that is kinematically disturbed or chemically processed (e.g. DeNoyer 1979a,b; Wootten 1981) motivated surveys of the galaxy for more examples (Frail *et al.*1996; Yusef-Zadeh *et al.*1996; Green *et al.*1997; Koralesky *et al.*1998; Yusef-Zadeh *et al.*1999a,b; Sjouwerman & Pihlström 2008), which have found that approximately 10% of galactic supernova remnants have 1720 MHz masers and are therefore interacting with molecular clouds.

The masers are collisionally pumped in warm, dense molecular gas with densities and temperatures in the range 50–125 K, $\sim 10^3$–10^5cm^{-3} and OH column densities 10^{16}–10^{17} cm^{-2} (Elitzur 1976; Lockett, Gauthier & Elitzur 1999). These conditions are not met in gas overrun by standard "J-type" shock waves which dissociate molecules: although molecules reform as the shocked gas cools (e.g. Hollenbach & McKee 1989), cooling is so rapid that the OH column that is produced in the relevant temperature range is too small (Lockett, Gauthier & Elitzur 1999). Instead the molecular shock must be "C-type"(Mullan 1971; Draine 1980), in which heating and compression of the gas within the shock front is achieved by magnetically-driven ion-neutral collisions. The low fractional ionisation of the molecular gas means that this process is relatively slow, yielding a broad

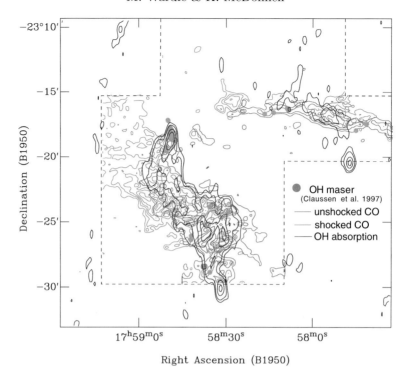

Figure 1. OH absorption from ATCA observations of SNR W28 (Green, Wardle & Lazendic 2000), overlaid on the locations of bright OH(1720 MHz) maser spots (Claussen *et al.* 1997), and contours of CO J=1–0 and J=3–2 emission (Arikawa *et al.* 1999) tracing unshocked and shocked gas, respectively. (Adapted from Arikawa *et al.* 1999).

shock front with an extended warm tail. However the requisite OH column is still higher than predicted by standard C-type shock models which rapidly convert OH into water (Draine, Roberge & Dalgarno 1983; Kaufman & Neufeld 1996). Instead an additional agent, such as X-rays or cosmic rays associated with the supernova remnant, is needed to dissociate water into OH as the gas cools behind the shock, (Wardle 1999).

The clear association of 1720 MHz masers with shocked gas (see e.g. Fig. 1), underpinned by a theoretical production mechanism hinging on molecular shock waves, provides a firm basis for assuming that the presence of these masers signals the existence of a shock wave driven by a supernova remnant into an adjacent cloud (Wardle & Yusef-Zadeh 2002). This has been confirmed by follow-up studies of individual remnants. A good illustration of this is provided by the supernova remnant G349.7+0.2, about which little was known apart from its having a radio continuum shell and four OH(1720 MHz) masers. A search for CO (J=1–0) emission from an adjacent molecular cloud, H_2 (v=1–0) emission at 2.12 μm from shocked molecular gas, and OH absorption associated with the masers found all three (Lazendic *et al.* 2010). The population of 1720 MHz-bearing supernova remnants show other hallmarks of interaction. For example "mixed-morphology" remnants are characterised a radio shell, or partial shell, filled with centrally-peaked emission in soft thermal x-rays (Long *et al.* 1991; Rho & Petre 1998) believed to be provided by the evaporation of dense interstellar gas in the interior of the remnants (White & Long 1991; Cox *et al.* 1999). There is a strong statistical association between remnants hosting 1720 MHz masers and mixed-morphology remnants (Yusef-Zadeh *et al.* 2003), and also with those with associated gamma-ray sources (Hewitt *et al.* 2009), which are likely produced by interactions of cosmic-rays accelerated in the remnant with the neighbouring molecular gas.

Apart from signalling locations where remnants are driving shock waves into molecular clouds, 1720 MHz masers are important probes of the interaction. The masers require significant columns of shocked molecular gas and good velocity coherence, and so tend to occur on the limbs of supernova remnant shells at velocities close to the systemic LSR velocity of the remnant, allowing distance estimates based on models of galactocentric rotation. Follow-up Zeeman observations at 1720 MHz have determined that the magnetic field in the post-shock gas is $\sim 1\text{--}3$ mG (Claussen *et al.* 1997, 1999; Brogan *et al.* 2000; Hoffman *et al.* 2005). The post-shock magnetic field acts as the piston driving a C-type shock wave, and a field of this strength provides sufficient post-shock pressure needed to drive a shock at $25\,\mathrm{km\,s^{-1}}$ into gas of density $\sim 10^4\,\mathrm{cm^{-3}}$, so these measurements provide an important sanity check on the basic picture. In addition the magnetic field measurements provides a good estimate of the pressure in the interior of the supernova remnant that is driving the shock into the molecular cloud.

2. Searches for 6049 MHz masers

While 1720 MHz masers are invaluable in pinpointing where supernova remnants interact with molecular clouds, their absence does not imply that an interaction is *not* occurring. Instead, one or more of the physical conditions required to collisionally invert the 1720 MHz transition may simply be absent – a sufficient column of OH, or line-of-sight velocity coherence within the OH column. Additional indicators of such interactions would therefore be useful.

The 1720 MHz inversion dies away for OH columns $\gtrsim 10^{17}\,\mathrm{cm^{-2}}$ due to photon trapping, but under similar physical conditions the analogous 6049 MHz transition in the first rotationally-excited state of OH becomes inverted (Pavlakis & Kylafis 2000; McDonnell, Wardle & Vaughan 2008). This prompted searches for masers at 6049 MHz associated with supernova remnants using Parkes (McDonnell, Wardle & Vaughan 2008), Effelsberg (Fish, Sjouwerman & Philstrom 2007) and the ATCA (McDonnell 2011) (see Table 1). These searches targeted SNR with 1720 MHz masers, mixed-morphology, or other evidence of interaction (note that there is substantial overlap between the SNR in the Parkes & Effelsberg and Parkes & ATCA searches), but all were unsuccessful.

Table 1. Searches for 6049 MHz masers

Telescope	N_{SNR}	beam	σ (mJy/beam)	Reference
Parkes	35	$200''$	~ 150	McDonnell, Wardle & Vaughan (2008)
Effelsberg	14	$130''$	~ 10	Fish, Sjouwerman & Philstrom (2007)
ATCA	17	$20'' \times 9''$	~ 22	McDonnell (2011)

The non-detections can be converted into upper limits on the 6049 MHz maser optical depth (i.e. maser amplification factor) using estimates of the brightness temperature of the background continuum and the likely beam filling factor of any putative maser emission. In turn, an upper limit on the maser optical depth can be translated to an upper limit on OH column density through the OH excitation calculations. It turns out that the relatively poor sensitivity of the Parkes observations does not place strong constraints on the OH column; by contrast, the derived limits on OH column density implied by the lack of detected 6049 MHz masers in the Effelsberg and ATCA searches are $\lesssim 10^{16.5}$ cm^{-2} (McDonnell 2011). These upper limits are at odds with the OH columns exceeding $\sim 10^{17}$ cm^{-2} that were inferred in nine supernova remnants by modelling the of OH absorption in the 1612, 1665, 1667 MHz lines observed with the Green Bank Telescope (Hewitt *et al.* 2006; Hewitt, Yusef-Zadeh & Wardle 2008). These nine remnants are listed in Table 2 – all but G349.7+0.2 were searched for 6049 MHz emission with Effelsberg and/or ATCA, and so one might have expected several detections.

One potential source of error is the 7.4 arc minute beam of the GBT observations. While a beam filling factor was included in the modelling by Hewitt *et al.* (2008) this does not account for different sub-structure in the background continuum and overlying OH absorption within the beam, and given that the inferred beam filling factors are typically $\lesssim 10^{-2}$ this omission may be significant. To check whether the inferred OH columns are systematically overestimated we conducted 1667 MHz OH absorption observations of 3 remnants (3C 391, W44, and IC 443) using the VLA in D array, yielding a beam size of $\sim 55'' \times 90''$. We followed the approach of Hewitt *et al.* (2008) in modelling the absorption at 1667 MHz in conjunction with 1720 MHz VLA data (the latter kindly provided by Jack Hewitt).

Notably, the derived solid angles subtended by the OH absorption components – which are typically marginally resolved in the VLA observations – are in reasonable agreement with the GBT results. However, the comparison of the derived OH columns gives mixed results: the peak column density derived from the VLA observations for 3C391, W44 and IC443 are much less, a little less, and somewhat greater than the GBT-derived columns, respectively.

Fig. 2 shows the 1667 MHz absorption against the continuum in IC 443. As expected, OH absorption surrounds the three known 1720 MHz OH masers, which are associated with molecular clumps clumps D and G and a new feature, labeled OH1. Focussing on clump G, which has the greatest column density, we derived model parameters quite similar to the GBT. This reflects that the OH emission is dominated by a single compact component and the background continuum is quite smooth. For 3C 491 by contrast, the continuum shows strong structure on the scale of the dominant absorption in the NW of the remnant (see Fig. 3). In this case the GBT observations, which have a beam area of about the size of the image in Fig. 3, underestimate the strength of the background continuum underlying the absorption by averaging over the beam, resulting in an over-estimate of the optical depth in the 1667 MHz line. The derived OH column density is therefore too large.

In summary, the VLA observations show that the GBT modelling of the does pretty well, but the poor resolution of single dish observations leads to significant overestimates of the OH column density when the background continuum and OH absorption have poorly correlated substructure within the beam. Nevertheless IC 443 and potentially some of the other remnants in Table 2 have OH column densities exceeding 10^{17} cm^{-3} and yet were not detected in the ATCA and Effelsberg 6049 MHz searches.

3. Line overlap

An alternative possibility is that the treatment of overlapping spectral lines in the OH excitation calculations is at fault. Line overlap in OH occurs even for velocity widths of order $1 \, \mathrm{km \, s^{-1}}$ because of the small splitting of the rotational levels arising through lambda doubling and hyperfine splitting. This enables photons emitted in one infrared transition between sub levels of different rotational states in one OH molecule to to be absorbed by a transition between different sub-levels in the same pair of rotational rates. This tends to scramble the level populations within the upper or lower pairs of sub levels

Figure 2. Grey-scale shows the continuum emission of IC 443 at 1667 MHz (Jy per $76'' \times 51''$ beam). The beam size and orientation are indicated by the ellipse in the lower left corner. Contours show the velocity-integrated OH absorption from 0.2 to 1.6 $\mathrm{Jy\,beam^{-1}\,km\,s^{-1}}$ in steps of 0.2 $\mathrm{Jy\,beam^{-1}\,km\,s^{-1}}$. The three crosses mark positions of 1720 MHz OH maser emission (Hewitt 2009). The OH absorption clumps B, C, D, F, and G are labelled following the identification of shocked CO clumps by Huang, Dickman & Snell (1986). The clump of OH absorption labelled OH1 A has not been previously identified.

Table 2. SNR with $N_{\mathrm{OH}} \gtrsim 10^{17}\,\mathrm{cm^{-2}}$ inferred from GBT observations.

SNR	N_{OH} (dex) from GBT	6049 MHz searches[1]	N_{OH} (dex) from VLA
W28	17.14 ± 0.01	P E A	
G16.7+0.1	16.9 ± 0.6	P E A	
3C391	17.1 ± 0.4	P E	< 16.5
W44	17.14 ± 0.01	P E	< 17.1
IC443	17.11 ± 0.01	P E	~ 17.5
CTB37A	17.1 ± 0.4	P A	
G349.7+0.2	17.04 ± 0.02	P	
G357.7-0.1	17.12 ± 0.01	P A	
G359.1-0.5	17.13 ± 0.01	P A	

[1] P=Parkes, E=Effelsberg, A=ATCA

inside each rotational level and suppress population inversion in the 1720 and 6049 MHz transitions, quenching the masers.

Line overlap is at present treated *locally* in the OH excitation code, i.e. the OH-bearing cloud is treated as a uniform slab with a velocity width that is everywhere the same.

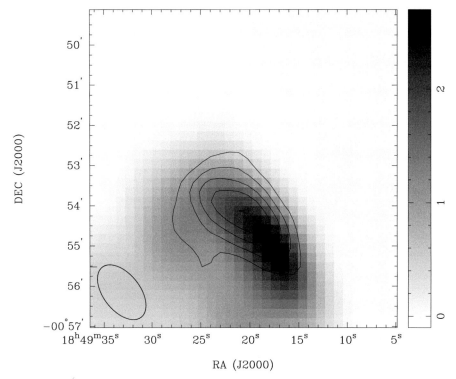

Figure 3. Grey-scale shows the continuum emission of the NE corner of 3C 391 at 1667 MHz between -0.1 and 2.7 Jy per $96'' \times 56''$ beam. Absorption in the 1667 MHz line of OH is indicated by the contours, with levels at -1, -0.8, -0.6, -0.4, -0.2 Jy beam^{-1} km s^{-1}. This image is $\sim$7.44 arcmin on a side, approximately matching the GBT beam area of the 3C 391 spectrum from Hewitt *et al.* (2008).

Its effects are illustrated in Fig. 4. The upper panel shows the maser optical depth as a function of OH column density in a uniform slab when overlap is neglected. In this case, line optical depths (ie. photon escape probabilities) scale as $N_{\rm OH}/\Delta v$ so the curves simply shift to the right as Δv is increased. The right hand column shows the effect of including local line overlap: once the line width exceeds 0.5 km s^{-1}, line overlap starts to shuffle the populations within rotational levels and the inversions are suppressed.

Observed 1720 MHz maser lines have $\Delta v \sim 1$ km s^{-1}, at which point local line overlap would suppress the masers, so this treatment overestimates the effect of line overlap. In reality the OH column does not have a single, uniform, velocity profile throughout – instead, the column of OH has local line widths that are much smaller than the net velocity width created by the systematic velocity gradient along the line of sight. This situation requires a "non-local" treatment of line overlap effects, in which photons emitted by an infrared transition are absorbed in a neighbouring transition at a different layer in the slab (e.g. Doel, Gray & Field 1990). This form of overlap is less effective, potentially enabling the 1720 MHz line to remain inverted for larger (~ 1 km s^{-1}) line widths; the 6049 MHz line however, may then only be weakly inverted at column densities $\sim 10^{17}$ cm^{-2}. Resolution of this will have to await calculations including non-local overlap.

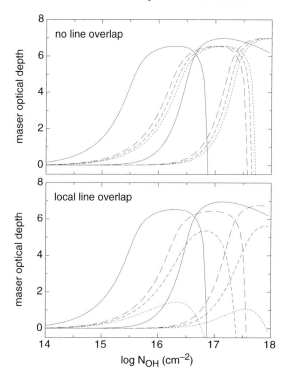

Figure 4. 1720 MHz (red) and 6049 MHz (blue) maser optical depth as a function of OH column density for a uniform slab with $n(H_2) = 10^4 cm^{-3}$ and $T = 50 K$ as the total (thermal + micro-turbulent) FWHM line width is increased: $0.1 km s^{-1}$ (*solid*), $0.5 km s^{-1}$ (*long-dashed*), $0.6 km s^{-1}$ (*short-dashed*), $0.7 km s^{-1}$ (*dotted*). The upper and lower panels show the results obtained neglecting or including local line overlap, respectively (see text).

4. Discussion

The lack of detected OH(6049 MHz) masers in the SNR searches conducted to date may be attributed to a combination if two factors. First, the OH column densities inferred from GBT observations may be systematically overestimated because of the tendency for OH absorption to overlay regions where the background continuum is brightest. If the continuum has structure within the beam the optical depth (and so the OH column) will tend to be overestimated, as appears to be the case in 3C 391 and W44. However, IC 443 clump G appears to have a column density $3 \times 10^{17} cm^{-2}$, more than twice that estimated from the single-dish observations. Therefore several additional remnants listed in Table 1 may still have OH column densities exceeding the $\sim 10^{17} cm^{-2}$ threshold suggested for 6049 MHz inversion. Second, non-local line overlap might suppress 6049 MHz inversion more effectively than the 1720 MHz transition, or simply push the threshold OH column density required for significant maser amplification to columns exceeding $3 \times 10^{17} cm^{-2}$.

Acknowledgements

We thank Jack Hewitt for providing his 1720 MHz VLA data.

References

Arikawa, Y., Tatematsu, K., Sekimoto, Y., & Takahashi, T. 1999, *PASJ*, 51, L7

Brogan, C. L., Frail, D. A., Goss, W. M., & Troland, T. H. 2000, *ApJ*, 537, 875

Claussen, M. J., Frail, D. A., Goss, W. M., & Gaume, R. A. 1997, *ApJ*, 489, 143

Claussen, M. J., Goss, W. M., Frail, D. A., & Desai, K. 1999, *ApJ*, 522, 349

Cox, D. P., Shelton, R. L., Maciejewski, W., Smith, R. K., Plewa, T., Pawl, A., & Rózyczka, M. 1999, *ApJ*, 524, 179

DeNoyer, L. K. 1979a, *ApJ* (Letters), 228, L41

DeNoyer, L. K. 1979b, *ApJ* (Letters), 232, L165

Doel, R. C., Gray, M. D., & Field, D. 1990, *MNRAS*, 244, 504

Draine, B. T. 1980, *ApJ*, 241, 102

Draine, B. T., Roberge, W. G., & Dalgarno, A. 1983, *ApJ*, 264, 485-507

Elitzur, M. 1976, *ApJ*, 203, 124

Fish, V. L., Sjouwerman, L. O., & Pihlström, Y. M. 2007, *ApJ* (Letters), 670, L117

Frail, D. A., Goss, W. M., & Slysh, V. I. 1994, *ApJ*, 424, L111

Frail, D. A., Goss, W. M., Reynoso, E. M., Giacani, E. B., Green, A. J., & Otrupcek, R. 1996, *AJ*, 111, 1651

Goss, W. M. & Robinson, B. J. 1968, *Astrophys. Lett.*, 2, 81

Green, A. J., Frail, D. A., Goss, W. M., & Otrupcek, R. 1997, *AJ*, 114, 2058

Green, A. J. & Wardle, M, Lazendic, J. S. 2000, IAU XXIV General Assembly, JD1 , abstract.

Hewitt, J. W., Yusef-Zadeh, F., Wardle, M., Roberts, D. A., & Kassim, N. E. 2006, *ApJ*, 652, 1288

Hewitt, J. W., Yusef-Zadeh, F., & Wardle, M. 2008, *ApJ*, 683, 189

Hewitt, J. W., Yusef-Zadeh, F., & Wardle, M. 2009, *ApJ* (Letters), 706, L270

Hewitt, J. W., 2009, PhD thesis, Northwestern University

Huang, Y.-L., Dickman, R. L., & Snell, R. L. 1986, *ApJ* (Letters), 302, L63

Hoffman, I. M., Goss, W. M., Brogan, C. L., Claussen, M. J., & Richards, A. M. S. 2003, *ApJ*, 583, 272

Hollenbach, D. J. & McKee, C. F. 1989, *ApJ*, 342, 306

Koralesky, B., Frail, D. A., Goss, W. M., Claussen, M. J., & Green, A. J. 1998, *AJ*, 116, 1323

Kaufman, M. J. & Neufeld, D. A. 1996, *ApJ*, 456, 250

Lazendic, J. S., Wardle, M., Whiteoak, J. B., Burton, M. G., & Green, A. J. 2010, *MNRAS*, 409, 371

Lockett, P., Gauthier, E., & Elitzur, M. 1999, *ApJ*, 511, 235

Long, K. S., Blair, W. P., Matsui, Y., & White, R. L. 1991, *ApJ*, 373, 567

McDonnell, K. E., Wardle, M., & Vaughan, A. E. 2008, *MNRAS*, 390, 49

McDonnell, K. E., 2011, PhD thesis, Macquarie University

Mullan, D. J. 1971, *MNRAS*, 153, 145

Pavlakis, K. G. & Kylafis, N. D. 2000, *ApJ*, 534, 770

Reach, W. T. & Rho, J. 1999, *ApJ*, 511, 836

Rho, J. & Petre, R. 1998, *ApJ* (Letters), 503, L167

Sjouwerman, L. O. & Pihlström, Y. M. 2008, *ApJ*, 681, 1287

Wardle, M. 1999, *ApJ* (Letters), 525, L101

Wardle, M. & Yusef-Zadeh, F. 2002, *Science*, 296, 2350

White, R. L. & Long, K. S. 1991, *ApJ*, 373, 543

Wootten, A. 1981, *ApJ*, 245, 105

Yusef-Zadeh, F., Roberts, D. A., Goss, W. M., Frail, D. A., & Green, A. J. 1996, *ApJ* (Letters), 466, L25

Yusef-Zadeh, F., Roberts, D. A., Goss, W. M., Frail, D. A., & Green, A. J. 1999a, *ApJ*, 512, 230

Yusef-Zadeh, F., Goss, W. M., Roberts, D. A., Robinson, B., & Frail, D. A. 1999b, *ApJ*, 527, 172

Yusef-Zadeh, F., Wardle, M., & Roberts, D. A. 2003, *ApJ*, 583, 267

Yusef-Zadeh, F., Wardle, M., Rho, J., & Sakano, M. 2003, *ApJ*, 585, 319

Cosmic Masers - from OH to H$_0$
Proceedings IAU Symposium No. 287, 2012
R.S. Booth, E.M.L. Humphreys & W.H.T. Vlemmings, eds.
© International Astronomical Union 2012
doi:10.1017/S1743921312007508

Class I Methanol Masers in the Galactic Center

Loránt O. Sjouwerman[1] & Ylva M. Pihlström[2]†

[1]National Radio Astronomy Observatory, P.O. Box O, 1003 Lopezville Rd., Socorro, NM 87801
[2]Department of Physics and Astronomy, University of New Mexico, MSC07 4220, Albuquerque, NM 87131

Abstract. We report on 36 and 44 GHz Class I methanol (CH$_3$OH) maser emission in the Sagittarius A (Sgr A) region with the Expanded Very Large Array (EVLA). At least three different maser transitions tracing shocked regions in the cm-wave radio regime can be found in Sgr A. 44 GHz masers correlate with the positions and velocities of 36 GHz CH$_3$OH masers, but the methanol masers correlate less with 1720 MHz OH masers. Our results agree with theoretical predictions that the densities and temperatures conducive for 1720 MHz OH masers may also produce 36 and 44 GHz CH$_3$OH maser emission. However, many 44 GHz masers do not overlap with 36 GHz methanol masers, suggesting that 44 GHz masers also arise in regions too hot and too dense for 36 GHz masers to form. This agrees with the non-detection of 1720 MHz OH masers in the same area, which are thought to be excited under cooler or denser conditions. We speculate that the geometry of the bright 36 GHz masers in Sgr A East outlines the location of a SNR shock front.

Keywords. Galaxy: center — Masers — Shock waves — Supernovae: individual(Sgr A East)

1. Introduction

The Sagittarius A complex is one of the best studied regions in the sky and encompasses several interesting phenomena like the nearest supermassive nuclear black hole (Sgr A*), the circumnuclear disk (CND), star forming regions (SFRs) and supernova remnants (SNRs). The line of sight toward the Sgr A complex consists of the SNR Sgr A East in the back and the CND (whose ionized part is known as Sgr A West) in the front partly overlapping with Sgr A East. Molecular gas is abundant and well distributed; the CND consists of irregularly distributed clumps of molecular gas and there are two giant molecular cloud cores (GMCs) called the +20 and +50 km s^{-1} clouds (M−0.13−0.08 and M−0.02 − 0.07 respectively). These GMCs form the molecular belt stretching across the Sgr A complex, providing the interstellar medium (ISM) that interacts with Sgr A East. A nice and recent comprehensive overview is presented by Amo-Baladrón, Martín-Pintado & Martín (2011).

Bright maser lines are useful probes of physical conditions within molecular clouds, especially when mapped in detail by interferometers. One example is the collisionally pumped 1720 MHz OH maser which is widely recognized as a tracer for shocked regions, observed both in SFRs and SNRs. In SNRs, Very Large Array (VLA) observations have shown that the OH masers originate in regions where the SNR shock collides with the interstellar medium (e.g. Claussen *et al.* 1997, Yusef-Zadeh *et al.* 2003, Frail & Mitchell 1998). Such OH masers are numerous in Sgr A East, thus probing the conditions of the interaction regions between the +50 and +20 km s^{-1} clouds and the SNR Sgr A East.

Dense gas structures in the Galactic center region, including Sgr A East, are traced by ammonia and methanol thermal emission (Coil & Ho 2000, Szczepanski *et al.* 1989,

† Ylva Pihlström is also Adjunct Astronomer at the National Radio Astronomy Observatory

Szczepanski, Ho & Gusten 1991). Methanol abundances are high enough to produce maser emission. Like 1720 MHz OH masers, Class I methanol masers such as the 36 and 44 GHz transitions are excited through collisions. Theoretical modeling of collisional OH excitation predicts that the 1720 MHz OH should be found in regions of $n \geqslant 10^5$ cm^{-3}, $T \sim 75$ K (Gray, Doel & Field 1991, Gray, Field & Doel 1992, Wardle 1999, Lockett, Gauthier & Elitzur 1999, Pihlström et al. 2008). The number density and temperature required for 36 GHz methanol masers are near those modeled for 1720 MHz OH masers, with $n \sim 10^4 - 10^5$ cm^{-3} and $T < 100$ K (Morimoto, Kanzawa & Ohishi 1985, Cragg et al. 1992, Liechti & Wilson 1996). At least in SFRs, higher densities and temperatures, $n \sim 10^5 - 10^6$ cm^{-3} and $T = 80 - 200$ K, the Class I 44 GHz line will have optimized maser output, while the 36 GHz maser eventually becomes quenched (Pratap et al. 2008, Sobolev et al. 2005, Sobolev et al. 2007). These methanol masers may therefore constrain the density in the shocked SNR regions. In turn upper limits can be used to estimate the importance of compression by shocks in the formation of stars near SNRs.

That Class I methanol maser lines are detectable in SNR/cloud interaction regions was shown by Sjouwerman, Pihlström & Fish (2010), using the 7 first antennas outfitted with 36 GHz receivers at the Expanded VLA (EVLA). Several bright masers were found near the 1720 MHz OH masers in the Sgr A East molecular cloud - SNR interaction region; a feature also observed by many others (e.g. Tsuboi, Miyazaki & Okumura 2009). To see whether the relation between the collisionally excited 36 and 44 methanol masers and 1720 MHz OH holds in general, we here present the result of a search for Class I 44 GHz methanol maser emission in the Sgr A region.

Meanwhile we have surveyed the entire $\sim 6' \times 8'$ Sgr A complex for 36 and 44 GHz methanol emission for which the complete results will be reported elsewhere; here we concentrate on the early detections in the $\sim 4' \times 4'$ Sgr A region.

2. Observations and Results

The EVLA was used in its C configuration to observe the transition of CH$_3$OH at 44 069.41 MHz with a bandwidth of 8 MHz in dual polarization at a velocity resolution of approximately 0.2 km s^{-1}. The primary beam at 44 GHz is about 56$''$; initially we selected five pointing positions ("A" through "E", see Figure 1) with central velocities based on previous results on 1720 MHz OH masers and 36 GHz methanol masers. Later we covered the whole Sgr A region as part of a larger $\sim 6' \times 8'$ survey. Position A corresponds to a region of high-velocity 1720 MHz OH masers belonging to the circumnuclear disk, covering LSR velocities between 106 and 157 km s^{-1}. In position B we previously detected 36 GHz methanol masers at velocities around 23 km s^{-1} (LSR coverage -3 to $+39$ km s^{-1}). This pointing position partly overlaps on the sky with pointing position C which has a central velocity of 48 km s^{-1} (LSR coverage $22 - 74$ km s^{-1}) based on the 1720 MHz OH masers. Finally, positions D and E correspond to a region where the 50 km s^{-1} molecular cloud interacts with Sgr A East, and where 1720 MHz masers are numerous. The LSR velocity coverage for these pointing positions was $22 - 74$ km s^{-1}.

The data were reduced using standard procedures in AIPS, using 3C 286 as the flux density calibrator resulting in a typical absolute flux density uncertainty of 15%. In the fields were masers were found, a strong maser channel was used for self-calibration. Each cube was CLEANed with robust weighting down to a level of five times the theoretical rms over a field approximately twice the primary beam. This was necessary since some masers appeared in the sidelobes and needed to be accounted for in the CLEANing process. The resulting typical channel rms noise is $15 - 20$ mJy/beam, and the restoring beam is $1.3 \times 0.5''$.

The image cubes were searched for masers, and parameters for the individual features were extracted using the AIPS task JMFIT, with peak fluxes corrected for primary beam attenuation using PBCOR. A few weaker masers exist in the cubes, but they are all located close to the brighter masers in position and velocity, and will not change any of the discussion in Sect. 3.

A few features show more than one spectral peak at a given position, implying there is structure on scales smaller than the EVLA beam. Some spectral features have wings of weaker emission extending over $6 - 8$ km s^{-1}. Since the observations were taken when the EVLA was in C-configuration, we suspect that some of this broad and weak emission is of thermal origin. The peak flux density of the spectral features corresponds to brightness temperatures exceeding 10^3 K, but as both the 44 GHz and 36 GHz masers are thought to be excited by the same process, it is likely that at least the 44 GHz detections co-located with 36 GHz masers are masers too.

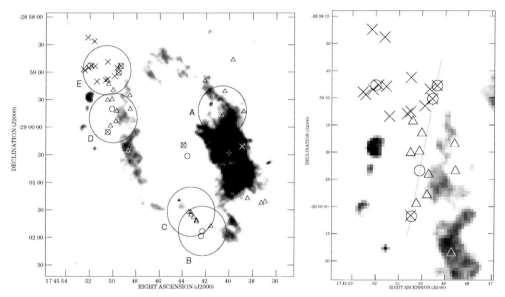

Figure 1. Relative sky positions of the 44 GHz methanol masers (crosses), 36 GHz methanol masers (circles) and 1720 MHz OH masers (triangles). The white plus symbol marks the position of Sgr A*. The large circles show the five field-of-view positions covered. A blow-up of the upper left region is shown at the right hand side, showing an apparent systematic offset in the location of the three maser species, with the 44 GHz methanol masers concentrated to the northeast (upper left), the 1720 MHz OH masers more to the southwest (lower right) and a NNW-SSE line of (four) very bright 36 GHz methanol masers, three co-spatial with 44 GHz, roughly dividing the two regions. The dashed line shows the alignment of 36 GHz masers with the shock front in the northnorthwest-southsoutheast direction.

Three masers were detected outside the primary beam (large circles in Figure 1) and their flux density is therefore less certain than for masers located within the primary beam. The position and velocity of maser 2 at 50 km s^{-1} agrees very well with the position of a 36 GHz maser at 51 km s^{-1} (Sjouwerman, Pihlström & Fish 2010), and we therefore trust that this detection is real. This 44 GHz maser has previously been reported on by Yusef-Zadeh *et al.* (2008), and is associated with a molecular clump 'G'. Similarly, we confirm their maser 'V' (our maser 1) in our data taken for the phase-reference calibrator J1745−2900 (i.e. Sgr A*), and both 'G' and 'V' in archival VLA data taken on 2009 April 23. The 44 GHz maser associated with clump 'F' reported by Yusef-Zadeh *et al.* (2008) is not confirmed by our, nor the archival, observations. The additional two masers detected outside the primary beam also do not have 36 GHz masers directly associated with them.

Figure 1 plots the position of the detected 44 GHz methanol sources compared to the previously detected 36 GHz and 1720 MHz OH masers. The right hand side panel of Figure 1 shows the northeast region of Sgr A East where most 44 GHz masers are located. From these plots a few main results can be concluded. Firstly, the positions and velocities of several 44 GHz masers (Figure 1) agree to within the errors with the values reported by Sjouwerman, Pihlström & Fish (2010) for the 36 GHz masers. Secondly, there is a systematic difference between the overall distributions of the 1720 MHz OH, 36 GHz methanol and 44 GHz methanol masers. In the northeast (upper left) region, in overlap with the densest part of the $50\,\mathrm{km\,s^{-1}}$ cloud, the 44 GHz masers are offset to the northeast with respect to a narrow, almost linear southsoutheast to northnorthwest distribution of the 36 GHz masers. The 1720 MHz OH masers are found on the other side of the 36 GHz masers, near the radio continuum of the SNR to the southwest (lower right). A positional offset between the OH and methanol is also observed in the southeastern interaction region, where the SNR G359.02–0.09 overlaps the Sgr A East continuum (Coil & Ho 2000, Herrnstein & Ho 2005). These positional offsets are discussed further in Sect. 3.

3. Discussion

3.1. *36 GHz versus 44 GHz Methanol Masers*

Modeling of methanol masers suggest that the 36 GHz transition occurs under somewhat cooler and less dense ($T \sim 30 - 100\,\mathrm{K}$, $n \sim 10^4 - 10^5\,\mathrm{cm^{-3}}$) conditions than the 44 GHz transition ($T \sim 80 - 200$ K, $n \sim 10^5 - 10^6\,\mathrm{cm^{-3}}$; see, e.g. Pratap *et al.* 2008). The range of physical conditions do however overlap, and some spatial overlap could therefore be expected. Seven 44 GHz masers show an almost perfect overlap in both position and velocity with 36 GHz masers (Sjouwerman, Pihlström & Fish 2010). Here, according to the modeling, the densities and temperatures should be close to $10^5\,\mathrm{cm^{-3}}$ and 100 K to produce both methanol maser lines.

It is striking that the brightest 36 GHz masers, all to the northwest in positions D and E (Sjouwerman, Pihlström & Fish 2010), are narrowly distributed along a line roughly from north to south, of which three coincide with 44 GHz masers in position and velocity within the errors. This NNW-SSE division more or less coincides with the sharp gradient in low-frequency radio continuum emission of the Sgr A East SNR (Pedlar *et al.* 1989) and appears to be located in the sheath in the CS emission as mapped by Tsuboi, Miyazaki & Okumura (2009). The mean velocity of each transition is 46 $\mathrm{km\,s^{-1}}$, implying that they arise in similar regions of the molecular cloud where the velocities still are less disturbed by the SNR shock (see Sect. 3.2). Apart from two individual exceptions located far from this area, we do not find any 44 GHz (nor 36 GHz) masers westward of this line in our pointings. It is therefore tempting to speculate that the line delineates the arrival of the shock front, where enough material has been swept up to provide the density for the creation of (perhaps due to geometry very bright 36 GHz) methanol masers, but not yet enough energy has dissipated to dissociate all methanol or to significantly disturb the velocity structure by means of a reverse shock (Section 3.2).

In the northeastern part of Sgr A East toward the core of the $50\,\mathrm{km\,s^{-1}}$ cloud, there is a group of 44 GHz masers with a distinct positional offset from the 36 GHz masers. The narrower distribution of 36 GHz masers suggests that the conditions required to produce masers in this transition are not fulfilled further to the northeast. The position of the 36 GHz emission is consistent with being just in the SNR/cloud interaction region, while the 44 GHz masers may be found deeper inside the denser parts of the cloud. These 44 GHz masers, which are typically found to be brighter than the 36 GHz masers, may originate near sites of massive star formation instead (e.g. Pratap *et al.* 2008, Fish *et al.* 2011). The lack of companion 36 GHz masers in this putative star-forming region (e.g. Tsuboi, Miyazaki & Okumura 2009) therefore may be due to the limited sensitivity of the 36 GHz observations. This picture, at least for this region in the Galactic center, in which 44 GHz masers are primarily associated with cloud cores and 36 GHz masers are found at the boundaries of the SNR interaction regions, is consistent with theoretical models indicating that 44 GHz masers can be produced at higher densities than 36 GHz masers. The existence of 36 GHz masers without accompanying 44 GHz masers in positions B and C may then indicate the interaction region of two SNRs without the presence of a dense cloud.

3.2. *OH versus Methanol Masers*

As is the case with Class I methanol masers, 1720 MHz OH masers are used as tracers of shocked regions. The presence of 1720 MHz OH masers indicates the presence of C-shocks (e.g. Lockett, Gauthier & Elitzur 1999). Modeling of OH and CH_3OH shows that the three maser transitions discussed here require similar densities and temperatures. This agrees well with the detection of all three masers in Sgr A East. However, we observe a distinct offset in positions between the methanol and OH masers (Figure 1). In the northeast interaction region between the $50\,km\,s^{-1}$ cloud and Sgr A East the OH masers are found more to the southwest. In addition, the 1720 MHz OH masers have a higher mean velocity of $\sim 57\,km\,s^{-1}$ versus $\sim 46\,km\,s^{-1}$ for the methanol. However, the 1720 MHz OH does overlap in the sky with the line of 36 GHz masers.

A similar offset is observed in the southeastern interaction region in pointing positions B and C, where the methanol masers are offset southwest from the OH. The OH mean velocities here are $58\,km\,s^{-1}$ to be compared to the $24.5\,km\,s^{-1}$ for the 36 GHz methanol. No 44 GHz methanol was detected in this region.

The association between 1720 MHz OH and 36 GHz methanol emission may be due to the processes that form these molecules. OH is created by dissociation of H_2O (and maybe also CH_3OH), and the propagation of a C-shock creates densities and temperatures suitable for 1720 MHz OH inversion. Thus, OH masers should preferentially be found in the SNR post-shock region. This agrees with the OH masers being co-located with positions of radio continuum, outlining regions where electrons have been accelerated by the shock. The production of methanol is less well understood, but it is believed that methanol is released from grains, either by sputtering from a shock or by evaporation when temperatures reach above $100\,K$ (Hidaka *et al.* 2004, Menten *et al.* 2009, Bachiller & Perez Gutierrez 1997, Voronkov *et al.* 2006, Hartquist *et al.* 1995).

References

Amo-Baladrón, M. A., Martín-Pintado, J., & Martín, S. 2011, *A&A*, 526, A54
Bachiller, R. & Perez Gutierrez, M., 1997, *ApJ*, 487 L93
Claussen, M. J., Frail, D. A., Goss, W. M., & Gaume, R. A. 1997, *ApJ*, 489, 143
Coil, A. L. & Ho, P. T. P. 2000, *ApJ*, 533, 245
Cragg, D. M., Johns, K. P., Godfrey, P. D., & Brown, R. D., 1992, *MNRAS*, 259, 203
Fish, V. L., Muehlbrad, T. C., Pratap, P., Sjouwerman, L. O., Strelnitski, V., Pihlström, Y. M., & Bourke, T. L., 2011, *ApJ*, 729, 14
Frail, D. A. & Mitchell, G. F. 1998, *ApJ*, 508, 690
Gray, M. D., Doel, R. C., & Field, D. 1991, *MNRAS*, 262, 30
Gray, M. D., Field, D., & Doel, R. C. 1992, *A&A*, 264, 220

Hartquist, T. W., Menten, K. M., Lepp, S., & Dalgarno, A., 1995, *MNRAS*, 272, 184

Herrnstein, R. M. & Ho, P. T. P. 2005, *ApJ*, 620, 287

Hidaka, H., Watanabe, N., Shiraki, T. m Nagaoka, A., & Kouchi, A., 2004, *ApJ*, 614, 1124

Liechti, S. & Wilson, T. L., 1996, *A&A*, 314, 615

Lockett, P., Gauthier, E., & Elitzur, M. 1999, *ApJ* (Letters), 511, L235

Menten, K. M., Wilson, R. W., Leurini, S., & Schilke, P., 2009, *ApJ*, 692, 47

Morimoto, M., Kanzawa, T., & Ohishi, M., 1985, *ApJ* (Letters), 288, L11

Pedlar, A., Anantharamaiah, K. R., Ekers, R. D., Goss, W. M., van Gorkom, J. H., Schwarz, U. J., & Zhao, J.-H. 1989, *ApJ*, 342, 769

Pihlström, Y. M., Fish, V. L., Sjouwerman, L. O., Zschaechner, L. K., Lockett, P. B., & Elitzur, M., *ApJ*, 676, 371

Pratap, P., Shute, P. A., Keane, T. C., Battersby, C., & Sterling, S., 2008, *AJ*, 135, 1718

Szczepanski, J. C., Ho, P. T. P., Haschick, A. D., & Baan, W. A., 1989, *IAU Symp.* 136, 383

Szczepanski, J. C., Ho, P. T. P., & Gusten, R. 1991, *ASP Conf. Series*, Vol. 16, 143

Sjouwerman, L. O., Pihlström, Y. M., & Fish, V. L. 2010, *ApJ* (Letters), 710, L111

Sobolev, A. M., Cragg, D. M., Ellingsen, S. P., Gaylard, M. J., Goedhart, S., Henkel, C., Kirsanova, M. S., Ostrovskii, A. B., Pankratova, N. V., Shelemei, O. V., van der Walt, D. J., Vasyunina, T. S., & Voronkov, M. A., 2007, *IAU Symp.*, 242, Vol. 242, 81

Sobolev, A. M., Ostrovskii, A. B., Kirsanova, M. S., Shelemei, O. V., Voronkov, M. A., & Malyshev, A. V., 2005, *IAU Symp.* 227, Vol. 227, 174

Tsuboi, M., Miyazaki, A., & Okumura, S. K. 2009, *PASJ*, 61, 29

Voronkov, M. A., Brooks, K. J., Sobolev, A. M., Ellingsen, S. P., Ostrovskii, A. B., & Caswell, J. L., 2006, *MNRAS*, 373, 411

Wardle, M. 1999, *ApJ* (Letters), 525, L101

Yusef-Zadeh, F., Braatz, J., Wardle, M., & Roberts, D. 2008, *ApJ* (Letters), 683, L147

Yusef-Zadeh, F., Wardle, M., Rho, J., & Sakano, M. 2003, *ApJ*, 585, 319

Cosmic Masers - from OH to H_0
Proceedings IAU Symposium No. 287, 2012
R.S. Booth, E.M.L. Humphreys & W.H.T. Vlemmings, eds.

© International Astronomical Union 2012
doi:10.1017/S174392131200751X

Radio Recombination Line Maser Objects: New Detections with the SMA

Izaskun Jiménez-Serra

Harvard-Smithsonian Center for Astrophysics,
60 Garden Street, 02138 Cambridge, MA, USA
email: ijimenez-serra@cfa.harvard.edu

Abstract. Hydrogen radio recombination line (RRL) masers are a rare phenomenon in star forming regions. Since RRL masers were first detected in 1989 toward the emission line star MWC349A, several single-dish surveys at millimeter wavelengths have been carried out to detect other RRL maser objects. However, although RRL maser amplification is expected to appear at wavelengths <2mm, MWC349A still remains as the only RRL maser object known to date. In this contribution, I will present our recent findings of two new RRL maser objects with the Submillimeter Array (SMA) toward the massive star forming regions Cepheus A HW2 and MonR2-IRS2. Sub-millimeter observations with interferometers such as the SMA and the Atacama Large Millimeter Array (ALMA) open the possibility to detect a much larger sample of RRL maser objects, where very detailed information about the kinematics and physical structure of the innermost ionized regions can be obtained toward these objects.

Keywords. masers, stars: formation, (ISM:) HII regions, techniques: interferometric

1. Introduction

Calculations of the level populations of atomic hydrogen in HII regions, very early showed that global population inversions could exist across the Rydberg levels of the hydrogen atom (Cillié 1936; Baker & Menzel 1938). These population inversions lead to the formation of hydrogen recombination lines under non-LTE (Local Thermodynamic Equilibrium) conditions, so that stimulated emission can occur. This idea has been confirmed by detailed theoretical modeling of the non-LTE populations in recombining hydrogen (Walmsley 1990; Strelnitski, Ponomarev & Smith 1996; Martín-Pintado 2002). In particular, this modeling shows that for ultracompact (UC) HII regions the measured electron densities ($N_e \sim 10^4$-3×10^5 cm^{-3}; Wood & Churchwell 1989) are high enough to significantly invert the level populations involved in RRL transitions at millimeter, submillimeter and far-IR wavelengths. Therefore, RRL maser amplification at wavelengths <2mm is expected to be a common phenomenon in massive star forming regions.

By carrying out single-dish observations of several RRLs toward the massive Young Stellar Object (YSO) MWC349A, Martín-Pintado, Bachiller & Thum (1989) reported the first detection of strong RRL maser emission at 1.3mm (i.e. for the H29α, H30α, and H31α lines). The line profiles of the observed 1.3mm RRLs are double-peaked, and their peak intensities are factors of ⩾50 brighter than the RRLs at 3mm. Since the derived line-to-continuum flux ratios toward MWC349A clearly exceed those predicted under LTE conditions†, Martín-Pintado, Bachiller & Thum (1989) concluded that the RRLs detected at 1.3mm toward MWC349A are masers.

† The line-to-continuum flux ratios provide a measurement of the departure from LTE for the observed RRLs, and are calculated as $\Delta v T_L / T_C$, with $\Delta v T_L$ the RRL integrated intensity and T_C the free-free continuum flux at the frequency of the RRL, ν. Under LTE and optically thin emission, $\Delta v T_L / T_C \propto \nu^{1.1}$ (Martín-Pintado, Bachiller & Thum 1989).

Since the discovery of RRL maser emission toward MWC349A, extensive searches of RRL masers have been carried out with single-dish telescopes toward other massive YSOs (Martín-Pintado 2002). However, MWC349A still remains as the only RRL maser object firmly reported to date. In this contribution, I will present the detection of two new RRL maser objects toward the Cepheus A HW2 and MonR2 star forming regions, carried out with the SMA.

2. Extremely broad RRL maser emission toward Cepheus A HW2

Cepheus A East is a very active region of massive star formation that shows a plethora of UC HII regions, water masers, and molecular outflows (e.g. Hughes & Wouterloot 1984; Narayanan & Walker 1996; Torrelles *et al.* 2011). The brightest radio continuum source, HW2, is a B0 star that has a rotating (and possibly photo-evaporating) molecular disk (Jiménez-Serra *et al.* 2007; Jiménez-Serra *et al.* 2009). In addition, HW2 powers a collimated, high-velocity ionized jet (Rodríguez *et al.* 1994), whose proper motions are consistent with expanding velocities $\geqslant 500\,\mathrm{km\,s^{-1}}$ (Curiel *et al.* 2006). This implies that if RRLs were detected toward this source, they would show extremely broad line profiles with linewidths $\geqslant 1000\,\mathrm{km\,s^{-1}}$.

Figure 1 reports the full spectra measured toward Cepheus A HW2 at the frequencies of the H40α, H34α, and H31α RRLs. The observed spectra show three different features: i) a strong slope due to the increase of the dust continuum emission with frequency ($S_\nu \propto \nu^\alpha$); ii) a forest of narrow molecular lines mostly arising from the HC source at 0.3" east HW2 (Jiménez-Serra *et al.* 2009); and iii) three faint and extremely broad features extending in velocity from $\sim -500\,\mathrm{km\,s^{-1}}$ to $\sim 600\,\mathrm{km\,s^{-1}}$ (zero-intensity linewidths of $\sim 1100\,\mathrm{km\,s^{-1}}$). Jiménez-Serra *et al.* (2011) have proposed that these extremely broad features correspond to RRL emission arising from the high-velocity HW2 radio jet.

The RRLs detected toward Cepheus A HW2 show asymmetric double-peaked line profiles, with the blue-shifted component slightly brighter than the red-shifted gas (Figure 1). Like MWC349A, the derived line-to-continuum flux ratios for the RRLs at 2 mm and 1.3 mm are significantly larger than those predicted under LTE, indicating that the RRLs are likely affected by maser amplification (see Table 1 and Jiménez-Serra *et al.* 2011).

By using the 3D radiative transfer code of Martín-Pintado *et al.* (2011), we have modeled the intensity and distribution of the free-free continuum and RRL emission toward Cepheus A HW2. We have considered an isothermal ($T_e \sim 10^4$ K), collimated (semi-opening angle $\sim 18°$), and bi-conical jet, where the ionized gas is accelerated constantly to reach a velocity of $500\,\mathrm{km\,s^{-1}}$ at 35 AU from the protostar. Our model not only reproduces the fluxes and morphology of the free-free continuum emission measured at centimeter wavelengths with the Very Large Array (VLA; Rodríguez *et al.* 1994), but also the extremely broad asymmetric profiles of the H40α, H34α, and H31α RRLs.

Our model, however, fails to reproduce two features of the RRLs observed toward Cepheus A HW2: i) the peak intensities of the H34α and H31α lines, expected to be more affected by maser amplification; and ii) the velocity shift of $160\,\mathrm{km\,s^{-1}}$, with respect to the radial velocity of the source of $-10\,\mathrm{km\,s^{-1}}$, required to reproduce the lines. For the former feature, the discrepancies between the observed and predicted intensities are due to uncertainties in the LTE departure coefficients used in the model (see the contribution by A. Báez-Rubio in this volume). For the latter, inhomogeneities in the electron temperature, density and kinematics of the ionized gas in the inner regions of the jet could give rise to large asymmetries in the RRL profiles. From all this we suggest that the extremely broad features detected toward Cepheus A HW2 are associated with RRL masers generated in the high-velocity HW2 radio jet (Jiménez-Serra *et al.* 2011).

3. A new RRL maser object toward the MonR2-IRS2 source

MonR2 is a well-known massive star forming cluster located at a distance of 950 pc. The brightest sources in the region are IRS1, IRS2 and IRS3. IRS1 ($L_{bol} \sim 3 \times 10^3$ $L_\odot$; Henning, Chini & Pfau 1992) is the exciting source of the evolved HII region detected with the VLA toward this region (Wood & Churchwell 1989; Massi, Felli & Simon 1985). IRS2 ($L_{bol} \sim 6500$ $L_\odot$; Henning, Chini & Pfau 1992) is a very compact source responsible for the illumination of the inner walls of the HII region (Aspin & Walther 1990). And IRS3 ($L_{bol} \sim 1.4 \times 10^4$ $L_\odot$; Henning, Chini & Pfau 1992) is a multiple source where the A and B objects show outflow activity (Preibisch *et al.* 2002). The MonR2 cluster is also associated with a large-scale CO outflow (total extent of 6.8 pc; Bally & Lada 1983) seen in the northwest-southeast direction.

By using the SMA in the very extended (VEX) and compact configurations (COM), we have recently imaged the H30α and H26α RRL emission toward the MonR2 cluster. Figure 2 shows the VEX spectra (beam $\sim$0.3"–0.5") of these lines extracted from the position of the MonR2-IRS2 continuum peak. The RRLs have double-peaked line

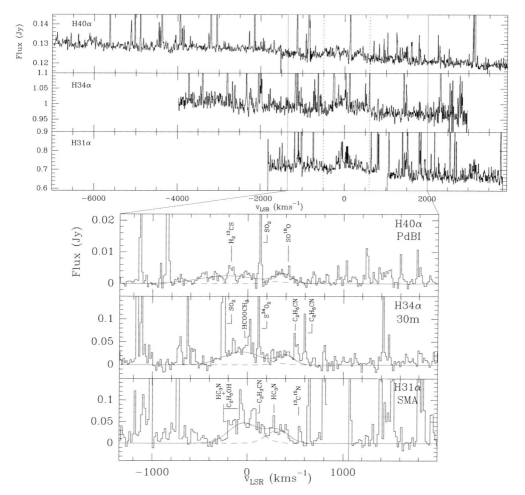

Figure 1. *Upper panels:* Full spectra of the H40α, H34α, and H31α RRLs measured toward Cepheus A HW2 with the IRAM Plateau de Bure Interferometer (PdBI), the IRAM 30 m telescope, and the SMA. *Lower panels:* Zoom-in of the H40α, H34α, and H31α spectra, showing the two-component Gaussian fits of the extremely broad RRLs (Jiménez-Serra *et al.* 2011).

Table 1. Line-to-continuum flux ratios measured toward Cepheus A HW2 and MonR2-IRS2, and compared with those derived toward MWC349A.

RRL	Cepheus A HW2			RRL	MonR2-IRS2		
	LTE^a (km s^{-1})	HW2 (km s^{-1})	MWC349 (km s^{-1})		LTE^a (km s^{-1})	IRS2 (km s^{-1})	MWC349 (km s^{-1})
H40α	~30	43	39^b	H30α	~65	90	298^b
H34α	~43	229	110^c	H26α	~103	174	1660^d
H31α	~58	280	215^b	...	...	...	...

Notes:
[a] Calculated for optically thin continuum emission, $T_e^*=10^4$ K and $N(He^+)/N(H^+)=0.08$.
[b] From Martín-Pintado, Bachiller & Thum (1989).
[c] From Thum, Martín-Pintado & Bachiller(1992).
[d] From Thum *et al.* (1994).

profiles resembling those measured toward MWC349A. The observed linewidths are ~20-30 km s^{-1} and the RRL peaks are centered at velocities very different from that of the ambient cloud (v_{LSR}=10 km s^{-1}; Figure 2). While the H30α red- and blue-shifted components have similar intensities, the intensity of the blue-shifted H26α gas is a factor of >3 brighter than that of the red-shifted peak. This indicates that the RRL emission toward MonR2-IRS2 is affected by maser amplification. This is confirmed by the derived line-to-continuum flux ratios (90 and 174 km s^{-1} for the H30α and H26α RRLs, respectively), which are clearly higher than those expected under LTE conditions (see Table 1).

The integrated intensity images of the red- and blue-shifted components of the H26α RRL toward MonR2-IRS2, reveal a small spatial shift (by ⩽0.1") in the east-west direction. This spatial shift could be interpreted as an ionized jet, like Cepheus A HW2, or as an ionized disk, as found toward MWC349A. Our SMA ^{12}CO $J = 2 \rightarrow 1$ COM data, however, do not show broad ^{12}CO emission arising from outflowing gas in the vicinity of IRS2. Therefore, the ~0.1" spatial shift between the red- and blue-shifted H26α gas could be associated with a Keplerian rotating disk around a source with mass ~9 M$_\odot$. This mass is consistent with the luminosity of the IRS2 source (L_{bol}~6500 L$_\odot$). We propose that the RRL masers detected toward the YSO MonR2-IRS2 likely arise from an ionized circumstellar disk, in the same fashion as observed toward the massive YSO MWC349A.

We finally note that, in contrast with the VEX data, the H30α RRL data in COM configuration (beam ~2.5"–3.5") show line emission with single-Gaussian profiles. This is probably due to the additional contribution arising from the large-scale and evolved IRS1 HII region within which the IRS2 source is embedded. This reveals that very high-angular resolution observations are needed to pinpoint the regions where RRL maser

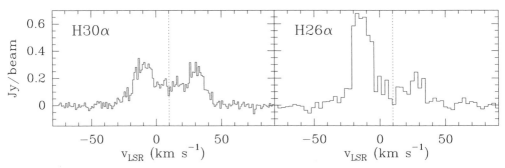

Figure 2. VEX spectra of the H30α and H26α RRLs extracted from the position of the MonR2-IRS2 continuum peak. The RRLs are double-peaked as in MWC349A. Vertical dotted lines show the central radial velocity of the MonR2 cloud (v_{LSR}=10 km s^{-1}).

amplification takes place, and it explains why single-dish surveys of RRL masers at 1.3 mm were unsuccessful in the past (Martín-Pintado 2002).

4. Conclusions

We report the detection of two new RRL maser objects toward the massive star forming regions Cepheus A HW2 and MonR2-IRS2. Our observations reveals that high-angular resolution interferometric observations at sub-millimeter wavelengths are strongly required to detect weakly amplified RRL masers such as those detected toward Cepheus A HW2 and MonR2-IRS2. Interferometric facilities such as the SMA and ALMA will be key in the detection of a larger sample of massive YSOs with RRL masers.

References

Aspin, C. & Walther, D. M. 1990, *A&A*, 235, 387

Baker, J. G. & Menzel, D. H. 1938, *ApJ*, 88, 52

Bally, J. & Lada, C. J. 1983, *ApJ*, 265, 824

Cillié, G. G. 1936, *MNRAS*, 96, 771

Curiel, S., *et al.* 2006, *ApJ*, 638, 878

Henning, Th., Chini, R., & Pfau, W. 1992, *A&A*, 263, 285

Hughes, V. A. & Wouterloot, J. G. A. 1984, *ApJ*, 276, 204

Jiménez-Serra, I., Martín-Pintado, J., Rodríguez-Franco, A., Chandler, C., Comito, C., & Schilke, P. 2007, *ApJ*, 661, L187

Jiménez-Serra, I., Martín-Pintado, J., Caselli, P., Martín, S., Rodríguez-Franco, A., Chandler, C., & Winters, J. M. 2009, *ApJ*, 703, L157

Jiménez-Serra, I., Martín-Pintado, J., Báez-Rubio, A., Patel, N., & Thum, C. 2011, *ApJ*, 732, L27

Martín-Pintado, J., Bachiller, R., & Thum, C. 1989, *A&A* 222, L9

Martín-Pintado, J. 2002, in: V. Mineese & M. Reid K. K. (eds.) *Cosmic Masers: From Proto-Stars to Black Holes, IAU Symposium 206* (San Francisco: Astronomical Society of the Pacific), p. 226

Martín-Pintado, J., Thum, C., Planesas, P., & Báez-Rubio, A. 2011, *A&A*, 530, L15

Massi, M., Felli, M., & Simon, M. 1985, *A&A*, 152, 387

Narayanan, G. & Walker, C. K. 1996, *ApJ*, 466, 844

Preibisch, T., Balega, Y. Y., Schertl, D., & Weigelt, G. 2002, *A&A*, 392, 945

Rodríguez, L. F., Garay, G., Curiel, S., Ramírez, S., Torrelles, J. M., Gómez, Y., & Velázquez, A. 1994, *ApJ*, 430, L65

Strelnitski, V. S., Ponomarev, V. O., & Smith, H. A. 1996, *ApJ*, 470, 1118

Thum, C., Martín-Pintado, J., & Bachiller, R. 1992, *A&A*, 256, 507

Thum, C., Matthews, H. E., Martín-Pintado, J., Serabyn, E., Planesas, P., & Bachiller, R. 1994, *A&A*, 283, 582

Torrelles, J. M., *et al.* 2011, *MNRAS*, 410, 627

Walmsley, C. M. 1990, *A&ASS*, 82, 201

Wood, D. O. S. & Churchwell, E. 1989, *ApJS*, 69, 831

Cosmic Masers - from OH to H_0
Proceedings IAU Symposium No. 287, 2012
R.S. Booth, E.M.L. Humphreys & W.H.T. Vlemmings, eds.

© International Astronomical Union 2012
doi:10.1017/S1743921312007521

Unveiling the kinematics of the disk and the ionized stellar wind of the massive star MWC349A through RRL masers

Alejandro Báez-Rubio[1] and Jesús Martín-Pintado[2]

Centro de Astrobiología (CSIC-INTA),
Ctra de Torrejón a Ajalvir, km 4, 28850 Torrejón de Ardoz, Madrid, Spain
email: [1]baezra@cab.inta-csic.es [2]jmartin@cab.inta-csic.es

Abstract. The kinematics of photoevaporating disks and their associated ionized outflows around massive stars are fundamental to understand how these stars are formed and they evolve in their early phases of their evolution. To date, the important advances have been provided by studying the UC-HII region of MWC349A thanks to their strong maser emission at Hydrogen radio-recombination lines (RRLs). This B[e] star is one of the best prototypes of massive star with an ionized outflow expanding at nearly constant velocity. A 3D radiative transfer model applied to the H30α line has allowed to constrain the disk kinematics, which seems to follow pure Keplerian rotation in its outer parts. The model has also allowed us to constraints the launching radius of the outflow. Our results are supported by the agreement of our model predictions with the observations for other observed RRLs. Recent high-frequency observations of RRL masers with the Herschel Space Telescope (HIFI) show that the kinematics of the disk inner regions is not well understood. Modeling of these lines will constrain the formation of the ionized winds.

Keywords. accretion, accretion disks, masers, line: profiles, radiative transfer, stars: winds, outflows, radio continuum: stars, radio lines: stars, submillimeter

1. Introduction

Although massive stars play a central role in the evolution of the Cosmos, their short lifetime, their rarity and their large distances make their detailed study difficult. In particular, there is a lack in the understanding of how they form and their early and late stages of evolution. Recent studies have shown that at least in some cases, the formation of massive stars seems to proceed through accretion of circumstellar disks, as in low-mass stars (Jiménez-Serra *et al.* 2009, Kraus *et al.* 2010). On the other hand, the late stages of evolution contain a rich zoo of different kind of stars (i.e. LBV, WR). Some of them, such as the supergiant B[e] stars, also have circumstellar disks. However, the kinematics of both the circumstellar disk and the ionized outflow are still poorly understood for both the pre-main or post-main sequence stars. One of the most relevant discoveries is the fact that UC-HII regions do not expand into the circumstellar medium as we could expect from the high pressure exerted by these hot ionized regions (Churchwell *et al.* 1990). Its explanation is still under debate. To unveil the kinematics of the ionized region around massive stars with circumstellar disks, we have studied the star of MWC349A due to its unique characteristics.

MWC349A is a star of spectral type B[e], characterized by its forbidden emission lines in the optical spectrum and its NIR excesses (Lamers *et al.* 1998). As often happens with B[e] stars, we do not know its evolutionary stage. It could be a pre-main sequence star or a supergiant B[e] star. Nevertheless, it is one of the few massive stars with a circumstellar disk well confirmed observationally and with an edge-on orientation very well established

(Danchi *et al.* 2001). For years it has been the only source known with maser emission at Hydrogen RRLs as proved by the integrated line-to-continuum ratios (Martín-Pintado *et al.* 1989). The maser lines along with the fact that it is the strongest radio-continuum source provides spectra with a high angular and spatial resolution to study in detail the kinematics of its ionized wind. We have constrained the physical characteristics and kinematics of the ionized circumstellar disk by comparisons of the radio-continuum and RRL observations with the results of a non-LTE 3D radiative transfer model. Our results provide the first self-consistent model that explain the bulk of the available data.

2. Radio-continuum emission

First of all, we have constrained the physical characteristics and geometry for the ionized wind of MWC349A on the basis of the radio-continuum images and its spectral energy distribution. Its spectral index, ~ 0.6 (Tafoya *et al.* 2004), is explained as due to an ionized wind expanding at constant velocity, v_{exp}. We have explained those data by considering a double-cone geometry with semi-opening angle of $\theta_a \sim 57°$, an electron density distribution depending on the radius and the zenith angle such as the density is larger at angles closer to the faces of the double-cone like $N_e(r,\theta)=3.85 \cdot 10^9 e^{\frac{\theta_a - \theta}{20}} \cdot r^{-2.14}$ cm^{-3} (with the angle θ as shown in Fig. 1) and an electron temperature of 12000 K.

3. RRL emission: constraining the kinematics of the ionized gas

We have constrain the kinematics by using the whole data set of RRLs. The double-peaked profiles and the H30α centroid map show two kinematic components (see Fig. 1): an ionized Keplerian-rotating disk located next to the neutral disk with opening angle θ_k and an outflow that must be expanding at nearly constant velocity as previously mentioned. By modelling the source with these two kinematic components, we have obtained the following input parameters for the model: an inclination angle, θ_i, of 8° (tilted-up), an opening angle for the ionized rotating disk, θ_k, of 6.5° which extends up 130 AU, a central mass for the star of 40 $M_\odot$, an electron temperature of 10000K for the disk, a terminal velocity of 60 km s^{-1} for the outflow and a turbulent velocity of 15 km s^{-1}.

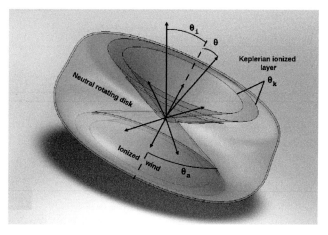

Figure 1. Sketch of the double-cone geometry used for the modelling the Keplerian-rotating ionized disk and the outflow of MWC349A.

3.1. *H30α centroid map*

The interpretation that the UC-HII region of MWC349A is composed by an ionized rotating disk and an outflow is supported by the behaviour of the H30α centroid map (see Fig. 2). The centroid map provides the spatial location where the line emission arises for every radial velocity with a relative accuracy of ∼2 to 5 milliarcseconds (Martín-Pintado *et al.* 2011).

In the centroid map we clearly distinguish two different regions. Firstly, a straight line overlapping with the NIR disk image (Danchi *et al.* 2001) for the radial velocities between the maser spikes. Our model clearly shows that the line emission for that range of radial velocities is mainly originated inside of the disk. On the other hand, the expanding outflow is responsible for the north-south loops observed in the centroid map for radial velocities larger than those of the maser spikes.

Since the intensity of the maser emission strongly depends on the electron density and electron temperature (Strelnitski *et al.* 1996), it provides tight constraints on the kinematics of the region where the maser spikes arise. By using the velocity peak separations and the H30α centroid map, we have deduced that the circumstellar disk is rotating following a Keplerian law around a central star of about ∼ 40 M$_\odot$.

On the other hand, since the emission of the outflow should be symmetric with respect the East-West plane for an edge-on disk, the north-south loops observed in the H30α centroid map must be due to the fact that the disk is slightly tilted. The fit to the height of the loops provides strong constraints on the inclination of the plane of the disk with respect the line-of-sight: between 4.5 and 15° for a tilted-up disk. In addition, the loops are a clear signature that the outflow is also rotating since otherwise, the emission would be symmetrical with respect to the north-south plane. In such a case, the loops would occur at right ascensions very close to 0 (Martín-Pintado *et al.* 2011), opposite to that observed. Thus, our model predicts the rotation for the outflow following the Keplerian law of the ionized circumstellar disk.

Another important finding is that the H30α centroid map is only explained if we assume that the ionized outflow is originated in the inner disk at a radius of <25 AU. This fact clearly disagrees with the more than 100 AU expected from the analytic photoevaporating model of Hollenbach *et al.* (1994) and it could provide constraints on which hydrodynamical models are viable candidates to explain the formation of the ionized

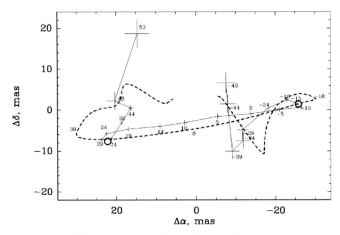

Figure 2. Observational (solid lines) and predicted (dashed lines) centroid map of the H30α. The observations were carried out with the IRAM Plateau de Bure Interferometer (Martín-Pintado *et al.* 2011).

Finally we have to stress that our final results were obtained by using the set of departure coefficients, b_n, provided by Storey & Hummer (1995) since no one set of input parameters reproduce the intensity of the maser spikes of mm RRLs if we use the set of b_n provided by Walmsley (1990). In particular, our model reproduces for the first time the intensities of the maser spikes at mm wavelengths and their integrated line fluxes.

3.2. *Integrated line fluxes*

The non-LTE model predictions matches well the observed integrated line fluxes for Hnα with $n > 26$ as shown in Fig. 3. This supports the proposed kinematic model and gives hints on the dominant processes involved in the RRL emission for the different ranges of frequencies.

Our predictions show that even for low frequencies RRLs (those with n>41), there is a constant shift between the non-LTE and the LTE predictions. This seems to indicate that even for such RRLs, the emission is out of the LTE since the observations can only be fitted by considering the non-LTE case. This is also consistent with the profile found for the observed RRL of largest frequency, the H76α (see Escalante *et al.* 1989). Its asymmetric profile is easily explained due to stimulated emission. In particular, the stronger emission of the blue-shifted wing is due to the amplification of the larger background continuum emission by the ionized material approaching to us. All these evidences seem to indicate that the radiation is amplified by stimulated emission even for low-frequency RRLs.

Secondly, the integrated line fluxes for Hnα RRLs with 21<n<41 significantly increases with the frequency, specially for RRLs with n<31. In this range of RRLs, the maser emission is the dominant process. Our non-LTE predictions agree very well with the predictions down to the H26α. However, we clearly overestimate the integrated line flux of the submm RRLs with n<26. This is because in our model the maser amplification increases in an exponential way for such RRLs, while one expects that such strong maser emission should be saturated. Thus, we would need to take into account in our model the saturation effects to explain the intensities of those lines. Finally our LTE predictions show that the RRL emission are emitted under LTE conditions for n<7.

3.3. *The inner disk kinematics. Keplerian rotation?*

The increase of the velocity peak separation of the maser spikes with frequency is well explained by our model for millimeter RRLs for frequencies larger than that of the H30α.

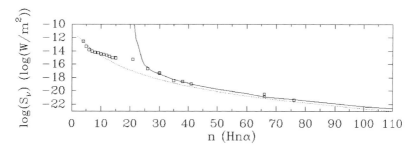

Figure 3. Observational (squares) and predicted integrated line fluxes for the non-LTE and LTE case (solid and dashed lines respectively). References for the observational data: H4α to H15α, Thum *et al.* (1998); H21α, Thum *et al.* 1994b; H26α and H30α, Thum *et al.* 1994a; H30α, Martín-Pintado *et al.* 1989; H35α and H38α, Martín-Pintado *et al.* 1994; H41α, Martín-Pintado *et al.* 1989; H66α, Loinard & Rodríguez 2010; H66α, Martín-Pintado *et al.* 1993; H76α, Escalante *et al.* 1989.

Nevertheless, the H26α and H21α profiles obtained with the IRAM-30m (Thum *et al.* 1995, Thum *et al.* 1994 respectively) and the very recent observations carried out provided by the HIFI instrument on board the Herschel Space Observatory show that the velocity peak separation does not increase as expected from the model for a Keplerian-rotating disk. In contrast, it remains approximately constant (~ 50 km s^{-1}) from the H24α to the H16α. The explanation of this behaviour will help to understand the inner regions of the disk, where the outflow is expected to originate. It could be a signature of the distortion of the disk kinematics of the inner regions likely due to launching of the ionized outflow.

4. Summary and perspectives

For the first time, we have used the powerful maser emission of hydrogen RRLs to derive the detailed kinematics of the ionized circumstellar disk and the ionized outflow around a massive star. By using the information derived from the maser RRLs along with the data of other RRLs, we have presented the first self-consistent model that describes the bulk of the observations carried out in a wide range of frequencies. Future submm high angular studies will help understanding the kinematics of the inner regions and the possible the mechanisms involved in the launching of the outflow. In addition, the recent detection of pre-main sequence stars with RRL maser emission (Jiménez-Serra *et al.* 2011, Jiménez-Serra *et al.* in prep.) open the possibility that RRL maser emission turns out to be an essential tool to disentangle the kinematics of circumstellar disks around massive stars.

Acknowledgements

We thank the Spanish MICINN for funding support through grants ESP2007-65812-C02-C01 and AYA2010-21697-C05-01 and AstroMadrid (CAM S2009/ESP-1496). A. Báez-Rubio acknowledges support from grant JAE program, CSIC, Spain.

References

Churchwell, E. 1990, *A&AR*, 2, 79
Danchi, W. C., Tuthill, P. G., & Monnier, J. D. 2001, *ApJ*, 562, 440
Escalante, V., Rodríguez, L. F., Moran, J. M., & Cantó, J. 1989, *Rev. Mexicana AyA*, 17, 11
Hollenbach, D., Johnstone, D., Lizano, S., & Shu, F. 1994, *ApJ*, 428, 654
Jiménez-Serra, I., Martín-Pintado, J., Caselli, P., Martín, S. *et al.* 2009, *ApJ*, 703, L157
Jiménez-Serra, I., Martín-Pintado, J., Báez-Rubio, Al. *et al.*, 2011 *ApJ*, 732, L27
Kraus, S., Hofmann, K., Menten, K. M., Schertl, D., Weigelt, G. *et al.* 2010, *Nature*, 466, 339
Lamers, H. J. G. L. M., Zickgraf, F.-J., de Winter, D. *et al.* 1998, *A&A* 340, 117
Loinard, L. & Rodríguez, L. F. 2010 *ApJ*, 722, L100
Martín-Pintado, J., Bachiller, R., Thum, C., & Walmsley, M. 1989, *A&A*, 215, L13
Martín-Pintado, J., Gaume, R., Bachiller, R., Johnston, K., & Planesas, P. 1993, *ApJ*, 418, L79
Martín-Pintado, J., Neri, R., Thum, C., Planesas, P., & Bachiller, R. 1994, *A&A*, 286, 890
Martín-Pintado, J., Thum, C., Planesas, P., & Báez-Rubio, A. 2011, *A&A*, 530, L15
Storey, P. J. & Hummer, D. G. 1995, *MNRAS*, 272, 41
Strelnitski, V. S., Ponomarev, V. O., & Smith, H. A. 1996, *ApJ*, 470, 1118
Tafoya, D., Gómez, Y., & Rodríguez, L. F. 2004, *ApJ*, 610, 827
Thum, C., Matthews, H. E., Martín-Pintado, J. *et al.* 1994, *A&A*, 283, 582
Thum, C., Matthews, H. E., Harris, A. H., Tacconi, L. J. *et al.* 1994, *A&A*, 288, L25
Thum, C., Strelnitski, V. S., Martín-Pintado, J. *et al.* 1995, *A&A*, 300, 843
Thum, C., Martín-Pintado, J., Quirrenbach, A. & Matthews, H. E. *A&A*, 333, L63
Walmsley, C. M. 1990, *A&A*, 82, 201

Cosmic Masers - from OH to H_0
Proceedings IAU Symposium No. 287, 2012
R.S. Booth, E.M.L. Humphreys & W.H.T. Vlemmings, eds.
© International Astronomical Union 2012
doi:10.1017/S1743921312007533

Intrinsic Sizes of the W3 (OH) Masers via Short Time Scale Variability

Tanmoy Laskar[1], W. M. Goss[2] and B. Ashley Zauderer[1]

[1]Harvard University, 60 Garden St., Cambridge, MA - 02138, USA
email: `tlaskar@cfa.harvard.edu`

[2]National Radio Astronomy Observatory, PO Box 0, Socorro, NM - 87801, USA
textitail: `mgoss@aoc.nrao.edu`

Abstract. We present a study of short time-scale variability of OH masers within a contiguous 15-hour Very Long Baseline Array observation of the high-mass star-forming region, W3 (OH). With an angular resolution of ~ 7 mas and a velocity resolution of $53\,\mathrm{m\,s^{-1}}$, we isolate emission from masers in the field into individual Gaussian-shaped components, each a few milliarcseconds in size. We compute dynamic spectra for individual maser features with a time resolution of 1 minute by fitting for the flux density of all sources in the field simultaneously in the uv-domain. We isolate intrinsic maser variability from interstellar scintillation and instrumental effects. We find fluctuations in the maser line shape on time scales of 5 to 20 minutes, corresponding to maser column lengths of 0.5 to 2.0 Astronomical Units.

1. Introduction

Galactic OH masers have been found to be variable on a variety of time scales: from years (Caswell & Vaile 1995) and months (e.g., Slysh *et al.* 2010) to days (Clegg & Cordes 1991). The origin of this long-time scale variability is unknown, although proposed mechanisms have included changes in the physical conditions of the maser column (e.g., Zuckerman *et al.* 1972), sudden onset of pumping (Salem & Middleton 1978), and radiative instabilities in the maser column (Scappaticci & Watson 1992). Maser variability studies on time scales on the order of minutes are useful in probing the length of the maser column. Such studies have proved challenging, in part due to the effects of interstellar diffractive scintillation and the need to study spectrally-isolated masers.

Recently, Ramachandran *et al.* 2006 (hereafter R06) studied variability in the galactic star-forming region, W3(OH) on short ($\sim$ minute) timescales with archival VLBA data (Wright *et al.* 2004). They employed a statistical technique to separate the effects of interstellar diffractive scintillation, based on the expectation for variability arising from scintillation to have a wider correlation bandwidth compared to the maser line width. After accounting for scintillation, they found significant residual variability (modulation indices up to 100%) on time scales of about 15-20 minutes in the line wings of three strong, spectrally-isolated masers at 1612, 1665 and 1720 MHz. Although this provided tantalizing evidence for intrinsic variability of OH masers, certain aspects of the data used made it unfavorable for a final and conclusive analysis: the data had non-uniformly spaced gaps whenever a calibrator was observed, making standard Fourier analysis on the time series difficult. Additionally, the velocity resolution of $90\,\mathrm{m\,s^{-1}}$ resulted in spectral blending, making it hard to find isolated maser sources for variability analysis.

We present the results of a variability study of OH masers in W3(OH) using Very Long Baseline Array (VLBA) data obtained with a factor of 2 better velocity resolution compared to the data used by R06. We fit for all maser sources present in the field

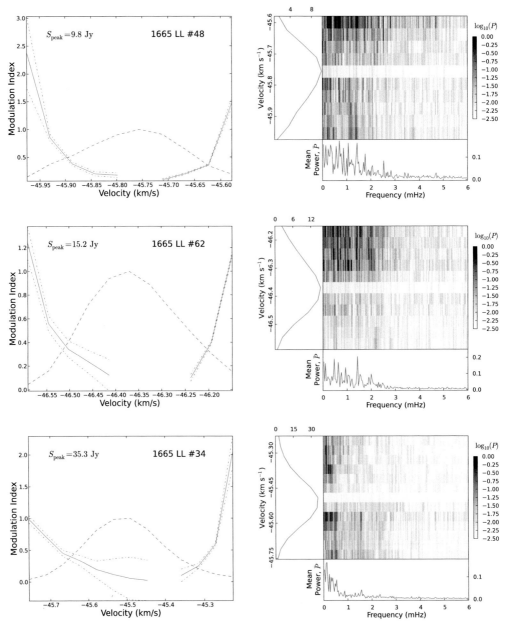

Figure 1. Left: Modulation index (solid green lines) and $\pm 1\,\sigma$ uncertainties (dotted red lines), together with normalized spectral line profiles (blue dashed lines) for eight maser sources in W3(OH). The details of the masers used are presented in Table 1. All sources exhibit statistically significant variability in the line wings (modulation indices $\gtrsim 50\%$). The line core appears not to be variable in all cases, confirming the results of Ramachandran *et al.* 2006. Right: Fluctuation power spectra for each source, obtained by Fourier transforming each channel in the dynamic spectrum and taking the square modulus. In each sub-figure, the 2d frame is the velocity-resolved fluctuation spectrum in logarithmic units, the bottom panel is the velocity-averaged fluctuation spectrum in linear units, and the left panel is the average line profile in Jy/beam. The power spectra demonstrate a cut-off at high frequencies, corresponding to time scales of $5 - 20$ minutes. The 1665 MHz Stokes LL sources (arranged here in order of increasing flux density) appear to show decreasing power at high fluctuation frequencies, indicating a possible increase in the time scale of variability with flux density.

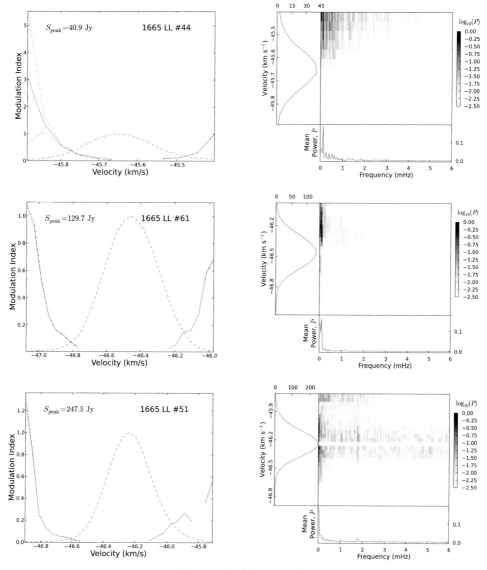

Figure 1. (Continued)

simultaneously directly in the *uv*-domain, lifting the restriction of sources under study to be spectrally isolated, which greatly increases the number of masers available for this study. Using the statistical tools of R06, we find variability on time scales of 5 $\sim$ 20 minutes with modulation indices between 50 and 100% for eight maser sources in W3(OH), implying maser column lengths of 0.5 to 2 astronomical units.

2. Observations and Data Analysis

We obtained a 15-hour contiguous observation of W3 (OH) with the VLBA (all ten VLBA stations) on 2005 July 07 (VLBA program BR107) in all four ground state transitions with a velocity resolution of $\sim 53\,\mathrm{m\,s^{-1}}$. RFI was excised and the data were phase self-calibrated. In $4'' \times 4''$ images (synthesized beam $6.7 \times 7.7\,\mathrm{mas}$), we detected

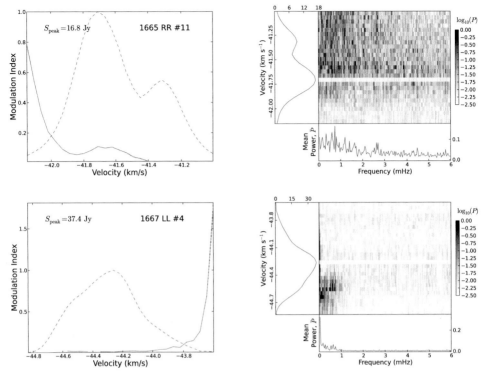

Figure 2. Similar to Figure 1, for a 1665 RR maser and a 1667 LL maser.

approximately 20 distinct maser sources at 1612 MHz, 10 at 1720 MHz, 180 at 1665 MHz and 40 at 1667 MHz brighter than 10 σ. The noise level in our maps ranged from ~ 30 mJy/beam in line-free parts of the spectrum to ~ 60 mJy/beam in channels with strong masers.

Dynamic spectra for every source brighter than 1 Jy were generated by simultaneously fitting for the flux density in the uv domain in the entire 256-channel spectral cube at 1 minute intervals. The fitting was carried out in the uv domain since the sparse uv-coverage at short integrations results in very poor image fidelity with large side-lobe responses whenever multiple sources are present in a single spectral channel, as is the case for the majority of channels in the data cube. With this simultaneous fit for all maser sources, we are not restricted to analyzing only spectrally-isolated sources and are able to extend the work by R06 to all strong sources in the entire data set. For the 1-minute fits, we fixed the positions and shape properties of each maser component to the values determined by imaging the full 15-hour observation.

We used statistical techniques from R06 to determine modulation indices across the line profile for 8 maser sources (Figure 1, left panels). These modulation indices imply significant ($\sim 100\%$) variability in the line wings. In each case, the modulation index is consistent with zero near the peak of the line, indicating that the line core is not significantly variable. Both of these results are in agreement with the conclusions of R06.

3. Fluctuation Analysis

To estimate the time scale corresponding to narrow band variability, the contribution of broad band variability to the dynamic spectrum must be removed. We achieve this by dividing the observed dynamic spectrum by the flux density in the reference channel

Table 1. List of masers used for the present analysis.

Transition (MHz)	Polarization	Source Number[1]	Velocity[2] (km s^{-1})	RA offset[3] arcsec	Dec offset[3] arcsec
1665	LL	34	−45.51	−0.595	−1.918
1665	LL	44	−45.65	−0.711	−1.673
1665	LL	48	−45.77	−0.851	−0.618
1665	LL	51	−46.24	−1.024	−0.123
1665	LL	61	−46.45	−0.960	−0.103
1665	LL	62	−46.37	−0.168	−0.044
1665	RR	11	−41.63	−0.141	−1.185
1667	LL	4	−44.31	−0.386	−1.886

Notes:
[1] The source numbers given here are internal to our analysis, and are likely to change in a future publication.
[2] LSR Velocity of the line centroid, determined by fitting a single Gaussian to the spectral line.
[3] Right Ascension and Declination offsets (J2000.0) from the phase center, corresponding to the location of the flux-weighted centroid of the maser line. Since our data were self-calibrated, the exact location of the phase center in RA and Dec is uncertain. Preliminary analysis suggests that our offset coordinate system is comparable (within about 10 mas) to that used by Fish *et al.* 2006.

defined by the peak of the average line profile. This is justified since the peak of the line does not exhibit statistically significant narrow band variations, as apparent from the modulation indices (Figure 1, left panels). We compute the fluctuation power spectrum for each channel by Fourier transforming the time series in each channel individually, followed by taking the square modulus. Our preliminary results are shown in Figure 1 (right panels). Excess power is evident at low frequencies, below about 3×10^{-3} Hz, corresponding to a minimum variability time scale of about 5 minutes. The power spectra for the 1665 MHz Stokes LL masers also hint at a possible decrease in power at high fluctuation frequencies (corresponding to an increase in the variability time scale), with increasing flux density of the maser line.

In summary, we confirm the results of R06 and find fluctuations in the maser line shape on time scales of 5 to 20 minutes, corresponding to maser columns of 0.5 to 2.0 AU. In a future publication, we aim to extend our analysis to other maser sources in the field and to compare variability between masers in Zeeman pairs.

The National Radio Astronomy Observatory is a facility of the National Science Foundation operated under cooperative agreement by Associated Universities, Inc.

References

Baudry, A., Desmurs, J. F., Wilson, T. L., & Cohen, R. J. 1997, *A&A*, 325, 255
Caswell, J. L. & Vaile R. A. 1995, *MNRAS*, 273, 328
Clegg, A. W. & Corders J. M. 1991, *ApJ*, 374, 150
Fish, V. L., Brisken W. F., & Sjouwerman L. O. 2006, *ApJ*, 647, 418
Ramachandran, R., Deshpande, A. A., & Goss, W. M. 2006, *ApJ*, 653, 1314
Rickard, L. J., Palmer, P., & Zuckerman, B. 1975, *ApJ*, 200, 6
Salem, M. & Middleton, M. S. 1978, *MNRAS*, 183, 491
Scappaticci, G. A. & Watson, W. D. 1992, *ApJ*, 400, 351
Slysh, V. I., Alakoz, A. V., & Migenes, V. 2010, *MNRAS*, 404, 1121
Wright, M. M., Gray, M. D., & Diamond, P. J. 2004, *MNRAS*, 350,1253
Zuckerman, B. Yen, J. L., Gottlieb, C. A., & Palmer, P. 1972, *ApJ*, 177, 59

Cosmic Masers - from OH to H_0
Proceedings IAU Symposium No. 287, 2012
R.S. Booth, E.M.L. Humphreys & W.H.T. Vlemmings, eds.

© International Astronomical Union 2012
doi:10.1017/S1743921312007545

OH Maser sources in W49N: probing differential anisotropic scattering with Zeeman pairs

Avinash A. Deshpande[1], W. M. Goss[2] and J. E. Mendoza-Torres[3]

[1]Raman Research Institute, C. V. Raman Avenue, Sadashivanagar, Bangalore, 560080 India
email: desh@rri.res.in

[2]National Radio Astronomy Observatory, P.O. Box O, Socorro, NM 87801 USA
email: mgoss@aoc.nrao.edu

[3]Instituto Nacional de Astrofisica Optica y Electronica, 72840, Mexico
email: mend@inaoep.mx

Abstract. Our analysis of a VLBA 12-hour synthesis observations of the OH masers in W49N has provided detailed high angular-resolution images of the maser sources, at 1612, 1665 and 1667 MHz. The images, of several dozens of spots, reveal anisotropic scatter broadening; with typical sizes of a few tens of milli-arc-seconds and axial ratios between 1.5 to 3. The image position angles oriented perpendicular to the galactic plane are interpreted in terms of elongation of electron-density irregularities parallel to the galactic plane, due to a similarly aligned local magnetic field. However, we find the apparent angular sizes on the average a factor of 2.5 less than those reported by Desai *et al.*, indicating significantly less scattering than inferred earlier. The average position angle of the scattered broadened images is also seen to deviate significantly (by about 10 degrees) from that implied by the magnetic field in the Galactic plane. More intriguingly, for a few Zeeman pairs in our set, we find significant differences in the scatter broadened images for the two hands of polarization, even when apparent velocity separation is less than 0.1 km/s. Here we present the details of our observations and analysis, and discuss the interesting implications of our results for the intervening anisotropic magneto-ionic medium, as well as a comparison with the expectations based on earlier work.

Keywords. masers; ISM: molecules, magnetic fields, individual (W49N); radio lines: ISM

1. Introduction

W49N is a well-known and extensively-studied massive star-forming complex in the Milky Way. The OH, H_2O masers in this region represent some of the most luminous of such sources in our Galaxy. In our efforts to study intrinsic short time-scale variability in W3OH (Ramachandran *et al.* 2006; also Laskar *et al.* in this volume) we were able to estimate and remove possible variability due to interstellar scintillation using the fact that decorrelation bandwidth is larger than line-widths. Such velocity-resolved analysis of W3OH data has suggested intrinsic variability on 15-20 minute time scale (Ramachandran *et al.*, 2006), whereas W49N data show variations on time-scales of 1 hour or longer (Goss *et al.*, 2007, talk at IAUS 242, Alice Springs).

The W49N complex, in the Galactic direction $l = 43°.17$; $b = 0°.01$, is located on the far-side of the Solar circle. Its large distance (~11.4 kpc) and low Galactic latitude together makes this object attractive for studying various propagation effects due to the intervening medium. The interstellar scattering in this case is comparable to that in the Vela direction. In an earlier study, Desai, Gwinn & Diamond (1994; hereafter DGD94) have found significant anisotropic scattering, attributable to electron-density irregularities that are preferentially elongated in the Galactic plane. In this paper, we describe our probe based on the observed manifestations of the interstellar scattering in the W49N direction, compare our results with those from DGD94, and present intriguing cases of differential anisotropic scattering apparent for the two hands of circular polarization.

2. Our data

Our visibility data from 12+ hour synthesis observations with VLBA (BR107) were calibrated & processed in the standard way using AIPS. The data were self-calibrated, and absolute coordinates are not available due to an absence of phase reference source obervations. The spectral-line image cubes, obtained separately at 1612, 1665 & 1667 MHz, provided high angular-resolution images of the set of sources for each of the transitions, and for each of the circular polarizations. The imaged angular extent of about 8 or so arcseconds corresponds to a transverse span of ∼ 0.5 pc at the distance of the source, although a majority of the sources are within central 2-arcsecond wide region. As a result of exclusion of the outer 2 antennas of VLBA from Self-Cal and this imaging, our synthesized beam size was compromised to ∼20 mas and ∼15 mas in RA and Declination, respectively (corresponding spatial resolution being ∼200 AU and ∼150 AU). Each data set consisting of cleaned and restored images across 240 spectral channels spanning 22 km/s velocity range, providing velocity resolution of ∼0.1 km/s, was used to identify discrete maser sources (avoiding cases with significant velocity gradients). For each of clearly identifiable discrete (or isolated) maser sources, JMFIT-based estimates of the deconvolved source shape and size were obtained, in addition to the estimates of the mean location and velocity, along with estimates of the respective uncertainties. The resultant data on a total of 205 sources (most of these at 1665 & 1667 MHz) were examined among other things, for positional proximity of LCP and RCP source pairs ($\leqslant$ 10 mas), and a few dozen Zeeman pairs were thus identified. Figure 1 shows the distribution of all of the discrete sources in the RA-Dec plane, where the size of the symbol is proportional to the mean velocity associated with the observed line.

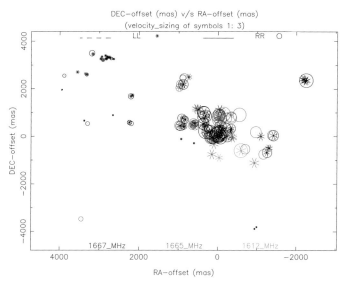

Figure 1. The observed distribution of W49N OH maser sources in RA-Dec. The symbol sizes are scaled proportional to the velocity (which are in the range +2 to +21 km/s). The velocity-position correlation, and the implied bipolar nature, is apparent from the majority of the sources in the sample.

3. Anisotropic scattering: apparent source sizes and shapes

Scatter-broadened shapes and orientations of the maser spots provide valuable information on the nature of the scattering medium. The elongations or the axial ratios (ratio of major to minor axis of the best-fit ellipse) are found in the range 1.5 to 3, indicating significant anisotropic scattering, consistent with the finding of DGD94. However, we find the overall apparent sizes of the sources to be significantly smaller (by a factor of $\geqslant 2$) than those reported by DGD94. Our size estimates are nonetheless consistent with the suitably scaled values of the scatter-broadening of H_2O masers reported by Gwinn (1994). Possible reasons for what appears to be an overestimation of the sizes in DGD94 are unclear, though remain intriguing. One of the most important aspects discussed by DGD94 (also see references therein) was that the apparent image shapes resulting from anisotropic diffraction are expected to have elongation orthogonal to that of the scattering irregularities. A presence of a magnetic field would induce such anisotropy in the electron-density irregularities; then the implied field direction would be perpendicular to the resultant image position angle. Based on their limited sample, of the apparent image shapes for 27 spectral components from 6 sources, DGD94 had suggested that the apparent anisotropy is induced by a magnetic field in and parallel to the Galactic plane.

In Figure 2, we present a distribution of our estimates of the position angles for our significantly ($\sim$ 30 times) larger sample. Some random spread in PA is apparent in both, our and DGD94's, reported PA values, and such a spread is not unexpected. However, we find that our PA distribution has a significant offset from the PA of $\sim$117 degrees, the expected PA if the density irregularities were to be elongated parallel to the Galactic plane. Our sample gives a mean PA of 107$\pm$3 degrees, implying a significant mean deviation, of $\sim$10 degrees. The DGD94 sample may have been too limited to make such a offset detectable. What might be the reason for this systematic offset ? In this context,

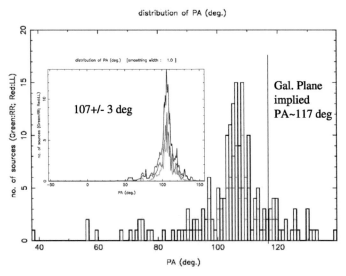

Figure 2. Distribution of the position angles (PA) of the scatter-broadened images. The main panel shows the histogram of the PA values for the set of OH maser sources in W49N. The vertical line at 117 degrees, shown for reference, marks the mean PA expected from the anisotropic scattering due to electron density irregularities, if these were to be elongated parallel to the Galactic plane. The inset shows the distribution of PAs after accounting for uncertainties in individual measurements (green and red profile correspond to LCP and RCP, respectively). This distribution is used to estimate the mean PA of the observed scatter-broadened images (107$\pm$3 degrees).

a closer look at the magnetic field structure along the W49N direction would be instructive. Interestingly, our sight-line to W49N is through the well-known North Polar Spur feature. From the study, by Wolleben (2007) and others, of the North Polar Spur, it is evident that significantly different magnetic field structure as well as enhanced scattering would be expected for the medium within half a kpc of the Sun. We note that the field orientation in this region, corresponding to the NPS, might be at a large inclination, if not almost orthogonal, to the Galactic plane. Also, given that the scattering medium closer to the observer is expected to make relatively higher contribution to the angular broadening (e.g., Gwinn *et al.* 1993, Deshpande and Ramachandran 1998), the above mentioned PA deviation can be caused by the more local scattering in the NPS region, with density anisotropy at a possibly large angle with respect to the Galactic plane. Careful modelling of the magneto-ionic medium would reveal the relative strengths and sense of anisotropic scattering, implied magnetic field strengths, for the different regions along the sight-line. Details and results of such modelling will be discussed elsewhere.

4. Differential Anisotropic scattering: probe with Zeeman pairs

As mentioned earlier, a few dozen Zeeman pairs are identified from our large sample of sources, using positional-proximity criterion. These Zeeman pairs allow us to probe yet unobserved aspect of scattering caused by the magneto-ionic component of the interstellar medium. A magneto-ionic medium would, in principle, render different refractive indices for the two hands of circular polarization. Diffractive scintillation and scatter-image shapes should therefore differ for LHC, RHC due to line-of-sight component of the magnetic field in the intervening medium. Hence scattering-dominated images of even a randomly polarized source might show circular polarization in unmatched parts of the images, when the Faraday rotation is significant. Macquart & Melrose (2000) indeed consider this possibility, but estimate the effect to be too small (circular fraction $\sim 10^{-8}$) to be observable. Contrary to that expectation, the scatter-broadened images of some of the W49N OH maser sources seem to significantly differ in L&R-hand circular polarizations, i.e. within a given Zeeman pair ! For the rest of the Zeeman pairs, the PA differences are either small or within the respective uncertainties. Figure 3 shows the observed differences in the image shape parameters for a subset of our sample of Zeeman pairs. The subset includes only those cases for which significant difference ($\geqslant 6\sigma$) in image PA is observed, when compared to corresponding uncertainty σ. Although there are only a few such Zeeman pairs, the PA differences range between 6 to 30 degrees. Difference in the line-velocities within most of these pairs, and hence in frequencies, is too small to account for the apparent differential scattering. Position differences are also within a few mas, and are unlikely to contribute to the observed PA differences.

To explore the issue further, we have made preliminary attempts to simulate magneto-ionic medium with a mild anisotropy. A random column-density distribution of free electrons, following a power-law spatial spectrum (with Kolmogorov index $-11/3$), is used to simulate a 2-d scattering screen across a transverse extent of a few Fresnel scales. For simplicity, a uniform magnetic field is assumed, so that the phase screens for the two circular polarizations are merely scaled versions of each other. The net circular polarization, if viewed with coarse resolution would be negligible, consistent with the conclusion of Macquart & Melrose (2000). However, when viewed with 10 or so milliarcsecond resolution, these preliminary simulations do indeed reveal noticeable differential diffractive effects, and hence the differing shapes and sizes of scatter-broadened images, for the two circular polarizations. Of course, more detailed simulations, with thick screens, will be needed to assess these issues in detail.

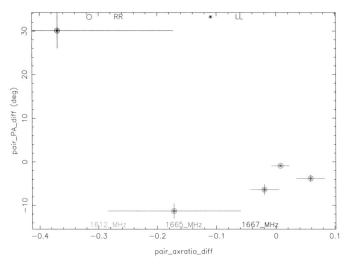

Figure 3. The PA differences within Zeeman pairs versus the respective differences in axial ratios for the cases where significant PA differences are apparent. The error-bars indicate $\pm 1 - \sigma$ uncertainties. All the axial ratios are all consistent with equal values for the two hands of circular polarization.

5. Summary

The OH maser sources in W49N do show anisotropic scattering, but the apparent scatter broadening is much less than reported earlier. The position angles of the source images deviate significantly from the value expected if scattering density irregularities were to be "stretched" due to magnetic field strictly aligned parallel to the Galactic plane, indicating significant scatter-broadening contribution from differently aligned density irregularities, possibly associated with the North Polar Spur. Some of our data also reveal differential scattering during propagation of the two circular polarizations, caused most likely by Faraday rotation. Through various aspects discussed above, the attractive, but yet unexplored, potential of the high-resolution maser observations for probing the intervening magneto-ionic medium is certainly evident.

Acknowledgements

It is a pleasure to acknowledge the contributions from R. Ramachandran and Sarah Streb at different stages of the reported work. The National Radio Astronomy Observatory is a facility of the Natiional Science Foundation operated under a cooperative agreement by Associated Universities, Inc.

References

Desai, K. M., Gwinn, C. R., & Diamond, P. J. 1994, *Nature*, 372, 754 (DGD94)
Deshpande, A. A. & Ramachandran, R. 1998, *MNRAS*, 300, 577
Gwinn, C. R., Bartel, N., & Cordes, J. M. 1993, *ApJ*, 410, 673
Gwinn, C. R. 1994, *ApJ*, 429, 253
Macquart, J.-P. & Melrose, D. B. 2000, *ApJ*, 545, 798
Ramachandran, R., Deshpande, A. A., & Goss, W. M. 2006, *ApJ*, 653, 1314
Wolleben, M. 2007, *ApJ*, 664, 349

T10:

Masers and the impact
of new facilities *Chair: Wouter Vlemmings*

Cosmic Masers - from OH to H$_0$
Proceedings IAU Symposium No. 287, 2012
R.S. Booth, E.M.L. Humphreys & W.H.T. Vlemmings, eds.

© International Astronomical Union 2012
doi:10.1017/S1743921312007557

Maser observations with new instruments

Alwyn Wootten

North American ALMA Science Center, NRAO†, 520 Edgemont Rd., Charlottesville,
Virginia 22903, USA
email: awootten@nrao.edu

Abstract. The Atacama Large Millimeter/submillimeter Array (ALMA)†, and the Jansky Very Large Array (JVLA) have recently begun probing the Universe. Both provide the largest collecting area available at locations on a high dry site, endowing them with unparalleled potential for sensitive spectral line observations. Over the next few years, these telescopes will be joined by other telescopes to provide advances in maser science, including NOEMA and the LMT. Other instruments of note for maser science which may commence construction include the North American Array, the CCAT, and an enlarged worldwide VLB network outfitted to operate into the millimeter wavelength regime.

Keywords. masers, radio lines: ISM, instrumentation: high angular resolution

1. Introduction

At the last IAU Symposium on masers, construction had just begun on the Atacama Large Millimeter/Submillimeter Array (ALMA), the upgrade of the Very Large Array to the Jansky Very Large Array was being planned, and millimeter instrumentation was being introduced on the Green Bank Telescope (GBT). CARMA had just commenced its first semester of routine science and would shortly populate its more extended arrays, enabling high resolution observations, particularly well adapted to maser observations. The Submillimeter Array was forging new paths with high resolution observations up to the edge of the atmospheric windows near 700 GHz. The Nobeyama Millimeter Array worked in conjunction with the 45 m telescope as the Rainbow Array. The Plateau de Bure continued to improve its bandwidth, sensitivity and resolution. Many of the results presented at this meeting were obtained over the intervening years with these telescopes.

In this review, the status is surveyed of new instruments which will provide new science during the time to the next Symposium beyond.

† The National Radio Astronomy Observatory is a facility of the National Science Foundation operated under cooperative agreement by Associated Universities, Inc.

† The Atacama Large Millimeter/sub-millimeter Array (ALMA), an international astronomy facility, is a partnership of Europe, North America and East Asia in cooperation with the Republic of Chile. ALMA is funded in Europe by the European Organization for Astronomical Research in the Southern Hemisphere (ESO), in North America by the U.S. National Science Foundation (NSF) in cooperation with the National Research Council of Canada (NRC) and the National Science Council of Taiwan (NSC) and in East Asia by the National Institutes of Natural Sciences (NINS) of Japan in cooperation with the Academia Sinica (AS) in Taiwan. ALMA construction and operations are led on behalf of Europe by ESO, on behalf of North America by the National Radio Astronomy Observatory (NRAO), which is managed by Associated Universities, Inc. (AUI) and on behalf of East Asia by the National Astronomical Observatory of Japan (NAOJ). The Joint ALMA Observatory (JAO) provides the unified leadership and management of the construction, commissioning and operation of ALMA.

2. Current Status

ALMA (see Wootten and Thompson 2009 for an instrument description) began its early science 'Cycle 0' operations in late September of 2011. ALMA construction has begun its final stages; there will be at least one further Early Science call, 'Cycle 1'.

ALMA Construction continues for somewhat more than a year.

At this writing, ALMA has transported 34 antennas delivered from the three vendors to the Array Operation Site (AOS), half of the final complement of 66 telescopes. At any given time the complement of antennas at the AOS is somewhat less than this as antennas return to the Operations Support Facility (OSF) for periodic maintenance, others are involved in commissioning or in the distinct Atacama Compact Array (ACA) or 12 seven meter antennas and 4 twleve meter antennas.

Three Front End Integration Centers, located in Charlottesville, near Oxford and Taipei assembled, tested and delivered 25 additional new Front Ends, or receiver packages, during 2011 to outfit those antennas (see, for example, Ediss _et al._ 2010). The units in the antenna cabin through which the Front Ends communicate with the equipment in the Array Operations Site Technical Building (AOS TB) are called Antenna Articles. Sixty-six of these were delivered by the end of the year, one for each of the planned ALMA antennas.

The Central Local Oscillator was also delivered and installed during 2011. It forms an important part of the ALMA nervous system, distributing and synchronizing signals across the array. A major technical problem of the ALMA Local Oscillator (LO) system is to maintain accurate phase stability in the signals used in the receiver components. The central reference part of the LO system is located at the AOS in the Technical Building. From here the signal is used to lock electronic oscillators and derived frequencies in antennas located at distances of many miles across the site via optical fiber. The outgoing frequency is transmitted as the difference between two laser frequencies in the infrared. The required phase stability of the LO signal is equivalent to less than 12 femtoseconds of time, less than 38 femtoseconds of phase noise. This time equivalent phase stability is to a second as a second is to 2.6 million years! The CLOA is now capable of supporting the 66 antennas and four subarrays built for the ALMA construction project; it may be expanded to provide signals for up to six subarrays and eighty antennas.

The antenna stations for the central cluster are in the final stages of preparation, ready for the deployment of the Early Science Array into its extended configuration early in 2012. Thus far, ALMA has operated on a network of generators. The Permanent Power System erection was completed in 2011 and is due for deployment on the array during early 2012, improving the quality and reliability of the array electrical power.

Production lines for ALMA components are shutting down as final components are delivered. The last of the four correlator quadrants was accepted. Two correlator quadrants were deployed at the AOS TB and a third was installed, while the fourth remained in Charlottesville for software testing. That quadrant will be installed by September 2012, when correlator installation and verification will finish. Receiver cartridges for the 3 mm (ALMA Band 3) and 0.45 mm (ALMA Band 9) frequencies have also completed production. See Table 1 for a current receiver summary. Steel fabrication for the Vertex antennas was complete; the final antenna will be shipped in April 2012 for delivery a few months later. The final third of the antennas constructed by the AEM consortium are nearing completion; most of the antennas are at the ALMA site. All sixteen antennas for the Atacama Compact Array (ACA) were delivered to Chile from Japan. The ACA with its initial complement of antennas employed its correlator to produce the first interferometric and total power data. The penultimate software release was deployed on the

Table 1. Front Ends on all Antennas[a]

Band	λ mm	ν GHz	Type	T_{rx}^{b}	Continuum Sensitivity (mJy)[c]
Early Science Cycles 0-1					
Band 3	3	84-116	2SB	$45K^{b}$	0.04
Band 6	1.3	211-275	2SB	$55K^{b}$	0.11
Band 7	0.9	275-373	2SB	$75K^{b}$	0.19
Band 9	0.45	602-720	DSB	$110K (DSB))^{b}$	1.0
Full Operations					
Band 4	2	125-169	2SB	<51K	0.06
Band 5	1.5	163-211	2SB	<65K	
Band 8	0.6	385-500	2SB	<196K	0.70
Band 10	0.35	787-950	DSB	<230K DSB	0.70
Water Vapor Radiometer	1.6	183	DSB Schottky	...	

[a] All have dual polarization with noise performance limited by atmosphere. [b] See the ALMA sensitivity calculator for T_{sys} under typical atmospheric conditions. For Early Science, actual realized receiver temperatures are given, often much better than the specification. T_{rx} for full operations is the specification value. [c] For sensitivity, 50 antennas are assumed in all cases. [c]

array and new releases of CASA software were released to the community for the further reduction of the delivered ALMA data.

During commissioning, carried out by a team of astronomers led by Project Scientist Richard Hills and his deputy Alison Peck (Hills *et al.* 2010, Mauersberger *et al.*. 2011), a typical early array consisted of a few antennas tightly packed onto several of the antenna stations designated to eventually host the Atacama Compact Array (ACA) 7 m antennas, as this small set of stations could be most quickly outfitted to host the few antennas in the earliest array. This commissioning array observed maser sources often, as these bright narrowband point emission sources provide excellent targets for assessment of the delays, pointing, baselines and a host of other instrumental qualities. The Orion SiO J = 1–0 masers, observable in most weather conditions and well-placed during the first commissioning periods in early 2010, were most often observed to provide test data to evaluate the performance of the array. As the year pressed on and weather improved, some early maser observations included 325 GHz observations of VY CMa with 3-5 antennas. An analysis by A. Richards of this test data showed a good correspondence between these masers and those mapped at 22 GHz by other arrays, to within the limited imaging ability of the small array. As the array grew further, to eight antennas by October, other masers were observed to provide test data, including the 658 GHz water masers toward a number of stars. With the increase in imaging capabilities of the array, testing moved forward to the imaging of more complex objects. NGC253 was imaged in all of the ALMA bands, as was the Beta Pictoris disk and other targets. The focus of the commissioning team moved on to calibration and verification of ALMA results through a comparison with existing data from other arrays. A call for science verification targets provided a favorable response, with over a hundred suggestions received.

A goal of science verification was the release of representative datasets to the community for inspection by investigators prior to the beginning of Early Science. Several science verification datasets were released on the ALMA portal to demonstrate the quality of ALMA data and its consistency with previous observations. Eight science verification datasets and one datum have now been released. Several publications based on Science

Verification data have been published, though none has yet dealt with maser science (Herrera *et al.* 2012, Oberg *et al.* 2012). The datasets released so far include:

- TW Hya: Band 7, high spectral resolution (casaguide). Band 3, Band 6.
- NGC3256: Band 3, low spectral resolution (casaguide).
- Antennae galaxies: Band 7, high spectral resolution, mosaic (casaguide). Band 6.
- M100 Band 3, low spectral resolution.
- SgrA* Band 6, recombination lines.
- Proof of Concept of Response to Targets of Opportunity: The GRB 110715A.

As commissioning continues, several additional science verification datasets will be released to demonstrate additional capabilities.

ALMA Operations has begun with Early Science as the array grew to 16 elements (Nyman *et al.* (2010), Rawlings *et al.* (2010)). The Early Science Cycle 0 Call for Proposals resulted in 2898 registrations at the ALMA portal. 919 proposals were received for the 2011 June 30 deadline. Investigators were notified in September of the status of their proposals; the limited time available was heavily ovsersubscribed. Scheduling and execution of the first batch of the 112 highest ranked proposals began on 30 September. By early December, quality assurance and packaging of the first project datasets had finished and delivery of the first data packages to principal investigators followed. By the February 2012 shutdown for implementation of ALMA's permanent power supply, nine periods of Early Science had been executed on the growing array.

ALMA construction will continue through 2012, which will see delivery of most of the remaining hardware. The most exciting prospect for 2012 is, of course, the expected publication of the first papers from Early Science. Some early results are making the rounds of science meetings already, whetting astronomers appetites for the announcement of the Cycle 1 Call for Proposals, expected within the next few months. During that period, of course, some of the Early Science observations will emerge into the literature,

By 2014 ALMA will be in full operation, having begun transforming science in 2011. Having invested $\sim\$1.3$B to realize the biggest advance ever in groundbased astronomy, it is vital to plan to keep the facility upgraded to maintain and expand its capabilities. When ALMA commenced its program of Early Science, it already eclipsed any other millimeter/submillimeter array in its sensitivity and resolution. The ALMA Operations Plan envisaged an ongoing program of development and upgrade. ALMA's design allows for expansion of the 50 antennas in the 12 m Array to a complement of 64. ALMA's wavelength coverage may be extended to cover 1 cm to 200 μm, or a factor of 50 and an increase of more than 50% from its first light capability. With a modest investment of less than 1% of capital cost per year divided among the three funding entities ALMA will lead astronomical research through the 2010 decade and beyond. Several programs have been identified by the scientific community which spearhead a development plan. The ALMA Development Plan will be coordinated by the Joint ALMA Observatory (JAO), which will issue calls for ALMA upgrades whose implementation will be assigned on a competitive basis. The details of this process are currently under discussion.

Fully operational ALMA will provide beamsizes of 0".01 at 300 GHz, appropriate for the sizes of some interstellar shock features seen in for example, water maser emission (see image in Wootten *et al.* (2002)). For a putative 1 kms^{-1} spectral resolution in an hour, the brightness temperature sensitivity of ALMA is a few K, allowing sensitive spatially and spectrally resolved observations of shock structures. The shocks shown in the image cited extend for several AU, which is several beams for nearby maser regions.

GBT (Prestage *et al.* (2009), Hunter *et al.* (2011)) brings the largest monolithic and steerable surface area on the planet to provide sensitive probes of distant weak masers. A particularly strong program there uses water masers to provide distances to maser host galaxies (see articles by Humphreys, Impellizzeri and Henkel in this volume). The first tests of a new 68-93 GHz receiver have proved promising; a multibeam array receiver may be implemented in the millimeter bands in the near future. Addition of the large collecting area to a sensitive VLB network will allow proper motions of masers in nearby galaxies to be measured, monitoring the kinematics of those spiral disks.

JVLA (Perley *et al.* (2009), Perley *et al.* (2011)) has begun providing exciting science results as demonstrated in data presented by many at this Symposium even before it reaches its full upgraded potential. The several order of magnitude increase in correlator capacity promises delivery of sensitive spectroscopy at high spectral and spatial resolution. As masers are often shock phenomena, investigations of the physical and chemical characteristics of the shocks will provide new insights into the origins of the masers. In the largest configuration, the JVLA provides tenth arcsecond resolution with a brightness temperature sensitivity of a few hundred Kelvin in an hour at a spectral resolution of 1 kms^{-1}, appropriate to most maser emission environments.

3. Next Symposium Instruments

NOEMA, the Northern Extended Millimeter Array at the Plateau de Bure(NOEMA Project, was agreed by the IRAM Executive Council to be expanded to four additional antennas in a first phase to be completed by 2016. A further two antennas are envisioned for construction in Phase II. The partners agreed to extend their successful participation in IRAM through 2024. The expansion will increase the collecting area of the array by 67% and imaging quality by 300% through tripling of the available baselines. Resolution is expected to double with the extension of baselines to 1.6 km. Continuum sensitivity will be further enhanced through a planned quadrupling of the bandwidth to 32 GHz. The enhanced array will provide a sensitive northern hemisphere baseline for intercontintental VLBI.

LMT, the Large Millimeter Telescope (Hughes *et al.* 2010), began operations at its 4600 mm mountaintop, Sierra Negra, in Mexico within the past year. Initially outfitted with a 32 meter surface, expansion to the full 50 m aperture is expected soon. The antenna surface will eventually be set to an accuracy sufficient to support operations in the 300 GHz spectral window. The antenna's aperture and location make it an important element of any worldwide VLB array.

4. Future

The North American Array (NAA) is a new initiative aimed at the realization of the high-frequency component of the Square Kilometer Array (SKA) program within North America. The primary activities of the NAA in the next decade are a technical development program and a prototype antenna station, leading up to SKA-high construction sometime after 2020. The key science drivers identified for the NAA cover a broad range of modern astrophysics, from studying the formation and evolution of planets, stars, and galaxies, to probing the overall structure of the Universe near and far. The primary technical goal of NAA is to ultimately exceed the sensitivity of the JVLA by a factor of 10 or greater, and thus survey speed by 100 or more, covering at least the core frequency range 3–45 GHz. The SKA-high is envisioned as the final component of the low/medium/high frequency triad of the International SKA Program.

CCAT (Radford *et al.* (2009)) will be a 25 m diameter telescope for submillimeter astronomy on a Cerro Chajnantor centrally located on the ALMA site. Primarily focussed on submillimeter continuum survey work, it is designed to reach frequencies above the

ALMA Band 10 window, which stops at 950 GHz. With a 6" beam at that frequency, it would be useful for submillimeter maser work. The telescope may be incorporated within the ALMA array at some future date for particular experiments, providing sensitive superterahertz capabilities.

Event Horizon Telescope (EHT, Fish & Doeleman (2010)) is aimed primarily at observations of the black holes at the center of M87 and the Milky Way, but it could also provide exceedingly high resolution images of bright masers. Incorporating the phased ALMA into its network, sensitive measurements of SiO or other masers in nearby galaxies, importantly the Magellanic Clouds, might be made which could determine the dynamics of those galaxies.

References

Ediss, G. A., Crabtree, J., Crady, K., *et al.* 2010, *Journal of Infrared, Millimeter, and Terahertz Waves*, Volume 31, Issue 10, pp. 1182–1204, 31, 1182
Fish, V. L. & Doeleman, S. S. 2010, *IAU Symposium*, 261, 271
Herrera, C. N., Boulanger, F., Nesvadba, N. P. H., & Falgarone, E. 2012, *A&A*, 538, L9
Hills, R. E., Kurz, R. J., & Peck, A. B. 2010, *Proc. SPIE*, 7733,
Hughes, D. H., Jáuregui Correa, J.-C., Schloerb, F. P., *et al.* 2010, *Proc. SPIE*, 7733,
Hunter, T. R., Schwab, F. R., White, S. D., *et al.* 2011, *PASP*, 123, 1087
Mauersberger, R., Villard, E., Peck, A. B., *et al.* 2011, *IAU Symposium*, 280, 402P
NOEMA Project (www.iram.fr/GENERAL/NOEMA-Phase-A.pdf)
Nyman, L.-Å., Andreani, P., Hibbard, J., & Okumura, S. K. 2010, *Proc. SPIE*, 7737,
Oberg, K. I., Qi, C., Wilner, D. J., & Hogerheijde, M. R. 2012, arXiv:1202.3992
Perley, R., Napier, P., Jackson, J., *et al.* 2009, *IEEE Proceedings*, 97, 1448
Perley, R. A., Chandler, C. J., Butler, B. J., & Wrobel, J. M. 2011, *ApJ*, 739, L1
Prestage, R. M., Constantikes, K. T., Hunter, T. R., *et al.* 2009, *IEEE Proceedings*, 97, 1382
Radford, S. J. E., Giovanelli, R., Sebring, T. A., & Zmuidzinas, J. 2009, *Submillimeter Astrophysics and Technology: a Symposium Honoring Thomas G. Phillips*, 417, 113
Rawlings, M. G., Nyman, L.-Å., & Vila Vilaro, B. 2010, *Proc. SPIE*, 7737,
Wootten, A. 2007, *IAU Symposium*, 242, 511
Wootten, A. & Thompson, A. R. 2009, *IEEE Proceedings*, 97, 1463
Wootten, A., Claussen, M., Marvel, K., & Wilking, B. 2002, *Cosmic Masers: From Proto-Stars to Black Holes*, 206, 100

Cosmic Masers - from OH to H_0
Proceedings IAU Symposium No. 287, 2012   © International Astronomical Union 2012
R.S. Booth, E.M.L. Humphreys & W.H.T. Vlemmings, eds.   doi:10.1017/S1743921312007569

MeerKAT and its potential for Cosmic MASER Research

Roy Booth[1,2], Sharmila Goedhart[1] and Justin Jonas[2]

[1]SKA South Africa, The Park, Park Road, Pinelands, 7005, South Africa,
[2]Rhodes University, P.O Box 94, Grahamstown 6140, South Africa
email: rbooth@ska.ac.za

Abstract. The MeerKAT radio telescope array is the South African precursor instrument for the proposed Square Kilometer Array's mid-band frequency range. It will be the most sensitive centimetre-wavelength telescope in the southern hemisphere until the SKA is built. It will cover a broad range of astronomical science from the evolution of galaxies to tests of Einstein's theory of General Relativity, using Pulsars. The chosen frequency bands will enable sensitive southern Galactic maser surveys in the main lines of hydroxyl, 12 GHz methanol and perhaps14.5 GHz formaldehyde lines as well as searches for redshifted water masers from red-shifts greater than about 0.5. Proposals for Large Surveys using MeerKAT were solicited in late 2009 and resulted in some 20 proposals from teams comprising 500 scientists, world-wide. The successful proposals relevant to maser research will be discussed below.

Keywords. radio telescope design, arrays, senstivity, surveys, masers

1. Introducion

MeerKAT is the South African SKA precursor telescope array. Its development began with a prototype 15 m paraboloid antenna, XDM, of innovative, light-weight design, the dish being cast in a single piece from composite (fibre-glass) material on a mould built on site at the Hartebeesthoek Radio Astronomy Observatory. The mould had been flame sprayed with a thin layer of aluminium before depositing the composite. The antenna performed well, having a surface rms of better than 2 mm, and set the scene for the first 7-element test array (KAT-7) for which all the (12 m diameter) dishes were cast at a base near the remote site in the Northern Cape region known as the Karoo, removing the need for long distance transportation. KAT-7 operates in the frequency range 1.2 - 1.95 GHz and is currently undergoing science commissioning and producing useful data. The receivers are cooled with Stirling-cycle coolers to about 70 K. Data signal processing is conducted using another innovative development, to which the KAT engineers have made a serious contribution. This is the ROACH (reconfigurable open architecture computing hardware) board which has become the primary building block for digital signal processing in several next-generation radio telescopes. As this conference draws to a close, the first narrow band correlator modes, suitable for (OH) MASER research are being tested.

2. The MeerKAT Concept

The experience gained from XDM and KAT-7 has been factored into the MeerKAT array design. The goal is to build the world's most sensitive centimetre wavelength radio telescope in the southern hemisphere, until the SKA becomes a reality and, after a design review in 2010, an ambitious concept of 64 x 13.5 m offset Gregorian antennas was conceived. The array will perform high dynamic range, high-fidelity imaging over nearly an order of magnitude in resolution, with a resolving power of 6 arcseconds at 1420 MHz. It will be optimised for deep, high fidelity imaging of extended low brightness emission, the detection of micro-Jansky radio sources, the measurement of polarization and the detection and monitoring of transient radio sources. Ultimately, the array will operate over a frequency range of 580 MHz to 14.5 GHz but in the first phase (2016) a band of 900-1670 MHz is planned. The other receiver bands will be phased as shown below.

The high sensitivity criterion implies a large effective collecting area and a low system temperature, or maximisation of A_e/T_{sys} - the goal is circa 300 m^2 per kelvin at L-band with cooled (20k) receivers. This, together with a high dynamic range, are the hallmarks of a powerful interferometer array. To achieve this goal the individual parameters should not be compromised. So, for example, rather than go for the widest band feeds, which compromise T_{sys} at the band edges, it is preferable to use somewhat narrower bandwidths. To maximise A_e we have chosen an offset Gregorian antenna design, giving an unblocked aperture and high aperture illumination without introducing high, or polarised sidelobes.

A further advantage of the offset antenna is that the arm carrying the secondary reflector can be designed to carry the individual cooled receivers thus alleviating the need to change receivers when switching between observing bands. A receiver indexer at the secondary focal position will have capacity for up to 4 receivers.

Finally, in another bold move, the MeerKAT digital engineers are working to eliminate the intermediate frequency receiver stage by direct digitisation of the entire RF band, after initial amplification in a low noise receiver. This is currently feasible at L-band but represents a significant challenge for the highest frequency (8-14.5 GHz) band. However they are confident that it wil be possible by 2016-17.

3. The Array configuration and receiver Phasing

As an array telescope, MeerKAT can and will be built in phases and early science will be possible before the full 64 antennae are erected; the receivers covering the different frequency bands will be deployed in a serial manner, as summarised below.

3.1. Phase 1

The array configuration will have a maximum baseline of 8 km and 70% of the collecting area will be within a 1 km diameter core to provide high brightness temperature sensitivity and reduce computational load for transient source data processing. The Phase 1 receiver will cover the frequency range 0.9-1.67 GHz and facilitate early continuum and red-shifted H1 and OH (maser) observations. Direct sampling at RF will be implemented, with all signal processing being performed in the digital domain using hardware and algoritms developed as part of the CASPER collaboration.

3.2. Future phases

The capabilities of MeerKAT will be expanded, as funding allows, and the frequency coverage of the instrument will be extended with the installation of up to three additional receivers:

(a) 580 - 1000 MHz: This receiver will define the redshift range of the deep HI/OH surveys and will be used to observe steep-spectrum continuum sources.

(*b*) 8 - 14.5 GHz (goal: 4-16 GHz): This wide bandwidth receiver will allow the detection of highly redshifted CO emission, red-shifted water masers from z = 0.5, as well as Galactic maser species due to methanol, formaldehyde and excited OH, and will facilitate the detection and monitoring of pulsars in the Galactic centre, and the mapping of the emission from a variety of molecules.

(*c*) 1.5 - 3 GHz: The primary use of this receiver would be the precision timing of pulsars for the detection of gravitational waves.

Finally, it is imperative that funding is found for several additional antennas to extend the maximum baselines of the array out to some 20 km or more. Such an extension will improve the resolution of the array so that the deep continuum survey will be able to probe to lower flux densities without being affected by source confusion and newly detected sources may be located to within about 1 arcsecond to allow follow-up observations at visible and other wavelengths. Further expansion of the array to increase the collecting area (and perhapsresolution) will depend on the decisions related to the implementation of the SKA.

4. MeerKAT Science: The Large Survey Projects

MeerKAT is one of a number of new or re-scoped radio astronomy arrays under construction/test at the moment. These include the ALMA in Chile, EVLA in the USA, E-MERLIN in the UK, LOFAR and APERTIF in The Netherlands, and ASKAP, in Australia. There is considerable synergy among these projects and at least four have chosen to allocate a large fraction of their observing time in the form of Large Survey Projects, which have generated considerable interest among members of the astronomical community. In the case of MeerKAT, we decided to allocate 70% of the observing time in this way, with the remaining 30% being reserved for PI projects ($\approx$ 20%) and Directors discretionary/urgent observations ($\approx$ 10%).

The MeerKAT call for Large Survey Project proposals (Booth *et al.* 2009) returned 20 excellent and timely project ideas from a total of 700 astronomers world-wide (some 500 individuals, including 68 South African astronomers). The total time requested amounted to about 10 years of full-time observing and so we set up a Time Allocation Committee (TAC) consisting of ten international experts to adjudicate and rank the proposals. Ten of the proposals were rated as outstanding and allocated observing time (amounting, in total, to 5 years). The most highly rated proposals were Pulsar timing observations related to gravitational radiation (Observations in the highest frequency band where interstellar scattering is unlikely to be a serious problem) (PI Matthew Bailes) and a deep HI survey to detect HI (and OH) at the highest possible red-shift: $z = 1.4$ at 580 MHz (Acronym LADUMA). (see below). The other highly rated proposals were related to Pulsars and Transient Radio Sources, Deep continuum observations, spectroscopy and VLBI. Of these we will discuss only those of relevance to MASER research.

4.1. *HI (and OH) Galaxy Surveys*

4.1.1. *LADUMA*

- PIs: Blythe (UCT), Baker (Rutgers) and Holweda (ESA)

This group will study HI and its properties at red-shifts out to the maximum MeerKAT can reach ($z = 1.4$) towards the Chandra Deep Field (CDFS). They will need to use the technique of "stacking" to detect the weak HI emission and this requires that they know the red-shift from surveys at other wavelengths (e.g. optical and IR). HDF-south is a well studied region at many wavelengths and probably one of the best regions for this kind of work. The detection of OH molecules and masers will be incidental but certainly a bi-product (sometimes a confusing bi-product) of this survey

4.1.2. *Galaxy Clustering*

- PI: Paolo Serra (ASTRON)

Fornax is the second most massive cluster within 20 Mpc from the Sun and the largest cluster in the southern hemisphere. Its low X-ray luminosity makes it representative of the environment where most galaxies live. Furthermore, Fornax's ongoing growth makes it an excellent laboratory for the study of structure formation. Ideally located for MeerKAT observation, it provides an excellent opportunity to study the assembly of clusters, the physics of the accretion and stripping of gas in galaxies falling in the cluster, and to probe for the first time the connection between these processes and the neutral medium in the cosmic web.

4.1.3. *MHONGOOSE*

- PI: de Blok (ASTRON)

This is a essentially a continuation of an earlier study of HI (and OH) in nearby galaxies of different types, with the VLA - THINGS. Among thelarge array of interesting studies in this proposal is the connection between star formation and HI dynamics and accretion. Of particular interest here is the study of the outer discs of galaxies and the Cosmic web, and galaxy halos and dark matter. The data will be compared with CO observations from ALMA and other Millimetre arrays, as well as Spitzer (IR) data, building up a picture of the relative distribution and importance of atomic and molecular hydrogen in the evolution of galaxies.

4.1.4. *Absorption Line Survey*

- PIs: Gupta (ATNF) and Srianand (IUCAA)

Since it is easier to detect small quantities of cool gas in absorption against a continuum radio source, than in emission, this is another way to detect HI, even the hydroxyl radical (OH) at high red-shift. Given sufficient frequency resolution, it is also an excellent way to detect the line splitting caused by the Zeeman effect and to measure magnetic fields. A further interesting study that will be conducted as part of this project is the constancy of Fundamental Constants as a function of z using OH lines and their different dependencies onthe fine structure constant and the electron/proton mass ratio .

4.2. *Spectroscopic observations with the high frequency receiver (8- 14.5 GHz)*

4.2.1. *MESMER*

- PI: Heywood (Oxford)

Carbon monoxide (CO) is the next most abundant molecule after molecular hydrogen (H_2). H_2 is difficult to observe in the ground state because it has no dipole moment and therefore has no rotational spectral lines. H_2 is usually detected only in the IR or UV, when highly excited.

CO has been detected in a galaxy with a red shift of 6.4 (SDSS J1148+52), and is commonly detected at $z = 4$, or less, showing that molecular hydrogen, perhaps unexpectedly, has a relatively high abundance in the early Universe. Theoretical studies have shown that molecular hydrogen may be more abundant in the early universe where galaxies may be more tightly wound, and the pressure and density may be relatively high. It is considered important to follow up on these, perhaps unexpected, discoveries and the highest frequency of MeerKAT was chosen for precisely this reason. The upper frequency of MeerKAT, 14.5 GHz is equivalent to the CO line red-shifted to $z \approx 7$.

4.2.2. *MeerGAL*

- PIs: Thompson (Hertfordshire) and Goedhart (SKA SA Project)

MeerKAT and its high frequency receiver will be unique in the southern hemisphere for studies of the southern Galaxy and the Magellanic Clouds. The MeerGAL study will measure the properties of the Galaxy, its stellar formation and evolution through continuum and line observations of HII regions, recombination lines and simple molecules. The recombination line work will give improved data on the Galactic rotation curve in the southern Galaxy. Interstellar methanol masers at 12 GHz are giving us important data on Galactic astrometry (e.g. the distance to the Galactic centre) and the discovery and utilisation of new southern hemisphere masers will be very important in this, and other respects. Finally, the 8-14.5 GHz spectral region has not been studied in detail for new molecules or masers, and especially astro-biologically important molecules.

5. The way forward

The MeerKAT organization is anxious to ensure that the Survey Teams will work with the project to ensure a coherent approach to issues like Project Development, Special Software, Data Format (we propose VO compatibility), Team Organization and Dynamics (working groups), Publication/Data Release Policy and Outreach.

We will hold regular (annual in the first instance) meetings of PIs (the first was in February 2011) and request that the Project teams conduct annual meetings with our scientific/technical liaison personnel to report progress, problems and possible changes in scientific priorities as other projects progress. We are also encouraging projects to set up a coherent organization within their projects.

The MeerKAT project will adapt its HR policy towards hiring special postdocs and PhD positions in fields related to the accepted Large Survey Programmes, and has requested that a reciprocal policy is adopted by the team leaders.

Finally, in due course we will issue calls for proposals for the 20% of time to be devoted to PI proposals and MASERS will indoubtedly be the theme of many of these.

References

Booth, R. S., de Blok, W. J. G., Jonas, J. L., & Fanaroff, B. 2009, *arXiv* 0910.2935B

Cosmic Masers - from OH to H_0
Proceedings IAU Symposium No. 287, 2012
R.S. Booth, E.M.L. Humphreys & W.H.T. Vlemmings, eds.

© International Astronomical Union 2012
doi:10.1017/S1743921312007570

KVN Single-dish Water and Methanol Maser Line Surveys of Galactic YSOs

Kee-Tae Kim, Do-Young Byun, Jae-Han Bae, Won-Ju Kim, Hyun-Woo Kang, Chung Sik Oh, and So-Young Youn

Korea Astronomy and Space Science Institute, 776 Daedeokdae-ro, Yuseong-Gu,
Daejeon 305-348, South Korea
email: ktkim@kasi.re.kr

Abstract. We have carried out simultaneous 22 GHz H_2O and 44 GHz Class I CH_3OH maser line surveys of more than 1500 intermediate- and high-mass YSOs in the Galaxy using newly-constructed KVN 21-m telescopes. As the central (proto)stars evolve, the detection rates of the two masers rapidly decrease for intermediate-mass YSOs while the rates increase for high-mass YSOs. These results suggest that the occurrence of the two masers is closely related both to the evolutionary stage of the central objects and to the circumstellar environments. CH_3OH masers always have very similar velocities (<10 km s^{-1}) to the natal dense cores, whereas H_2O masers often have significantly different velocities. The isotropic luminosities of both masers are well correlated with the bolometric luminosities of the central (proto)stars.

Keywords. maser, water, methanol, star formation, young stellar object

1. Korean VLBI Network (KVN)

The Korean VLBI Network (KVN) consists of three 21-m radio telescopes, which are located in Seoul (Yonsei University), Ulsan (University of Ulsan), and Jeju island (Tamna University) (Kim *et al.* 2004; Kim *et al.* 2011; Lee *et al.* 2011). The three baselines are 305, 359, and 478 km. KVN is a compact VLBI network, compared to VLBA and EVN, both of which have the longest baselines > 5000 km. In order to obtain more and longer baselines, KVN will be often run in combination with Japanese and Chinese VLBI facilities at 22 and 43 GHz. The so-called East Asian VLBI Network (EAVN) is expected to be comparable to VLBA or EVN in spatial resolution, sensitivity, and imaging capability (e.g., Yi & Jung 2008). In addition, KVN will be operated in four (22/43/86/129 GHz) different frequency bands simultaneously using multi-frequency band receiver systems (Han *et al.* 2008). Simultaneous multi-frequency observations will make it possible to use the 22 GHz data for performing phase calibrations of higher-frequency data (Jung *et al.* 2011). By this multi-frequency phase referencing technique, KVN could play an important role in millimeter VLBI observations of maser sources in galactic star-forming regions and evolved stars and active galactic nuclei.

Figure 1. KVN Yonsei, Ulsan, and Tamna stations.

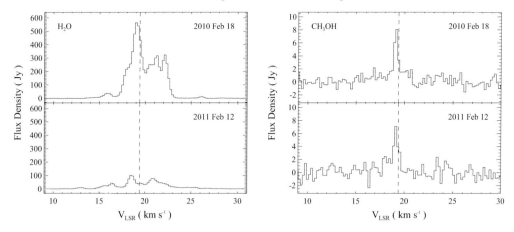

Figure 2. (a) H_2O maser spectra and (b) 44 GHz Class I CH_3OH maser spectra of UCHII G15.04−0.68. The observing date is shown at the top right corner in each panel. The vertical dotted line represents the systemic velocity of the parental dense core.

All KVN 21-m radio telescopes were equipped with 2-channel (22/43 GHz) receiving systems in 2009. We have been making single-dish observations as well as VLBI test observations. As of 2012 March, 4-channel receiving systems were installed at all KVN telescopes. We plan to start regular VLBI operation at 22/43 GHz from late 2012 and at 22/43/86/129 GHz from late 2013.

2. KVN Single-dish Maser Line Surveys of YSOs in the Galaxy

From 2009 September to 2011 June we have carried out simultaneous 22 GHz H_2O and 44 GHz Class I CH_3OH maser line surveys of more than 1500 YSOs in the Galaxy using KVN 21-m telescopes. Our sample is composed of intermediate- and high-mass young stellar objects (YSOs) in different evolutionary stages (e.g., Wood & Churchwell; Sridharan *et al.* 2002). The primary goals of these surveys are (1) to investigate the relationship between the two masers, and (2) to study the relationship of the two masers with the central (proto)stars and the natal dense cores, (3) to build a new database for higher-resolution observations of the two masers with EVLA and KVN(+VERA).

The overall detection rates of H_2O and CH_3OH masers in high-mass YSOs are about 40% and 35% with a flux limit of ∼0.5 Jy (1σ), respectively. The detection rates significantly increase for high-mass YSOs as the central objects evolve from the protostellar stage to the main sequence stage (see also Fontani *et al.* 2010). This is in contrast with the trend observed in low- and intermediate-mass YSOs, that the detection rates drop with the evolution of the central (pro)stars (Furuya *et al.* 2003; Bae *et al.* 2011). These results suggest that the occurrence of the two masers is closely related to the circumstellar environments as well as the evolutionary stage of the central (proto)stars. Figure 2 shows sample maser spectra detected in G15.04−0.68, which is a spherical ultra-compact HII region (UCHII) (Kim, W.-J. in preparation). CH_3OH maser lines do not usually show any significant variations in velocity, intensity, and shape over one-year period, but H_2O maser lines show significant variations. CH_3OH masers always have similar velocities to the parental dense cores within 10 km s^{-1}, while H_2O masers often have many velocity components significantly shifted from the systemic velocities.

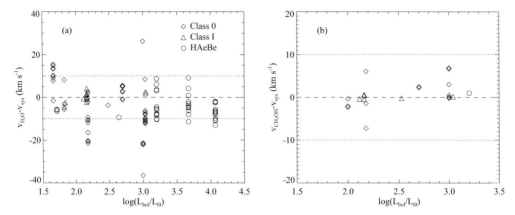

Figure 3. Relative velocity vs. bolometric luminosity of the central (proto)star for (a) H_2O masers and (b) 44 GHz Class I CH_3OH masers.

3. The Survey Results of Intermediate YSOs

In Bae *et al.* (2011) we reported a multi-epoch, simultaneous 22 GHz H_2O and 44 GHz class I CH_3OH maser line survey toward 180 intermediate-mass $(4-10\ M_\odot)$ YSOs. The sources consists of 14 Class 0, 19 Class I objects, and 147 Herbig Ae/Be (HAeBe) stars, which are widely believed to be intermediate-mass pre-main sequence stars.

We detected H_2O masers in 16 objects and CH_3OH masers in 10 objects with one new H_2O maser source (HH 165) and six new CH_3OH maser sources (CB 3, IRAS 00338+6312, OMC3 MMS9, IRAS 05338−0624, V1318 CygS, and IRAS 23011+6126). HH 165 is the ninth HAeBe star with detected H_2O maser emission, and V1318 CygS is the first HAeBe star with detected 44 GHz CH_3OH maser emission.The overall detection rates of H_2O and CH_3OH masers are 9 % and 6 %, respectively. The rates rapidly decrease as the central (proto)stars evolve, as for low-mass YSOs. The detection rates of H_2O masers are 50 %, 21 % and, 3 % for Class 0, Class I, and HAeBe objects, respectively. Those of CH_3OH masers for Class 0, Class I, and HAeBe objects are 36 %, 21 %, and 1 %. As mentioned above, the detection rates in high-mass star-forming regions increase as the central objects evolve.

The relative velocities of H_2O masers with respect to the ambient molecular gas are $9\ \mathrm{km\,s^{-1}}$ on average, whereas those of CH_3OH masers are tightly distributed around $0\ \mathrm{km\,s^{-1}}$. No CH_3OH maser velocity deviates more than $10\ \mathrm{km\,s^{-1}}$ from the systemic velocity. Large relative velocities are mainly shown in the Class 0 objects: $|v_{H_2O} - v_{sys}| > 10\ \mathrm{km\,s^{-1}}$ and $|v_{CH_3OH} - v_{sys}| > 1\ \mathrm{km\,s^{-1}}$. This is consistent with previous suggestions that H_2O masers originate from the inner parts of outflows while class I CH_3OH masers arise from the interacting interface of outflows with the ambient dense gas (e.g., Kurtz *et al.* 2004).

The intensities and shapes of the observed H_2O maser lines were quite variable. Half of the maser-detected sources showed velocity drifts. The integrated line intensities varied by up to two orders of magnitude. On the contrary, the observed CH_3OH lines do not reveal any significant variability in intensity, shape, or velocity. The line integrals were maintained within ~ 50 % over the observations. These different variability behaviors of the two masers may be connected with different emitting environments.

The isotropic luminosities of both masers are well correlated with the bolometric luminosities of the central objects (Fig. 4). The linear fits result in $L_{H_2O} = 1.71 \times 10^{-9}(L_{bol})^{0.97}$ and $L_{CH_3OH} = 1.71 \times 10^{-10}(L_{bol})^{1.22}$ when only data points of this

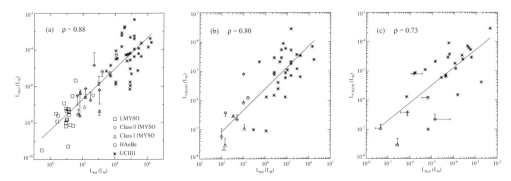

Figure 4. a) H_2O maser isotropic luminosity vs. bolometric luminosity of the central (proto)star, (b) CH_3OH maser luminosity vs. bolometric luminosity, and (c) CH_3OH maser luminosity vs. H_2O maser luminosity. In each panel, the solid line is the fitted relation to the data points and the error bars represent the variability of maser luminosity through the entire observations.

survey are considered, while those yield $L_{H_2O} = 4.10 \times 10^{-9}(L_{bol})^{0.84}$ and $L_{CH_3OH} = 2.89 \times 10^{-9}(L_{bol})^{0.73}$, after the data points of low- and high-mass regimes are added (Furuya *et al.* 2003; Kim, W.-J. 2012 in preparation).

References

Bae, J.-H., Kim, K.-T., Youn, S.-Y., Kim, W.-J., Byun, D.-Y., Kang, H.-W., & Oh, C. S. 2011, *ApJS*, 196, 21

Fontani, F., Cesaroni, R., & Furuya, R. S. 2010, *A&A*, 517, A56

Furuya, R. S., Kitamura, Y., Wootten, H. A., Claussen, M. J., & Kawabe, R. 2003, *ApJS*, 144, 71

Kim, H.-G., *et al.* 2004, in: R. Bachiller *et al.* (eds.), *Proceedings of the 7th Symposium of the European VLBI Network*, (Toledo:OAN), 281

Kim, K.-T., *et al.* 2004, *J. Korean Astron. Soc.*, 44, 81

Kurtz, S., Hofner, P., & Carlos Vargas, A. 2004, *ApJS*, 155, 149

Han, S.-T., *et al.* 2008, *Int. J. Infrared and Millimeter Waves*, 29, 69

Lee, S.-S., *et al.* 2011, *PASP*, 123, 1398

Sridharan, T. K., Beuther, H., Schilke, P., Menten, K. M., & Wyrowski, F. 2002, *ApJ*, 566, 931

Wood, D. O. C. & Churchwell, E. 1989, *ApJS*, 69, 831

Yi, J. & Jung, T.-H. 2008, *P. Korean Astron. Soc.*, 23, 37

Cosmic Masers - from OH to H₀
Proceedings IAU Symposium No. 287, 2012
R.S. Booth, E.M.L. Humphreys & W.H.T. Vlemmings, eds.

© International Astronomical Union 2012
doi:10.1017/S1743921312007582

Methanol masers in the Herschel era
Putting them in the star formation context

Michele Pestalozzi

IAPS - INAF
via del Fosso del Cavaliere 100, 00133 Roma, Italy
email: michele.pestalozzi@gmail.com

Abstract. Methanol masers are known to be among the most reliable tracers of high-mass stars in early stages of evolution. A number of searches across the Galaxy has yielded to date, a complete census of those masers in two thirds of the Milky Way, providing a catalogue of some 800 sources to be studied in depth. In particular, it is important to characterise the physical properties of the objects hosting methanol masers, and this is possible today using data from the Herschel Space Observatory (HSO). The exceptional spatial resolution of HSO and its wavelength coverage are perfectly tuned to put the methanol maser phase into its star formation context. This paper presents results on the characterisation of methanol maser hosts using Herschel data from the Hi-GAL project, an Open Time Key Project to survey the inner Galactic plane at 5 wavelengths between 70 and 500 μm.

Keywords. Star formation, high-mass stars, masers, methanol

1. Introduction

Methanol masers at 6.7 GHz are known to mark, exclusively, high-mass star forming regions in early stages of evolution. After their discovery (Menten 1991) a number of searches have been undertaken in the last two decades, using several strategies: on the one hand, targets were selected on the basis of their Infrared (IR) colours, at that time limited to the wavelength coverage of the IRAS satellite (see e.g. Schutte *et al.* 1993; van der Walt *et al.* 1995; Walsh *et al.* 1998; Szymczak *et al.* 2000 and several others, a complete list is given in Pestalozzi *et al.* 2005); on the other hand, large regions of sky have been systematically covered with observations to detect potentially all masers within a certain sesntivity limit (see e.g. Szymczak *et al.* 2002; Pestalozzi *et al.* 2002b; Pandian & Goldsmith 2007; Green *et al.* 2008). While the first strategy yielded a detection rate of some 15%, the latter technique, although more time consuming, was more efficient in finding new sources with e.g. unexpected IR colours. Examples of this are multiple, some of them are described in Pestalozzi *et al.* (2002a). The ultimate search for 6.7 GHz methanol masers in the Milky Way is the Methanol MultiBeam (MMB) Survey†, through which two thirds of the Galaxy has been covered, yeilding some 800 sources, 30% of which are new detections (see e.g. Caswell *et al.* 2011).

The main idea behind targeted searches was to define the characteristics of the objects that host methanol maser emission, something that, in view of the low detection rates, was only partially achieved. Characterisation of 6.7 GHz methanol maser hosts was also done using Spitzer GLIMPSE and MSX data, as reported by Ellingsen (2006). The latter

---

† The MMB Survey was conducted using the Parkes Telescope in Australia. The area covered to date comprises the longitude range $60° > \ell > -174°$. See http://www.jb.man.ac.uk/research/methanol/

study expressed MIR criteria to characterise methanol maser hosts, but apart from a test on a sample of known sources, no searches for new masers were performed on that basis.

The Hi-GAL Survey (High-mass Galactic Plane Survey)† is an Open Time Key Project that has been awarded some 300 hours of observation with the HSO, to survey the inner Galactic Plane in five contiguous bands of the Far Infrared (FIR) dust continuum between 70 and 500 μm. The unprecedented spatial resolution due to its 3.5 m mirror, as well as high-dynamical range instruments onboard (in our case two of the three, PACS for wavelengths shorter than 200 μm and SPIRE for the wavelength range between 200 and 600 μm, see Poglitsch *et al.* 2010; Griffin *et al.* 2010) make Hi-GAL data the best place to charaterise Galactic cold material (dust) that pinpoints the early stages of star formation. Technical details of the survey are summarised in Molinari *et al.* (2010).

In this paper we investigate the IR properties of methanol maser hosts that are detected in Hi-GAL. The hypothesis behind this study is that, due to the short duration of the methanol maser phase (between 2.5×10^4 and 4.5×10^4 years, see van der Walt 2005), all methanol maser hosts should basically be coeval. Therefore all differences in their IR appearance are not due to evolution but e.g. as a different level of embededness of the maser pumping object in its primordial dust cocoon. We attempt to disentangle these two effects (evolution and embededness) by studying and characterising of the SEDs of the hosts, allowing, in this way, a precise designation of the methanol maser phase in the context of (high-mass) star formation.

2. Data: Hi-GAL maps, source extraction, masers, SED fitting

The data used for the present paper consist of the public MMB catalogue and the latest Hi-GAL maps in all five Herschel bands, as obtained by the pipeline devolped at IAPS-INAF in Rome and described in detail in Traficante *et al.* (2011). The overlap between available Hi-GAL maps and public positions of methanol masers detected by MMB corresponds to coordinates $20.0° > \ell > -71.0°$ and $1.0° > b > -1.0°$. In this portion of the Galactic Plane there are 635 6.7 GHz methanol masers. All of them have a Hi-GAL counterpart, albeit not at all Herschel wavelengths.

Compact sources in the Hi-GAL data were extracted around every methanol maser position in a subimage with varying size, depending of the wavelength. In all cases the closest, in angle, on the sky was assigned to the maser as its counterpart. Compact source detection and photometry extraction was performed using CuTEX‡, a package developed explicitly to work on images with bright and highly varilable background levels, a characteristic of all Herschel images of regions on the Galactic Plane (see Molinari *et al.* 2011 for details).

In order to work with an homogeneous set of data, a series of filters were applied to the original source list. For the present work this means only sources for which counterparts at five Herschel wavelengths were found and that were closer than 5 arcseconds to the original maser position. All sources that showed complex SEDs have been excluded; these have to be treated on a sigular basis. The final list contains 421 methanol maser bearing Hi-GAL sources (66% of the starting 635 sources), the ones that are plotted with filled black circles in Fig. 1.

For all sources in the final list, grey body SED fitting was performed, obtaining masses and luminosities for near- and far heliocentric distances. The distribution of several physical properties of methanol maser bearing Hi-GAL sources is shown in Fig. 2.

---

† See: https://hi-gal.ifsi-roma.inaf.it/higal/
‡ CuTEX analyses the second derivative of Herschel images to detect soures and extracts photometric information with Gaussian photometry.

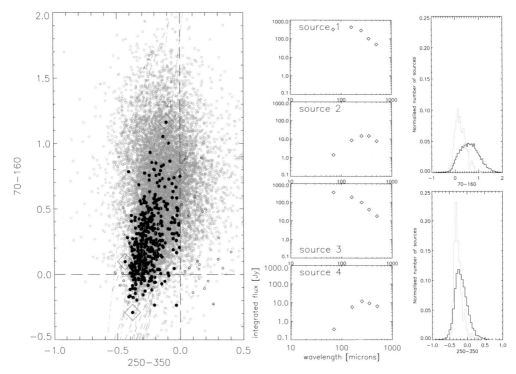

Figure 1. left: Color-color diagram of Hi-GAL sources extracted in the galactic longitude range $20° > \ell > -71°$ with a clear detection at all 5 Herschel continuum wavelengths. Black symbols indicate the methanol maser bearing sources, orange open circles all other sources. Grey body models are overlaid in blue (equal temperature) and magenta (equal β). Red diamonds indicate the "extreme" sources, for which the SEDs are shown in the four panels in the middle. The two most right panels show the histograms in the colors plotted in the left panel; green lines refer to methanol maser bearing sources, black lines to all other Hi-GAL sources. In both panels histograms have been normalised to the total number of sources.

3. Results: methanol masers in the context of star formation

3.1. *Colors*

The left panel of Fig. 1 shows the color-color diagram of all methanol maser bearing sources in the final list (filled black circles), overlaid on all other Hi-GAL sources detected in all 5 Herschel bands. Most of the sources fall in the region with positive 70-160 color and negative 250-350 color, indicating that their SED must peak between 160 and 250 μm. This is even more true for methanol maser bearing sources, which seem to concentrate in the lower part of that region and mainly to the left of the thick blue line. The latter fact indicates that methanol maser bearing sources are mostly modelled with grey bodies with varying values of β (blue lines show grey body models with constant β) and with temperatures between 15 and 25 K (see magenta lines that show grey bodies with constant temeprature).

The two right-most panels of Fig. 1 show the distribution of sources with and without associated methanol maser (green and black lines, respectively) in the two colors of the left panel. Again, what is immediately clear is the difference in shape between the green and black distribution, the green being more peaked, strongly suggesting that the methanol maser phase is a short one - in the evolution from large dust clump or core to stars. Another important aspect is that methanol maser bearing sources concentrate

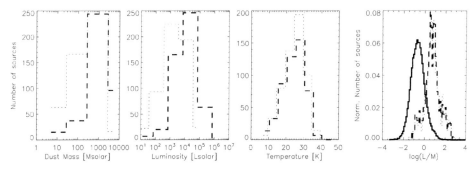

Figure 2. Physical properties of the methanol maser bearing Hi-GAL sources. From left to right: dust mass ($M_\odot$), bolometric luminosity ($L_\odot$) and dust temperature (K). Dotted and thick dashed lines refer to quantities for the near- and far heliocentric (kinematic) distances, respectively. The right most panel shows the distribution of the L/M ratio for methanol maser bearing sources (dotted and dashed lines) and Hi-GAL sources without methanol maser (solid line), normalised to their total. The peaks of the distribution are distinct, meaning a different overall placement in the L vs M diagram (see Molinari *et al.* 2008). Also, methanol maser bearing sources seem to have an overall more peaked L/M distribution, suggesting that the methanol maser phase is short.

in the "warmer" part of the diagram, suggesting that methanol masers probably do not appear at the very beginning of the formation of stars, but later on.

Finally, red diamonds indicate the area enclosing Hi-GAL methanol maser bearing sources with simple SEDs. The SEDs of these are displayed in the four panels in the middle of Fig. 1. The quick observation here is that there is a larger spread of colors in the 70-160 than in the 250-350 color.

Note that all temperatures mentioned here refer to what is visible by Herschel, i.e. with a spatial resolution between 5 and 20". At distances of a few kpc, this scale is much larger than the single, maser pumping protostar, explaining the discrepancy in dust temperature between the data presented here and the values obtained in methanol maser pumping models (see e.g. Sobolev *et al.* 1997). In fact, to allow methanol to evaporate from icy dust grains, a dust temperature of some 100-150 K is required in the immediate surroundings of the protostar. Only interferometry at millimeter and sub-millimeter wavelengths will allow a connection of the physical properties on the large scales seen by Herschel, with the immediate surroundings of the maser pumping source seen e.g by ALMA.

3.2. *Physical properties and L/M ratio*

Figure 2 shows some physical properties of methanol maser bearing Hi-GAL sources as a result of grey body SED fitting. While mass and luminosity histograms show some difference depending on which distance is adopted (dotted line is near- dashed thick line is far heliocentric distance), temperatures seem to be more stable around 25 K, with a spread between 15 and 35 K. This was also seen in the color-color diagram. As mentioned above, this fact serves to put methanol maser bearing sources into the context of star formation and not to extract the physical properties of the maser pumping source.

A distance independent parameter is the ratio of luminosty to mass, L/M. Comparing the distribution of this ratio among methanol maser bearing sources and all other Hi-GAL sources (Fig. 2, right most panel) one notices that the former peak at values of L/M indicating that the luminosity is one order of magnitude higher than the mass, in contrast to all other Hi-GAL sources. In a luminosity versus mass plot, the two distributions can indicate two different evolutionary stages. Models by Molinari *et al.* (2008) indicate that methanol maser bearing sources lie closer to the end of the accretion phase shortly before, or landing on the ZAMS, while all other Hi-GAL sources are mostly concentrated in regions where accretion from the large scale dust envelope is still strongly ongoing. Methanol masers then, seem to appear in an "advanced" stage of evolution, and might not be the very early stages indicator as believed in the past.

A further consideration comes from the shape of the distribution of L/M ratios. General Hi-GAL sources seem to have a broader distribution than methanol maser bearing sources. Circumstantially, this supports what was found in van der Walt (2005) regarding the length of the methanol maser phase. In that paper, methanol masers are calculated to last a few 10^4 years, a much shorter time than the time resolution of SED-based age determination. This would support the idea that methanol maser bearing objects actually host coeval sources. The ultimate statement on this hypothesis will not arrive before extensive observational campaigns with an iterferometer are performed. This will be a challenge for ALMA.

On the basis of all observables presented above, one can conclude that the difference in the IR appearance of several methanol maser bearing sources is probably not an indication of age but rather of different level of embededness of the methanol maser pumping source. This idea is also supported by an example of methanol maser bearing objects in the high-mass star forming region NGC 7538 (see e.g. Pestalozzi *et al.* (2006)). The three masers in that region are associated with apparently very different sources. Despite this evident difference, these could be hosting very similar objects in their interior, objects that are able to pump a maser.

Acknowledgements

This work was possible thank to the infatigable work of many people involved in the projects from which data were used here. I would like to mention the MMB team in Manchester and at ATNF Sydney (in particular Gary Fuller, James Caswell and Jimy Green), as well as the Hi-GAL team in Rome (in particular the Hi-GAL P.I. Sergio Molinari as well as Eugenio Schisano and Davide Elia).

References

Caswell, J. L., Fuller, G. A., Green, J. A., et al. 2011, *MNRAS*, 1330
Ellingsen, S. P. 2006, *ApJ*, 638, 241
Green, J. A., Caswell, J. L., Fuller, G. A., et al. 2008, *MNRAS*, 385, 948
Griffin, M. J., Abergel, A., Abreu, A., et al. 2010, *A&A*, 518, L3
Menten, K. M. 1991, *ApJ*, 380, L75
Molinari, S., Pezzuto, S., Cesaroni, R., et al. 2008, *A&A*, 481, 345
Molinari, S., Schisano, E., Faustini, F., et al. 2011, *A&A*, 530, A133+
Molinari, S., Swinyard, B., Bally, J., et al. 2010, *PASP*, 122, 314
Pandian, J. D. & Goldsmith, P. F. 2007, *ApJ*, 669, 435
Pestalozzi, M. R., Humphreys, E. M. L., & Booth, R. S. 2002a, *A&A*, 384, L15
Pestalozzi, M. R., Minier, V., & Booth, R. S. 2005, *A&A*, 432, 737
Pestalozzi, M. R., Minier, V., Booth, R. S., & Conway, J. E. 2002b, in Cosmic Masers: from Proto-Stars to Black Holes, IAU Symposium 206, 139
Pestalozzi, M. R., Minier, V., Motte, F., & Conway, J. E. 2006, *A&A*, 448, L57
Poglitsch, A., Waelkens, C., Geis, N., et al. 2010, *A&A*, 518, L2
Schutte, A. J., van der Walt, D. J., Gaylard, M. L., & MacLeod, G. C. 1993, *MNRAS*, 261, 783
Sobolev, A. M., Cragg, D. M., & Godfrey, P. D. 1997, *MNRAS*, 288, L39
Szymczak, M., Hrynek, G., & Kus, A. J. 2000, *A&AS*, 143, 269
Szymczak, M., Kus, A. J., Hrynek, G., Kepa, A., & Pazdereski, E. 2002, *A&A*, 392, 277
Traficante, A., Calzoletti, L., Veneziani, M., et al. 2011, ArXiv e-prints
van der Walt, D. J. 2005, *MNRAS*, 360, 153
van der Walt, D. J., Gaylard, M. J., & MacLeod, G. C. 1995, *A&AS*, 110, 81
Walsh, A. J., Burton, M. G., Hyland, A. R., & Robinson, G. 1998, *MNRAS*, 301, 640

Cosmic Masers - from OH to H$_0$
Proceedings IAU Symposium No. 287, 2012
R.S. Booth, E.M.L. Humphreys & W.H.T. Vlemmings, eds.
© International Astronomical Union 2012
doi:10.1017/S1743921312007594

Early results from a diagnostic 1.3 cm survey of massive young protostars

Crystal L. Brogan[1], Todd R. Hunter[1], Claudia J. Cyganowski[2], Remy Indebetouw[1,3], Rachel Friesen[1] and Claire Chandler[4]

[1]National Radio Astronomy Observatory (NRAO),
520 Edgemont Rd, Charlottesville, VA 22903, USA
email: cbrogan@nrao.edu

[2]Harvard-Smithsonian Center for Astrophysics
60 Garden St., Cambridge, MA 02138, USA

[3]University of Virginia, Charlottesville, VA 22903, USA

[4]National Radio Astronomy Observatory (NRAO), Socorro, New Mexico, 87801, 8USA

Abstract. We have used the recently-upgraded Karl G. Jansky Very Large Array (JVLA) to conduct a K-band (~ 24 GHz) study of 22 massive young stellar objects in 1.3 cm continuum and a comprehensive set of diagnostic lines. This survey is unique in that it samples a wide range of massive star formation signposts *simultaneously* for the first time. In this proceeding we present preliminary results for the 11 sources in the 2-4 kpc distance bin. We detect compact NH$_3$ cores in all of the fields, with many showing emission up through the (6,6) transition. Maser emission in the 25 GHz CH$_3$OH ladder is present in 7 of 11 sources. We also detect non-thermal emission in the NH$_3$ (3,3) transition in 7 of 11 sources.

Keywords. masers, stars: formation, ISM: jets and outflows, ISM: molecules, radio lines: ISM, radio continuum: ISM, infrared: ISM, submillimeter

1. Introduction

The formation of massive stars remains a poorly-understood phenomenon, primarily because they typically form in complex clusters, at distances of several kiloparsecs (see review by Zinnecker & Yorke (2007). Indeed, the tools traditionally used to classify low mass YSOs (primarily near-IR imaging) are largely inapplicable due to extreme dust obscuration, making it a challenge to gauge the evolutionary state of massive young stellar objects (MYSOs). Useful samples of MYSOs have been compiled on the basis of $> 1'$ resolution IRAS far-infrared colors coupled with a lack of bright centimeter wavelength emission (Molinari *et al.* 1996; Sridharan *et al.* 2002). However, detailed studies of individual MYSOs from these samples have consistently shown that they contain not a single protostar but a cluster (Hunter *et al.* 2006; Cyganowski *et al.* 2007, 2011; Brogan *et al.* 2008). Furthermore, members of a single cluster can exhibit a wide range of massive star formation signposts (e.g., Brogan *et al.* 2009; Hunter *et al.* 2008), suggesting a range of mass and/or evolutionary states. Thus the term MYSO is really synonomous with massive proto-cluster as opposed to the previous concept of a single massive protostar.

A wide variety of signposts for MYSOs have been established: H$_2$O and CH$_3$OH masers, ultracompact (UCHII) and hypercompact (HCHII) regions, dense dust cores, warm (> 30 K) NH$_3$ cores, massive outflows, hot core line emission, and infrared dark clouds (IRDCs). Several as yet unproven but plausible evolutionary sequences amongst these signposts have been proposed (e.g., Churchwell 2002). For masers in particular, Class I and II CH$_3$OH masers are thought to be early-stage, H$_2$O intermediate-stage, and OH late-stage tracers (e.g., Ellingsen *et al.* 2007). However, for the most part, this zoo of phenomena has been compiled from a heterogeneous set of observations with varying angular resolution, and by and large ignores the likely clustered nature of the observed star formation. Thus, subsequent correlation analyses are plagued by sensitivity and angular resolution mismatches, astrometric uncertainties or time variability. To advance this field of study, simultaneous high angular resolution observations of a majority of these diagnostic signposts are essential, and must be performed at wavelengths long enough to penetrate the high column densities of dust and with a high enough angular resolution to distinguish individual massive protostars.

Under the auspices of the NRAO Resident Shared Risk Program (RSRO), we have carried out a K-band ($\sim$ 24 GHz) study of 22 MYSOs in 1.3 cm continuum and a comprehensive set of diagnostic lines. This survey is unique in that it samples a wide range of massive star formation signposts *simultaneously* for the first time. Ultimately our goal is to identify possible observational discriminators of evolutionary state, and answer the questions: (1) Is there a correlation between gas temperature and mid-IR luminosity? (2) Is there a correlation between gas temperature, density, or compactness and presence of hot core line emission? (3) When do hot cores develop – is detectable free-free emission required? (4) Does the level of source multiplicity correlate with other diagnostics like temperature, and density? To answer these questions we assembled and observed an NH_3-selected sample of 22 MYSOs as described below.

2. The Observed Sample

Cyganowski *et al.* (2008) compiled a promising new catalog of MYSO candidates ($\sim$ 300), based on the presence of extended 4.5 μm emission in the *Spitzer* GLIMPSE Legacy Survey – emission that is thought to arise predominantly from shocked H_2 lines (from outflows) within the 4.5 μm continuum bandpass. About half of the cataloged EGOs reside in IRDCs – thought to be the birth sites of massive stars (e.g., Rathborne *et al.* 2006). More recently, Cyganowski *et al.* (2009) have confirmed that these sources do contain massive, actively accreting protostars from VLA observations of 20 EGOs in 6.7 GHz Class II CH_3OH masers (signpost of MYSOs) and 44 GHz Class I CH_3OH masers (signpost of outflows), and single dish detections of thermal CH_3OH, $H^{13}CO^+$, and SiO. About 2/3 of the cataloged EGOs in the northern sky have also been observed in NH_3 (1,1) to (3,3) using the Nobeyama 45-m with a 94% detection rate (66 sources, 65″ resolution; Cyganowski *et al.*, in prep, also see proceedings in this volume). From the NH_3 observed EGO subsample, we selected 20 of the strongest [$T_A(NH_3 (1,1)) > 0.7$ K; all detected in NH_3 (1,1) to (3,3)] to include in the current survey (14 of these overlap with the Cyganowski *et al.* (2009) VLA EGO subsample). We also include 2 IRDCs that are not EGOs, but are otherwise similar, and were observed previously by the VLA in NH_3 (1,1) and (2,2) (Devine *et al.* (2011) and show saturated NH_3 rotation temperatures and outflows. Most prior MYSO samples have been selected based on having at most weak free-free emission, in order to concentrate on (presumably) younger regions (e.g., Molinari *et al.* (1996); Sridharan *et al.* (2002). To avoid this potential bias, and provide a contrasting sample for the (presumably) younger EGO/IRDC sources, 9 of the objects are known to harbor UCHII or weak HCHII/wind counterparts. All 22 MYSOs in the sample also have 1.2 mm counterparts in the CSO BOLOCAM Galactic Plane Survey (Rosolowsky *et al.* 2010).

Salient facts about the sample sources are summarized in Table 1, with the sources separated into two distance categories: 2–4 kpc and 4–6 kpc. The distances have been consistently derived from the NH_3 LSR velocities using the new Galactic rotation curve of Reid *et al.* (2009); we assume the sources lie at the near distance. This assumption is almost certainly valid for the IRDC sources (18/22) since they are observed in silhouette against the Galactic background. Thus, we have created a sample that is unified by the presence of dense gas, dust, and outflows and lie at distances between 2–6 kpc. In contrast, the sample exhibits a wide range of mid-IR morphology, 24 μm derived luminosity, as well as a range in free-free ionized continuum properties. In addition to the targeted sources, we may also make a number of serendipitous detections from other mid-IR bright sources in the 2′ diameter FWHP primary beam. Massive star forming cores have typical sizes of 0.1 pc, and at a distance of 2 kpc (6 kpc), 4″ (1.3″) is equal to a physical scale of 0.04 pc or 8,000 AU. This is the physical scale that is essential to probe with this survey, so we observed the 2–4 kpc sample in the D (or DnC) configuration from September to December 2010 and the the 4–6 kpc sample in the C-configuration (or CnB) configuration from January to April 2011.

Table 1. The Observed Sample

Source	D kpc	H$_2$O maser	Methanol Masers 6.7 GHz	44 GHz	EGO	IRDC	Ionized gas	Methanol 25 GHz maser	Ammonia (3,3) maser
D-configuration sources (D = 2–4 kpc)									
G10.29−0.13	2.2	Y	Y	Y	Y^V	Y		Y	N
G10.34−0.14	2.1	Y	Y	Y	Y^V	Y		Y	Y
G11.92−0.61	3.8	Y	Y	Y	Y^V	Y	Weak	Y	Y
G14.33−0.64	2.8	Y	Y	Y	Y	Y		Y	Y
G14.63−0.58	2.3	Y	−	−	Y	Y		N	N
G19.36−0.03	2.4	Y	Y	Y	Y^V	Y		Y	Y
G22.04+0.22	3.6	Y	Y	Y	Y^V	Y		Y	N
G24.92−0.16	3.3	N	−	−	N	Y		N	Y
G24.94+0.07	3.0	N	Y	Y	Y^V	N	Weak	N	Y
G28.28−0.36	3.3	N	Y	N	Y^V	N	Weak	N	N
G35.03+0.35	3.4	Y	Y	Y	Y^V	Y	UCHII	Y	Y
DNC-configuration sources (D = 2–4 kpc)									
G12.42+0.50	2.6	Y	−	−	Y	Y	UCHII		
C-configuration sources (D = 4–6 kpc)									
G12.91−0.03	4.4	Y	−	−-	Y	Y			
G16.59−0.05	4.2	Y	Y	Y	Y	Y			
G18.67+0.03	4.9	N	Y	Y	Y^V	N			
G18.89−0.47	4.5	Y	Y	Y	Y^V	Y			
G25.27−0.43	4.0	N	Y	Y	Y^V	Y			
G28.83−0.25	5.0	N	Y	Y	Y^V	Y	Weak		
G45.47+0.05	4.8	Y	Y	Y	Y	Y	UCHII		
G49.27−0.34	5.6	Y	N	Y	Y^V	Y	UCHII		
CNB-configuration sources (D = 4–6 kpc)									
G08.67−0.35	4.5	Y	Y	Y	N	Y	UCHII		
G12.68−0.18	4.7	Y	Y	−	Y	N			

EGO column: Y=EGO from Cyganowski *et al.* (2008) catalog, Y^V = EGO also in Cyganowski *et al.* (2009) VLA CH$_3$OH maser survey sample.
Ionized column: Weak means < 1 mJy beam^{-1} at 3.6cm in the Cyganowski *et al.* (2011) continuum survey

3. Key Diagnostics

To sample a wide range of potential diagnostics of MYSO evolutionary state and answer the questions posed above, we have used the WIDAR correlator with sixteen 8 MHz subbands to simultaneously observe the key diagnostics described below at a spectral resolution of ~ 0.4 km s^{-1}. The correlator setup is shown in Figure 1.

Ammonia: The NH$_3$ ladder from (1,1) to (6,6) with E_l=23–400 K allows us to fit for the rotation temperature (T_R) of the hot dense gas, as well as NH$_3$ column density, and core diameter. In the past, only the (1,1) and (2,2) transitions were typically observed, which saturate for $T_R > 25$ K. Indeed, a recent VLA study of several IRDCs in NH$_3$ (1,1) and (2,2) has demonstrated that toward dense NH$_3$ cores with mid-IR counterparts, T_R is too high to be constrained by the two lowest transitions alone (Devine *et al.* 2011). For sources in confused regions, the (1,1) line can be difficult to image, but the emission in (2,2) and higher is generally very compact.

Hot Core Lines: Thermal emission from the CH$_3$OH-E ladder from E_l=35 - 150 K can provide an independent check on T_R (e.g. Menten *et al.* 1986), and establish whether the source harbors "hot core" line emission. Comparing the NH$_3$ and CH$_3$OH T_R is interesting because CH$_3$OH is often used to calculate hot core temperatures in the (sub)millimeter band (Brogan *et al.* 2007, 2009). Several of the lower lying transitions have been observed as Class I masers (e.g. Menten *et al.* 1986).

Continuum Emission: For sources with jets, HCHII, or UCHII regions, we have obtained images of the free-free emission with resolution matching the molecular gas. For sources lacking either an HII region designation or Y^V EGO designation in Table 1, these are the first sensitive high resolution cm-λ observations.

Radio Recombination Lines: H63α and H64α also lie within our bandpass, and when combined with free-free continuum, these RRLs can be used to measure the density, electron temperature, and mass of the ionized gas.

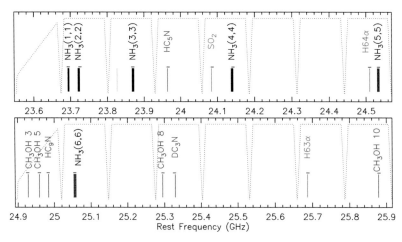

Figure 1. Sample JVLA correlator setup showing the sixteen 8 MHz spectral windows superposed on the inherent filter shape (dotted line).

4. Results

Processing of the D-configuration (2–4 kpc) sample has been completed, and the images reveal complexity in terms of core multiplicity, gas temperatures and kinematics, shocks, masers, and continuum properties. The results on the first source, G35.03+0.35, appeared in Brogan *et al.* (2011) and are summarized in Fig. 2. We detect compact

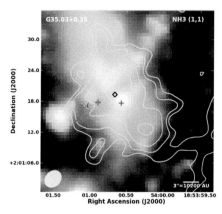 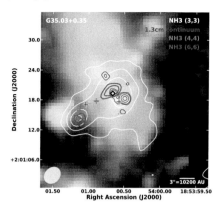

Figure 2. *Spitzer* GLIMPSE 3-color image with RGB= 8, 4.5, and 3.6 μm. Left: NH_3 (1,1) emission in yellow contours. Right: NH_3 (3,3) emission in white contours (most extended), NH_3 (6,6) in red, and 1.3 cm continuum in blue. The synthesized beam is shown in the lower left. Also see Brogan *et al.* (2011)

NH_3 cores in all of the fields, some seen in (1,1) and (2,2) only, and others up through NH_3 (6,6) - indicating the presence of "hot cores". 25 GHz CH_3OH maser emission is present in 7 out of 11 sources in close proximity in position and velocity to 44 GHz masers. Weak thermal emission in CH_3OH is also detected in some sources. We also detect non-thermal emission in the NH_3 (3,3) transition in 7 out of 11 sources, excited by outflow shocks (e.g. Zhang *et al.* 1999). Our preliminary results already demonstrate the powerful new capability offered by the WIDAR correlator and the JVLA. Additionally, millimeter follow-up observations have been obtained for more than half the sample with the remaining sources hopefully observed soon. When analysis of the survey is complete we expect to have a much greater understanding of the physical properties, and multiplicity of massive forming proto-clusters.

References

Brogan, C. L., Hunter, T. R., Cyganowski, C. J., Friesen, R., *et al.* 2011, *ApJL*
Brogan, C. L., Hunter, T. R., Cyganowski, C. J., Indebetouw, R., *et al.* 2009, *ApJ*, 707, 1
Brogan, C. L., Hunter, T. R., Indebetouw, R., *et al.* 2008, *Ap&SS*, 313, 53
Brogan, C. L., Chandler, C. J., Hunter, T. R., *et al.* 2007, *ApJ*, 660, L133
Churchwell, E. 2002, *ARAA*, 40, 27
Cyganowski, C. J., Brogan, C. L., & Hunter, T. R. 2011, *ApJ*, 743, 56
Cyganowski, C. J., Brogan, C. L., Hunter, T. R., & Churchwell, E. 2009, *ApJ*, 702, 1615
Cyganowski, C. J., *et al.* 2008, *AJ*, 136, 2391
Cyganowski, C. J., Brogan, C. L., & Hunter, T. R. 2007, *AJ*, 134, 346
Devine, K., Chandler, C. J., Brogan, C. L., *et al.* 2011, *ApJ*, 733, 44
Ellingsen, S. P. *et al.* 2007, *IAU Symposium*, 242, 213
Hunter, T. R., Brogan, C. L., Indebetouw, R., & Cyganowski, C. J. 2008, *ApJ*, 680, 1271
Hunter, T. R., Brogan, C. L., Megeath, S., *et al.* 2006, *ApJ*, 649, 888
Menten, K. M., Walmsley, C. M., Henkel, C., & Wilson, T. L. 1986, *A&A*, 157, 318
Molinari, S., Brand, J., Cesaroni, R., & Palla, F. 1996, *A&A*, 308, 573
Reid, M. J., *et al.* 2009, *ApJ*, 700, 137
Rathborne, J. M., Jackson, J. M., & Simon, R. 2006, *ApJ*, 641, 389
Rosolowsky, E., Dunham, M. K., Ginsburg, A., *et al.* 2010, *ApJS*, 188, 123
Sridharan, T. K., Beuther, H., Schilke, P., Menten, K. M., & Wyrowski, F. 2002, *ApJ*, 566, 931
Zhang, Q., Hunter, T. R., Sridharan, T. K., & Cesaroni, R. 1999, *ApJ*, 527, L117
Zinnecker, H. & Yorke, H. W. 2007, *ARAA*, 45, 481

Cosmic Masers - from OH to H$_0$
Proceedings IAU Symposium No. 287, 2012
R.S. Booth, E.M.L. Humphreys & W.H.T. Vlemmings, eds.
© International Astronomical Union 2012
doi:10.1017/S1743921312007600

EVLA imaging of the water masers in the massive protostellar cluster NGC6334I

Todd R. Hunter and Crystal L. Brogan

NRAO, 520 Edgemont Rd, Charlottesville, VA 22903, USA;
email: `thunter@nrao.edu`

Abstract. We have used the recently-upgraded Karl G. Jansky Very Large Array (VLA) in A-configuration to observe the water masers in the massive protostellar cluster NGC6334I with broad bandwidth and high spectral resolution. Four groups of maser spots are found. The two groups with the broadest velocity span (40 km/s) are towards the UCHII region and the hot core SMA1. The spatial kinematics of the SMA1 masers are consistent in sense and orientation with the large-scale CO outflow and appear to trace the base of the outflow from a protostar at the dust peak of SMA1. Additional masers at the southern end of SMA1 provide evidence for a second protostar. The highest intensity maser lies about 2″ north of SMA1. Interestingly, no water masers are seen on the equally impressive hot core SMA2. Finally, we have detected maser emission toward the enigmatic source SMA4, which shows no millimeter molecular lines despite having strong, compact submillimeter continuum and may trace another protostar.

Keywords. masers, stars: formation, ISM: jets and outflows, ISM: molecules, radio lines: ISM, radio continuum: ISM, infrared: ISM, submillimeter

1. Introduction

The association of water masers with massive star formation has been clear for many decades. Surveys of molecular outflow sources show a strong correlation with the presence of water masers (e.g., Tofani *et al.* 1995). The detailed kinematics of these masers have been interpreted as arising either in protostellar accretion disks or outflows. Much of the uncertainty arises because massive protostars form in clusters with multiple outflows overlapping simultaneously (e.g., Zhang *et al.* 2007) which can make it difficult to disentangle the individual outflow axes. Thus it is important to identify well-collimated massive bipolar outflows and observe them at high angular resolution in both water maser and thermal molecular lines. A good example is found in the massive protocluster NGC6334I (Hunter *et al.* 2006; Beuther *et al.* 2005), whose bipolar outflow has been imaged by single-dish telescopes (Leurini *et al.* 2006; Qiu *et al.* 2011). Here we present the first interferometric CO(2-1) observations along with the 22 GHz water maser emission.

2. Observations

We performed 22.235 GHz observations of NGC6334I in September 2011 with the VLA in A-configuration to demonstrate the flexibility of the upgraded IF system and correlator and confirm the operation of high resolution modes. In both of the independently tunable 1 GHz basebands, we configured 8 adjacent subbands each with a bandwidth of 4 MHz (54 km/s) and channel width of 0.42 km/s. The IFs were shifted by 2 MHz in order to provide uniform sensitivity across the subband edges for a total bandwidth of 34 MHz. Similarly, we observed the central 2.1 MHz of emission using subbands of 250 kHz bandwidth (3.4 km/s) with 0.026 km/s channels. CO(2-1) was observed with the Submillimeter Array (SMA) in February 2004 in extended config and in August 2008

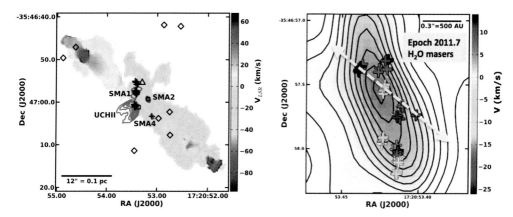

Figure 1. Left: Image of the CO (2-1) velocity field ($2.5'' \times 1.0''$ beam), with $870\mu m$ continuum overlaid in blue contours (Brogan *et al.* in prep.), VLA 8.4 GHz continuum in magenta contours (Hunter *et al.* 2006), 22 GHz H_2O masers (black crosses), 44.069 GHz CH_3OH 7_0-$6_1 A^+$ masers (crosses) (Gómez *et al.* 2010), 12.2 GHz CH_3OH $2_0 - 3_{-1} E$ masers (squares) (Breen *et al.* 2012), and the 23.121 GHz CH_3OH 9_2-$10_1 A^+$ maser (triangle) (Kurtz, Hofner & Álvarez 2004). **Right:** $870\mu m$ continuum contours ($0.5'' \times 0.3''$ beam) overlaid with the water maser positions.

in very-extended config. The $870\mu m$ continuum was observed in the very-extended config in August 2008.

3. Results

In Figure 1, the CO(2-1) velocity field reveals the outflow to be highly-collimated on 0.5 pc scales. Of the two hot cores (SMA1 & SMA2), SMA1 lies closest to the outflow axis and the interface between the redshifted and blueshifted CO. Similar to its neighboring source NGC6334I(N) (Brogan *et al.* 2009), multiple groups of maser spots are found. The masers with the broadest velocity span (40 km/s) are towards the UCHII region and the hot core SMA1. The spatial kinematics of the SMA1 masers match the large-scale CO outflow in sense and orientation and appear to trace the origin of a protostellar outflow at the dust peak of SMA1. Additional masers at the southern end of SMA1 provide evidence for either an outflow or a disk toward another protostar. The highest intensity masers lie $\sim 2''$ north of SMA1. Interestingly, no water masers are seen on the equally impressive hot core SMA2. Finally, we detect maser emission toward the enigmatic source SMA4, which shows no thermal molecular lines despite having strong, compact submillimeter continuum. The compact spatial velocity gradient here could trace another outflow or a disk with dynamical mass of $2M_\odot$, suggesting the presence of another protostar.

References

Breen, S., Ellingsen, S., Caswell, J., *et al.* 2012, arXiv:1201.1330

Beuther, H., Thorwirth, S., Zhang, Q., *et al.* 2005, *ApJ*, 627, 834

Brogan, C. L., Hunter, T. R., Cyganowski, C. J., Indebetouw, R., *et al.* 2009, *ApJ*, 707, 1

Hunter, T. R., Brogan, C. L., Megeath, S., Menten, K., *et al.* 2006, *ApJ*, 649, 888

Gómez, L., Luis, L., Hernández-Curiel, I., *et al.* 2010, *ApJS*, 191, 207

Kurtz, S., Hofner, P., & Álvarez, C. V. 2004, *ApJS*, 155, 149

Leurini, S., Schilke, P., Parise, B., *et al.* 2006, *A&A*, 454, L83

Qiu, K., Wyrowski, F., Menten, K. M., *et al.* 2011, *ApJ*, 743, L25

Tofani, G., Felli, M., Taylor, G. B., & Hunter, T. R., 1995, *A&AS*, 112, 299

Zhang, Q., Hunter, T. R., Beuther, H., *et al.* 2007, *ApJ*, 658, 1152

Cosmic Masers - from OH to H_0
Proceedings IAU Symposium No. 287, 2012
R.S. Booth, E.M.L. Humphreys & W.H.T. Vlemmings, eds.

© International Astronomical Union 2012
doi:10.1017/S1743921312007612

OH maser observations using the Russian interferometric network "Quasar" in preparation for scientific observations with the space mission RadioAstron.

I. D. Litovchenko[1], A. V. Alakoz[1], V. I. Kostenko[1], S. F. Lihachev[1], A. M. Finkelstein[2] and A. V. Ipatov[2]

[1]Astro Space Center of the Lebedev Physical Institute, Moscow, Russia
email: grosh@asc.rssi.ru

[2]Institute of Applied Astronomy of RAS, Saint Petersburg, Russia

Abstract. We present results of a VLBI experiment at a wavelength of 18 cm, which simulates the ground-space interferometer with space link to RadioAstron. An array of five antennas was used, four of them are located in the Russian Federation, plus the the 32-m radio telescope in Medicine (Italy). The 22-m radio telescope in Pushchino (Moscow Region) acted in place of the space arm. It has an effective area of 100 square meters. The three other Russian 32-m antennas are operated by the Institute of Applied Astronomy RAS; they are located at Badary, Svetloe and Zelenchukskaya (interferometer network "Quasar"). The maximum base-line, Badary-Svetloe, was about 4402 km, providing an angular resolution of about 0.009 arc seconds at a wavelength of 18 cm. The duration of the experiment was 10 hours on 02/03 February 2011. The program of observations included quasars 3C273, 3C279, 3C286 and the maser source - W3(OH). W3(OH) was observed only by the Russian telescopes and was investigated at the frequency of the 1665 MHz main line. The data were recorded on the MK5 recorder (32-m radio telescopes) and the RDR system (RadioAstron Digital Recorder) in Pushchino. The low SEFD (system equivalence of flux density) of Pushchino emulated the RadioAstron antenna. Correlation was performed with the universal software correlator of the AstroSpace Center of Lebedev Physical Institute. The correlator output format is compatible with that used by the AIPS package, which was used for data analysis. After analyzing the correlated data we obtained relative coordinates of the maser components. The main results are tabulated and presented in the figures. The data quality is sufficient for astrophysical analysis and comparison with previous observations of maser source W3(OH) on VLBI networks EVN and VLBA.

Keywords. interferometry, masers, hydroxyl, W3(OH)

1. Radio telescopes that observed the maser radio source W3(OH)

Three 32-m fully steerable reflectors of the Institute of Applied Astronomy RAS were used and we list their locations and other parameters below.

1. Svetloe, $\phi = 60°32'$, $\lambda = 29°47'$, $h = 86m$.
2. Zelenchukskaya, $\phi = 43°47'$, $\lambda = 41°34'$, $h = 1175m$.
3. Badary, $\phi = 51°46'$, $\lambda = 102°14'$, $h = 813m$.

All these antennas have five dual circular polarization receivers for the wavelengths: 1.35-cm (K-band), 3.5-cm (X-band), 6.2-cm (C-band), 13-cm (S-band) and 18-21-cm (L-band). Mark5B VLBI recorders were used. The SEFD at L-band is about 280 Jy.

4. The fourth antenna was the 22-m reflector in Pushchino, $\phi = 54°49'$, $\lambda = 37°38'$, $h = 239m$.

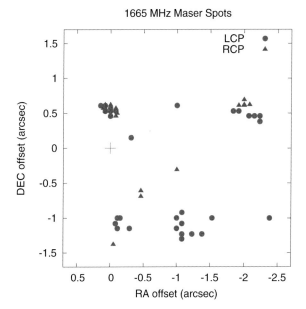

Figure 1. The 1665-MHz maser spot map of W3(OH). Dots indicate emission in LCP and triangels indicate emission in RCP. The cross marks the phase centre with coordinates: RA $02^h 27^m 03^s.825$ and DEC $61°52'2''.653$.

It has receivers for 1.35-cm (K-band), 6.2-cm (C-band), 18-21-cm (L-band) and 92-cm (P-band).

In this experiment we used RadioAstron recording system RDR (RadioAstron Digital Recorder). The SEFD at L-band is about 1000 Jy.

2. Correlation

Correlation was performed with the universal software correlator of the AstroSpace Center of Lebedev Physical Institute. The correlator output format is compatible with that used by the AIPS package. The final data had 6148 frequncy channels, bandwidth 3.002 MHz with channel separation 0.4883 kHz.

3. Data analysis

To obtain the positions of maser spots of W3(OH) we used the AIPS fringe-rate mapping task FRMAP. Then we determined the positions of the masers from lines of intersection on the fringe-rate maps. Results for all channels and both polarizations are shown on Fig. 1 which shows the relative postions of individual maser features. These are well matched to the map obtained by Wright *et al.* 2004.

References

Wright, M. M., Gray, M. D., & Diamond P. J. 2004, *MNRAS*, 350, 1253

Cosmic Masers - from OH to H_0
Proceedings IAU Symposium No. 287, 2012
R.S. Booth, E.M.L. Humphreys & W.H.T. Vlemmings, eds.

© International Astronomical Union 2012
doi:10.1017/S1743921312007624

IAU (Maser) Symposium 287 Summary

Karl M. Menten

Max-Planck-Institut für Radioastronomie,
Auf dem Hügel 69, 53121 Bonn, Germany
email: kmenten@mpifr.de

1. Introduction

Almost exactly twenty years ago, the first of a series of conferences dedicated to cosmic masers took place in Arlington, Virginia in the USA (March 9–11, 1992). Two more followed, each on a different continent, in Mangaratiba, near Rio de Janeiro, Brasil (March 5–10, 2001) and in Alice Springs, Australia (March 12–16, 2007). As at all others, a large part of the international maser community convened from January 29 to February 3, 2012 in splendid Stellenbosch, South Africa, to discuss the state of the art of the field.

Here I'm trying to summarize the many contributions made in 70 oral presentations and by 45 posters. Adhering, roughly, to the order of themes as defined by the sessions at the meeting. I'm trying to be comprehensive, but I apologize beforehand for any undue omissions and misrepresentations.

2. Maser Theory and Supernova Remnants

Like many other a maser meeting, this one started with a theory session. An excursion to the thermodynamical basics of maser action [→ STRELNITSKI] was followed by a demonstration of how complex things can get by the description of class II methanol (CH_3OH) maser excitation via mid-infrared pumping, cycling through the molecule's torsionally excited levels [→ SOBOLEV channeled by GRAY].

New ways of addressing the excitation of well known maser emission from the 1720 MHz $F = 2^+ - 1^-$ hyperfine structure (hfs) satellite line in the $^2\Pi_{3/2}, J = 3/2$ ground state of hydroxyl (OH) in supernova remnants (SNRs) were presented [→ GRAY]. Closely related, a careful theoretical and multi-line observational (absorption) study of OH excitation predicts maser emission in the 6049 MHz $J = 3^- - 2^+$ hfs satellite line of the rotationally excited $J = 5/2$ state over a range of conditions thought to be prevalent in SNRs†, which was, however, not detected despite a sensitive search. The results appear in conflict with the OH column density distributions predicted for some chemical models of SNRs [→ WARDLE].

1720 MHz OH masers are important tracers of the sub-mG magnetic fields of the 10^4 cm^{-3} density interstellar medium in which they are excited. Density- and also temperature-wise ($T \sim 100$ K) this post shock gas might also be expected to produce class I methanol masers *if* the elevated methanol abundance needed to produce high enough gain for observable maser emission can be attained (see §5.2). However, so far searches for 36 and 44 GHz CH_3OH maser emission toward SNRs have had little success with the Sgr A East SNR near the Galactic center being a spectacular exception. Observations of a Fermi satellite γ-ray selected sample of SNRs only produced a couple of

† Both the 1720 and the 6049 MHz line connect, in their respective rotational state, the hfs level with the highest energy and that with the lowest energy.

detections in these lines [→ SJOUWERMAN]. Very interestingly, vice versa, narrow spectral features in the 1720 MHz OH line have been found at the velocities of more than 50 narrow class I methanol maser features in a survey of more than a hundred sources containing such masers; these are *not* SNRs [→ VAL'TTS].

3. Bill Watson R. I. P.

With great sadness, the maser community has observed the passing of Bill Watson, one of its most eminent theorists. Prof. William D Watson, born January 12, 1942 in Memphis, TN, died on October 12, 2009 in Urbana, where he had been doing research and teaching since 1972 at the University of Illinois. In the early seventies, after an important contribution on the formation of molecules on dust grain surfaces, Professor Watson independently from and contemporaneously with another team established a chemistry driven by molecular ions as the basic paradigm of the gas phase chemistry of the interstellar medium. Ion-molecule chemistry explained the observed abundances of then newly detected molecules and predicted the observed strong deuterium enhancement as one of its natural consequences. After working on several other topics related to interstellar chemistry, Bill Watson went on to become a world expert on astronomical masers and, particularly, on the radiative transfer of maser radiation and its polarization properties. He completed a large number of studies on this very complex problem, significantly expanding fundamental work done in the early years of astronomical maser research. His work was always in unison with the forefront of observations. As an example, he correctly explained the origin of the high velocity features in the archetypical nuclear AGN maser NGC 4258 as originating in a ~ 0.1 pc sized ring, which was *subsequently* confirmed by Very Long Baseline Interferometry.

4. Polarization and Magnetic Fields

Models of molecular clouds in which magnetic fields are dynamically important (via ambipolar diffusion) predict that the relation between the magnetic field, B, and the density, n, is $B \propto n^\kappa$ with $\kappa \lesssim 0.5$. Observations (of the Zeeman effect) show that this relation holds over several orders of magnitude for $n > 100$ cm^{-3}, below which B is observed and predicted to be independent of n and has a value of ≈ 5 μG. In particular, it holds for 1720 MHz OH masers ($n \sim 10^4$ cm^{-3}, $B \sim 0.5$ mG; see §2) and for the elevated densities required for OH maser emission in high mass star forming regions (HMSFRs) ($n \sim 10^{6-7}$ cm^{-3}, $B \sim$ a few to 10 mG) and even for 22.2 GHz H$_2$O masers ($n \sim 10^{8-9}$ cm^{-3}, $B \sim$ tens of mG). Masers are good B-field probes since, even given the usually small splitting coefficient, i.e. the proportionality factor between the value of B and the frequency (or equivalent velocity) shift between right and left circularly polarized signal, B fields under the extreme conditions of maser regions (high n implies high B) are relatively easy to measure; the narrow line widths help. It is important to keep in mind that in many cases, whether a line is masing or not is a very sensitive function of density. For example, for many OH maser lines just a factor of two increase in density over the value at which optimal maser action occurs removes the inversion and, if a radio continuum background is present, absorption may be observed. In other words, within a certain range, Zeeman observations of all masers requiring a certain density for operation should deliver the same magnetic field strength. In particular, the values reported in the recent past for class I and class II methanol masers were orders of magnitude higher than expected, given the densities needed for their inversion (see §5.2) and fail to obey this requirement. Talks at the meeting make poorly known or unknown splitting coefficients responsible for this [→ VLEMMINGS, SARMA]. Theoretical calculations and laboratory measurements are urgently needed.

An increasing amount of Very Long Baseline Interferometry (VLBI) results map B-field configurations in the vicinity of, mostly, high mass young stellar objects (YSOs) [→ VLEMMINGS, SURCIS, CHIBUEZE], but now also of a solar-like, low mass YSO. The latter (VLA) observations, of H_2O maser emission in the famous IRAS 16293–2422, show that the protostellar evolution of this object appears to be magnetically dominated [→ DE OLIVEIRA ALVES].

Interesting polarization results for evolved stars include the finding that a dynamically significant and ordered B- field is maintained over the whole of a circumstellar envelope [→ AMIRI]. VLBI imaging of poloidal magnetic field configurations at the launching site of very high velocity water maser emission in so-called "water fountain" sources indicate a magnetic origin of the observed bipolarity. This result could have far-reaching consequences on the long-standing question whether magnetic fields play a role in the symmetry breaking mechanism that transforms a circularly symmetric envelope into an axially symmetric planetary nebula. [→ AMIRI, IMAI, also CLAUSSEN, SUAREZ].

The question which of two different models that had been proposed to interpret circular polarization of SiO masers for the applicable case of weak Zeeman splitting is the correct one can be addressed by observations of different maser transitions [→ RICHTER].

5. Star Formation Masers

The star formation session was dominated by presentations on methanol masers, which were the topic of a total of ten presentations in this and other sessions and ten posters. Below, I therefore dedicate some extra space on this molecule.

Other results on star forming regions include multi-epoch H_2O maser VLBI of two different deeply embedded objects in Cepheus A HW, tracing the motions of a bipolar outflow in one [→ CHIBUEZE] and in the other revealing what had appeared to be a spherical expanding shell, but seems to be showing more complex filamentary structure and dynamics down to few AU scales [→ TORRELLES].

5.1. *Spinning the Big Spin – Masers in Disks*

Masers in accretion disks have a long and chequered history. Consider, for example, one determines, via Gaussian fits, the centroid positions of two well-separated maser spots, A and B, peaking at distinctly different velocities, v_m and v_n, in the velocity channels m and n of the correlator that have been observed (e.g., with the VLA) with an angular resolution that is much coarser than the separation of the spots. Then fits to the emission in velocity channels i, with $m < i < n$, will yield centroid positions lying between the real positions of the spots that seem to move monotonously from A to B, mimicking a linear structure with a linear velocity gradient suggesting solid body rotation. This mis-interpretation of data gave rise to the *Saga of the Rotating Methanol Maser Disks*†. Of course, such "disks"' are easily debunked by VLBI, which resolves the two maser spots into separate entities,

VLBI indeed delivered the exciting picture of a rotating disk structure whose material giving rise to SiO maser emission appears also to be flowing in polar direction. The whole scenario, documented by a multi-Very Long Baseline Array-epoch movie, is that of an

---

† It remains a mystery why none of the disk saga's proponents noticed that, in order to show solid body (and *not* Keplerian) rotation the mass of the protostar would have to be negligible compared to the mask of the disk.

equatorially expanding "excretion" disk surrounding the peculiar radio source I in the Orion-KL region, whose inner part is photo-ionized [→ GREENHILL].

Other bona fide maser emission from a disk, this time a completely photo-evaporating one, is that seen in (sub) millimeter hydrogen recombination line emission around the peculiar emission line B[e]star MWC 349 A, which excites a bipolar radio nebula. Fifteen year long monitoring of this object's polarization characteristics have yielded information on its magnetic field, while modeling its velocity distribution reveals Keplerian rotation in the disk's outer parts and more complicated kinematics in its inner parts [→ THUM, BÁEZ RUBIO].

For a very long time, the MWC 349 A (sub)mm/far infrared recombination line maser/laser was one of its kind. It comes as a great relief that a second such source has now been found in Mon R2 IRS 2 [→ JIMENEZ-SERRA].

5.2. *The Class I – Class II Methanol Maser Dichotomy*

For more than 25 years it has being realized that strong interstellar methanol masers come in two varieties: Class I methanol masers arise from outflows from high mass protostellar objects (HMPOs) often significantly removed from the driving source, whereas Class II methanol masers (cIIMMs) arise from the nearest vicinity of the HMPO; sometimes forming conspicuous rings revealed by VLBI imaging [→ BARTKIEWICZ presented by VAN LANGEVELDE].

To build up the gain for observable maser flux, a common requirement for both maser varieties is a substantial CH_3OH abundance. For cIIMMs masers this is achieved by the heating of dust grains in the dense HMPO envelope that evaporates methanol-containing ice mantles and increases the gas phase methanol abundance by several orders of magnitudes. The warm grains, which attain an equilibrium temperature around 100–200 K emit intense mid/far infrared (IR) radiation that pumps the maser via torsional excitation†. Interestingly, the dust emission from these so-called *hot cores* is difficult or impossible to detect at near- or mid-IR wavelengths even when superior astrometry is achieved [→ DE BUIZER]. Submillimeter continuum emission is frequently detected, but often with too coarse resolution to establish a clear association with the masers. Here, the shortest wavelength band of the Photodetector Array Camera and Spectrometer (PACS) on Herschel at 70 μm promises to deliver crucial information [→ PESTALOZZI] on the maser host sources' spectral energy distribution and luminosity.

Combining cIIMM proper motions from VLBI with interferometric observations of non-maser CH_3OH (from the torsional ground and first excited state ($\sim$ 400 K above ground) will yield, in addition to n and T, precious 3 dimensional velocity information on the closest vicinity of HMPOs [→ TORSTENSSON, DE LA FUENTE], allowing searches for infall, which is indicated in one object (AFGL 5142) from VLBI observations alone [→ GODDI].

In complete opposition to cIIMMs, class I methanol masers (cIMMs) work in the *absence* of a strong IR field, which means at significant offsets from HMPOs. An association of these masers with protostellar outflows has been established a long time ago and has recently been illustrated for very many sources by their coincidence with so-called extended green objects (EGOs), which are shocked regions [→ CYGANOWSKI]‡

---

† Note that the Planck function, $B_\nu(T)$, for $T = 200$ K, has its maximum at 11.75 THz, corresponding to a wavelength of 25.5 μm, very close to that of the main cIIMM pumping transition identified by modelers.

‡ EGOs or "green fuzzies" are regions of enhanced emission in images made with the InfraRed Array Camera (IRAC) on the Spitzer Space Observatory using the camera's 4.5 μm filter. Imaged in the course of the Galactic Legacy Infrared Mid-Plane Survey (GLIMPSE), the emission in

Also for a very long time, it was known that the inversion of the observed maser lines follows naturally, over a range of physical conditions, from an interplay of both E- and A-type methanol's arrangement of energy levels in k-ladders and certain levels' transition probabilities (Einstein A-values). This is confirmed by statistical equilibrium/radiative transfer calculations. Whether some cIMM lines show maser action in some regions and not in others depends (for temperatures of up to a few hundreds K) on the region's density, which has to be between 10^4 and 10^5 cm^{-3}, significantly lower than required for cIIMMs. Actually, under conditions conducive for cIMM action, the strongest cIIMM lines, the 12.2 GHz $2_0 - 3_{-1}$ E and 6.7 GHz $5_1 - 6_0$ A^+ transitions, are predicted to show enhanced absorption, which is actually observed in some regions¶.

The physical conditions in the post shock gas hosting cIIMMs, n and T (for B see below), can actually be quite well constrained just by the fact which of the 25 known cIMM lines (including a new one [→ VORONKOV, WALSH, BROGAN]) are masing and which are not. Here extensive new surveys in the most prominent transitions, at 44 and 25 GHz, deliver an abundance of new data [→ KURTZ, BYUN, BRITTON]. In particular, the H_2O southern Galactic Plane Survey (HOPS), which apart from the 22.2 GHz H_2O maser transition, covers several cI and cIIMM lines (plus multiple thermally excited lines from NH_3 and other species [→ WALSH]) has been most successful. Also, such surveys finally found the first long-sought after methanol masers in low mass star forming regions: Maser emission from cIMM lines was found in high CH_3OH abundance clumps of well known outflows [→ KALENSKII].

The elevated CH_3OH abundances required to produce an observable cIMM signal may also result, as for cIIMMs, from grain ice mantle desorption. However, in this case the necessary energy would be provided by a shock wave *and not* by central heating from the HMPO. A high CH_3OH abundance might even result from endothermic gas phase reactions, which require high temperatures (of order 10000 K), which may be reached in shocks. If cIMMs arise in shock fronts, why do they virtually never show high velocity emission and have velocity spreads of just a few km/s? In contrast, H_2O masers, which are unequivocally associated with shocked outflows have much larger LSR velocity spreads of many tens, even up to hundreds of km/s. The answer is likely that the cIMMs do barely have high enough methanol column densities to produce an observable signal. Therefore, the geometry of a swept up shell seen edge on, i.e. from a direction perpendicular to that of the outflow motion would produce the longest coherent gain path in the direction of the observer. The radial velocity of emission from such a configuration would *naturally* only have a small offset from the systemic velocity of the outflow source since the bulk of the outflow motion is in the plane of the sky. This scenario also implies that cIMM spots, partaking in the outflow, have large transverse motions, which for H_2O masers can be measured with VLBI. Unfortunately, cIMMs maser spots have all been found to be too large for successful VLBI. However, future proper motion measurements with the Jansky Very Large Array (JVLA) may prove the above picture directly.

In this context it is noteworthy that magnetic field strength determinations of cIMMS via the Zeeman effect will provide measurements in an ISM density regime barely covered by existing data and moreover deliver supremely important input for

---

this band is dominated by highly (shock-)excited lines of molecular hydrogen. The name derives from the fact that this emission was chosen to be color coded as green in false color presentations of multi-IRAC-band images.

¶ This enhanced absorption ("over-cooling") follows from the fact that for E-(A)-type methanol *all* levels in the $k = -1$ ($K = 0$) energy ladder are overpopulated relative to levels in the neighboring $k = 0$ ($K = 1$) ladder. The same gives rise to the prominent $4_{-1} - 3_0$ E and $5_{-1} - 4_0$ E ($7_0 - 6_1$ A^+ and $8_0 - 7_1$ A^+) *maser* lines near 36 and 84 GHz (44 and 95 GHz).

magnetohydrodynamic modeling of interstellar shocks. Unfortunately, as reported above (§4), the *B*-field values reported so far (also at this meeting) are marred by our ignorance of the Zeeman splitting factors.

5.3. *Periodic Methanol Masers*

With periods from 20 to > 500 days, 6.7 and 12.2 GHz cIIMMs periodicity, firmly established by observations with South Africa's own Hartebeesthoek Radio Observatory by our conference organizer Sharmila Goedhart in her dissertation is one of the most peculiar phenomena in all maser science [→ GOEDHART]. VLBI appears to rule out a periodic infrared radiation pump, leaving a periodically varying 6–12 GHz radio continuum background as a possibility. A model for this has been worked out in the framework of a colliding wind binary scenario providing ionizing radiation [→ VAN DEN HEEVER, VAN DER WALT]. If a variable continuum background were at the heart of cIIMM variability, this would raise the question whether *all* cIIMMs need continuum photons as seeds to operate. While the first detections of these masers found the most prominent ones associated with ultracompact HII regions, in contrast, subsequent interferometric imaging surveys actually found that most cIIMMs had no associated continuum emission at the few mJy level (at 8.4 GHz). Future, much more sensitive JVLA continuum surveys will address this question.

Finally, a comment on the chronology of masers in star forming regions, another perennial "chicken and egg" topic of maser and star formation conferences [→ BREEN]. Clearly, H_2O and cIMMs are found in outflows and are thus accretion powered. CIMMs are frequently located quite far away from the outflow source (travel times of several ten thousands of years). Here we have the caveat that the transverse velocities are very uncertain and, given above arguments, may be quite a bit larger than the radial velocity spread, resulting in shorter time scales. In contrast, H_2O masers arise from within a few thousand AU (travel times a few hundred yr). The expansion time of an ultracompact HII region is thousands of yr, that of a hypercompact HII region hundreds of yr, comparable to the H_2O maser time scale. Critical questions are as follows: Before a HMPO has formed (started fusion): Can a H_2O maser outflow be driven, possibly by magnetically assisted disk-to-outflow angular momentum conversion? And, can accretion luminosity be sufficient to power a hot core and its associated cIIMMs ? Unequivocal evidence for disks and, in particular, high resolution imaging of outflow launching sites will address the first question, and evidence for the presence of radio emission at cIIMM positions the second.

6. Stellar Masers

The molecules producing circumstellar maser emission are either formed in or near the stellar photosphere (SiO, H_2O) or (OH) by photo dissociation in the other parts of the expanding circumstellar envelope surrounding mass losing evolved stars. They are observed, preferably, with interferometers [→ RICHARDS]. Nevertheless, single dish studies give complementary interesting results, e.g., on the physical conditions in the extended atmosphere (by high excitation SiO maser line monitoring [→ RAMSTEDT]) or on the magnetic field via polarization measurements. For OH lines, such measurements, of over a hundred AGB and post-AGB stars, imply $B = 0.2$–2.3 mG [→ WOLAK]. Monitoring the phase lag of OH maser variability between signals arriving from the blue- and redshifted parts of the circumstellar shell, combined with angular shell size diameters from interferometry will deliver distances for ∼ 20 objects [→ ENGELS].

IR interferometry and radio wavelength VLBI have enjoyed great synergy when IR imaging found the molecular regions just outside the photospheres of oxygen-rich Asymptotic Giant Branch stars (AGBs) (so-called MOLapheres) [→ WITTKOWSKI] at the same distance from the stellar surface as SiO masers, which in many cases form beautiful rings, strongly indicating tangential amplification [→ COTTON, AL MUTAFKI, DESMURS]. SiO masers, have the potential of probing the magnetic fields, via polarization measurements [→ COTTON], and also the dynamics in these interesting regions, as demonstrated by the spectacular movie made from 112 epochs of SiO maser Very Long Baseline Array (VLBA) imaging [→ GONIDAKIS].

7. Cosmology and the Hubble Constant: AGN and Megamasers

After a concise, but comprehensive introduction to the *Standard Model of Cosmology*, a status report on the Megamaser Cosmology Project (MCP) was given [→ HENKEL]. The goal is to measure the Hubble constant, H_0, with a precision of a few percent to complement the existing cosmic microwave background radiation data in placing constraints on the nature of Dark Energy. The MCP's targets are active galaxies with 22.2 GHz H_2O maser mission originating from a < 1 pc region around the central supermassive Black Hole. Best suited are systems seen nearly edge on, such as the "Golden Source" NGC 4258 (see §3). Comparing the maser distributions' rotational speeds and radii, determined from VLBA imaging, with the measured centripetal acceleration (from spectral monitoring) directly delivers the systems' distances, D, which can be compared with the measured redshifts, yielding H_0 and also the Black Hole mass [→ WARDLE]. Naturally, systems with $D > \sim 50$ Mpc are desirable, which are well partaking in the Hubble flow (and whose redshifts are not significantly influenced by local dynamics). Recently, a beautiful new example of such as system was found, UGC 3789, for which a distance of 50 Mpc was determined and an even farther system, NGC 6262 at $D = 152$ Mpc. Together, these two yield a Hubble constant of 67 ± 6 km s^{-1}Mpc^{-1}. [→ HENKEL]. Another nice system is Mrk 1419 [→ IMPELLIZZERI].

For NGC 4258 itself, we were shown what sophisticated modeling of 18 epochs of VLBI observations and 10 years of single dish monitoring can do for you: a detailed analysis of disk warping and elliptical orbits of maser clumps with differential precession [→ HUMPHREYS] and, on top of all this, a highly precise distance.

To date, several thousand galaxies have been surveyed for H_2O megamasers, yielding just ≈ 130 detections. Obviously, increasing the sample is highly desirable as is any criterion that could help to increase the success rate. Cross-matching galaxies with H_2O megamaser detections with systems found in the Sloan Digital Sky Survey yields an increased maser detection rate for galaxies showing strong [OIII] $\lambda 5700$ emission. [→ ZAW].

Apart from nuclear disks, H_2O megamasers have been found in outflows from AGN and in the jets' interaction zones with the interstellar medium and new detections were reported. [→ TARCHI].

Interesting progress was also reported for H_2O "kilomasers", which are found in star burst galaxies. Spectacular JVLA imaging revealed such masers in several active locations in the merging Antennae system (NGC 4038/4039), likely marking the birth sites of super star clusters. In particular, its 80 times higher bandwidth (compared to the VLA) make the JVLA a tremendously efficient survey machine for extragalactic maser emission [→ DARLING].

OH megamasers in the central starbursts of Ultra Luminous Infra Red Galaxies probably mark the most extreme star forming conditions in the local Universe. These masers even "contaminate" blind surveys for 21 cm emission from HI. Finding such systems and also formaldehyde (H_2CO) megamasers at higher redshift is highly desirable, among others in a H_2CO Deep Field [DARLING, BAAN], but have so far been unsuccessful (for OH at $z > 1$) [→ WILLETT]. Here the Five hundred meter Aperture Spherical Telescope, currently being built in south western China, holds great promise for the foreseeable future [→ J. ZHANG]. FAST will have almost three times the collecting area of the Arecibo 300 m telescope, the existing OH megamaser detection machine, and a *much* larger sky coverage.

8. Maser Astrometry

Already at the previous maser conference a whole series of contributions reported high precision multi-epoch VLBI astrometry (mostly) of masers in HMSFRs, yielding accurate distances and proper motions. In the meanwhile this field has greatly expanded and about 50 parallaxes obtained with the Japanese Very Long Baseline Exploration of Radio Astronomy Array (VERA)† and the Bar and Spiral Structure Legacy survey (BeSSeL)‡ using the VLBA have established the location of Outer Galaxy spiral arms and even resulted in a revision of the Galactic rotation parameters [→ REID, HONMA, SAKAI, MATSUMOTO]. So far, mostly 22.2 GHz H_2O masers and 12.2 GHz cIIMMs [→ XU] have been employed. Astrometry with the much stronger 6.7 GHz cIIMMs has started with the European VLBI Network (EVN) and VERA and will soon (in mid-2012) be possible with the VLBA.

Sources discussed at our meeting include well and (up to now) not so well studied HMSFRs, for example ON 1 and 2 [→ NAGAYAMA], W33 [→ IMMER]. and the prominent W51Main/South region, for which H_2O and 6.7 GHz cIIMM astrometry allows a comparison of the sources from which emission in the different species arises, outflows (with measured expansion motions) traced by H_2O [→ SATO] and hot cores by CH_3OH [→ ETOKA].

In addition, astrometry for the famous protoplanetary "rotten egg" nebula OH231.8+ 4.2 [→ CHOI] and other post AGB stars [→ IMAI] as well as the classical OH/IR hypergiant NML Cyg was reported [→ B. ZHANG]. So far, SiO masers have not yet played a major role in VLBI/Galactic structure astrometry efforts. However, in the future this may change a great deal thanks to the extensive surveys for vibrationally exited ($v = 1$ and 2), $J = 1 - 0$ masers conducted with the Nobeyama 45 meter telescope that have led to the detection of well over 1000 sources [→ DEGUCHI]. These masers, around 43 GHz, can be observed with VERA and the VLBA.

Other surveys will find many more methanol and water masers (see §5.2). In particular, an interferometric follow-up of the 6.0/6.7 GHz Parkes/ATCA/MERLIN multi-beam survey in the 22.2 GHz H_2O line will certainly detect many new water masers associated with high mass star formation in the general vicinity of the methanol masers. However, it also has the potential of detecting H_2O masers associated with low mass YSOs have formed together with the high mass YSOs, probably in clusters. In fact, the luminosity of known such masers can be high enough to make them detectable at distances of many kpc and make them the *only* signposts for low mass star formation outside of the Solar neighborhood [→ TITMARSH].

† http:// veraserver.mtk.nao.ac.jp/outline/index-e.html
‡ http://www.mpifr-bonn.mpg.de/staff/abrunthaler/BeSSeL/index.shtml

9. Odds and Ends: New masers, Propagation/Scattering, New Facilities

Maser action from molecules other than OH, H_2O and SiO has been discovered, at centimeter wavelengths, in many lines from (mostly) non-metastable levels of ammonia (NH_3) and formaldehyde (H_2CO) NH_3 and H_2CO masers have been exclusively found in the hot cores around HMPOs, in the vicinity, but generally not coincident with cIIMMs [$\rightarrow$ MENTEN, BROGAN]. Interestingly, the H_2CO maser line is the 4.8 GHz $1_{10} - 1_{11}$ K-doublet transition, which is ubiquitously found almost always in absorption throughout the Galaxy (and even in others, even in the diffuse ISM. The excitation of this maser is unclear, but it strikes me as peculiar that in all of the 10 known maser sources the emission is very weak. The flux densities of most sources are around 0.1 Jy or smaller. The most luminous one known, in Sgr B2, (≈ 0.5 Jy), has a luminosity that is roughly 100 times lower than that of the strongest 6.7 GHz cIIMM in that region.

Maser emission in (J,K = 3,3) inversion line of ortho NH_3† has been known for a while. In very few outflow sources [NGC 6334 I and DR21(OH)] it has been shown that this and sometimes also the (6,6) line, share properties of cIMM lines found in the same region, i.e. identical location of maser spots a and narrow single, component profile. In fact, this is the *only* known molecular line emission with a one-to-one correspondence to CIMM emission. In contrast, numerous (mostly weak) maser lines have found from non-metastable ammonia levels, even in the rare $^{15}NH_3$ isotopopologue, which is more than 200 times less abundance than $^{14}NH_3$. Non-stable levels, with $J > K$, decay rapidly down their K-ladders until they arrive at the lowest levels (with $J = K$). These are metastable and form a thermal distribution, which is the reason for ammonia's fame as a molecular cloud thermometer. Given the high transition probabilities of the FIR rotational lines connecting them, the non-metastable levels' populations are strongly influenced by the mid IR continuum, which, first, gives rise to maser action in certain lines and, second, places, as for cIIMMs, the maser's emission regions close to HMPOs.

In the (sub)millimeter range, HCN maser lines from within several vibrationally excited states have been found (plus the $J = 1 - 0$ line from the ground state). To these we may add H_2O masers in the vibrational ground state and the excited bending mode. The emission region of these vibrationally excited lines is pretty clear: The exceedingly high energies above the ground state (up to more than 4000 K) locates the origin of their emission very close to the stellar photospheres of mass losing stars with carbon-rich (HCN) and oxygen-rich (H_2O) chemistry, respectively.

Until a very short time ago, the only means to get information about the astrophysically very important water molecule was observing *maser* emission in lines emitted from high energies above the ground, which are not sufficiently excited in the Earth's atmosphere to make it complelely opaque at and around their frequencies. Earlier space missions either observed, with modest angular resolution, only the H_2O $1_{10} - 1_{01}$ ground-state line near 557 GHz (SWAS and ODIN) or far- and mid- IR lines also with limited spectral resolution (ISO). This situation has changed with Herschel, which produces a wealth of H_2O data between 500 and 1400 GHz with excellent sensitivity and spectral resolution. However, one should keep in mind that interferometric observations of the maser lines accessible from the ground are the *only* means to get information on this molecule with an angular resolution better than $10''$ [$\rightarrow$ MENTEN].

---

† Ortho-NH_3 assumes states, J, K, with $K = 0$ or $3n$, where n is an integer (all H spins parallel), whereas $K \neq 3$ for para-NH_3 (not all H spins parallel). The principal quantum numbers J and K correspond to the total angular momentum and its projection on the symmetry axis of the pyramidal molecule.

Exploring the nature of all these masers will greatly benefit from the comprehensive or greatly expanded frequency coverage and much larger number of spectral channels available with the JVLA and the Atacama Large Millimeter Array (ALMA) [→ WOOTTEN, BROGAN], e-MERLIN and ATCA, with MeerKAT playing a role as well [→ BOOTH].

Observations of short time scale maser variability caused in part by (even anisotropic) interstellar scattering may provide information on the intervening interstellar medium, but also on the intrinsic line of sight dimensions of individual maser spots [→ LASKAR, DESHPANDE, MCCALLUM].

10. Famous Last Words

Our field is like the (unsaturated) maser process – it stimulates itself and grows very rapidly! A few remarks:

• Big surveys are going on, but results need to be digested! Maser surveys need to be cross correlated with radio, IR and dust continuum surveys. In order to be useful, e.g., for planning VLBI observations, positions determined need to be listed with realistic uncertainty estimates.

• Think big! Compared to their predecessors, the new and upgraded facilities, JVLA, ALMA, e-MERLIN, and ATCA, offer awesome advances in *much* wider band correlator capability. Therefore, when planning your observation, make sure to use all the capabilities at your disposal. For, example, observe not just "your" target maser line, but cover as much bandwidth as possible, e.g., to get good continuum sensitivity or to cover other interesting lines. This is called "commensal" (= symbiotic) observing and hopefully will be a policy adopted and encouraged by observatories. Disk space is cheap and gets ever cheaper!

• Much more astrometry is needed! VERA and VLBA/BeSSeL run at full tilt! The EVN can also do astrometry. What about the Japanese VLBI and the East Asian VLBI Networks? We need VLBA-like capability in the southern hemisphere!

• If possible, all maser VLBI observations should use phase referencing, even if astrometry is not the goal. The resulting *absolute* position information is indispensable for comparative studies.

• It is worth to emphasize projects that have an impact on *all* of astronomy and even beyond, namely those that address Galactic structure, Local Group dynamics, precision H_0, protostellar collapse, magnetic fields: cIMMs could probe MHD shocks; the launching of protostellar outflows/shaping of planetary nebulae and other phenomena.

Finally, as a veteran of all the meetings in the series, I thank the organizers of this one very much for a perfect conference and the most pleasant time I had attending it.

I'm thankful to Christian Henkel and Mark Reid for their comments on the manuscript.

Author Index

Alakoz, A. V. – 294, 504
Alcolea, J. – 252
Alekseev, E. V. – 294
Alves, F. – **292**, 74
Amiri, N. – 54, 79
Anglada, G. – 258, 377
Assaf, K. A. – 235

Bae, J.-H. – 284, 488
Báez-Rubio, A. – 460
Barkiewicz, A. – **146**, 117, 151
Bayandina, O. – **280**, 294
Bendjoya, P. – 230
Bergman, P. – 161, 252
Blanton, M. – 316
Boboltz, D. A. – 209
Booth, R. – 483
Bourke, S. – 146
Braatz, J. A. – 301, 311
Breen, S. – **275**, 156, 433
Britton, T. R. – 176, 282, 433
Brogan, C. L. – 127, 497, 502
Brunthaler, A. – 340, 368, 407, 423, 427
Bujarrabal, V. – 252
Byun, D.-Y. – 284, 488

Castangia, P. – 340
Caswell, J. – **275**, 433
Cesaroni, R. – 180, 396
Chandler, C. – **497**, 166
Chapman, J. M. – 250
Chen, X. – 288
Chibueze, J. O. – 141
Cho, S.-H. – 290
Choi, Y. K. – 260, 407
Chong, S.-N. – 141
Chrysostomou, A. – 194
Churchwell, E. – 127
Claussen, M. – 225
Condon, J. J. – 301, 311
Conway, J. – 186
Cotton, W. – 245
Curiel, S. – 377
Cyganowski, C. J. – 127, 497

Darling, J. – 333
Deacon, R. M. – 250
De Buizer, J. M. – 151
Deguchi, S. – 265
Delaa, O. – 245
Deshpande, A. A. – 93, 470
Desmurs, J.-F. – 79, 217, 252

de Villiers, H. M. – 194
Diamond, P. J. – 79, 235, 240
Doi, A. – 190, 288
Dunham, M. K. – 286

Eliav, D. – 180
Ellingsen, S. – **275**, 156, 433
Engels, D. – 254, 256
Esimbek, J. – 178, 296
Estalella, R. – 377
Etoka, S. – 171

Finkelstein, A. M. – 504
Friesen, R. – 497
Fujisawa, K. – 98, 103, 188, 190, 288
Fuller, G. A. – 171
Furuya, R. S. – 180

Gaylard, M. – **85**, 108, 110
Gérard, E. – 59, 254
Girart, J. M. – 74
Goddi, C. – 166, 184, 396, 401
Goedhart, S. – 85, 483
Gómez, Y. – 249, 258
Gómez, J. F. – 230, 258, 377
Gonidakis, I. – 240, 250
Goss, W. M. – 465, 470
Gray, M. D. – 13, 23, 171, 209, 235
Green, A. – 250
Green, J. A. – 433
Greenhill, L. J. – 166, 184, 316
Guerrero, M. A. – 230

Hachisuka, K. – 288
Hallet, N. – 254
Henkel, C. – 301, 311, 340
Hideyuki, K. – 415
Hirota, T. – 103, 141, 288, 386, 415
Hoare, M. G. – 112
Hofner, P. – 161
Honma, M. – 98, 103, 188, 288, 386, 419, 421
Humphreys, E. – **184**, 292, 166, 209
Hunter, T. R. – 127, 497, 502

Imai, H. – 103, 141, 411
Immer, K. – 413
Impellizzeri, C. M. V. – **301**, 311, 340
Indebetouw, R. – 497
Ipatov, A. V. – 504

Jerkstrand A. – 186

516

Jiménez-Serra, I. – 455
Johansson, L. E. B. – 161
Jonas, J. – **483**, 81

Kalenskii, S. V. – 161
Kameya, O. – 141
Kang, H.-W. – 488
Karovicova, I. – 209
Kawaguchi, N. – 103, 386
Kemball, A. – **79**, 54, 81, 240
Kim, J. – 290
Kim, K.-T. – 284, 288, 488
Kim, M. K. – 103, 415
Kim, S. J. – 290
Kim, W.-J. – 488
Kobayashi, H. – 103
Koda, J. – 127
Konovalenko, A. A. – 294
Kostenko, V. I. – 504
Kramer, B. H. – 69
Kristensen, L. E. – 146
Kulishenko, V. F. – 294
Kuo, C. Y. – **301**, 311
Kurayama, T. – 417, 421
Kurtz, S. – **161**, 133

Larionov, G. – **280**, 294
Laskar, T. – 465
Leal-Ferreira, M. L. – 79
Li, D. – 350
Li, J. J. – 180, 368
Lihachev, S. F. – 504
Lindqvist, M. – 252
Litovchenko, I. D. – 294, 504
Lo, K. Y. – 301, 311

Martín-Pintado, J. – 460
Maswanganye, J. P. – 108
Matsumoto, N. – 386, 419
Matthews, L. D. – 166
McDonnell, K. – 441
McKean, J. P. – 340
Mendoza-Torres, J. E. – 470
Mennesson, B. – 245
Menten, K. M. – 180, 368, 407, 413,
    423, 425, 427, 506
Migenes, V. – 182, 192
Millan-Gabet, R. – 245
Miranda, L. F. – 230, 258
Mohamed, S. – 260
Morris, D. – 49
Morris, M. – 225
Moscadelli, L. – 180, 368, 396, 401
Moss, V. A. – 176
Motogi, K. – 98, 288
Mukha, D. V. – 294
Murata, Y. – 288

Nabatov, A. S. – 294
Nagayama, T. – 386, 391
Nakanishi, H. – 421
Niederhofer, F. – 184
Nikolaenko, V. S. – 294

Odincov, S. A. – 294
Ogawa, H. – 288
Oh, C. S. – 488
Olofsson, H. – 260
Omodaka, T. – 103, 141
Otto, S. – 110

Patel, N. A. – 377
Perrin, G. – 245
Pérez-Sánchez, A. F. – 64
Pestalozzi, M. – 180, 186, 492
Pihlström, Y. M. – 449
Pittard, J. M. – 112

Ramos-Larios, G. – 230
Ramstedt, S. – 260, 292
Reid, M. J. – 301, 311, 359, 368, 407,
    413, 423, 425, 427
Richards, A. M. S. – 199, 235
Richter, L. L. – 81
Rizzo, J. R. – 230
Rodríguez, I. T. – **182**, 192
Rogers, H. – 225
Rosolowsky, E. – 127
Roy, A. L. – 340
Rygl, K. L. J. – 368

Sahai, R. – 225
Sakai, N. – 386, 421
Sakanoue, H. – 421
Sanna, A. – 368, 396, 401
Sarma, A. P. – 41
Sato, M. – 423
Sawada-Satoh, S. – 188, 288
Schisano, E. – 180
Scholz, M. – 209
Shen, Z.-Q. – 288
Shibata, K. M. – 103
Shimoikura, T. – 103
Shino, N. – 190
Singh, N. K. – 93
Sjouwerman, L. O. – 449
Slysh, V. I. – 161
Sobolev, A. M. – 13, 433
Sorai, K. – 98
Soria-Ruiz, R. – 252
Strelnitski, V. – 3
Sugiyama, K. – 98, 188, 190, 288
Surcis, G. – 69
Suárez, O. – 230, 258
Szymczak, M. – 59

Tafoya, D. – 141, 258
Tang, X. D. – 178
Tarchi, A. – 323
Thompson, M. A. – 194
Thum, C. – 49
Titmarsh, A. – 275
Torrelles, J. M. – 74, 141, 258, 377
Torstensson, K. J. E. – 146
Towers, S. – 127
Trinidad, M. A. – 182, 192
Tsuboi, M. – 103

Uscanga, L. – 230
Usuda, T. – 180

Val'tts, I. – **280**, 294
van den Heever, S. P. – 112
van der Tak, F. F. S. – 146
van der Walt, D. J. – 112, 194
van der Walt, J. – 85
van Langevelde, H. J. – 54, 69, 117, 146
Vázquez, R. – 258
Vlemmings, W. – **64**, 260, 292, 31, 54, 69, 74, 79, 146
Voronkov, M. – 282
Voronkov, M. A. – 176, 275, 433

Wajima, K. – 188
Walmsley, C. M. – 161

Walsh, A. J. – 433
Wang, J. Z. – 350
Wardle, M. – 354
Whitney, B. – 127
Wiesemeyer, H. – 49
Willett, K. W. – 345
Wittkowski, M. – 209
Wolak, P. – 59
Wootten, A. – 477
Wu, G. – 178, 296
Wu, Y. W. – 425
Wucknitz, O. – 340

Xu, Y. – 172, 351, 408 – 180, 368, 425

Yonekura, Y. – 103, 288
Youn, S.-Y. – 488
Yusef-Zadeh, F. – 354

Zakharenko, V. V. – 294
Zauderer, B. A. – 465
Zaw, I. – 316
Zhang, B. – 351, 410, 368, 427
Zhang, J. S. – 350
Zhang, Q. – 127
Zheng, X. W. – 368, 425, 427
Zhou, J. J. – **178**, 296
Zhu, G. – 316

Printed in the United States
by Baker & Taylor Publisher Services